21st Process Plant Safety Symposium 2019

Topical Conference at the 2019 AIChE Spring Meeting
and 15th Global Congress on Process Safety

New Orleans, Louisiana, USA
31 March - 4 April 2019

ISBN: 978-1-5108-8686-5

Printed from e-media with permission by:

Curran Associates, Inc.
57 Morehouse Lane
Red Hook, NY 12571

Some format issues inherent in the e-media version may also appear in this print version.

Printed by Curran Associates, Inc. (2019)

For permission requests, please contact AIChE
at the address below.

AIChE
120 Wall Street, FL 23
New York, NY 10005-4020

Phone: (800) 242-4363
Fax: (203) 775-5177

www.aiche.org

Additional copies of this publication are available from:

Curran Associates, Inc.
57 Morehouse Lane
Red Hook, NY 12571 USA
Phone: 845-758-0400
Fax: 845-758-2633
Email: curran@proceedings.com
Web: www.proceedings.com

TABLE OF CONTENTS

Lifecycle Management of Protection Layers and Safeguards

Jack Chosnek, PhD, PE
Knowledge*One* LLC
PO Box 451629, Houston, TX 77245
jc@knowledge1.net

Prepared for Presentation at
American Institute of Chemical Engineers
2019 Spring Meeting and 15th Global Congress on Process Safety
New Orleans, LA
March 31 – April 3, 2019

Lifecycle Management of Protection Layers and Safeguards

Jack Chosnek, PhD, PE
Knowledge*One* LLC
PO Box 451629, Houston, TX 77245
jc@knowledge1.net

Keywords: Process Safety, Hazards Register, Risk management, Layers of Protection, PHA, LOPA, Safeguards

Abstract

From project inception to continuous plant operation a stream of safety studies will be conducted on the process, and the identified hazards will have to be either eliminated or reduced to an acceptable risk level. This risk level needs to be maintained at an acceptable level throughout the lifecycle of the plant. As changes occur and time passes the safeguards and protection layers start getting disconnected from the intent of the safety studies. This happens for many reasons but mainly because on proposing and implementing the layers of protection, the assumptions and intent made during the study are not explicitly attached to the specification of the equipment, instrumentation, or procedures. This is aggravated when the recommendations coming out of a study are implemented in a different manner as the recommended one or a totally different solution is adopted. Furthermore, the same equipment may be part of different safeguards in different studies.

A Hazards Register would contain all the pertinent information related to the risks assessed during all of the safety studies performed by the company, whether a Process Hazards Analysis, or a Management of Change review, or an incident investigation. The resolution of each hazard should be available in the Register, and not only the latest resolution but also its evolution (history) starting from the original study. In order to be effective, the Hazards Register should be easily accessible, be capable of simultaneous use by all plant personnel, have appropriate security, and be fully and effortlessly searchable. It should also automatically provide metrics that allow management of outstanding recommendations and automatically recalculate relevant risk information (e.g., cumulative probability of failure on demand, pfd, from a Layer of Protection Analysis study). It should be able to import data from any type of safety study and to export all or part of the data for other uses (e.g., instrument specifications). Such a system was successfully used in a very large project in which over 9,000 safeguards and their justifications were managed. At the end of the project the Hazards Register was transferred to the operating company for continued management of the process risks.

Introduction

Managing the risks of a process is a task that starts at the conceptual stage and continues throughout the design, startup, operation of the process, and decommissioning of the plant. Thus, we have to continuously manage the risks through the life cycle of the process. But the risks keep changing as we define in more detail the equipment and instrumentation of the process during the design, or introduce changes as a result of a desire for process improvement, or a change in operation. The nature and

severity of the risks will also change when we analyze them and implement measures for either avoiding them, or minimizing or mitigating them.
Whichever method we use for safety studies, we will start with identifying the hazards, defining a potential cause for a process deviation that will make the hazard become an event, determining the consequence of that event, and then we will identify what safeguards exist to prevent or mitigate the consequence. If we use risk-based process safety we will also want to describe the risk by assessing the severity of the consequence and the probability of that occurring given the identified safeguards. During the analysis we will make assumptions and develop some logic that supports the description of the found risk. We may want to reduce the risk and, therefore, we will recommend and implement additional safeguards. So, there is a lot of information that helps us understand the risk and find commensurate ways to control it.

As time passes this information evolves and the basis for its creation becomes faint. This may lead to adding, modifying or removing necessary safeguards adopted from safety studies, as the reason for these safeguards being there is hidden or forgotten. Thus, a tool is needed for the long-term maintenance of the assumptions and logic for all of the control schemes, instrumentation loops and equipment specifications that make up the safety infrastructure of the plant.

The Evolution of Risk Assessment in the Life Cycle of the Plant or a Project

Any process design will undergo safety studies to determine the hazards of the process and the means to control them. Most of the time this is done through Process Hazards Analyses (PHAs) using many different methodologies. If the company practices risk-based process safety, this may be followed by a Layer of Protection Analysis (LOPA) and/or a Quantitative Risk Analysis (QRA), if the consequence level demands it. When performing the PHA oft times recommendations are made to mitigate the risk resulting from a given cause that leads to a specific consequence (we will call it a cause-consequence pair or c-c pair for short). The recommendation will generally propose adding or modifying safeguards (or even changing equipment to make that part of the process inherently safer). If a LOPA is performed in order to evaluate the risk in more detail, the same c-c pair may produce a more detailed safeguard, which can be an independent layer of protection (IPL) with its own rules [1, 2, 3]. Those rules will dictate, depending on the desired risk reduction and independence from other instrumented protections, a Safety Integrity Level (SIL) for the IPL. We then specify instrumentation that is part of a Safety Instrumented System (SIS) and which gets documented in the SIS archives of the plant. The same applies to a non-SIS layer of protection or safeguard, except that the documentation is in another place. This applies as well as to procedural (administrative) safeguards, where only the required actions will become a part of a Standard Operating Procedure (SOP) and their basis may not be documented in any place except for a PHA report. Unless we document very well the reason behind a layer of protection (and are able to find it at any time), we are destined to defeat it at some point in time.

A change in risk also occurs when a change in the plant occurs. The change will go through the Management of Change (MOC) process and obviously an MOC safety study will be performed. In order to be successful, the safety study will need to review the hazards identified during the process PHA, and the protections that were identified at the time, plus the mitigations recommended by the PHA team. In addition, the safety study will have to review any previous modifications of the equipment and instrumentation, that is, search the MOC database to see if any of the existing layers of protection are not impacted.

An additional challenge is that recommendations made by any of the teams, be it a PHA or an MOC (or an incident investigation), sometimes don't get executed as the team proposed, as a better solution may be implemented once there is more time to think and come up with a potentially better design. When this happens, chances are that the implementation will only refer to the PHA report, or MOC number, and there will be a disconnect between the cause of the event and the implemented solution. It is very common to create a new document in order to track implementation of the recommendations and this document will increase the gap between the implementation and the understanding of why the recommendation was needed.

The effect of all this is that, as time goes by, the process safety reason for a certain process configuration gets lost. A new safety analysis may totally miss the underlying safety need for that configuration as the need may have been identified a couple of iterations before and it may not be obvious at the present time. This gets complicated even more when there are interlocks that activate various valves (for example a plant shutdown) but one of the valves is also part of a control loop that has a different safety function (SIF) from that of the shutdown. The plant shutdown may be obvious after some analysis, but the other SIF may not be that obvious. At some point in time a decision could be made to reduce the SIL rating of the loop because it is unnecessary for the shutdown, but by doing this the risk reduction for the other safety function may be compromised.

The challenge resides on how the safety documentation and the process documentation are interrelated. The following documents would need to be consulted for a proper risk estimation in case of a change in the process:

1. PHA report
2. MOC database
3. Safety study of the MOC (if not integrated into the MOC database)
4. Cause and Effect Table
5. Instrument list containing instrument specifications.
6. Incident investigation reports if there had been any incidents in the plant.
7. Action-tracking table to see if there are unimplemented safety recommendations.

And, of course, up-to-date Piping and Instrumentation Diagrams (P&IDs) that reflect all the actual control schemes and other protections, be it SIS, Distributed Control System (DCS), or other.

All of this information can be easily integrated in a Hazards Register, which is a dynamic database that can also serve as a tracking device for all safety recommendations. The Hazards Register ought to also include a historical record of all the resolutions that were ever made with respect to safety studies or changes affecting safety systems. Such a database was created and its capabilities are described below.

The Hazards Register

The Hazards Register is a database with all of the causes and consequences that were ever identified in a plant, including its grass-roots design project. These c-c pairs will have unique identifiers that allow them to be tracked throughout the life cycle of the plant. The history is preserved. In order to avoid errors created by compilation of data, all of the data is imported directly from most commonly used PHA/LOPA software, or from any spreadsheets used in safety studies. The only required manual entries are the name, date and type of the source safety study (PHA, LOPA, MOC, etc.). If desired, additional information can also be entered, such as study team composition, remarks, etc., without limitations.

The data elements that the Hazards Register includes for each c-c pair are:

- C-C unique identifier (created automatically on entering the data by importing or otherwise)
- Source study name, type and date
- System (e.g. Operating Unit, plant area)
- Subsystem (e.g. Compressor System, Cold Box, Final Purification, etc.)
- Process deviation/keyword used in PHA for the c-c pair
- Consequence (Hazard)
- Cause
- Consequence severity level
- Type of consequence severity (e.g. economic, safety, environmental, etc.)
- Safeguards or IPLs
- Frequency of initiating event (LOPA)
- Frequency and type of enabling event (LOPA)
- Frequency and type of modification event (LOPA)
- PFD (Probability of Failure on Demand) for each IPL
- Probability (calculated from PFDs or entered from Risk Matrix)
- Risk (from risk matrix)
- Mitigated risk
- Reference (e.g. P&ID or other)
- PFD gap
- Recommendation or Action Plan
- Owner (for tracking purposes)
- Target date of implementation (for tracking purposes)
- Resolution (the latest resolution for the item will show, but all the resolutions from the beginning can be seen by clicking a button)
- New safeguards/IPLs resulting from recommendation(s)
- SILs of each of the new IPLs
- Final cumulative PFD (calculated automatically)
- Documentation description
- Status (open, closed, deferred, in progress, etc.). Milestones are defined by the user. Past due items, that is items not closed by target date, will show in red.
- Comments

That is, all the information contained in a PHA and LOPA plus some. But everything is electronically imported to avoid errors. A main table (see Figure 1) is a list of the c-c pairs with all the data above and is fully searchable. If a c-c pair from a Hazard and Operability (HAZOP) study, for example, which was imported into the program, was taken further into a LOPA which was then also imported into the program, the database will create a link between the two identical c-c pairs. At the click of a button you can switch between one and the other. All the data can be easily searched, filtered, sorted and exported. Excel spreadsheets and Adobe Acrobat reports can be created at the click of a button. The main table will also automatically show statistics of all the table contents, both of those shown (after filtering) and of the total number of items. The statistics will reflect the milestones previously selected by the company (e.g., open, closed, in progress, deferred, etc.).

KnowOne Chemical Co.

HAZARDS REGISTER
CO Production

KnowledgeOne LLC

Total Items: 4454 [Shown: 2]
Open: 7 (0%) [1]
Deferred: 237 (5%) []
MOC: 14 (0%) []
Closed: 4196 (94%) [1]
Past Due: 258 (6%) [1]

Recommendations

New | Delete | Show All | Find

Items with Action Plans: 1704 [2]
With Action: 1654 (97%) [2]
Open: 183 (11%) [1]
Past Due: 183 (11%) [1]

ID	Source Study	Consequence (Hazard)	System	Sub-system		Cause	Cons	Type	Safeguards/ IPLs	IPLs PFDs	Prob	Risk	Mitigated Risk	Reference	PFD Gap	Action Plan (Recommendation)	Owner	Target Date	Action/Resolution
2646 Link	EPC CO Purification LOPA May 2017	Potential loss of containment with potential fire	CO Production Deviation	4	Cold Box	Thermal stress of the plate fin heat exchanger (HEX -76) during startup and shutdown or upsets resulting in fatigue failure, generating a leak Freq Initiating Event 1.0E-2 Freq Enabling Event 1 Freq Init Modifier 0.1 Vendor confirmed only	3	Safety and Health			.001	M		P&ID 2531	.03	**1270.** (7) See Cold Box HAZOP Recommendations 12 and 13 (differential temperature alarming and temperature trending). **1271.** (8) Verify fatigue life (PFD/MTBF) of heat exchangers, e.g., how many cycles until the exchanger fails. **1272.** (9) Assure O&M procedures address and that operators are trained to operate the exchangers in a way that prevents overstressing (i.e. heat up and cool down rates). See Liquefaction HAZOP	Process	10/18/17	TI-83 added to DCS and addition of TDI -243 and TDI-244 with rate of change alarms to alert panel operator.
2652 Link	EPC CO Purification LOPA May 2017	Potential loss of containment inside the cold box resulting in rupture of the cold box and dispersal of perlite insulation and cold vapor/liquid	CO Production Deviation	6	Cold Box	Thermal stress of the plate fin heat exchanger (HE -072) during startup and shutdown or upsets resulting in fatigue failure or mechanical defect, generating an external leak. Freq Initiating Event 1.0E-2 Freq Enabling Event 1 Freq Init Modifier 0.1 Vendor confirmed only	4	Economic	2) Gas detectors in the area tied to the SIS initiate area shutdown 1) Multiple shift operators monitoring restart procedures.	2) 1.0E -1 1) 1.0E -1	.0001	H		P&ID 2531	.01	**1282.** (7) See Cold Box HAZOP Recommendations 12 and 13 (differential temperature alarming and temperature trending). **1283.** (8) Verify fatigue life (PFD/MTBF) of heat exchangers, e.g., how many cycles until the exchanger fails. **1284.** (9) Assure O&M procedures address and that operators are trained to operate the exchangers in a way that prevents overstressing (i.e. heat up and cool down rates).	Process	10/18/17	TI-83 added to DCS and addition of TDI -243 and TDI-244 with rate of change alarms to alert panel operator. TDI-243 and TDI-244 with independent local indication are monitored by multiple outside operators during start up as a standard operating procedures

Figure 1. Hazards Register Main Table

Each c-c pair can be closely examined by clicking on the item ID which will take you to a detailed view as shown in Figure 2.

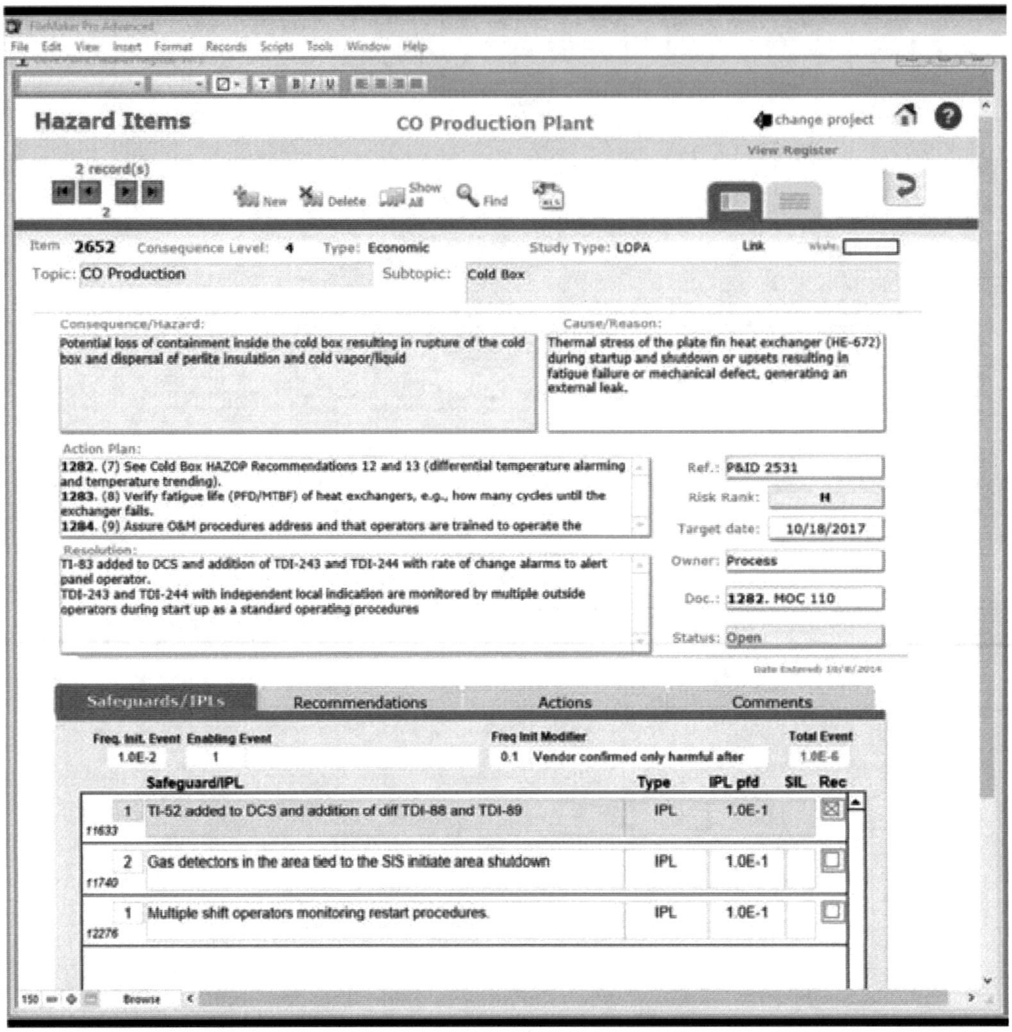

Figure 2. Item Detailed View Showing Safeguards/IPLs

In this view, all the safeguards/IPLs can be seen and SIL values assigned. If a PFD has been entered, the program will calculate the cumulative (total event) PFD. If an IPL PFD is changed, the cumulative PFD will be recalculated. If a safeguard or IPL came from a recommendation, it will be marked so, and it is possible to see the contribution of the recommendation to the total item risk.

There are other views, of all the recommendations and actions taken for each recommendation (with dates and name of implementer), and of all the resolutions, which are the explanations of how each item was finally resolved. There is a resolution for every item, even if a recommendation wasn't made and even more important, if a recommendation was not followed upon. Although the latest resolution is the one that will be shown in the main table, all of the history of the item from day one is available by clicking on the history button. Thus, the evolution of the reasoning of why the resolution stands as it is, is available, as well the justification for all the safeguards and IPLs that are part of the current process configuration. When making a change, it will be immediately obvious why a certain protection is in place. If a protection is removed or a new protection is added, the program will recalculate the risk.

Another very useful part of the Register is a list of all the safeguards/IPLs with their PFDs and SIL values (Figure 3). This list can be easily exported to Excel and serve as the basis for creating or checking the plant instrument list as well as inspecting and maintaining the Cause and Effect table. Since it is fully searchable any instrument or device can be found, and its participation in more than one Safety Instrumented Function (SIF) at a time scrutinized.

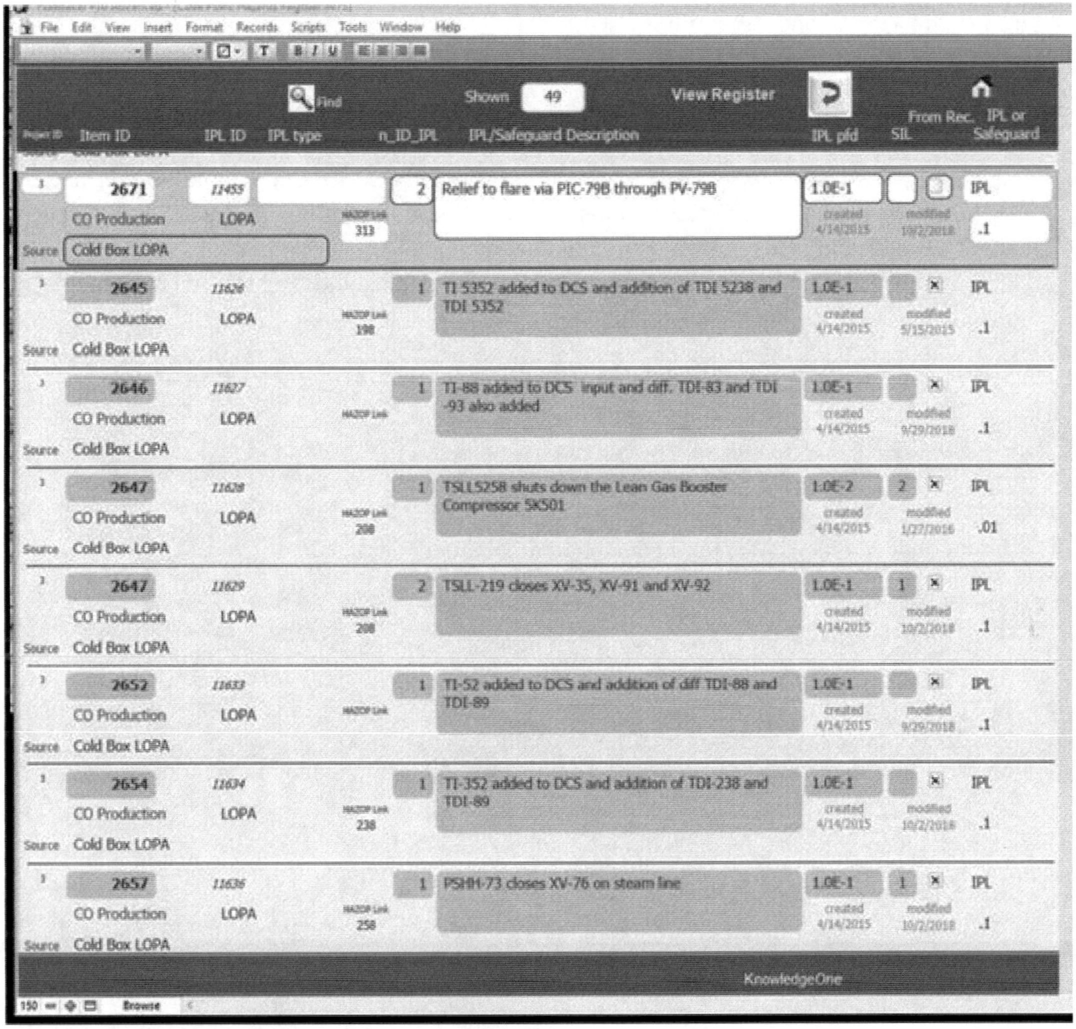

Figure 3. List of Safeguards/IPLs

SUMMARY

In summary, the Hazards Register that was created maintains in one place all the hazards and related risks of the facility. The basis and reasoning for resolving the hazards through time is available since when the first safety study was imported. This is essentially in order not to forget the intent of an instrument or protection layer which could be later be changed, unintentionally increasing the risk of the facility. Since it resides in a database with simultaneous access to all, and is fully searchable, all the data can be easily found and continually used to maintain a safe design throughout the life cycle of the plant.

This Hazards Register was successfully used in a very large Engineering, Procurement and Construction (EPC) project ($4 billion) that lasted over three years. The database contained about 2,300 c-c pairs and 9,000 safeguards/IPLs and it was used to track resolution of all these items. At the end of the project the data was incorporated in the new facility's information. The Register could be seamlessly transferred to the facility and continued to be used.

REFERENCES

1. CCPS, "Layer of Protection Analysis: Simplified Process Risk Assessment", October 2001.
2. CCPS, "Guidelines for Enabling Conditions and Conditional Modifiers in Layers of Protection Analysis", November 2013.
3. CCPS, "Guidelines for Initiating Events and Independent Protection Layers in Layer of Protection Analysis", February 2015.

How to Prioritize Safeguards When They Are All Critical

Ronjit Mukherjee
ACM Facility Safety Inc.
926 5 Ave SW #300, Calgary, AB T2P 0N7
rmukherjee@acm.ca

Richard Carter
ACM Facility Safety Inc.
926 5 Ave SW #300, Calgary, AB T2P 0N7
rcarter@acm.ca

Prepared for Presentation at
American Institute of Chemical Engineers
2019 Spring Meeting and 15th Global Congress on Process Safety
New Orleans, LA
March 31 – April 3, 2019

How to Prioritize Safeguards When They Are All Critical

Ronjit Mukherjee
ACM Facility Safety
926 5 Ave SW #300, Calgary, AB T2P 0N7
rmukherjee@acm.ca

Keywords: Safeguards, Safety Critical Equipment, Prioritization, Manage Risks, Process Safety

Abstract

With the industry norm of budget constraints and the continuous effort of trying to do more with less, the ability to correctly prioritize work has become increasingly critical. This becomes especially important when it relates to risk-reducing Safeguards and the prioritization of Safeguards for maintenance. With a growing list of equipment and corresponding Safeguards, the ability to stay on top of maintenance prioritization can be significantly impacted. This can lead to critical Safeguards having delayed or missed maintenance, and in turn lead to unintended and significant risk exposure.

In order to correctly identify critical Safeguards and prioritize them, it would require hundreds of hours of effort from cross-disciplinary Subject Matter Experts (SMEs). Fortunately, these are the same SMEs involved in developing Process Hazard Analysis (PHA) documentation. By using PHAs as a data source, the documentation of hazardous scenarios and Safeguards can be used to prioritize critical Safeguards using the following criteria:

- Frequency of a Cause
- Severity of a Consequence and its corresponding Tolerable Frequency
- Probability of Failure on Demand (PFD) of Safeguards intended to prevent that Consequence

These criteria are used to determine the risk reduction associated with each occurrence of a Safeguard, and by accumulating the risk reduction, the total risk-reducing benefit of each Safeguard can be determined to assist in prioritization. By having Safeguard prioritization based on SME-approved data, as well as using a simulation tool to consider plant realism of Safeguards failing and PHA recommendations being implemented, the confidence in Safeguard maintenance prioritization will increase. An optimized and reflective prioritization list will reduce the probability of missing maintenance on critical Safeguards, which in turn will minimize risk exposure to site personnel.

1 Introduction

In the process industries, effort is placed on trying to reduce the risk of hazardous events. The key items involved in reducing the risk of these scenarios are Safeguards. A safeguard can be an engineered system or an administrative control that either interrupts the chain of events following an initiating event or that mitigates the consequences. Safeguards are used to ensure the impact to people, equipment and the environment is minimized from many critical consequences such as overpressures of equipment to fires and explosions. Examples of Safeguards include:

- Relief and Vent Devices
- Control Systems
- Interlocks and Emergency Shutdown Systems
- Administrative Controls including Alarms, Procedures, Operator Rounds

The effectiveness of a Safeguard is highly dependent on the maintenance of that equipment. Without regularly scheduled maintenance the probability of safeguard failure will increase, and when a Safeguard fails it will no longer be effective in reducing risk. Unfortunately, the realism of the industry is that there will be always be a constraint on budgets and people, which means there will be equipment that we are not able to maintain at the desired level and associated reliability. The scenarios which are protected by the equipment maintained at a lower level are then more likely to occur, increasing the risk to site personnel, the environment, and facility production.

As with all problems related to resource constraints, the key solution is correct prioritization. Thus, the key solution to the problem of maintenance constraints is Safeguard prioritization. Although organizations in industry have made steps in the right direction to create prioritized Safeguard lists through Safety Critical Equipment identification and Alarm Rationalization, there are several current issues which may hinder the goal of ensuring critical Safeguards are maintained correctly to keep high risk scenarios at a tolerable level. By improving the methodologies for prioritizing Safeguard maintenance, the opportunity for unanticipated risk exposure can be minimized, and the following issues related to the current approach to critical Safeguard lists can be resolved:

- Within current lists of critical Safeguards such as the outputs of Alarm Rationalization, in which alarms are tiered into different priorities, or Safety Critical Equipment lists, in which equipment related to high risk scenarios are documented, it can be difficult to identify the relative importance between each unique Safeguard. The importance of distinguishing safeguards among a large group magnifies all the more when maintenance budgets and resources are further constrained.
- Current documentation of identification of critical Safeguards is not easily understood by personnel unfamiliar with the process. It is important that the reasoning for identifying Safeguards as critical is recorded clearly, so that anyone who reviews the document will be able to understand the justification. Through regular swings in the economy, it is commonplace for management to direct engineers and operations to find areas of cost savings. If engineers or operations review critical Safeguard lists to identify areas of

savings, and due to poor documentation unnecessarily declassify critical Safeguards, it can result in missed maintenance of critical Safeguards and increased risk exposure.
- Current Safeguard lists are static, however, in a processing facility Safeguards fail or are bypassed, and new Safeguards are implemented, changing the importance of the remaining functional Safeguards on site.

Fortunately, the solution to improve this methodology is available today to all members of the process industry. This paper presents how integration of PHAs with the creation of a prioritized Safeguard list can resolve these current issues and will show that:

- Although all Safeguards reduce risk, they do not all have equal weighting in their risk reduction and therefore they do not have equal importance for ensuring regular maintenance.
- PHAs already have enough documentation of what scenarios Safeguards are protecting against and provide an easy-to-understand reasoning of the Safeguard criticality.
- Using tools to simulate failing Safeguards and implementing recommendations can help keep an active and reflective list of Safeguard criticality.

The PHAs analyzed in this paper were Hazard and Operability (HAZOP) studies performed on existing processing sites. Data from the Center for Chemical Process Safety (CCPS) publication on Layer of Protection Analysis (LOPA) was used for the study when risk reduction credits were applied to existing Safeguards [2].

It should be noted that:

- Irrespective of how much maintenance is performed on a safeguard, there will always be a certain limit on safeguards PFD due to its design and features, and maintenance will simply ensure that the reliability of a safeguard does not reduce to a point that increases the risk of a scenario past its target tolerable frequency.
- There are some safeguards identified in a PHA that are purely personnel related, such as procedures. Although these safeguards do not relate to equipment maintenance, they do relate to refresher training and procedure knowledge verification.
- Multiple companies not only perform HAZOPs but also LOPAs. As LOPAs may not only go into more detail on scenarios analyzed in HAZOP, but also may identify new scenarios, it is important to conglomerate the learnings from PHA methodologies. This would create a data set of HAZOP scenarios without LOPA equivalents, LOPA scenarios that supersede certain HAZOP scenarios, and LOPA scenarios that were not identified in HAZOP.
- As with all knowledge gathering sessions that are to be driven by SMEs, they are only are good as the people involved. As some companies use the mentality of "use whoever is available" for attendance of PHA sessions, the quality of these risk assessments may be below average quality. This would also mean that there would be less value for the purpose of using this documentation for input into safeguard maintenance prioritization. The higher the quality of PHA sessions, the more valuable they are for prioritizing safeguard maintenance and PHA analytics in general.

2 Safeguard Risk Reduction

2.1 Calculating the Risk Reduction Benefit Provided by Safeguards

PHA studies are used, and in some countries are required by regulatory bodies, to identify process hazards and operational issues for facilities. These studies enable management, engineers and site personnel to make decisions that reduce the risk of hazards identified in PHAs.

Although the outputs of PHAs are often focused on the recommendations and future action items that are identified, hundreds of hours of Subject Matter Expert effort is put into breaking down hazardous scenarios and the existing Safeguards which protect them. This information can be used to identify the risk reduction associated with each Safeguard, which in turn will directly correlate with the maintenance prioritization of that Safeguard. Simply put, the more risk reduction a Safeguard provides to a facility, the more critical it is to ensure its maintenance.

Safeguards

Safeguard	Category	Max # Credits Taken	Place(s) Used
8. MEC - PSV-1000 set at 1000 kPag, sized for regulator failure, will relieve to safe location.	MEC	2	Consequences: 3.2.1.1
2. OCC - Personnel in the area less than 10% of the time.	OCC	1	Consequences: 1.1.4.2, 1.1.5.2, 1.2.1.1, 1.2.2.1, 1.2.2.2, 1.2.3.1, 2.1.2.1, 2.1.8.1, 3.2.1.1
1. Pneumatic - PSHH-1000 will close ESDV-0100.	Other	1	Consequences: 1.1.4.2, 1.1.5.2, 1.2.2.2
7. PM - UT scan on process gas piping.	PM	1	Consequences: 2.2.4.2
5. PRO - Chemical injection upon initial start-up of well.	PRO	1	Consequences: 1.16.1.1
9. ROUND - Operator round monitoring LG-2000 and initiate draining of Fuel Gas Scrubber	ROUND	1	Consequences: 3.5.1.2
3. PLC-T - FAHH-2020 will close ESDV-0100.	RTU	1	Consequences: 1.2.1.1, 1.2.2.1, 1.2.3.1
6. PLC-T - PAHH-1002 will initiate wellsite ESD.	-, RTU-T	1	Consequences: 2.1.7.1, 2.1.8.1
4. PLC-T - LSHH-0550 will close ESDV-0100.	RTU-T	1	Consequences: 1.2.4.1

Figure 1. Sample PHA Safeguard Summary

In order to utilize this PHA dataset to determine the actual risk reduction provided by Safeguards and assist in identifying critical Safeguards, the following need to be considered:

- The current frequency of the consequence and/or the frequency of the cause.
- The probability of failure on demand (PFD) of Safeguards on related consequences, which is a measure of the safeguard reliability.
- The severity level and associated tolerable frequency of each cause/consequence pair the Safeguard is associated with. These tolerable frequencies are usually identified on the Corporate Risk Matrix.

In the presented case studies, the following assumptions were made when calculating the risk reduction provided by each Safeguard. This was based on the Risk Matrix shown in Figure 2:

- Frequency of causes were assigned based on the correlations provided by the PHA Team. (Table 1.).
- A PFD of 0.1 was assigned to each Safeguard given 1 "credit" of risk reduction in the PHA. Safeguards given 2 "credits" were assigned a PFD of 0.01 [2].

- Tolerable Frequencies were determined based on the below assumption for each severity (Table 2.). Each consequence receptor (e.g. Health and Safety, Environment, etc.) were assumed to have the same relationship between severity and tolerable frequency, which is found from LOPA documentation. Note that in certain LOPA documentation it may be specified that various consequence receptors will have different relationships between severity and tolerable frequency (e.g. for the same severity, Health and Safety may have a tolerable frequency of 0.0001, whereas Financial may have a tolerable frequency of 0.001).

Companies can use specific IEFs and PFDs from their internal equipment databases for the risk reduction calculations to improve the precision of the calculations. Only safeguards that are credited for risk reduction in the PHA have been reviewed for this report.

Table 1. Frequency Correlation for Risk Matrix Likelihoods

Likelihood	Cause Frequency (/yr)
1	0.0001
2	0.001
3	0.01
4	0.1
5	1

Table 2. Target Tolerable Frequency for Risk Matrix Severities

Severity Code	Tolerable Frequency (/yr)
A	1
B	0.1
C	0.01
D	0.001
E	0.0001

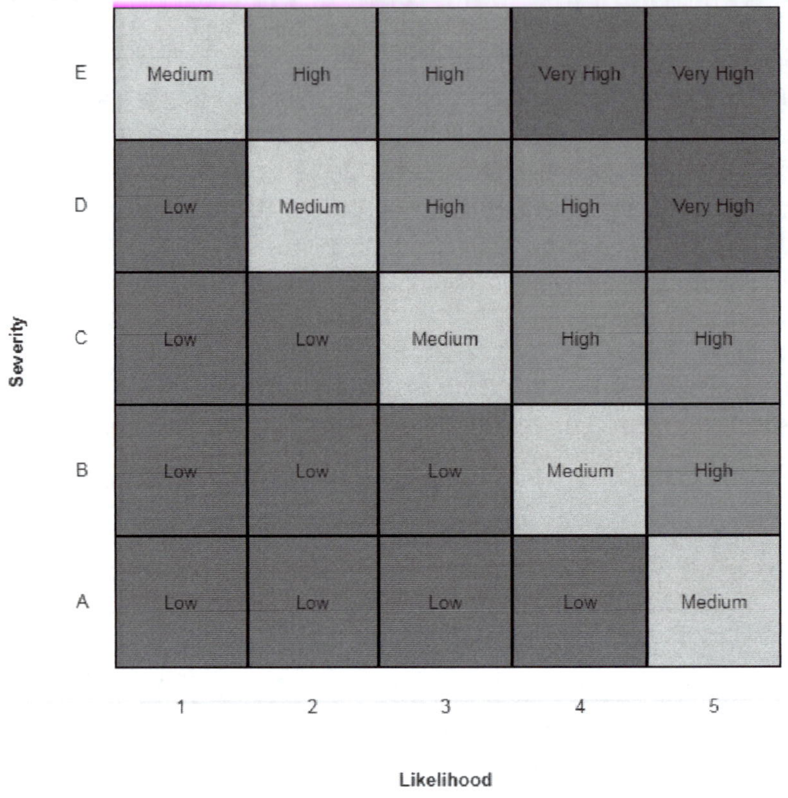

Figure 2. Case Study Risk Matrix

The methodology used to calculate risk reduction provided by Safeguards included the following criteria:

- Safeguards provide only reduction of the frequency of a consequence from occurring. The severities of the worst credible consequences identified in the PHA were assumed to be constant before and after the consideration of Safeguards.
- The assumed tolerable frequency of a consequence is indicative of its relative impact. The less tolerable a consequence is considered, the more severe its impact.
- If existing Safeguards are not as effective as described in the session, the corresponding risk reduction provided for the Safeguard may change.
- The assumed frequency of the cause and the PFD of the corresponding Safeguards can be used to determine the current projected frequency of the consequence scenario.
- The assumed frequency of the cause and the PFDs of the Safeguards remaining once the Safeguard of focus has been excluded can be used to determine the projected frequency of the consequence if the Safeguard of focus were to fail or be bypassed.
- The risk reduction of a Safeguard is the change in risk associated with the simulated failure or bypassing of that Safeguard.
- Risk reduction across different severities and consequence receptors were aggregated to determine the cumulative risk reduction associated with a Safeguard. This means that the more times a Safeguard was used in a PHA, the more risk reduction it will provide.

These cumulative values can be broken down into different scenarios, consequence receptors and severities as required.

- These calculated Safeguard risk reductions are snap shots in time that can be continually updated as Safeguards fail or are bypassed. This will be further elaborated in Section 3.

By using these assumptions and criteria, it is possible to calculate a "Criticality" value for each Safeguard that indicates the relative importance of that Safeguard in preventing hazardous scenarios. This Safeguard criticality considers how often the Safeguard is required for risk reduction in the PHA, the reliability of that safeguard, and the severity of the hazardous event or events it is protecting against.

2.2 Examples of Safeguard Risk Reduction Calculated from PHAs

Below are "snapshots" which present the initial static risk reduction of each Safeguard identified in three PHAs from different facilities. Figure 3 to Figure 8 below show:

- Safeguards sorted in descending order of the risk reduction, indicating the priority of that Safeguard for maintenance
- The risk reduction shown below is filtered to only show Health and Safety Consequence Receptor risk reduction as this is often the focus of Safety Critical Equipment lists
- Bowtie Visualizations associated with critical Safeguards to showcase how linking to the original PHA Data can make it easier for reviewers unfamiliar with the original study to understand why a Safeguard may be considered critical. On the bowtie visualization, blue rectangles represent the current risk-reducing Safeguards and black rectangles represent additional PHA recommendations for potential future Safeguards.

Figure 3 and Figure 4 showcase a 3-day PHA Session, in which 82 unique Risk-reducing Safeguards were identified. Observing the risk reduction for each Safeguard associated with Health and Safety related scenarios, we see that:

- The top risk-reducing Safeguard, a mechanical relief valve, is utilized 35 times for risk reduction and as can be seen on the Bowtie Visualization in Figure 4, as SG 18, is applied to high risk scenarios, contributing to its importance. The importance of the top Safeguard stems from protection of overpressure of the 1st stage discharge piping, which results in leaks or rupture and loss of containment of gas leading to fire. By visualizing this scenario, we can understand the logic of how this Safeguard may be identified as safety critical.
- The majority of risk reduction contribution comes from Mechanical Safeguards, such as Pressure Safety Valves (PSVs) and other relief devices.
- Occupancy a Modifier is within the Top 10 risk-reducing items identified in the PHA session, indicating the importance to communicate to site personnel the potential risks in the Flare Knock Out Drum area.
- The categories listed in the Donut Chart are Mechanical (MEC), BPCS Trips/Interlocks (BPCS-T), Compressor Panel Trips/Interlocks (CP-T), Procedure (PRO), Operator Rounds (ROUND), Occupancy (OCC), and Other

- The consequence receptors listed in the bottom right panel and the right hand side of the Bowtie are People (PPL), Environment (ENV), Asset (AST), Financial (FIN), Operational Costs (OPS), Regulatory (REG) and Reputation (REP).

Twenty (approximately 25%) unique Safeguards contribute to more than 90% of the risk reduction identified in the PHA. This indicates that a large concentration of the risk reduction is in only 25% of the identified Safeguards, making it logical to focus on this subset for ensuring regular maintenance. It should be noted that for larger facilities, its risk assessments are likely broken down into multiple PHAs. These PHAs can be combined together to give a complete picture of risk and risk reduction on a facility level.

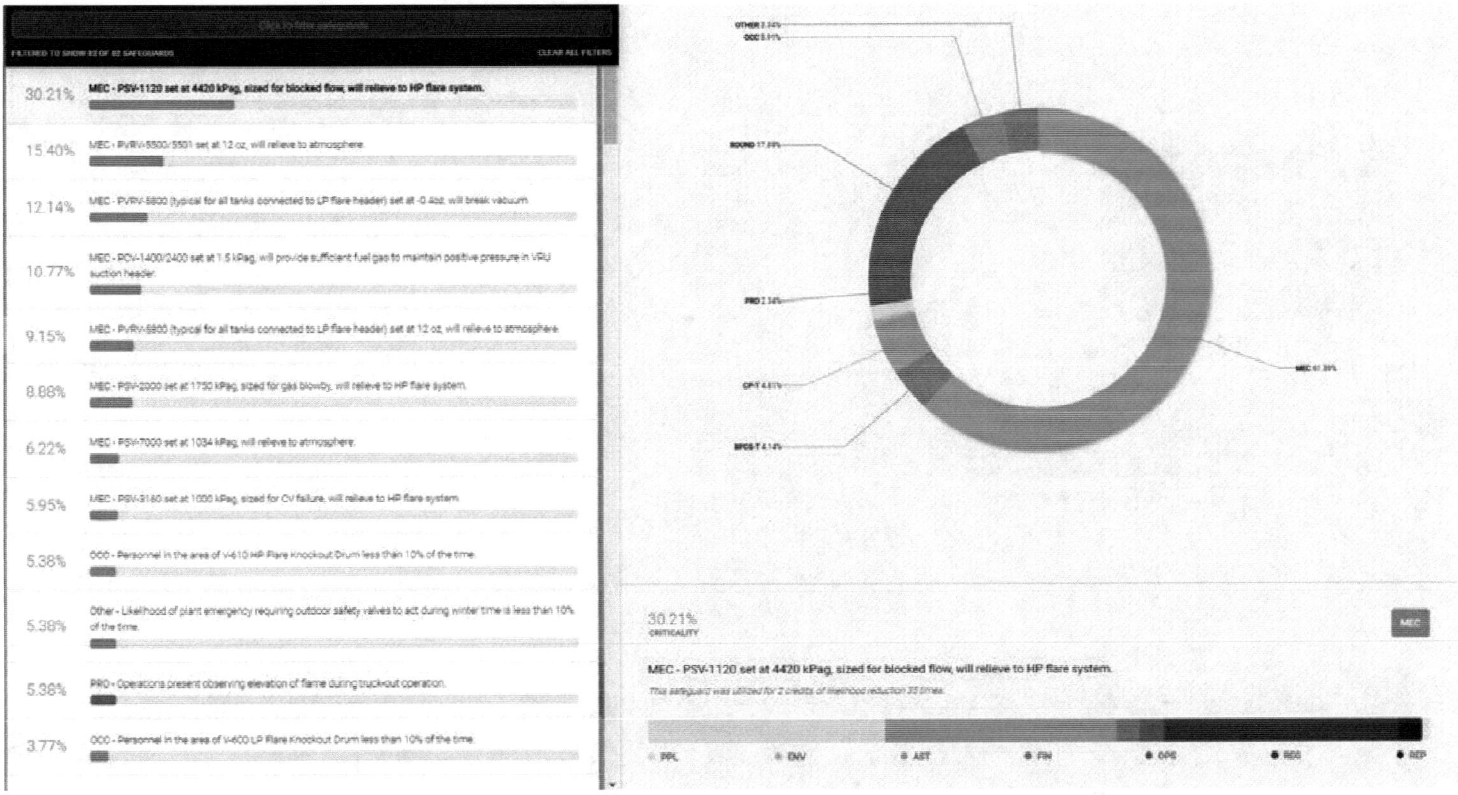

Figure 3. Safeguard Criticality for Compressor Station Filtered for H&S (Top 10)

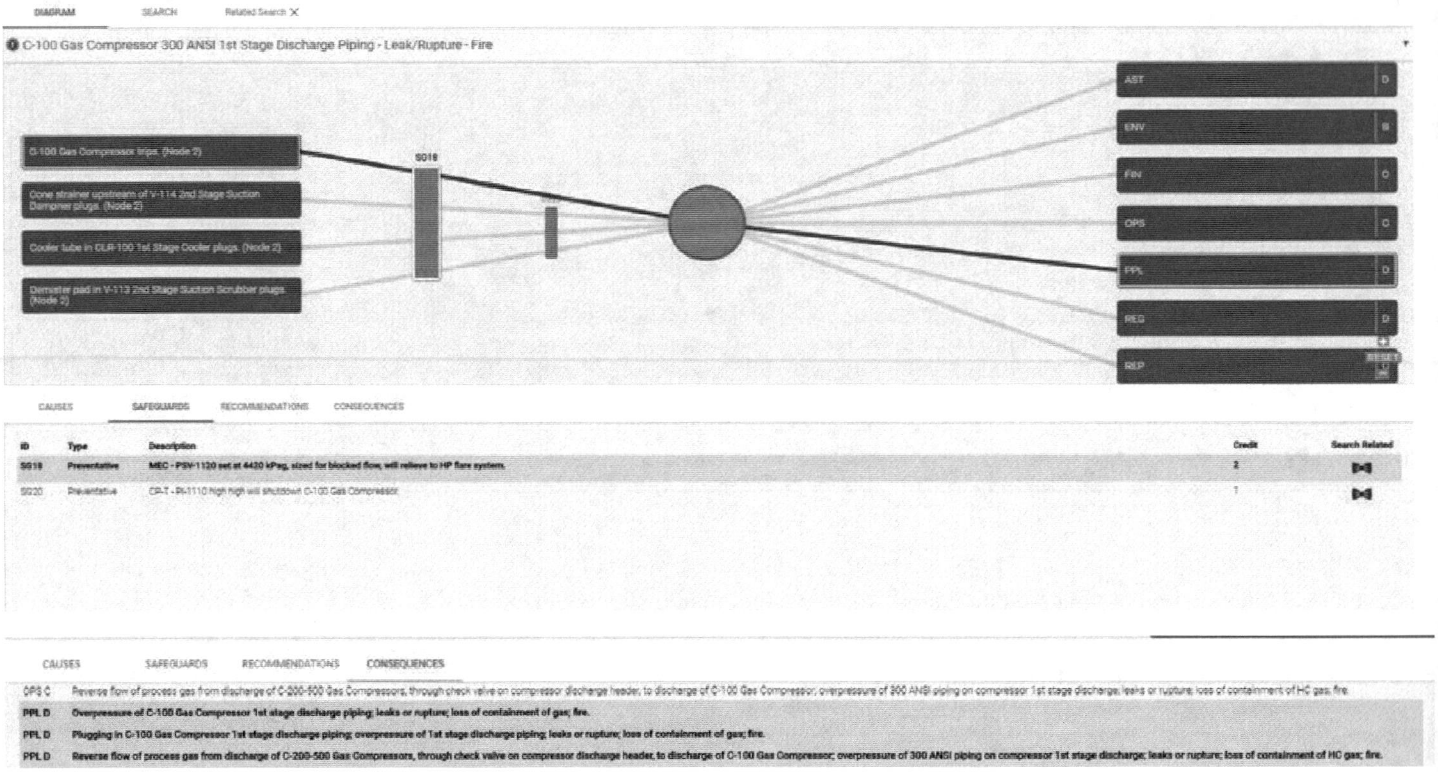

Figure 4. Bowtie Visualization for Top Safeguard of Compressor Station

Figure 5 and Figure 6 showcase a four-day PHA session, in which 40 unique risk-reducing Safeguards were identified. While analyzing the risk reduction for each Safeguard associated with Health and Safety related scenarios, it can be seen that:

- The top risk-reducing Safeguard, a BPCS alarm with operator action (BPCS), is utilized only 4 times for risk reduction, but as seen on the Bowtie Visualization is applied to high risk scenarios which have no other current Safeguards, contributing to its importance. The importance of the top Safeguard stems from protection of overpressure of a tank and the potential for fire or explosion. By visualizing the scenario, we can understand the logic of how this Safeguard may be identified as safety critical
- The majority of risk reduction contribution comes from BPCS Alarms with Operator Action, which implies the importance of not only maintaining the physical alarm equipment functionality, but also the training of operations to ensure they are aware of what actions to take when the alarm activates
- The top 5 (approximately12.5%) unique Safeguards contribute to more than 90% of the risk reduction identified in the PHA. This shows a high concentration of risk reduction in few Safeguards, providing a focus on which Safeguards require maintenance prioritization.
- The categories listed in the Donut Chart are Mechanical (MEC), BPCS Alarm with Operator Action (BPCS), Operator Rounds (ROUND)
- The consequence receptors listed in the bottom right panel and the right hand side of the Bowtie are Health and Safety (H&S), Environment (ENV), Economic (ECO), Regulatory (REG) and Reputation (REP).

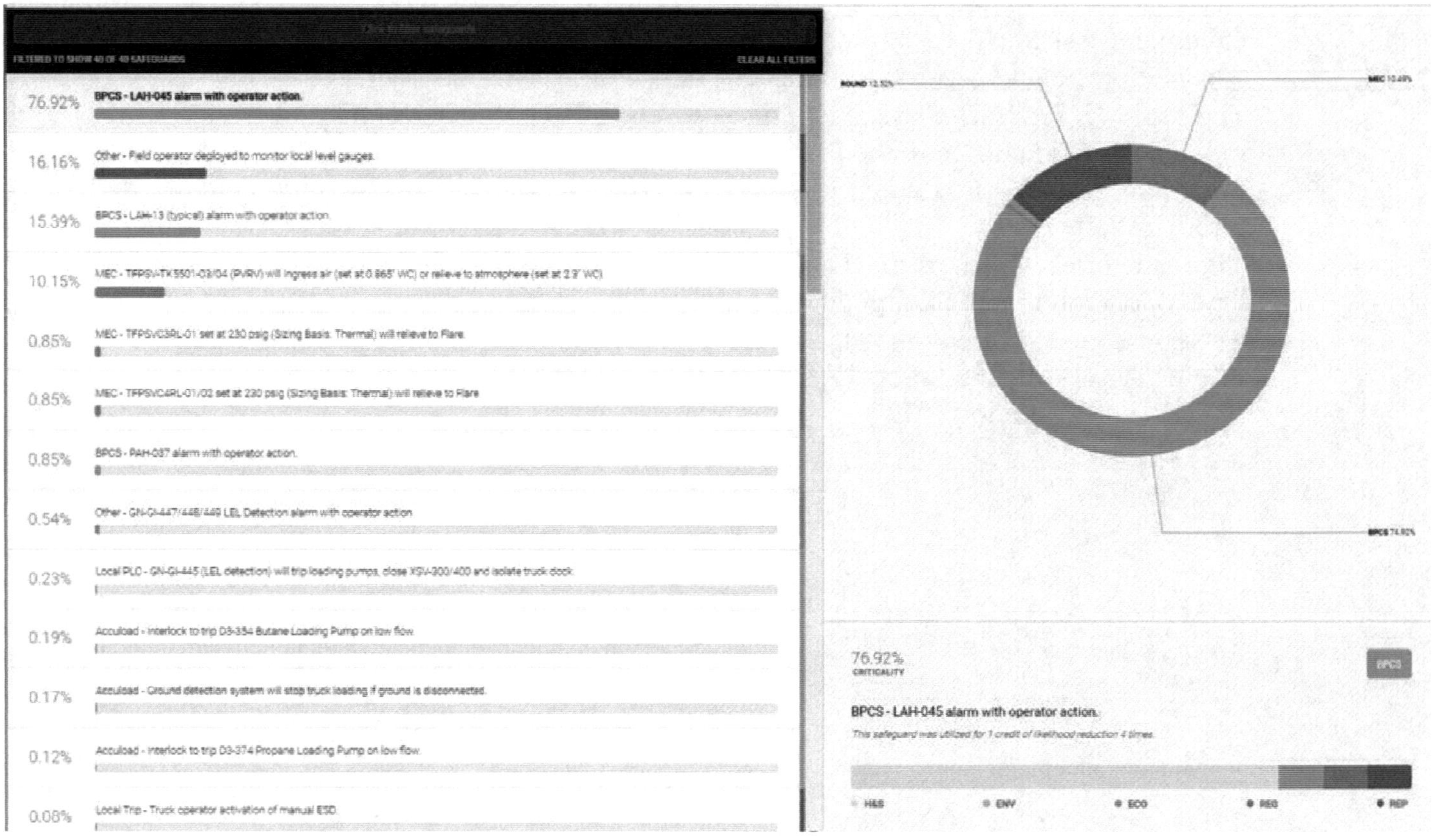

Figure 5. Safeguard Criticality for Butane and Propane Loading Station Filtered for H&S (Top 10)

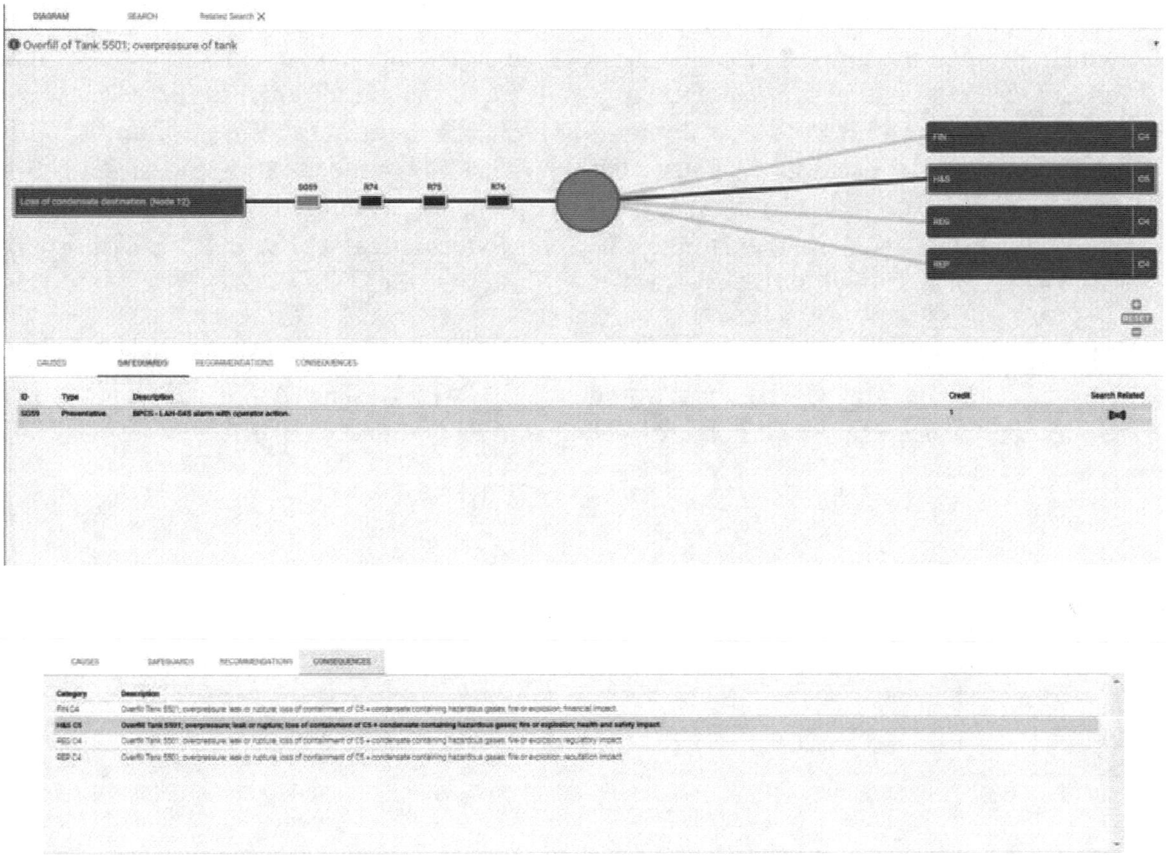

Figure 6. Bowtie Visualization for Top Safeguard of Butane and Propane Loading Station

Figure 7 and Figure 8 showcase a seven-day PHA session, in which 166 unique risk-reducing Safeguards were identified. Observing the risk reduction for each Safeguard associated with Health and Safety related scenarios, it can be seen that:

- The top risk-reducing item, excluding Modifiers of Ignition or Occupancy considered in the PHA, is from personnel using building entry procedures with personal O2 monitors, and as can be seen on the Bowtie Visualization it is applied to high risk scenarios. The importance of the top Safeguard stems from low level of oxygen in a building due to nitrogen leaks. By seeing this scenario, we can understand the logic as to why this Safeguard may be identified as safety critical.
- The majority of risk reduction contribution comes from Mechanical Relief Devices, which implies the importance of having a stringent preventative maintenance plan to ensure the relief valves are maintained to achieve the required reliability level.
- The top 30 (approximately 20%) unique Safeguards contribute to more than 90% of the risk reduction identified in the PHA. This shows a high concentration of risk reduction in relatively few Safeguards, allowing for easier focus to ensure maintenance prioritization.
- The categories listed in the Donut Chart are Mechanical (MEC), PLC Trip/Interlock (PLC-T), Local Trips/Interlocks (Local-T), Shutdown PLC (SD-PLC), Procedure (PRO), and Modifiers.
- The consequence receptors listed in the bottom right panel and the right hand side of the Bowtie are People (PPL), Environment (ENV), Asset (AST), Financial (FIN), Operational Costs (OPS), Regulatory (REG) and Reputation (REP).

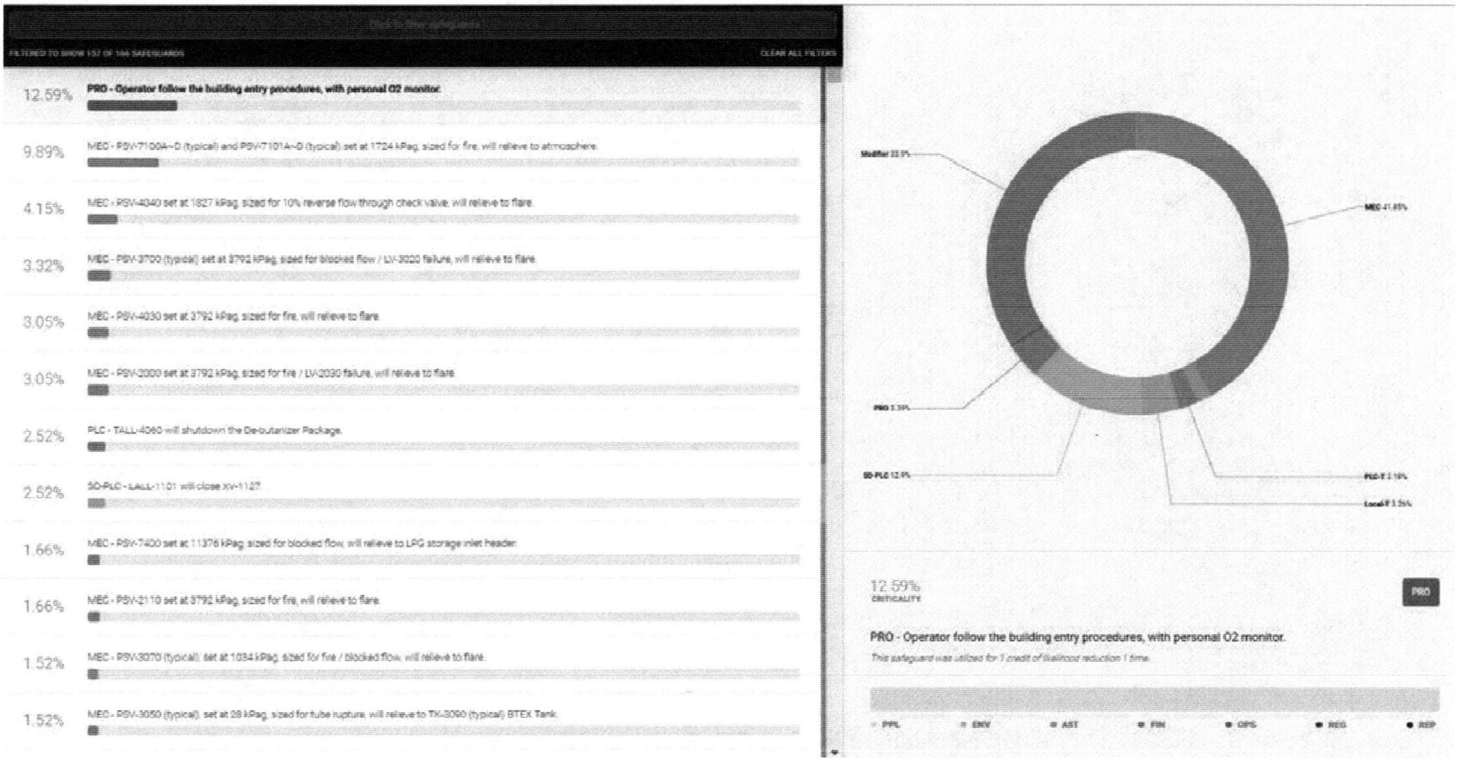

Figure 7. Safeguard Criticality for Refrigeration Plant Filtered for H&S (Top 10)

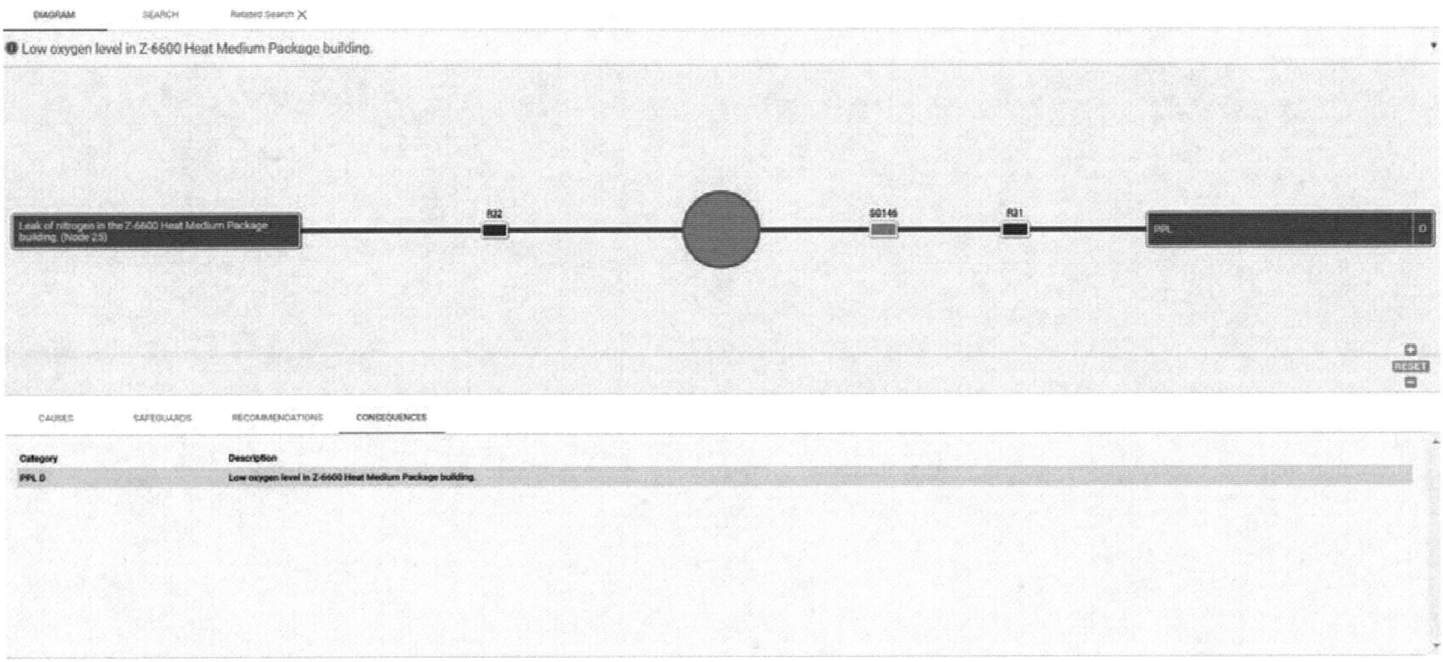

Figure 8. Bowtie Visualization for Top Safeguard of Refrigeration Plant

In the above examples, most facility risk reduction can be found within less than 25% of Safeguards, which will allow maintenance schedulers to focus in on the critical few. PHA data, either through utilizing Bowties or simply through PHA documentation reference, can provide an efficient way to identify critical Safeguards and communicate the reasons behind its criticality ranking.

3 Impact of Non-Effective Safeguards on Critical Safeguard Lists

3.1 Determining the Change in Critical Safeguard Lists

As mentioned in the prior sections, simply calculating the risk reduction of Safeguards based on the initial dataset may not be enough to reflect real processing facility scenarios, where Safeguards unexpectedly fail or are bypassed, and new recommendations are implemented to become Safeguards.

When the functionality of a Safeguard changes and it may no longer be relied upon to reduce risk, its impact on the frequency of the consequence and the criticality of the remaining Safeguards may need to be updated. For example, if a Safeguard fails, the current frequency and risk of a consequence scenario will increase, increasing the requirement and criticality of the remaining functional Safeguards to act when required. This simulation methodology will improve the ability of personnel to ensure critical Safeguard lists are up to date and reflect the current status of the processing facility, which in turn can minimize the risk exposure of site personnel.

The below figures showcase three simulation scenarios which demonstrate the impact on Safeguard criticality and the maintenance focus of remaining functional Safeguards when changes in a plant occur:

- Figure 9 and Figure 10 showcase the baseline case of a Butane and Propane Loading Facility and highlight an LEL detector
- Figure 11 and Figure 12 showcase the change that occurs if the LEL Detector were to fail or be bypassed and its effect on the criticality of remaining Safeguards
- Figure 13 and Figure 14 showcase the change that occurs to the baseline Safeguard criticality if a risk-reducing recommendation were to be implemented.

Figure 9 shows that the LEL Detection alarm is utilized 66 times within the facility, causing it to be the 9th most critical Safeguard identified in the PHA. The bowtie visual in Figure 10 shows that for one of the hazardous scenarios that the LEL alarm protects against, there is another Safeguard currently present, an interlock to trip the butane loading pumps on low flow to prevent damage to the pump seals causing loss of containment of butane.

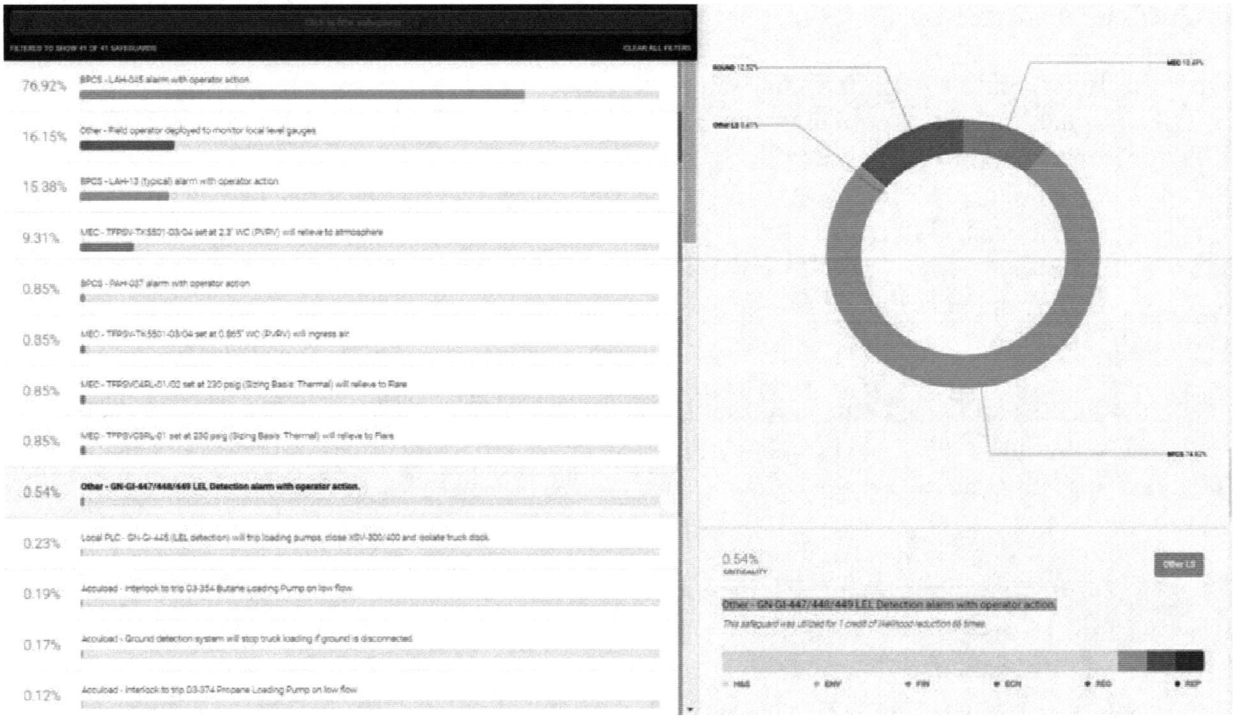

Figure 9. Butane and Propane Loading Station Safeguard Criticality (Baseline)

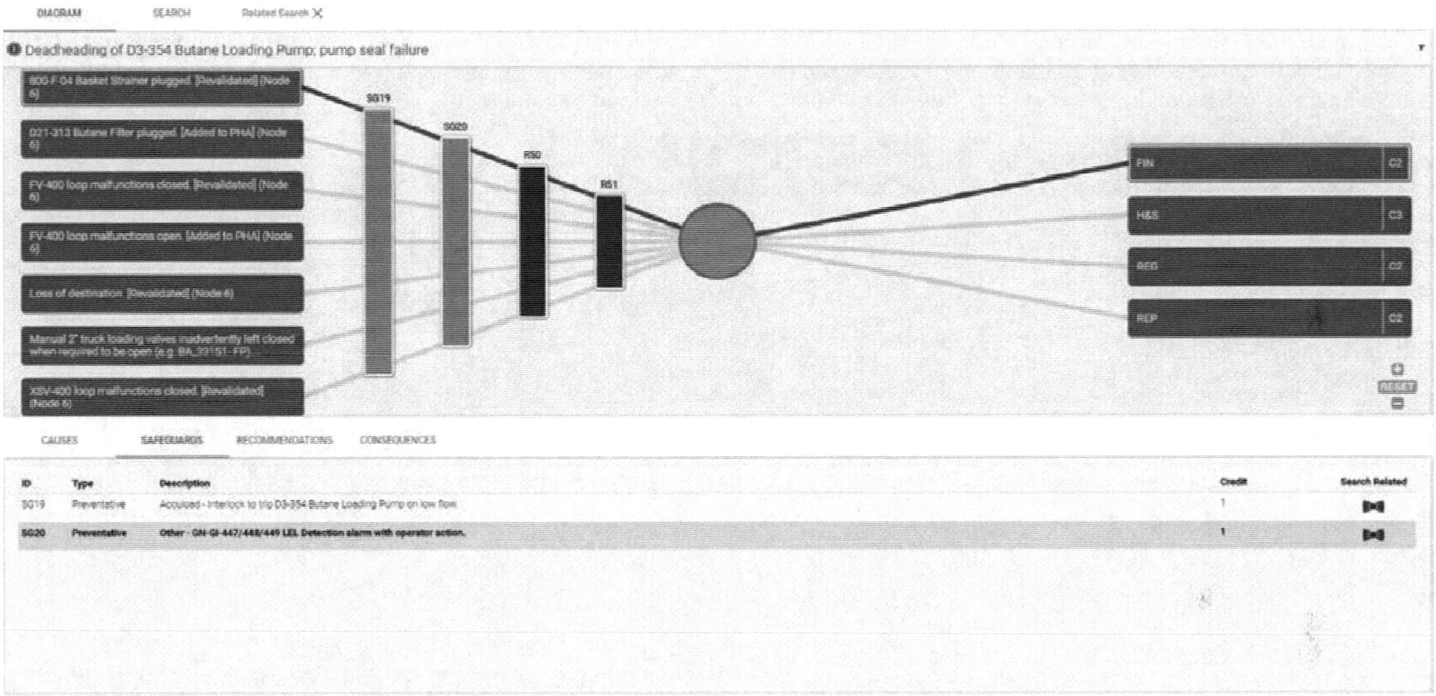

Figure 10. Bowtie Visualization for LEL Detection in Butane and Propane Loading Station

Figure 11 and Figure 12 show that once the LEL Detector has failed or been bypassed, the criticality of the remaining Safeguards has changed. For example, the Interlock to trip the Butane Pump has moved from Rank 11 to Rank 5, as its criticality has increased without the LEL detector available to help in reducing the risk of the scenario. This example showcases the importance of ensuring Safeguard Prioritization Lists are not static, but rather adjust to represent current conditions.

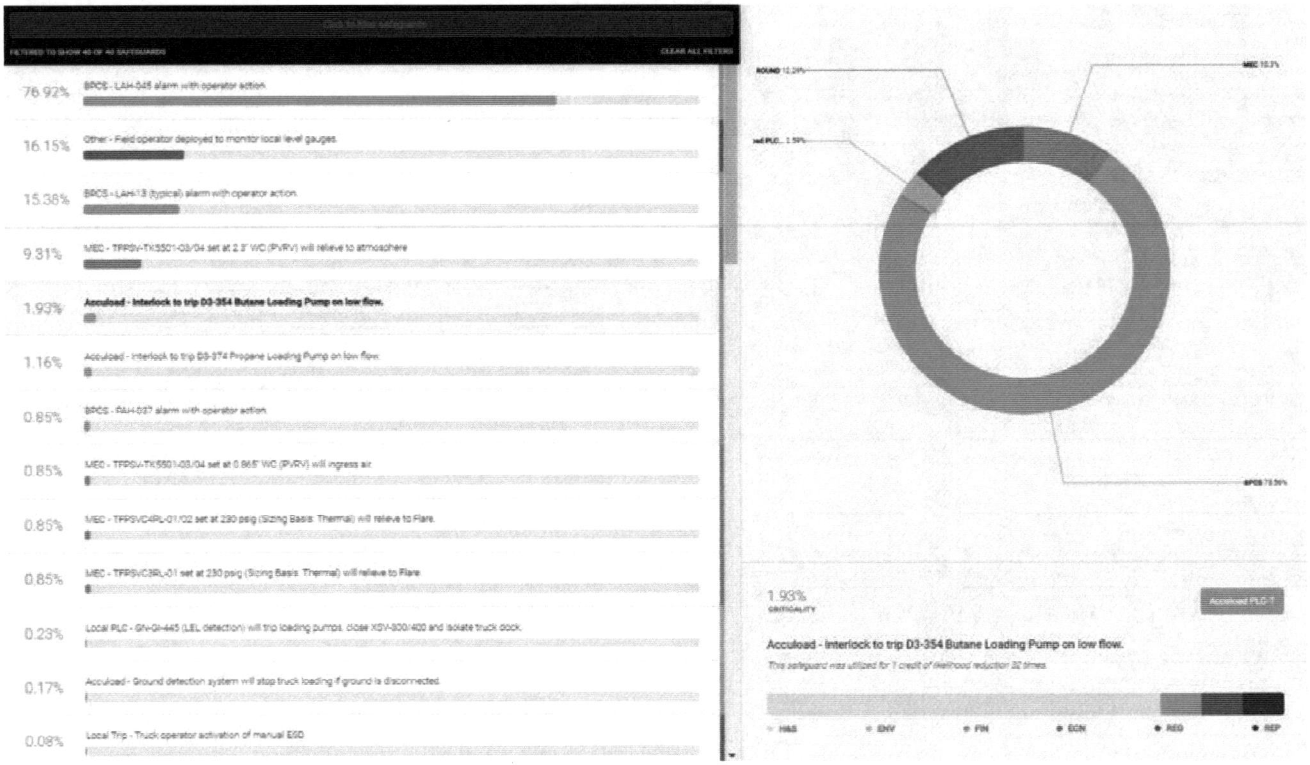

Figure 11. Butane and Propane Loading Station Safeguard Criticality After Failure of LEL Detector

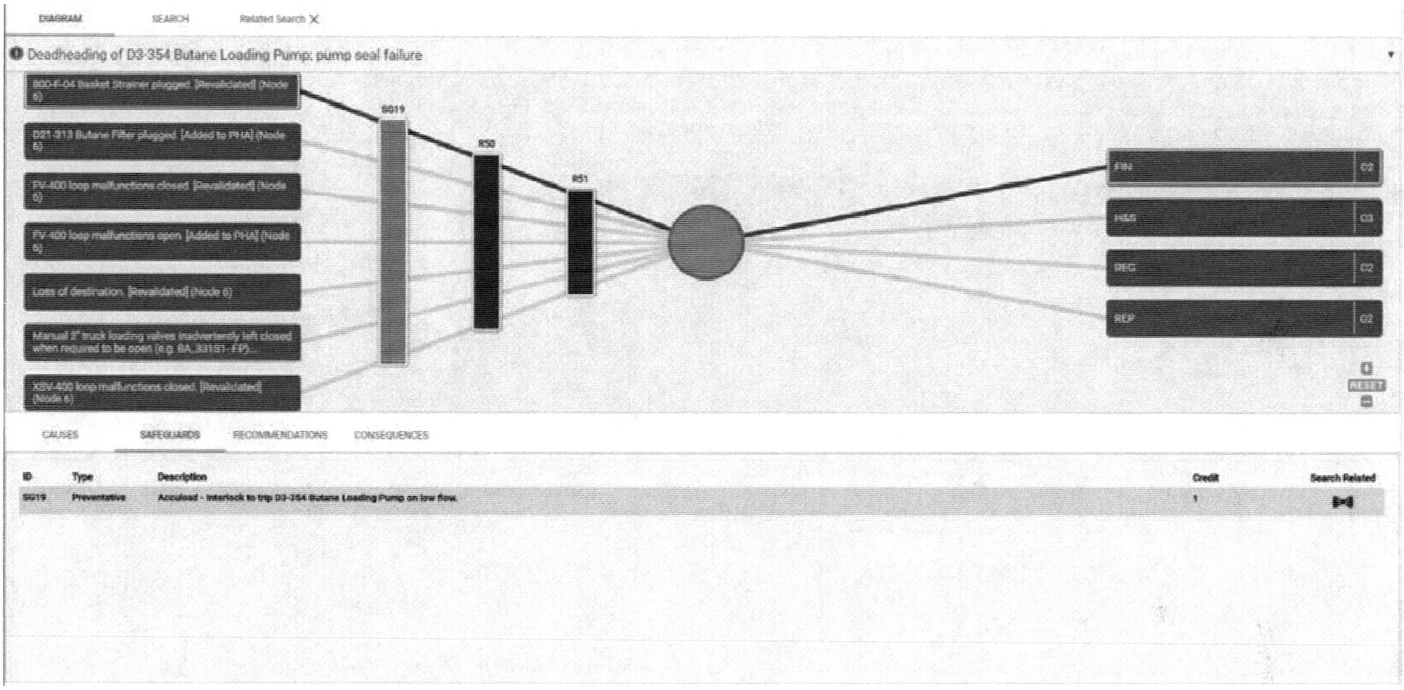

Figure 12. Bowtie Visualization of Butane and Propane Loading Station After Failure of LEL Detector

Figure 13 and Figure 14 showcase what would happen to Safeguard Criticality and prioritization for maintenance if a recommendation were to be implemented. When Recommendation # 36 is implemented, it becomes the 2nd most critical Safeguard for maintenance, while the Butane Pump Interlock drops in rank from # 11 to # 12. This shows that when recommendations are implemented, the risk on current scenarios may change, which in turn changes the risk reduction of remaining Safeguards. Without updating the Safeguard criticality list, the high criticality and importance of maintenance of the new Safeguard may not be identified.

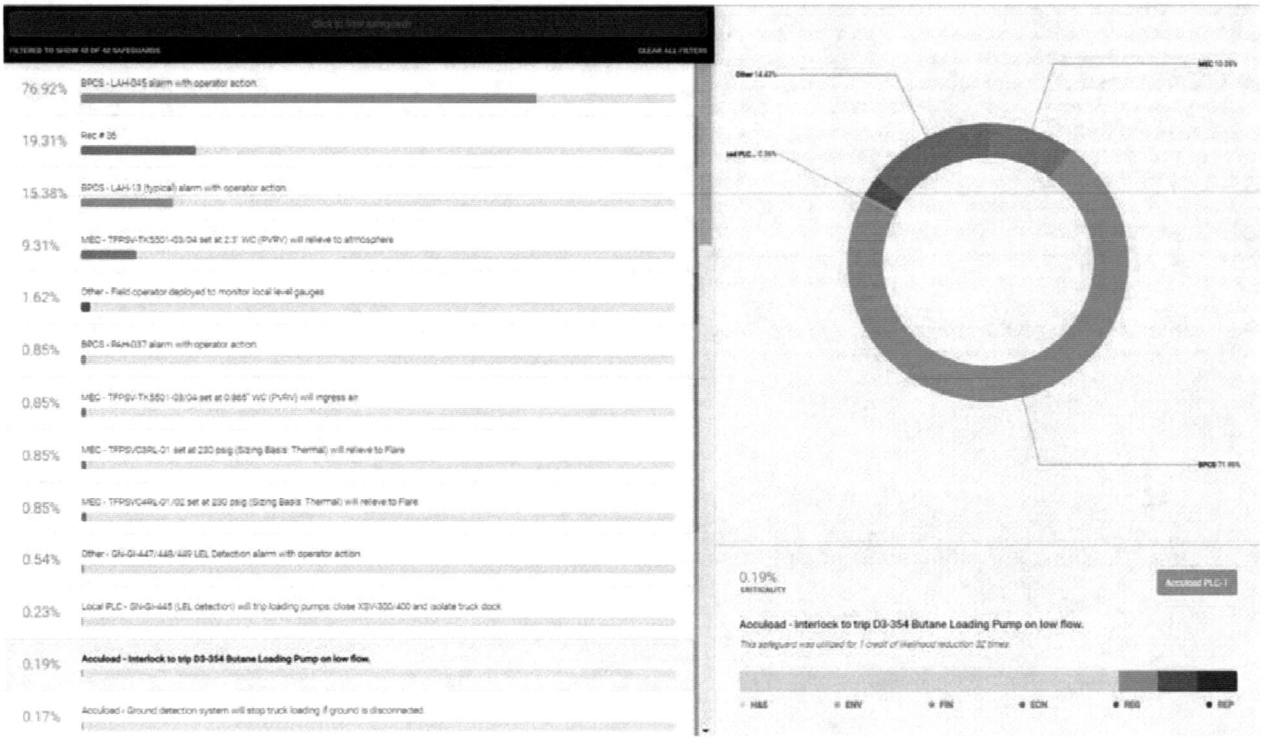

Figure 13. Butane and Propane Loading Station Safeguard Criticality After Implementation of PHA Recommendation # 36

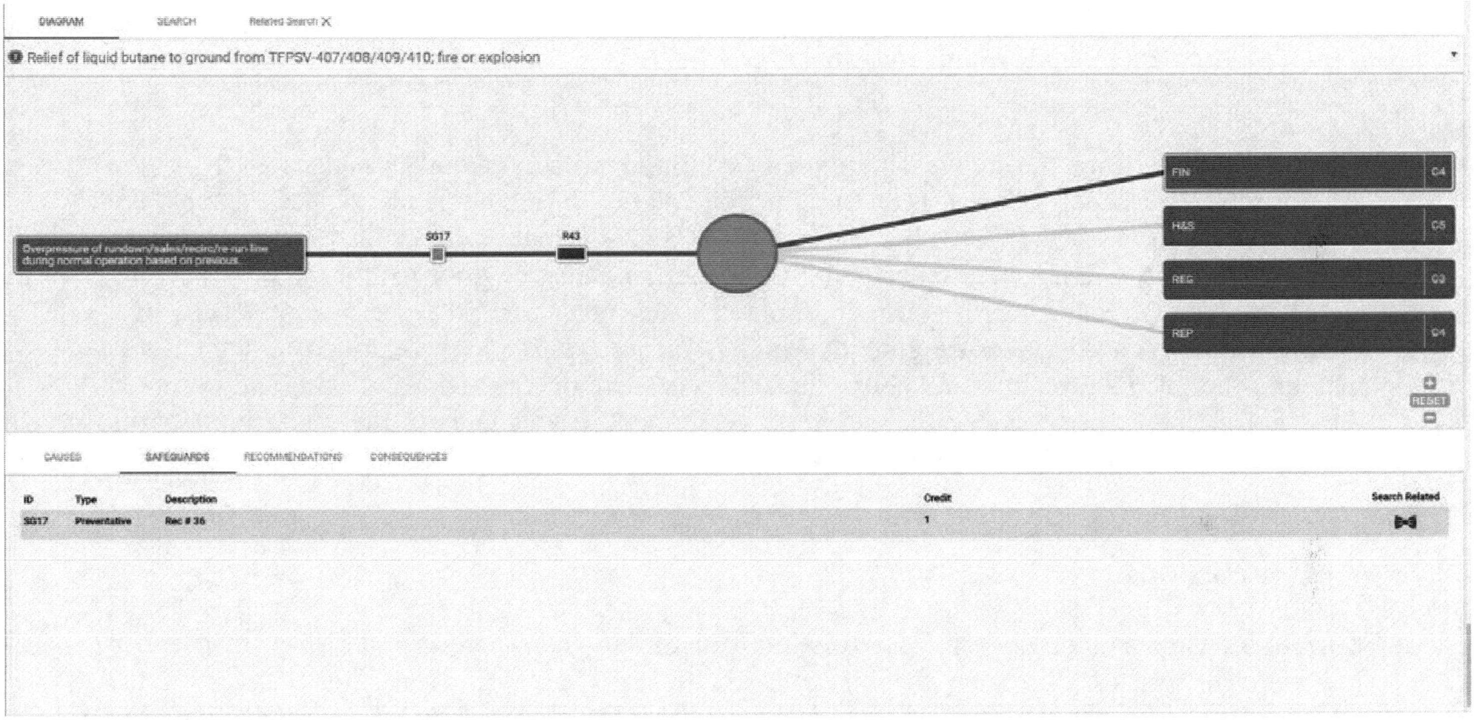

Figure 14. Bowtie Visualization of Butane and Propane Loading Station After Recommendation # 36 Implementation

Table 3 shows a summary of how the Butane Pump Interlock Safeguard changes in criticality in each of the above scenarios.

Table 3. Changes in Safeguard Criticality of Butane Loading Pump Trip over Simulations

Simulation	Rank
Baseline	11
Failure/Bypass of LEL Detector	5
Implementing Recommendation # 36	12

As can be seen from the above simulations, common process facility changes such as the effectiveness of current Safeguards and the implementation of recommendations can change the criticality and prioritization of other Safeguards. With current static critical Safeguard lists, teams will not be aware of the impacts of changes to maintenance prioritization, but by keeping an up-to-date list reflecting the current in-plant status of safeguards, the risk exposure in a facility can be reduced by ensuring maintenance is focused on the highest priority safeguards to maintain the intended required reliability levels. As mentioned previously, it should be noted that irrespective of how much maintenance is performed on a safeguard, there is a limit on its PFD due to its inherent design and features.

4 Conclusions

Current industry methodology for determining critical Safeguards produce a static list that may not be reflective of a process facility on a day-to-day basis, as there are changes in the effectiveness of Safeguards or recommendations implemented. Also, there is difficulty in understanding the relative criticality differences of various Safeguards, and documentation gaps which may lead independent reviewers or management to question why something is critical. These cumulative issues may lead to poor prioritization of maintenance, critical Safeguards with reduced availabilities and reliabilities, and a greater chance for process incidents with associated higher risk exposures.

By determining baseline Safeguard criticality from PHAs, industry can have a consistent way of determining the importance of Safeguards by taking advantage of the hundreds of SME hours that are put into developing PHA scenarios and the associated identified Safeguards. By utilizing simulation modelling, the Criticality List can also be easily updated to reflect risk changes from implemented recommendations and from Safeguards failing or being bypassed. This can provide an understanding in changes in criticality of the remaining functional Safeguards, revealing the change in the maintenance requirements of those Safeguards. Finally, by linking a critical Safeguard list to the corresponding PHA documentation, the reasoning behind the criticality of a Safeguard can be preserved. This improves the ability for engineers, operations and maintenance schedulers to make informed, effective decisions when considering modifying or removing existing Safeguards.

These documentation and simulation tools will allow management to better focus on Critical Safeguard maintenance at any point in the facility operation and ensure the long-term reliability of the most impactful Safeguards, as well as providing consistent input for Critical Safety Equipment and Alarm Rationalization.

Disclaimer:

The information in this paper is general in nature only and should not be relied upon without first obtaining advice from a qualified professional person. The advice and strategies herein may not be suitable for your situation. Any use which a third party makes of this paper, or any reliance on or decisions made based on it, are the responsibility of such third party. Neither the author nor ACM Risk Sciences and Development Inc. shall be responsible for damages, if any, suffered by any third party as a result of decisions made or actions taken based on this paper.

5 References

[1] Center for Chemical Process Safety (CCPS). *Guidelines for Enabling Conditions and Conditional Modifiers in Layer of Protection Analysis*. Hoboken, NJ, USA: John Wiley & Sons, Inc., 2012

[2] Center for Chemical Process Safety (CCPS). *Layer of Protection Analysis: Simplified Process Risk Assessment.* New York, New York, USA. American Institute of Chemical Engineers, 2001

Designing a Fluid Curtain System as an Independent Protection Layer

Marcel Amorós-Martí
ioMosaic Corporation
2100 Addison St. Suite 300, Berkeley, CA 94704
amoros.m.ca@iomosaic.com

Neil Prophet
ioMosaic Corporation
1900 St. James Place Suite 700, Houston, TX 77056
prophet@iomosaic.com

Gene Gorski
ioMosaic Corporation
93 Stiles Road, Salem, NH 03079
gorski@iomosaic.com

Prepared for Presentation at
American Institute of Chemical Engineers
2019 Spring Meeting and 15th Global Congress on Process Safety
New Orleans, LA
March 31 – April 3, 2019

Designing a Fluid Curtain System as an Independent Protection Layer

Marcel Amorós-Martí
ioMosaic Corporation
2100 Addison St. Suite 300, Berkeley, CA 94704
amoros.m.ca@iomosaic.com

Neil Prophet
ioMosaic Corporation
1900 St. James Place Suite 700, Houston, TX 77056
prophet@iomosaic.com

Gene Gorski
ioMosaic Corporation
93 Stiles Road, Salem, NH 03079
gorski@iomosaic.com

Keywords: Quantitative Risk Assessment (QRA), Risk Reduction, Fluid Curtains, Mitigation, Reliability, Safety Integrity Level (SIL)

Abstract

Fluid Curtains are an effective means of post-release mitigation, and are typically intended to dilute, disperse or neutralize toxic vapor cloud dispersions.

When designing a Fluid Curtain System, there are many parameters to consider regarding the effectiveness and reliability of the system. Focusing on the system effectiveness, these parameters include spray and nozzle characteristics, fluid solubility, contact time, fluid flowrate, and orientation; as well as considering all the Loss Of Containment scenarios (LOCs) that require mitigation.

If the Fluid Curtain System is to be considered as an Independent Protection Layer (IPL), then it must be capable of preventing a LOC from proceeding to the undesired consequence without being adversely affected by the initiating event, or the action of any other protection layer associated with the scenario. Applying this concept to the Fluid Curtain System design, it must guarantee a sufficient level of reliability, typically expressed by Safety Integrity Level (SIL).

This paper discusses the steps and requirements involved in designing a Fluid Curtain System, and the additional effort required to ensure it functions as an IPL.

1 Introduction

The Center for Chemical Process Safety (CCPS) [1] defines risk as the measure of human injury, environmental damage, or economic loss in terms of both the incident likelihood and the magnitude of the loss or injury. The loss, damage, or injury resulting from an undesired event are defined as the consequences. The likelihood of the undesired event measures how frequently this incident may occur. Accordingly, when determining risk, both the frequency of occurrence and the consequences that could lead to an undesired event such as an explosion, a fire, and a flammable and/or toxic vapor dispersion need to be considered.

The main purpose of a Quantitative Risk Assessment (QRA) is to answer the following key questions: (1) What can go wrong? (2) How likely is it? (3) What are the impacts? (4) What are the facility risk levels? (5) Is the risk tolerable? and (6) If not, which are the most appropriate safeguards for reducing risk to a tolerable level?

Figure 1 illustrates a simplified risk management program flowchart.

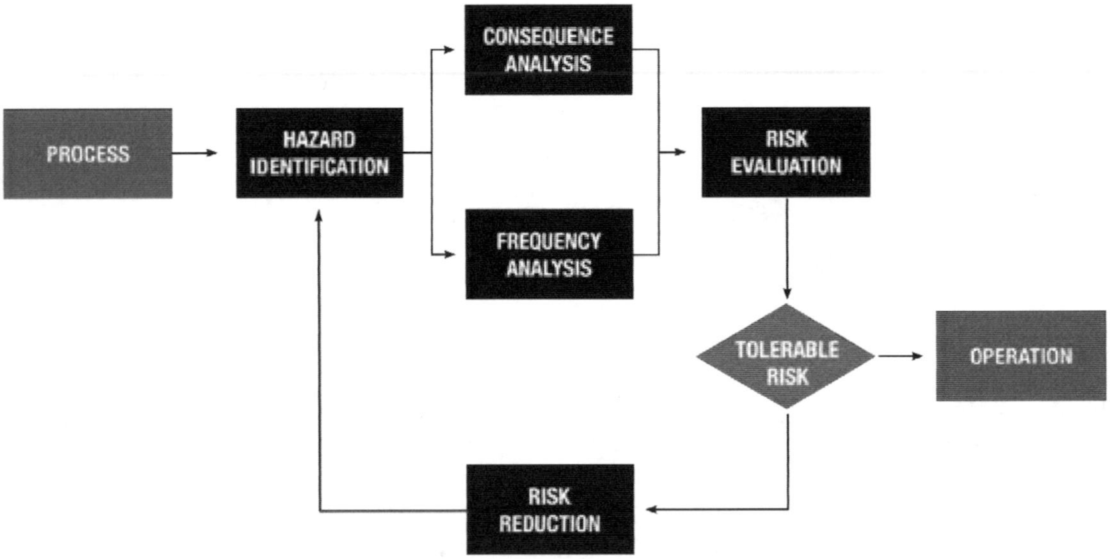

Figure 1: Risk Management Program Simplified Flowchart

The goal of the risk reduction step is to analyze the implementation of safeguards such as inherently safer process design, adding Independent Protection Layers (IPLs), prevention and/or mitigation activities in order to decrease the risk to at least a “tolerable if ALARP” (As Low As Reasonably Practicable) risk level.

The next section of this paper describes the basics of risk reduction.

1.1 Basics of Risk Reduction

During operation, things sometimes "go wrong". Normally, in a well-designed process, many layers of protection are present. These are intended to prevent a hazardous scenario from occurring. Generalized risk reduction measures found in the Chemical Process Industry (CPI) are illustrated in Figure 2 below.

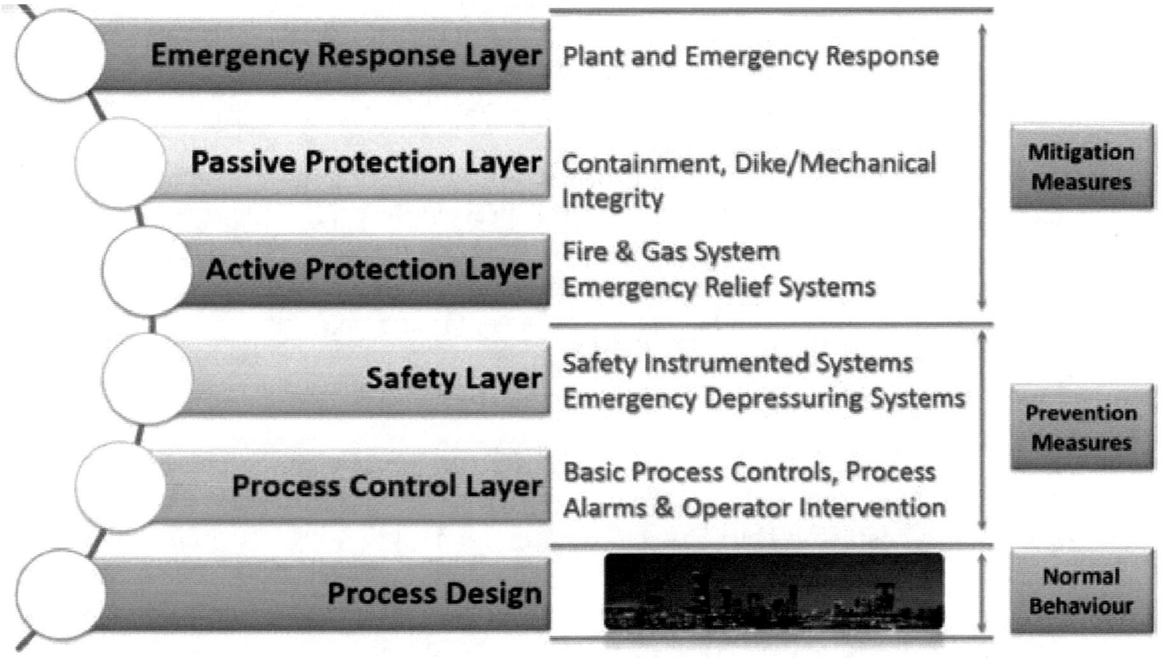

Figure 2: Generalized Layers of Protection

The generalized layers of protection that could be present in a typical CPI facility are categorized as follows:

Process Design Layer

- Process: the baseline design of a process, if done properly, may provide some risk reduction for certain scenarios, i.e., inherent safety.

Process Control Layer

- Basic Process Control System (BPCS): process control which includes continuous control loops, on/off control loops, and process control-based interlocks. The BPCS may be considered a protection layer only under certain circumstances, because the BPCS must not be the cause of the accident and must be independent of the cause and other independent protection layers.

- Process Alarm and Operator Intervention: layer of protection which requires operator intervention to prevent the accident, i.e., operator surveillance-based and/or alarm-based operator intervention.

Safety Layer

- Safety Instrumented System (SIS): set of equipment, i.e., Safety Instrumented Functions (SIFs), specifically designed to prevent the hazardous scenario from proceeding by taking an industrial process to a safe state when specific conditions are violated. If the SIS complies with applicable standards such as International Electrotechnical Commission (IEC) 61508 [2], IEC 61511 [3], and International Society of Automation (ISA) Technical Report (TR) 84.00.17 [4], it provides strong protection according to the design Safety Integrity Level (SIL) and associated Probability of Failure on Demand (PFD) of each SIF.

- Emergency Depressuring Systems; protection systems that usually consist of piping and valves that must be adequately sized to prevent the protected process equipment from experiencing an overpressure and/or over-temperature rupture. Sizing of these systems typically involves dynamic simulation of the pressure, temperature, and composition profiles from the start of the event under analysis.

Active Protection Layer

- Emergency Relief Systems (ERS): layer of protection which includes devices such as relief valves and rupture disks. It can be debatable if these layers of protection should be considered active or passive. In some situations, ERS are intended or capable of preventing the most severe consequence from occurring (i.e., overpressure prevention), but in other cases, the mentioned devices can eliminate or mitigate the consequence when their sizing and stable operation are ensured and tied into a well-designed flare system (i.e., mitigation) or other effluent handing system.

- Fire and Gas detector Systems (FGS): modern fire and gas detectors combined for example with an emergency shutdown system can pertain to this category of active protection layer as these can reduce or even eliminate the hazardous scenario. Specifically, FGS should comply with criteria established in standards such as ISA TR 84.00.07 [4].

- Fluid Curtains and Deluge Systems: FGS rationale can be applied to fluid curtains and deluge systems, which are automatically activated by detection systems. These systems provide protection against fires and/or flammable/toxic dispersions, and these systems are normally composed of open sprinklers attached to a piping system that is connected to an extinguishing material supply (e.g., water, steam, foam, dry powder, inert gas). When the system is activated, the discharge must occur rapidly and efficiently in response to the scenario.

Passive Protection Layer

- Layers of protection that do not need to be physically activated to provide risk reduction, e.g., dikes, fire walls, and blast walls. These layers of protection are similar to process design protection layers such as metallurgy and pressure rating, because they are a static characteristic of the process that provides risk reduction without involving sensors, logic solvers, or moving parts.

Emergency Response Layer

- Emergency Response: last layer of protection category mainly intended for addressing catastrophic consequences such as releases of toxic materials, fires and explosions; e.g., plant and community evacuation or shelter-in-place plans.

When discussing the generalized typical protection layers present in a CPI facility, it is also important to address enabling conditions and conditional modifiers. Enabling conditions and conditional modifiers, if/when considered, may also effectively reduce the calculated level of risk. Note that enabling conditions and conditional modifiers can be defined as follows [5]:

- Enabling Conditions: a condition that is not a failure, error or protection layer that makes it possible for an incident sequence to proceed to a consequence of concern. It consists of a condition or operating phase that does not directly cause the scenario, but that must be present or active in order for the scenario to proceed to a loss event; expressed as a dimensionless probability, and

- Conditional Modifiers: one of several probabilities included in scenario risk calculations, generally when risk criteria endpoints are expressed in impact terms (e.g. fatalities) instead of in primary loss event terms (e.g. release, vessel rupture). Conditional modifiers include, but are not limited to: probability of a hazardous atmosphere, probability of ignition, probability of explosion, probability of personnel presence, probability of injury or fatality, and probability of equipment damage or other financial impact.

2 Background

ioMosaic Corporation (ioMosaic) conducted a detailed QRA study for a facility with more than 100 process units using Process Safety Office® SuperChems™ [6]. The findings from this QRA indicated that the risk levels in a populated area surrounding the plant were higher than the Company's risk tolerability limits for land-use planning purposes. The main hazard contributing to the intolerable risk level was identified as toxic vapor cloud dispersions containing high quantities of hydrogen sulfide.

Since the risk in the populated public area was estimated to be higher than the risk tolerability criteria limits when considering the public to be vulnerable, the next step was the identification of potential risk reduction measures to ensure that the risk to the affected populated public area was within the risk tolerability limits. As a result, ioMosaic developed a methodology to identify and quantify the most suitable, practicable and economic risk reduction measures to achieve this goal [7]. This methodology consists of three (3) steps: (1) Identification of the most contributing process units and associated pieces of equipment impacting the target location, (2) quantification of Risk Reduction Factors (RRFs) for the most contributing hazardous scenarios; Table 1 below is an example listing the RRFs as a function of process unit, and (3) definition of risk reduction measures that met the calculated RRFs.

The RRF was defined as the ratio of the actual Rate Of Death (ROD) level for each piece of equipment and a constant target ROD level iteratively calculated. The RRF is expressed as follows (see Equation 1):

$$RRF_i = \frac{actual\ ROD_i}{constant\ Target\ ROD} \qquad \text{Equation 1}$$

where i is the piece of equipment being analyzed.

The ROD or Potential Loss of Life (PPL) is an index which describes the expected number of fatalities per year (or over the life time of a project). ROD is a type of risk integral, and is a summation of risk as expressed by the product of frequency and consequences; i.e., number of fatalities. The integral is summed over all potential events that can occur. It is mainly used to compare options and enables the inclusion of different risk types like process, transport, workplace hazards in one number.

Table 1. Highest Risk Process Units and Required RRFs

Risk Contributor / Process Unit	ID Name	RRF
Process Unit 02-Train 01/02	PU-02	994
Process Unit 04	PU-04	618
Process Unit 01	PU-01	253
Process Unit 03-Train 01/02/03	PU-03	117
Process Unit 05-Train 01/02	PU-05	45

Based on the approach explained in reference [7], it was concluded that well-designed fluid curtains should be capable of mitigating the Loss Of Containment scenarios (LOCs) from the identified pieces of equipment (and process units) and thereby reducing risk to the populated public area from "intolerable" to a "Tolerable if ALARP" level.

Figure 3 below provides an example comparison of the QRA individual risk contours before and after the implementation of the risk reduction measures with the associated RRFs. The land-use planning risk tolerability criteria are based on the United Kingdom Health and Safety Executive (UK HSE) criteria [8] and are provided in Table 2. Referring to the selected risk tolerability criteria listed in Table 2, the middle risk region, the so-called "Tolerable if ALARP" level is based on the ALARP principle[1], and can only be considered tolerable if it can be demonstrated that the risk has been averted to the point where additional risk reduction would result in a gross disproportion between the costs and benefits of doing so [8].

Table 2. UK HSE Risk Tolerability Criteria

Risk Region	**Non-Vulnerable Public [Fatalities yr^{-1}]**	**Vulnerable Public [Fatalities yr^{-1}]**
INTOLERABLE	Risk > 1.00E-05	Risk > 1.00E-06
TOLERABLE if ALARP	1.00E-05 ≥ Risk ≥ 1.00E-06	1.00E-06 ≥ Risk ≥ 0.33E-06
BROADY ACCEPTABLE	Risk < 1.00E-06	Risk < 0.33E-06

[1] ALARP: As Low As Reasonable Practicable; i.e., "*risk must be averted unless there is a gross disproportion between the costs and benefits of doing so*".

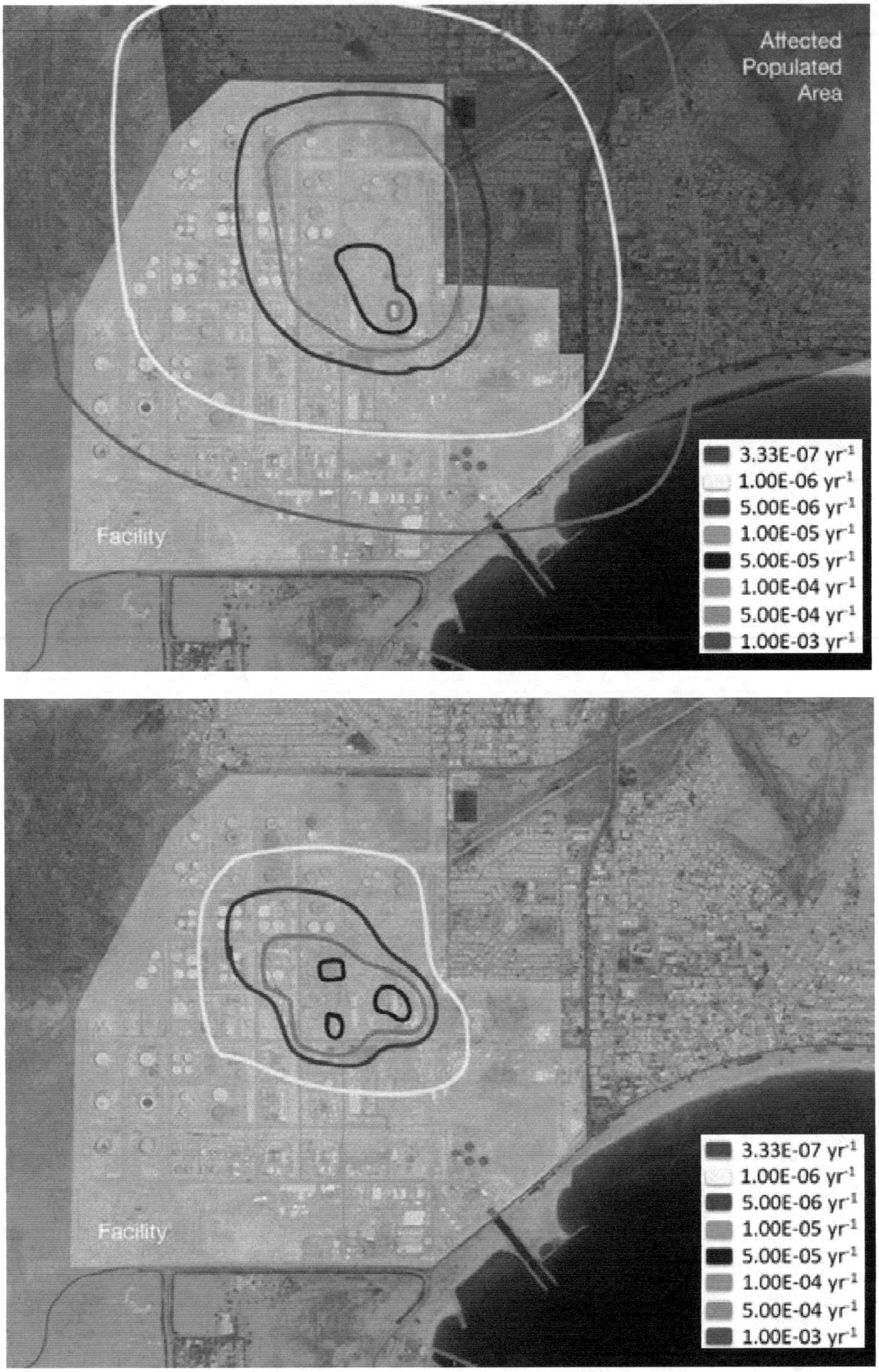

Source: Extracted from *Advanced QRA Methodology – Quantifying Risk Reduction Measures to Minimize Economic Investment [7]*

Figure 3: Example Comparison of QRA Risk Results

3 Fluid Curtain System Design Approach

Fluid curtain systems are an effective means of post-release mitigation, and are typically intended to dilute, disperse or neutralize toxic vapor cloud dispersions. With regard to the fluid curtains design approach, this Independent Protection Layer (IPL) needs to comply with a consequence-based design and a frequency-based design:

- Fluid curtain consequence-based design; i.e., the toxic cloud impact distance after fluid curtain treatment must not reach the affected populated area at a certain toxicity threshold.
- Fluid curtain frequency-based design; i.e., the fluid curtain including associated storage and supply must meet a certain reliability level.

Based on the fluid curtains design requirements explained above, it is important to note that existing fire water systems (e.g., deluges, sprinklers, isolated fire monitors, etc.) are generally designed for other purposes other than mitigation of toxic vapor dispersions or other consequences as a result of a LOC scenario. Additionally, these types of systems lack the required reliability to be considered an IPL. Thus, relying on existing fire water systems to protect against these types of events is not a good engineering practice; i.e., it is necessary to design new systems (fluid curtains) to ensure proper mitigation.

The next sections of this paper discuss the approach and steps necessary to ensure both the fluid curtain design (in terms of consequence mitigation) and level of reliability required to function as an IPL are met.

3.1 Consequence-Based Design

The consequence-based design approach explained in this paper consists of the following 3 steps: (1) selecting the applicable threshold (i.e. end-point criteria) for consequence modeling, (2) conducting a sensitivity analysis to identify the worst-credible LOCs, and (3) designing the fluid curtains considering the physical and chemical phenomena associated with spraying a toxic vapor cloud with a fluid.

3.1.1 Selection of Threshold Criteria

The design of fluid curtains for mitigating toxic vapor cloud dispersions depends on the threshold selected, because the amount of fluid (among other parameters) necessary to neutralize toxic dispersions will vary based on the toxic gas concentration end-point selected.

There are several recognized exposure limits for airborne concentrations, which are methods based on a combination of results from animal experiments, observations of long- and short-term human exposures and expert judgment. The published hydrogen sulfide thresholds include Emergency Response Planning Guidelines for Air Contaminants (ERPGs), Immediately Dangerous to Life and Health Levels (IDLHs), Emergency Exposure Guidance Levels (EEGLs), Acute Exposure Guidelines Levels (AEGLs), and Short Term Public Emergency Guidance Levels (SPEGLs).

The ERPGs are published by the American Industrial Hygiene Association (AIHA) [9]. Because of the comprehensive effort to develop acute toxicity values, ERPGs are becoming an acceptable industry norm. The ERPGs provide three (3) concentration levels, i.e., ERPG-1, ERPG-2, and ERPG-3, with each value providing a different level of expected damage for sixty (60) minutes of exposure to a specific toxic substance [10]:

- ERPG-1 is the maximum airborne concentration below which it is believed nearly all individuals could be exposed for up to 1 hour without experiencing effects other than mild transient adverse health effects or perceiving a clearly defined objectionable odor.
- ERPG-2 is the maximum airborne concentration below which it is believed nearly all individuals could be exposed for up to 1 hour without experiencing or developing irreversible or other serious health effects or symptoms that could impair their abilities to take protective action.
- ERPG-3 is the maximum airborne concentration below which it is believed nearly all individuals could be exposed for up to 1 hour without experiencing or developing life-threatening health effects.

Table 3 below provides those concentrations for hydrogen sulfide:

Table 3. Hydrogen Sulfide ERPGs

Threshold Level	**Concentration [ppm]**
ERPG-1	0.1
ERPG-2	30
ERPG-3	100

Based on the concentration thresholds as listed in Table 3 above, and according to the ERPG-2 threshold definition, 30 ppm could be deemed as a conservative and feasible concentration value to be used for designing fluid curtains.

3.1.2 Sensitivity Analysis to Identify Worst-Credible LOCs

The fluid curtains need to ensure that any toxic dispersion release within the identified process units will be mitigated. Thus, the next step is to conduct a detailed sensitivity analysis in order to calculate and identify the worst-credible LOC scenarios. These hazardous scenarios will be used later to analyze and determine the fluid curtain design basis, such as fluid flow requirements, type and number of nozzles, etc. In order to cover all possibilities, parameters such as release angle, release height and meteorological conditions should be studied.

With regard to the release angle, the fluid curtains need to be high enough to be able to capture the toxic vapor cloud dispersions; thus, the identification of worst-credible LOCs should account for the possibility of vertical releases. Additionally, QRA studies typically model horizontal releases, as these are expected to result in the worst toxic impact distances [10].

Addressing release heights, releases occurring at 40 meters and 10 meters above ground are expected to result in different toxic concentrations at 2-meter height; the typical QRA receptor elevation.

Finally, the meteorological conditions, generally defined as the combination of wind speed and atmospheric stability based on Pasquill criteria [11], can affect the dispersion modeling. Typically, meteorological conditions D5 and F2 are used when calculating toxic impact distances due to the release of a toxic vapor cloud. D5 refers to a wind speed of 5 m/s and a D atmospheric stability (i.e., day-time), while F2 refers to a wind speed of 2 m/s and an F atmospheric stability (night-time). Note that generally D5 meteorological conditions are considered to be the most credible conditions, while F2 are considered to be the worst-case conditions for continuous toxic releases.

3.1.3 Modeling Design Basis

When designing fluid curtain systems, the following physical and chemical phenomena need to be considered.

Release Momentum

Fluid curtain effectiveness claims by certain vendors are misleading; i.e., a curtain placed at the periphery of process equipment will only scrub a puff of a release. Studies that demonstrate fluid curtains to be effective assume low gas velocities. However, calculations show that pressurized contents when released from an orifice to the atmosphere exit containment with high velocity and momentum in the form of a gas or two-phase jet. The gas velocity should be as low as possible for proper scrubbing, otherwise, the jet may penetrate the fluid curtain without sufficiently scrubbing the toxic material. With no physical barrier other than air, the jet must travel quite a distance before the velocity drops enough to be effectively scrubbed by a fluid curtain (e.g.., 15-30 meters). This is especially true for high velocity releases where the distance is larger to attain a low velocity (e.g., 5 $m \cdot s^{-1}$) that can be successfully scrubbed. Therefore, a separation distance between the loss of containment point and the fluid curtain is needed for it to be effective and, as a result, the dimension of a fluid curtain manifold ring that surrounds the equipment could be quite large [12].

In addition, Ghosh and Hunt [13] studied the effects of a cross-flow on a spray jet by considering how fluid jets are deflected and deformed by cross-flows. When a liquid first emerges through a pressure atomizing nozzle, it is in the form of a jet or a sheet which quickly disintegrates into droplets due to aerodynamic instabilities in the “break up region” which interacts strongly with the atmosphere. The study concludes that to be effective, the gas velocity at the spray (or fluid curtain) boundary should be around 5 m/s and the droplet velocity should be around 15 m/s, considering an adequate droplet size. With regard to the droplet size, low droplet carryover is accomplished with a droplet size of 250-300 mm, which still shows an acceptable mass transfer interfacial area.

As a result, when designing fluid curtains, it is critical to accurately model the fluid momentum for each LOC scenario. For each LOC, the first step is to determine the distance from the release where the vapor/two-phase jet downwind distance and height reaches a capture (or target) velocity of 5 $m \cdot s^{-1}$. The series of points that define these radial distances from the release point can be used to determine the fluid curtain manifold layout, location, and height.

The impact distance at the target (or capture) velocity is referred to as the fluid curtain capture distance, while the height that meets the target velocity may be called the fluid curtain capture height. Figure 4 below illustrates an example of the fluid curtains capture distances for four LOC scenarios. Referring to Figure 4, the dark blue line shows the optimal location of the fluid curtain manifold for these four scenarios.

Once release phenomena are characterized, the next step is to decide what type of fluid curtain may be applied to the system in the study scope. CCPS [14] describes and explains the different types of fluid curtains; i.e., water curtains, steam curtains, air curtains, foam scrubbing and dry powder curtains. This paper discusses water curtains, as this type has been successfully installed for toxic releases such as ammonia and hydrogen fluoride. The two main phenomena associated with water curtains is dilution and absorption.

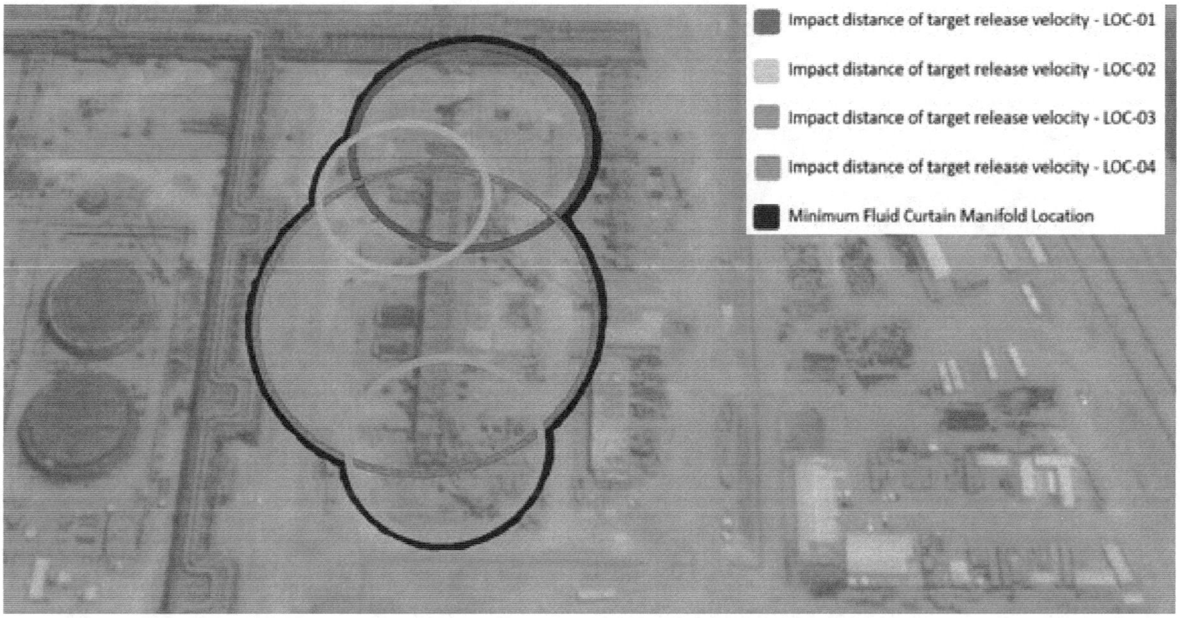

Figure 4: Fluid Curtain Capture Distances that Satisfy Target Velocity Criteria

Dilution Phenomena

Water sprays may dilute a toxic vapor cloud by entraining large volumes of air. Air entrainment is induced by momentum transfer from the water droplets to the ambient air and is a strong function of spray configuration, droplet size, droplet velocity, and spray location. Large droplets entrain smaller amounts of air than smaller droplets and induce better mixing and dispersion. Although smaller droplets entrain more air, they have smaller terminal velocities which are almost constant but cause poor mixing and dilution. A good estimate of air entrainment flow in a spray can be obtained by examining in detail the momentum exchange between the spray and the surrounding air. The forces acting on a single droplet include gravity (g), which acts vertically downward, and drag, which depends on the droplet shape and its relative velocity with respect to surrounding air (u_d). A force balance on a single droplet is illustrated in Equation 2. The force balance ignores the time dependence on droplet velocity [14].

$$\frac{du_d}{dz} = \frac{g}{u_d} - \frac{F_d}{m_d u_d} \qquad \text{Equation 2}$$

In Equation 2, m_d is the droplet mass, F_d is the aerodynamic drag on the drop, and z is the vertical distance from the droplet source.

Heskestad et al. [15] integrated Equation 2. The subsequent equations provided in reference [14] conclude that in order to estimate the air entrainment rate from a water curtain, parameters such as droplet velocity, the momentum coefficient of a nozzle, the spray angle, the volumetric flow rate of water, and the cross-section spray area need to be determined. Normally the analysis is conducted for a single spray and then extended to a curtain design by accounting for an aggregate surface area and water flow rate [14].

Absorption and Chemi-Absorption Phenomena

Absorption depends on fluid curtain flow rate, drop size, and gas solubility. Fthenakis [16] developed a model for evaluating mass transfer induced by a single water-spray nozzle for a point-source release (see Equation 3).

$$\frac{C(x)}{C(0)} = \exp\left[-\frac{K_g a}{G(1+\frac{K_g a m z}{2L})} x \right] \qquad \text{Equation 3}$$

Fthenakis [16] considers a nozzle located at an elevation z above the ground, spraying water at a rate of L (in gmoles$\cdot$m$^{-2}\cdot$s^{-1}) over an area of A (in m^2). All water droplets have the same diameter d and velocity u. If the source strength of the point source inside the spray cone is G (in gmoles$\cdot$m$^{-2}\cdot$s^{-1}), and if it is assumed that the gas is well mixed in the vertical direction, the gas phase concentration of a given chemical can be calculated as a function of radial distance within the spray. a is the interfacial spray area per unit volume, K_g is the mass transfer coefficient, x is the axial distance from the spray cone center, m is the equilibrium curve slope, and a can be calculated based on the spray volume, drop exposure time, molecular weight, and fluid density.

The equilibrium curve slope m can be estimated from Henry's law (see Equation 4).

$$m = C \exp[kT] \qquad \text{Equation 4}$$

Referring to Equation 4, T is the water temperature in Celsius units and C and k are constants which are a function of the toxic chemical being addressed.

Another key parameter when designing fluid curtains is the water-to-toxic chemical ratio as investigated by Blewitt et al [17] for hydrogen fluoride. Blewitt reported that increasing this ratio increases the absorption and chemical reaction with the capturing fluid (water) of the toxic chemical under study.

Successful fluid curtain applications have been implemented with hydrogen fluoride and ammonia using pure water, as these two chemicals have good affinity with water; i.e., high solubility. However, for hydrogen sulfide, it can be observed that it does not have affinity with water and, therefore, relying solely on water absorption is not practical.

Table 4 below lists the Henry's Law constant values and overall mass transfer coefficients for several toxic chemicals in water; i.e., hydrogen fluoride, ammonia, hydrogen sulfide, and hydrogen chloride.

Table 4. Henry's Law Constant and Mass Transfer Coefficient Values for Toxic Chemicals in Water

Toxic Chemical	Henry's Law Constant [mol m^{-3} Pa^{-1}]	Overall Mass Transfer Coefficient [kmol h^{-1} m^{-3}]
Hydrogen Fluoride (HF)	130	152
Ammonia (NH_3)	0.59	337
Hydrogen Sulfide (H_2S)	0.001	-
Hydrogen Chloride (HCl)	15	353

Source: Henry's Law Constant values extracted from [18], and overall mass transfer coefficients based on [19]

For fluid curtains intended to capture hydrogen sulfide, the addition of an alkali promotes the absorption phenomena (i.e., chemi-absorption). Astarita and Gioia [20] studied the absorption of hydrogen sulfide in aqueous hydroxide solution and aqueous carbonate-bicarbonate buffered solution. Additionally, Van Krevelen et al. [21] carried out a huge experimental campaign of absorption of hydrogen sulfide in sodium carbonate solutions, presenting a large set of data that consisted of 22 experimental runs. Note that the Astarita-Gioia model [20] incorporated Van Krevelen et al. [21] findings in his work. Based on the findings of Astorita and Gioia [20] and Van Krevelen et al. [21], for fluid curtains that need to capture toxic chemicals that do not have affinity with water, the addition of an alkali should be considered.

3.2 Frequency-Based Design

The frequency-based design addresses the level of reliability that the fluid curtains need to achieve in order to be considered as an IPL. The frequency-based design needs to ensure that the number and placement of fire and gas detectors is correct, as well as meet the Risk Reduction Factors (RRFs) identified during the Quantitative Risk Assessment (QRA) activity.

3.2.1 Fire and Gas Mapping

The main purpose of a Fire and Gas detectors Systems (FGS) mapping study is to identify and assess the location and performance of detectors and cover both gas (flammable and toxic gases) and fire detection. This analysis is normally conducted by developing the following tasks: (1) philosophy considerations, (2) zone definition and categorization, (3) performance targets, (4) detector coverage verification, and (5) functional safety availability.

The first step in a FGS mapping study is to confirm that the FGS philosophy considerations are based on corporate guidelines as well as industry standards such as ISA TR 84.00.07 [4].

The main purpose of the second step, defining zones, is to prevent the migration of combustible gas or toxic gas hazards from the process area to non-process areas where the fugitive gas can either impact humans or be ignited by electrical equipment. When addressing zone definition and categorization, it is important to recognize that each selected area may require different types of FGS detection and different levels of performance.

In the zone definition step, each zone can be segmented into smaller geographic divisions known as graded areas. This segmentation is used to establish the specific performance targets for all FGS. Categories to be considered are defined as a direct function of the type of hazards identified in the zone (e.g., hydrocarbon fire, non-hydrocarbon fire, toxic cloud) and is related to the nature of the zone (e.g., process areas, non-process areas with or without electrical equipment protection or occupied buildings, specific areas with turbine or engine enclosures, ventilation air intakes). Therefore, each area receives a performance target for fire, flammable gas and toxic gas hazards which takes the form of grades. Each of the grades serves to define a relative level of fire or gas risk with grade A referring to the highest risk areas and grade C being associated with the lowest risk areas requiring detection (see Table 5).

Table 5. Example of FGS Zone Categorization – FGS Grades

Grade Level	**Description**
A	Toxic cloud with high concentration of H_2S (high risk level area)
B	Toxic cloud with moderate concentration of H_2S (medium risk level area)
C	Toxic cloud with low concentration of H_2S (low risk level area)

FGS performance targets are based on the grade level required for each graded area. For example, for a high-risk level area (grade A), the FGS performance target may be ninety percent (90%), while for a low-risk level area (grade C) may be seventy percent (70%). Additionally, according to ISA TR 84.00.07 [4], the two primary modes of FGS failure are the following:

1. Coverage: insufficient number, type or location of fire or gas detectors resulting in a hazard that is not detected by the FGS. Two types of coverage are typically evaluated: geographic coverage and scenario coverage. The type of coverage to be used during the fire and gas mapping step will depend on the purpose of the study itself. Generally, scenario coverage (where the coverage relates to the location of the worst credible release scenarios) can be done if the goal is to cover a few specific critical pieces of equipment within a process unit. If all the process equipment within that unit is deemed to be critical, then geographic coverage may be a better approach.

2. Safety Availability: component failures of FGS hardware (e.g., detector itself, power supply, etc.) that result in the FGS being in an unavailable state when a demand condition arises. For setting performance targets for availability, performance targets must be established for the Probability of Failure on Demand (PFD) of the equipment that comprises the instrumented fire and gas function; i.e., the PFD must comply with the RRFs already calculated. Note that section 3.2.2 below describes how to establish the reliability of a fluid curtain system, which starts with establishing the reliability of the FGS (i.e., sensor subsystem).

With regard to detector coverage verification, this step consists of modeling and calculating the location and number of fire and/or gas detectors necessary to meet the established FGS performance targets. Commercial software such as Detect3D [22] can help the user determine the best location and optimize the number fire/gas detectors to minimize the economic investment.

3.2.2 Reliability Verification

The first task within the reliability verification is to develop Safety Requirements Specification (SRS) documents addressing the reliability lifecycle of the Fire and Gas detector System Safety Instrumented Systems (FGS-SIS). This information is key for defining the Safety Instrumented Functions (SIFs) that must be activated for the FGS-SIS; i.e., identification of the sensor subsystem, logic subsystem, and final element that comprise the fluid curtain system. The SRS is a document where all the safety requirements for a SIF are clearly stated.

An SRS must address the two main categories of requirements associated with each SIF:

1. Functional safety requirements: which are the SIFs to be implemented to prevent or act upon demands (hazardous scenarios).

2. Functional integrity requirements: how reliable the SIFs must be in order to achieve adequate risk reduction.

The functional safety requirements include defining relevant and realistic SIFs for each hazardous scenario (resulting in a demand) for areas or specific targets with intolerable risk levels. Accordingly, a SIF will be further defined after addressing its modes of operation. Three SIF modes of operation can be defined according to IEC 61511 [3]:

- Low demand mode: Mode of operation where the SIF is only performed on demand, in order to transfer the process into a specified safe state, and where the frequency of demands is no greater than one per year.

- High demand mode: Mode of operation where the SIF is only performed on demand, in order to transfer the process into a specified safe state, and where the frequency of demands is greater than one per year.

- Continuous mode: Mode of operation where the SIF retains the process in a safe state as part of normal operation.

Based on the definition of these three modes of operation, the Chemical Process Industry (CPI) mainly uses mostly low demand SIFs. A SIS element can be designed according to two different principles: Energize-to-trip or De-energize-to-trip. Many SIS elements are currently designed according to the de-energize-to-trip principle. This principle is also the basis for the fail-safe principle, which is a design property that causes an SIS element to go to a predetermined safe state in the event of a specific failure or malfunction. However, note that each SIS must be individually considered.

When addressing functional integrity requirements, it is necessary for the FGS-SIS to meet the required availability based on the calculated RRFs. This availability is characterized by the Safety Integrity Level (SIL), which is a function of the predicted RRF, which directly defines the target Probability of Failure on Demand average (PFD_{avg}) in the probabilistic verification. The PFD_{avg} is a term defined for low demand systems. The relationship between RRF and PFD_{avg} is provided in Equation 5 and the relationship between RRF, PFD_{avg} and SIL is listed in Table 6. Note that the RRF as listed in Table 6 only addresses the likelihood portion of the risk; i.e., consequences are not addressed.

$$PFD_{avg} = \frac{1}{RRF} \quad \text{Equation 5}$$

Table 6. Relationship between RRF, PFD_{avg}, and SIL

RRF	PFD_{avg}	SIL
$10{,}000 \le RRF < 100{,}000$	$1.00E\text{-}05 \le PFD_{avg} < 1.00E\text{-}04$	4
$1{,}000 \le RRF < 10{,}000$	$1.00E\text{-}04 \le PFD_{avg} < 1.00E\text{-}03$	3
$100 \le RRF < 1{,}000$	$1.00E\text{-}03 \le PFD_{avg} < 1.00E\text{-}02$	2
$10 \le RRF < 100$	$1.00E\text{-}02 \le PFD_{avg} < 1.00E\text{-}01$	1

Once the functional integrity requirements are addressed, it is necessary to ensure the architecture requirements; i.e., SIS/SIF redundancy. These requirements are typically expressed by the parameter Hardware Fault Tolerance (HFT), which indicates the ability of a hardware subsystem to continue to perform the required function in the presence of faults or errors. The HFT must be met based on criteria in standards such as IEC 61508 [2] and/or IEC 61511 [3]. Note that the HFT accounts for the required redundancy, which means having two or more items, such that if one item fails, the system can continue to function by using the other items.

The concept of redundancy introduces the voting concept. Figure 5 below illustrates the sensor, logic solver and final element subsystems of a SIS. Each subsystem may have one or more voted groups or channels. A channel is defined as a structure of one or more elements that can independently perform a safety function. For example, the block denoted as "pressure transmitter" (see Figure 5) is a channel intended to detect when the pressure goes beyond acceptable limits and send a signal to the logic solver subsystem. A group of *n* identical channels can be configured in several ways. One extreme is when the group is functioning only when all the *n* channels are functioning; the other extreme is when the group is functioning as soon as at least one channel is functioning. The first extreme case is called an *n-out-of-n* voted structure and the second a *1-out-of-n* voted structure. In a general case, the group may be configured such that it is functioning when at least *k* of its *n* channels are functioning (a *k-out-of-n* voted structure). Such a structure is often written *koon* and is said to be *koon* voting.

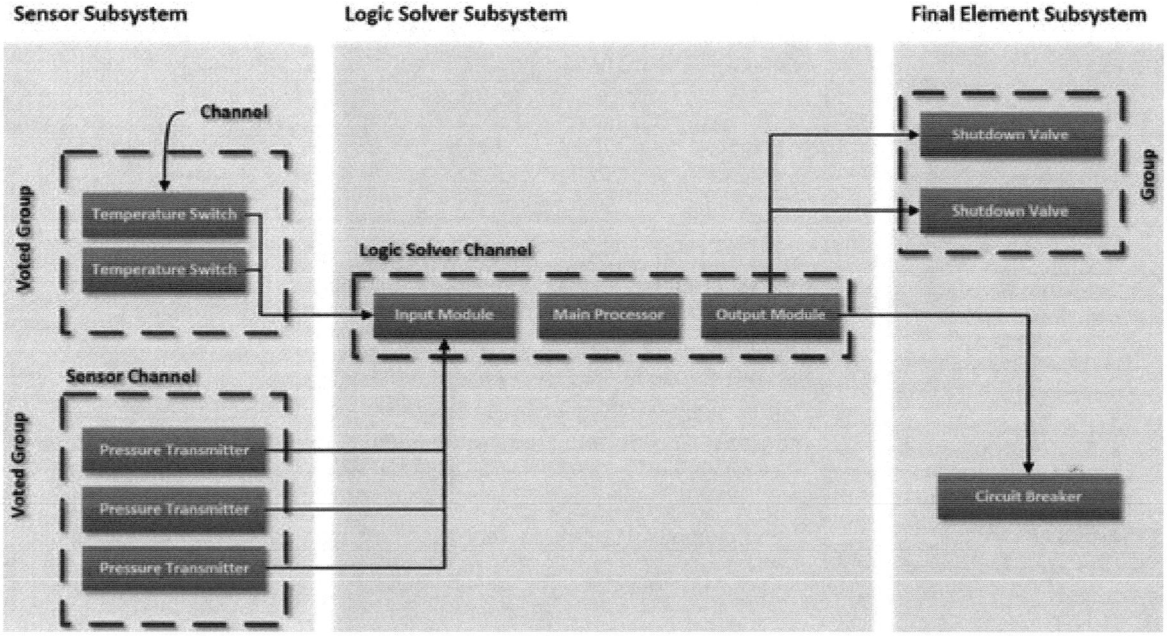

Figure 5: SIS Subsystems Including Groups and Channels

Based on the concepts introduced above, it is evident that the key step during reliability verification is to calculate the PFD_{avg}. Table 7 below lists the parameters that are necessary to calculate the PFD_{avg}. Firstly, it is critical to identify the failure rate data of each component of the SIS. Secondly, it is important to establish test and maintenance data. Detailed information about the mathematical equations involved in the calculation of the PFD_{avg} can be found in IEC 61508 [2] and ISA TR 84.00.02 [23].

Table 7. Reliability Verification Key Parameters

Parameter	Symbol	Definition
Failure Rate	λ	Frequency of the occurrence of failures. Failures can be classified based on consequence (i.e., Dangerous or Safe), and detectability (i.e., Detected or Undetected).
Mission Time	MT	Maximum time period for which a system is intended to be used. After this time period, the system needs to be replaced.
Proof Test Interval	TI	Time interval between two proof tests, i.e., periodic test performed to detect failures in a safety-related system so that, if necessary, the system can be restored to an "as new" condition or as close as practical to this condition.
Proof Test Coverage	C_{PT}	Indicates the effectiveness of a proof test. A 100% proof test coverage would mean that 100% of all dangerous failures would be detected in the test.
Proof Test Duration	PTD	Duration of the proof test while the safety function is put into bypass.
Mean Time to Repair/Restore	$MTTR$	Expected time required to detect that a failure has occurred as well as the time required to make a repair once the failure has been detected and identified, i.e., average value which includes diagnostic detection time and actual repair time.
Common Cause Failure	CCF	Condition that affects the operation of multiple devices that would otherwise be considered independent, i.e., redundant devices.
System Capability	SC	Measure of the effectiveness of quality management techniques applied to components, i.e., Operational-Maintenance Capability.

4 Conclusions

This paper discusses the different steps required to design fluid curtains as an effective means of post-release mitigation technique ensuring that these function as an Independent Protection Layer (IPL). The approach discussed focuses on providing safeguarding against certain Loss Of Containment scenarios (LOCs) identified to contribute the most to the intolerable risk levels in a populated public area based on the Quantitative Risk Assessment (QRA) results.

Fluid curtains need to meet a specific consequence-based design. In order to do that, the first task is to model and identify the worst-credible LOCs by conducting a sensitivity analysis of parameters that can affect the toxic impact distances calculated. The worst-credible LOCs will then be used to define the fluid curtain design basis. Additionally, to ensure that all toxic releases are captured, considerations for physical release phenomena such as fluid momentum needs to be modeled. Furthermore, absorption and dilution phenomena need to be considered when modeling and designing fluid curtains, which are a function of the fluid curtain type, and the toxic chemical under analysis. During this step, fluid flow requirements, fluid-to-toxic chemical mass ratio, type of nozzles, number of nozzles, fluid curtain layout and height among other system features are defined and characterized.

Once the consequence-based design is addressed, this type of safeguard needs to achieve a certain level of reliability to be considered as an IPL; i.e., frequency-based design. First, in order to ensure that a potential toxic release is detected, a detailed fire and gas mapping study must be conducted with the aim to optimize the location and layout as well as calculate the optimum number of detectors required. Then, a Safety Requirement Specification (SRS) document needs to be developed to address the system lifecycle. Finally, to meet the required level of reliability from the Risk Reduction Factors (RRFs) calculated during the QRA activity, the Probability of Failure on Demand (PFD) needs to be calculated for all Safety Instrumented Functions (SIFs) within the Safety Instrumented System (SIS).

Thus, based on the fluid curtains design requirements (consequence-based design and frequency-based design), existing fire water systems should not be considered as a mitigation technique of toxic vapor dispersions or other consequences as a result of a loss of containment; i.e., these types of systems are generally designed for other purposes and lack the level of reliability required.

In summary, fluid curtains when appropriately designed, maintained, and tested can effectively safeguard and provide proper protection while operating as an IPL.

5 References

[1] CCPS. *Guidelines for Developing Quantitative Safety Risk Criteria*. Center for Chemical Process Safety, American Institute of Chemical Engineers, New York, NY, 2009.

[2] IEC 61508. *Functional Safety of Electrical / Electronic / Programmable Electronic Safety-related Systems, Parts 1-7*. International Electrotechnical Commission, 2010.

[3] IEC 61511. *Functional Safety - Safety Instrumented Systems for the Process Industry Sector, Parts 1-3*. International Electrotechnical Commission, 2018.

[4] ISA TR 84.00. 07. *Guidance on the Evaluation of Fire and Gas System Effectiveness*. International Society of Automation, 2018.

[5] CCPS. *Guidelines for Enabling Conditions and Conditional Modifiers in Layer of Protection Analysis.* American Institute of Chemical Engineers, New York, NY, 2013.

[6] ioMosaic Corporation, Process Safety Office® SuperChems™, http://www.iomosaic.com/software/process-safety-office-.

[7] Prophet, N., Dunjó, J., Amoros, M., and Gorski, G. “Advanced QRA Methodology – Quantifying Risk Reduction Measures to Minimize Economic Investment”. Presentation at the 2018 Spring Meeting and 14th Global Congress on Process Safety, Orlando, FL, 2018.

[8] HSE. “Risk Criteria for Land-Use Planning in the Vicinity of Major Industrial Hazards”. Health and Safety Executive, HMSO, 1989.

[9] AIHA. “Emergency Response Planning Guidelines and Workplace Environmental Exposure Levels”. American Industrial Hygiene Association, Fairfax, VA, 2016.

[10] VROM, CPR 14E. *Methods for the calculation of Physical Effects Due to releases of hazardous materials (liquids and gases).* 2005.

[11] Pasquill, F. “Atmospheric Diffusion”. Wiley, 1974.

[12] Mukherjee, S., Dharmavaram, S., and Jaskolka, S. “Effectiveness of Water Sprays in Mitigating Toxic Releases”, Process Safety Progress, 37: 256-262, 2018.

[13] Ghosh, S., and Hunt, J. C. R. “Spray jets in a cross-flow”. Journal of Fluid Mechanics, Vol. 365, pp. 109-136, 1998.

[14] CCPS. *Guidelines for Postrelease Mitigation Technology in the Chemical Process Industry.* Center for Chemical Process Safety, American Institute of Chemical Engineers, New York, NY, 1997.

[15] Heskestad, G., Kung, H. G., and Todtenkopf, N.F. “Air entrainment into water sprays and spray curtains”. ASME, (76-WA/FE-40):1-14, 1976.

[16] Fthenakis, V. M. “The Feasibility of Controlling Unconfined Released of Toxic Gases by Liquid Spraying". Chemical Engineering Comm., 83: 173-189, 1989.

[17] Blewitt, D. N., J. F. Yohn, R. P. Koopman, T. C. Brown, and W. J. Hague. “Effectiveness of Water Sprays on Mitigating Anhydrous Hydrofluoric Acid Releases”. American Institute of Chemical Engineers, Proceedings of the International Conference on Vapor Cloud Modeling, pp. 155-171, New York, 1987.

[18] R. Sander. “Compilation of Henry's Law constants (version 4.0) for water as solvent”. Atmos. Chem. Phys., 15, 4399-4981, 2015.

[19] Perry, R. H., and Green, D. W. *Perry's Chemical Engineers' Handbook.* McGraw-Hill, New York, 7th Edition, 1997. ISBN: 0-07-049841-5.

[20] Astarita, G., Gioia, F. “Hydrogen sulphide chemical absorption”. Chemical Engineering Science, Vol. 19, pp. 963-971, 1964.

[21] Van Krevelen, D.W., Hoftijzer, P. J., van Hooren, C. J. "Studies of Gas Absorption. IV. Simultaneous Gas Absorption and Chemical Reaction". Recueil des Travaux Chimiques des Pays-Bas, Vol. 67, Issue 2, 1948.

[22] Insight Numerics, Detect3D software, http://insightnumerics.com/.

[23] ISA TR 84.00. 02. *Safety Integrity Level (SIL) Verification of Safety Instrument Functions*. International Society of Automation, 2015.

Think a BIT!

Process Safety Excellence through Incident Barrier Based Inspection and Training

Majed Al-Mutairi
Team Leader – HSE (Gas), Kuwait Oil Company
Ahmadi-61008, Kuwait
Almutairim@kockw.com

Mazharuddin R. Shaikh
Safety Engineer, Kuwait Oil Company
MaRShaikh@kockw.com

Sami Al-Enezi
Sr. Engineer – HSE (Gas), Kuwait Oil Company
SZEnezi@kockw.com

Prepared for Presentation at
American Institute of Chemical Engineers
2019 Spring Meeting and 15th Global Congress on Process Safety
New Orleans, LA
March 31 – April 3, 2019

Think a BIT! Process Safety Excellence through Incident Barrier Based Inspection and Training

Majed Al-Mutairi
Team Leader – HSE (Gas), Kuwait Oil Company
Ahmadi-61008, Kuwait
Almutairim@kockw.com

Mazharuddin R. Shaikh
Safety Engineer, Kuwait Oil Company
MaRShaikh@kockw.com

Sami Al-Enezi
Sr. Engineer – HSE (Gas), Kuwait Oil Company
SZEnezi@kockw.com

Keywords: Process Safety, Training, Inspection, Knowledge, Hazard Communication

Abstract

One common goal of an operating facility is to maintain production at a maximum while minimizing cost by preventing incidents. It is achievable if the pertinent risks are properly assessed, managed and communicated. Understanding and communication of Hazard/ Risk/ Control Measures and Monitoring of the same at an optimum frequency is very important. Identifying risks and providing simpler references eliminate having to act without proper knowledge within a limited reaction time especially for non-risk professionals. Covering all aspects of major accident hazard in a single simple document, and focusing on the same during regular inspection and for training, may make it easy for people to understand and act. Therefore, incident **B**arrier based **I**nspection and **T**raining (BIT) can be defined as the incident prevention tool that is used to simplify the approach to training people and monitoring barriers at regular intervals in an easy way.

This paper will shed light on the representation of risk for non-risk professionals in a barrier model, identify HSE critical tasks and roles which have the responsibility to maintain effectiveness and integrity of the barrier (preventive, escalation factor control, or mitigative measures) and monitoring of these tasks and roles throughout the process life cycle.

1 Introduction:

Process Safety management involves making sure that our facilities are well designed, safely operated, properly maintained and inspected for anything that has potential to cause Loss of Containment. We are in the Oil and Gas business and in order to run or operate a facility with hydrocarbons, proper inspection and action based on observation is key to safe operation.

To create and maintain safe processes, our actions are just as important as equipment safeguards. For this reason, we have to develop multiple barriers associated with the processes. Every single barrier counts in addressing a threat to critical equipment. Barrier thinking means small holes at each level are addressed. This keeps a series of small holes from aligning to create a major problem. Over a period of time, the effectiveness of the barrier deteriorates from its intended design. This requires that a facility uses proactive measures to ensure all barriers are in place and operating at their intended efficiency.

HSE Inspection is one of the proactive measures to maintain and improve HSE performance of any organization. A specific HSE Inspection Manual is designed to provide safety engineers with an understanding of their role in effective workplace inspections. It includes a list of hazards, the required barrier to control the top event and consequence, startup and shutdown management.

Therefore, Incident Barrier-based Inspection and Training (BIT) can be developed as the incident prevention tool as a simplified approach to train people and monitor barriers at regular intervals in an easy way. It is a part of the management plan for incident prevention. This means the inspection strategy is driven by risk, and major risks are monitored regularly.

BIT is built on a systematic assessment that covers:

- Incident Data gathering
- Analysis of causes
- Analysis of recommendations.
- Review of Safety Studies (HAZOP, HAZID, QRA, FERA, Risk Assessment etc.).
- Review of the Operation and Maintenance Manual.
- Review of Shut Down and Start Up Procedures.
- Review the MoC implemented for a particular piece of equipment. Consider reason and benefits.
- Identification of threats to each equipment.
- Determination of a Consequence due to each threat identified.
- Identify Barriers to prevent an event leading to a consequence.
- Development of Inspection and Training Manual.
- Train people for inspection and safe operation of each equipment and for the whole process.
- Assign Tasks to respective teams.
- Incorporation of findings and improvement of management plans.

2 Terminologies:

2.1 ***Bow Tie:*** It is a graphical representation of threats & potential consequences relating to a hazard, and the barriers that prevent or mitigate the consequences.

2.2 ***BIT:*** Barrier Based Inspection and Training

2.3 ***Hazard:*** It is an operation, activity or material with the potential to cause harm to people, property, environment, normal operation (can cause production loss), increase liability. Examples: Hydrocarbons, chemicals, radiation etc.

2.4 ***Top Event:*** It is an incident which occurs when hazard is realized due to control over Hazard is lost. Example: Loss of Primary Containment, structural failure etc.

2.5 ***LOPC:*** Loss of Primary Containment. Example: Unintended release of a hydrocarbon to the atmosphere from a storage tank due to leakage.

2.6 ***Threat:*** A possible cause that will potentially release a hazard and produce a Top event if controls are lost. Example: High temperature/ pressure, wear and tear, corrosion etc.

2.7 ***Consequence:*** Consequences are unwanted outcomes that could result from the top event and lead to damage or harm. Example: Fire (pool, jet, flash), explosion, fatality or injury, asset damage, oil spill etc.

2.8 ***Escalation Factor:*** It is a condition that will lead to increased risk by defeating/ overriding barriers or recovery preparedness/ mitigation measures and reducing the effectiveness of the barrier to which it is attached. An escalation factor does not directly cause a top event or consequence, but since it degrades the main pathway barrier, the likelihood of reaching undesired consequences will be higher. Examples: Loss of Utility Services; Operating outside the design Envelope.

2.9 ***Barrier:*** Barriers can be physical or non-physical measures to prevent/ reduce the probability of each threat resulting in a Top Event (Preventive Barrier) or limit/ provide recovery from unwanted outcomes (Mitigation Barrier). Barriers can be associated with engineering, Maintenance or Operational activities.

2.10 ***Escalation Factor Control Barrier:*** Controls that are put in place to manage conditions that lead to increase risk due to loss of threat barriers or loss of recovery preparedness measures. It helps to defeat the escalation factor.

2.11 ***HSE Critical Task:*** An HSE Critical Task is one that is in place to develop, implement or maintain the effectiveness and integrity of a Barrier, Escalation Factor Control or Recovery Measure Control in the bow-ties.

3 Development of Barrier Model:

3.1 Methodology:

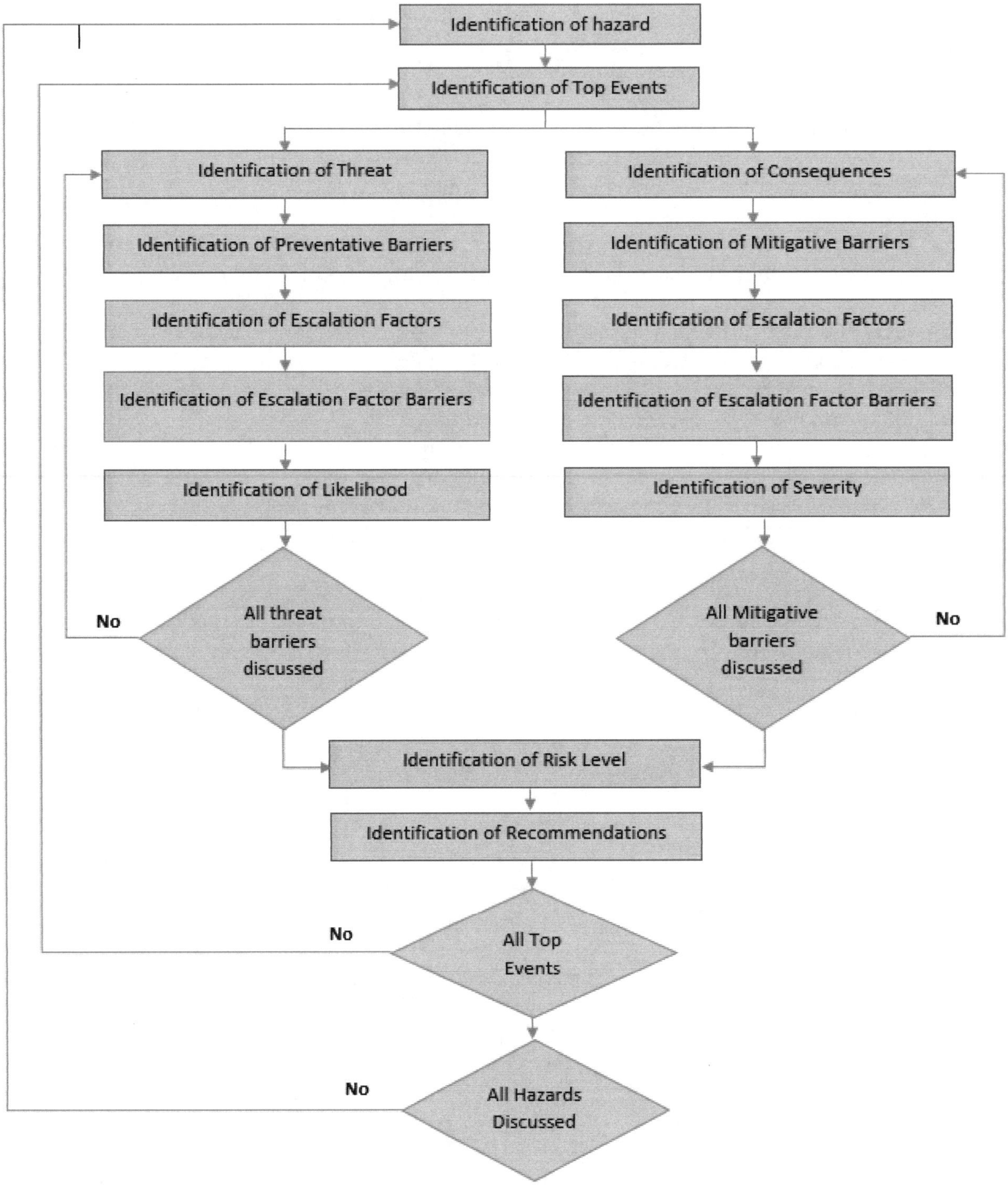

Figure 1. Methodology for Barrier Model Development

In this barrier model (see Fig. 1), we have concentrated on the hazards of equipment/ process. After selection of the equipment, a top event is defined which takes place once a control of hazard is lost. Each top event can have several consequences if the event is allowed to escalate. All consequences are identified and an initial screening assessment for each consequence is carried out to assess risk. Causes of the top event are identified and listed as threats. Based on the hazard control hierarchy of elimination, prevention, mitigation and recover, measure to block a threat from releasing a hazard and that limit the chain of consequences were defined.

3.2 Selection of Barriers

Barriers were selected after careful examination and review of safety studies, the Operation and Maintenance Manual, Start up and Shutdown philosophy, recommendations from MoC and most importantly analysis of incidents and their recommendations. Table 1 provides details of data and its source/ reference.

Table 1. Data for bow tie and Source referred for Identification of barriers

# No.	Data	Source/ Reference	Output
1	Incident Recommendations	Company Database (MyHSSE), External Resource (What Went Wrong, CSB)	Barriers
2	Hazard, assumptions, causes, consequences, safeguarding and Recommendation from safety studies	Study Reports – HAZOP, QRA, FERA, EERA, HAZID,	Hazards, Threats, Barriers
3	Precautions during Start up Shut Down	Plant Operation and Maintenance Manual. Start Up Shut Down Procedure	Barriers
4	Recommendations from Change Management	Management of Change Details	Confirming Safety studies are updated after change and no additional hazard is introduced due to change. Barrier

Kuwait Oil Company has developed online system for HSE Management and Data collection which is termed as MyHSSE. All incident are reported and investigations are uploaded in MyHSSE Incident database. This is very useful to refer incidents of last 10 years. Incidents can be filtered based on type, level, severity etc. (Figure 2).

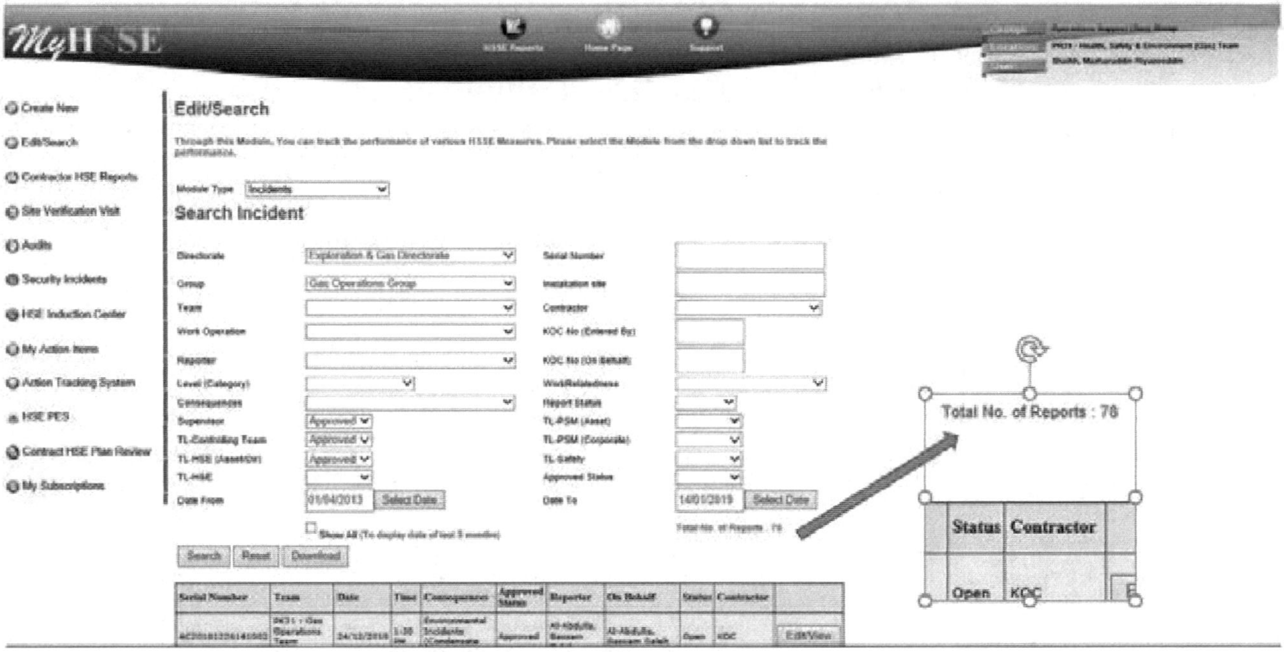

Figure 2. Company incident database screenshot

Barrier Model:

All identified elements fixed as shown in figure 2, to generate final barrier model for display.

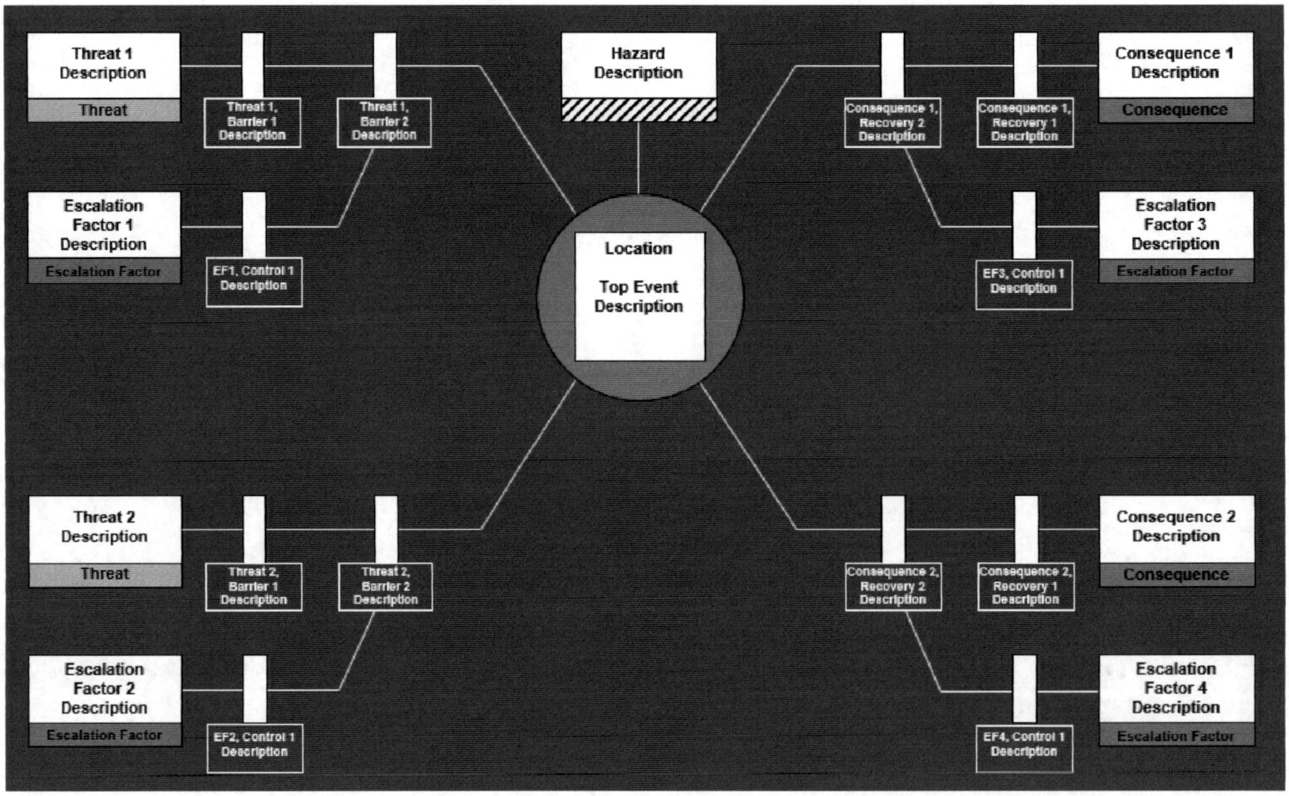

Figure 3. Sample Barrier Model

4 Inspection and Monitoring Program:

4.1 Activities:

4.1.1 Direct Activities: Activities within Direct Control of the Location
Example: Offshore Helicopter Landing/ Take-off controlled by offshore HLO

4.1.2 Interface Activities: Activities outside control of location but impact on it.
Example: Helicopter servicing, flight procedures, onshore landing/ take off interface activities by helicopter contractor.

4.2 HSE Critical Task:

Step 1: Identify tasks necessary to provide, enforce and maintain the barriers, recovery preparedness measures and escalation factor controls.
Step 2: Build the task
Step 3: Link task to the associated barrier, recovery preparedness measures or escalation factor control.
Step 4: Identify responsibilities.
Step 5: Implementation of activities to return barrier to its original performance.

Example:

Hazard –	Hydrocarbon Gas
Top Event –	Loss of Containment
Threat –	Over Pressurization
Barrier –	Safety Relief Valve
Task –	Inspection/ Testing/ Maintenance of SRVs

<u>Inspection Check:</u>
Safety relief valves checked and tagged on a regular basis (at least every two years). (Yes/ No)
Date of Inspection and Testing:
Due Date of Inspection and Testing:
<u>Inputs:</u> Maintenance System Instructions, Job Requests, Legislation, Maintenance Schedule
<u>Task Performer:</u> Maintenance Engineer.
<u>Verification:</u> Maintenance and Inspection Records, SRV Tags, Relief Valve Records.
<u>Competency:</u> PTW Training, 5 Years' Experience, Maintenance Tracing System User.

4.3 Prepare Inspection Checklist Manual:

List out all critical task with periodicity and prepare quarterly checklist. Segregate quarterly, half yearly and yearly activities to be checked and create four checklists for each equipment.

4.4 Prepare Inspection Plan:

The Inspection Plan is prepared as per the inspection requirements for each barrier. Some barriers need frequent checks; however, others are yearly or every two years as per their maintenance schedule and manufacturer requirements. Where guidelines are not available, risk assessment can be helpful to finalize inspection frequency of a particular barrier.

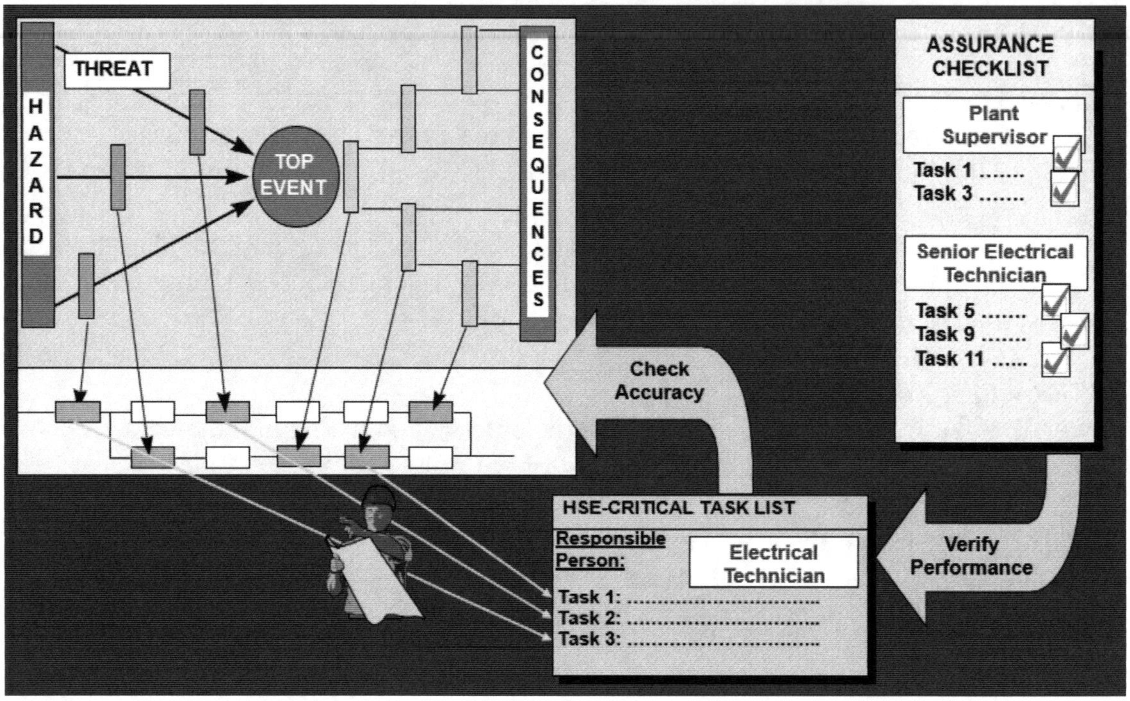

Figure 4. Implementation of HSE Critical Task

5 Development of the Inspection and Training Manual:

The purpose of this Inspection and Training Manual is to provide safety engineer with a systematic guideline to carry out the facility inspection properly and to ensure a standardized incident barrier-based approach for inspection. The manual describes a brief description of the process and potential hazards normally encountered in each operational section/ area. It includes threats for each equipment and tasks to be checked to ensure efficiency and effectiveness of the control barrier.

Inspection and training module is used particularly for two purposes as name suggests:

1) Plant inspection
2) Training

5.1 Plant inspection:

Inspection and training manual serves as a ready reference with all check points for periodic inspection. The inspection checklist focuses on all possible deviations/ imperfections normally observed in the facilities, which prompts the safety engineer to verify and ensure compliance.

Inspection is quarterly bases for each facility. Frequency of inspection, testing and maintenance of barrier with responsibility is provided in manual. During inspection, safety engineer verifies that activity to ensure intended function of barrier and escalation factor control is performed and barrier integrity is ensured.

Status from inspection is listed and followed up through HSE Action Tracking Committee on Monthly basis. If persistent errors are found, then remedial trainings are arranged.

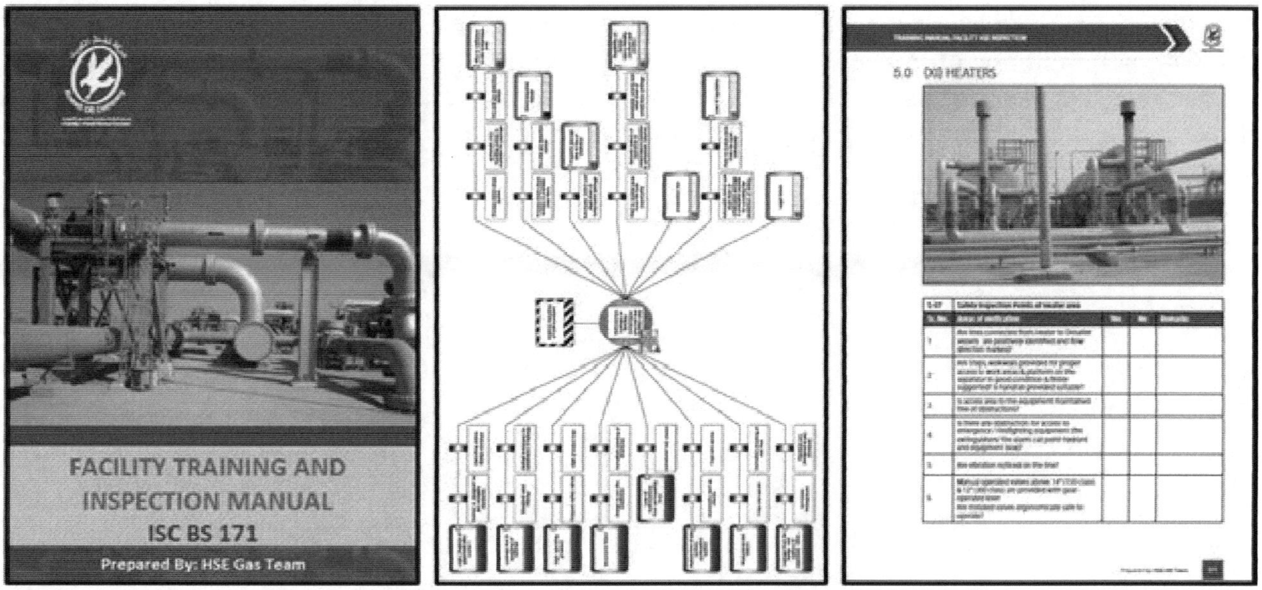

Figure 5. Facility Training and Inspection Manual

5.2 Training:

Inspection and training manual also covers description of process and bow tie diagram. Bow ties assist with the understanding of accident risk and demonstrate the role of barriers and escalation factor controls in managing the risk. The bow tie diagram shows how barrier can degrade and how the support of barriers links to the overall facility management system.

New employees are assisted by experienced engineer for their training and development. This manual provides ready reference for new employees and assists them with understanding of process/ equipment and hazards associated with them. It also provides details of existing control measures to prevent and mitigate undesired event. During joint inspection, they are well informed about how to maintain integrity of barrier.

The manual specifies the proper training and skills necessary for personnel to effectively complete the assigned task and strive for continuous improvement.

The Inspection and Training Manual covers the items below:

1) General Facility Layout, PFD and Introduction of plant
2) List of Equipment, Vessels, Manifolds and other operational.
3) Description of Equipment/ process
    a. Picture
    b. Description
    c. Bow Tie Diagram
4) Inspection checklist for the equipment (e.g. heaters)/ process area (e.g. manifold)
5) Emergency Response
6) Fire protection measures at a production facility
7) Evacuation Route and Assembly Points

Most barriers rely on training and competency. Many aspects of bow tie diagram can be linked to training and competency management. With development of this manual, training requirement for barriers including human related barriers were also identified. For example, in a "gas detection, human decision and ESD system", elements in a training program that addresses critical human decision and action to press the ESD button is added.

6 Conclusion:

Barrier based Inspection and Training program provides useful information for driving improvement in process safety and when acted upon, contributes to reducing risk by verification of activities carried out for barrier effectiveness. Detailed checkpoints in manual further reduce chances of missing verification for any important barrier. This is a better and effective alternative to traditional inspection for safety engineers. While acted upon, inspection program ensures that the risk and barriers are managed throughout the life of the facility. Addressing gaps in management of effectiveness of barriers help to ensure awareness of potential process safety risk. The outcome of inspection is useful for measurement of safety performance for particular facility. It enhances understanding of how incidents occurs.

Further, bow tie after structured assessment is ideal for communication of risks. Manual also provides a better understanding of plant, process, maintenance and operational risk associated and barriers to prevent primary event.

The process of developing bow ties and "BIT" manual itself benefit to improve process safety management as it involves review of process safety information and PHA studies which increases process knowledge, foster employee involvement, ensuring mitigation barriers are in place for emergency management. Excellent training and communication resource builds process safety competency amongst existing and new employees. It is useful for deriving immediate and root causes of incident.

7 References:

[1] Kuwait Oil Company. *Incident data base on MyHSSE.*

[2] Kuwait Oil Company, *PHA study reports of facilities.*

[3] Kletz, Trevor (2010). *What went wrong* – 5th Edition

[4] Kletz, Trevor (2009). *Still Going Wrong* – 1st Edition -

[5] Kletz, Trevor (2001). *Learning from accidents* – 3rd Edition

[6] Atherton, John and Gil, Frederic (2008). *Incidents that define process safety.* CCPS

[7] Center for Chemical Process Safety (CCPS) of the AIChE (2018), Bow Ties in Risk Management - *A concept book for risk management.*

546830 PHAs at Identical Plants: Why Aren't They the Same?

There is a common misunderstanding about the nature of HazOps and other process hazard analyses, that they are objective studies. That is, that different teams, working at different times and different places, but looking at the same process, will get the same results. In fact, HazOps are not objective. Their virtue is not in their absolute objectivity, but in the discipline they impose on a review team and how they document the relative subjectivity of the team.

There are several reasons that HazOp teams may get different results for nominally identical processes: structural differences in methodology, differences in assessments of likelihood, and differences in assessments of consequences. Even accounting for these differences, different PHA teams also can and do make different recommendations.

How different can they be? In addition to a review of why HazOps can yield different results, this study looks at two separate processes, each operated at several different sites, and quantifies the differences between the HazOps conducted at those sites. It is just as likely that only one facility will recognize a hazard as it is that all facilities will recognize a different hazard. The study then goes on to discuss what these differences tell us about the potential for overlooking hazards and mischaracterizing the risk of hazards that are discovered. The study concludes with recommendations for increasing the rate at which "identical" facilities discover the same hazards, while at the same time acknowledging that facilities are rarely identical.

Staying Alert: Incorporating Fatigue in Risk Management

Chason J. Coelho, Ph.D., CSP, CFI
Exponent | Managing Scientist | Human Factors
15375 SE 30th Place, Suite 250, Bellevue, WA 98007
ccoelho@exponent.com

Patrick Cullen, Ph.D.
Exponent | Scientist | Human Factors
4580 Weaver Parkway, Suite 100, Warrenville, Illinois 60555
pcullen@exponent.com

Sunil D. Lakhiani, Ph.D., P.E., CSP
Exponent | Managing Engineer | Human Factors
4580 Weaver Parkway, Suite 100, Warrenville, Illinois 60555
slakhiani@exponent.com

Delmar R. "Trey" Morrison, III, Ph.D., P.E., CFEI
Exponent | Principal Engineer | Thermal Sciences
4580 Weaver Parkway, Suite 100, Warrenville, Illinois 60555
tmorrison@exponent.com

Prepared for Presentation at
American Institute of Chemical Engineers
2019 Spring Meeting and 15th Global Congress on Process Safety
New Orleans, LA
March 31 – April 3, 2019

Z190115092205 - 6620

Staying Alert: Incorporating Fatigue in Risk Management

Chason J. Coelho, Ph.D., CSP, CFI
Exponent | Managing Scientist | Human Factors
15375 SE 30th Place, Suite 250, Bellevue, WA 98007
ccoelho@exponent.com

Patrick Cullen, Ph.D.
Exponent | Scientist | Human Factors

Sunil D. Lakhiani, Ph.D., P.E., CSP
Exponent | Managing Engineer | Human Factors

Delmar R. "Trey" Morrison, III, Ph.D., P.E., CFEI
Exponent | Principal Engineer | Thermal Sciences

Keywords: Fatigue, Risk Management, Human Factors, Human Error, Human Performance

Abstract

This paper concerns human fatigue in process safety and risk management. It reviews the impact of fatigue on workplace injuries and presents a risk-based method for addressing fatigue in process safety and risk management processes. The method entails developing elements of a data-driven fatigue risk management system. Specific issues include potential data sources and practical applications within existing Process Safety Management (PSM) schemes and other safety management systems. Special attention is paid to the fatigue risk assessment, which mirrors a common process safety risk assessment and affords systematic control of fatigue-related human error.

1 Introduction

Human fatigue is a peculiar issue in safety and risk management. On one hand, imagining fatigue as a factor that can contribute to adverse safety incidents seems rational and intuitive, while on the other, managing fatigue risks in practice can prove quite challenging. Specific difficulties include realistic assessment of fatigue, prioritization of work activities and fatigue hazards to be assessed, and development of risk-assessable scenarios that link fatigue hazards to adverse safety incidents. This paper reviews these and other challenges and presents practical ways to approach them. Guidance that follows applies to process safety professionals, managers, and field personnel tasked with promoting safety performance.

Fatigue is defined as a state of reduced mental or physical performance capability that results from (1) sleep loss, (2) circadian challenge, (3) difficult task factors, or (4) some combination thereof [1, 2]. Circadian challenge refers to any factor that disrupts the normal human preference for sleeping at night or the natural daily ebb and flow of alertness. Figure 1 presents a guiding framework for this paper. The most critical feature is that fatigue provides a mechanistic link between the three fatigue-hazard categories on the left and the adverse safety impact on the right.

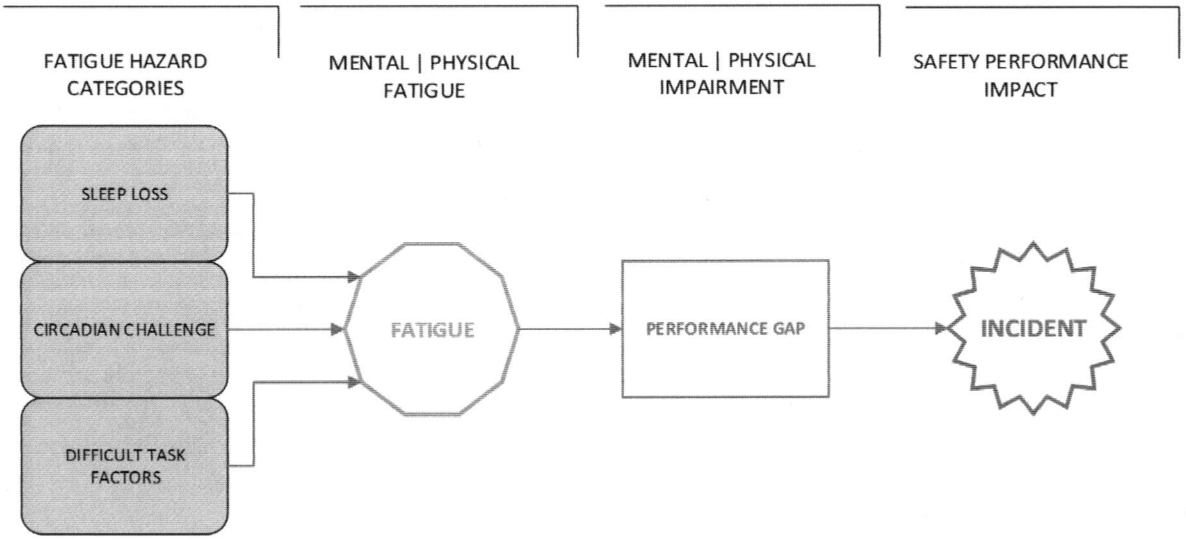

Figure 1. Guiding Framework for Linking Fatigue Hazards to Safety Incidents

There exist many fatigue hazards within each fatigue hazard category. Some examples of such hazards are long commutes, shiftwork, stress, and high cognitive task demands. Personnel are faced with fatigue risk when they are exposed to fatigue hazards. Thus, a fatigue risk is defined by the probability that the hazard leads to an adverse incident and the potential severity of that incident. Fatigue risks can precipitate mental and physical performance impairments that can lead, in turn, to an incident. The notion that fatigue can link hazards and incidents gleans support from the history of catastrophic industry incidents, safety statistics, and scientific literatures.

1.1 Human Fatigue and Catastrophe

The BP Texas City Refinery experienced an explosion and fire in 2005 which killed 15 people and injured 180 others when flammable material escaped from a large vessel and exploded [3]. An investigation by the US Chemical Safety and Hazard Investigation Board found that the board operator had worked 12-hour shifts for at least 29 consecutive days and suggested that fatigue-related human error contributed to the accident [3]. The company had neither a corporate nor a site-specific fatigue prevention policy, and a contract between the labor union and BP specified a minimum number of hours per week but not a maximum [3].

Fatigue is thought to have played a causal role in other notable accidents, such as the Three Mile Island nuclear power plant release and the Chernobyl nuclear meltdown [4], the Space Shuttle Challenger explosion [5], and the grounding of US tankship Exxon Valdez and subsequent spill of approximately 258,000 barrels of oil [6].

1.2 Fatigue-Related Safety Statistics

It is estimated that (a) 360,000 worldwide workplace fatalities occur per year, (b) 960,000 workplace injuries occur per day, and (c) 13% of these injuries can be attributed to fatigue in some way [7]. This statistic works out to around 125,000 fatigue-related injuries per day.

In the United States, about 23% of the population suffers significant sleep problems – the highest rate in the world [7]. Reduced sleep and longer work hours are associated with greater injury risk; the increase in risk is approximately three-fold in the United States [8]. Nighttime workers are almost three times more likely to be injured compared to daytime workers [9]. Jobs with overtime are associated with a 61% higher injury rate compared to jobs without overtime [10]. Working at least 12 hours per day is associated with a 37% higher injury rate, and working more than 60 hours per week is linked to a 23% increase in injury rate [10]. Finally, transportation data reveal 3,662 fatal crashes and 160,000 injury crashes involving fatigue-related drowsiness between 2011-2015 [11].

These findings indicate that fatigue poses significant risks for personal safety. It is logical that fatigue risks extend to process safety. For example, fatigue in a control board operator may increase the likelihood of an error that drives a process outside a safe operating limit.

1.3 Scientific Literature

A comprehensive review of scientific literature has shown that fatigue makes people perform worse than 90% of rested people on various tasks [12]. Performance on cognitive tasks, such as reasoning and mental arithmetic, motor tasks, such as walking and manipulating objects, and emotional tasks, such as regulating individual mood, is negatively affected by fatigue [12]. Human systems in all three categories – mental, physical, and emotional – are especially impaired when less than five hours of sleep are obtained in a given 24-hour period [12].

Sleep loss has been shown to decrease vigilance and impair reaction time [13]. Fatigue-related performance impairments on reaction-time, divided-attention, and memory-recall tasks can even be similar to performance deficits that result from alcohol intoxication [14]. Individuals who lose just two hours of sleep from a normal eight-hour sleep schedule perform vigilance, attention, and memory tasks similarly poorly to those who have consumed two to three beers within 2.5 hours prior to testing; people who lose four hours perform similarly to those who have consumed five or six beers [14]. Fatigue has also been shown detrimental to tasks involving judgment, planning, learning or storing new material, and maintaining situation awareness [15].

These findings show that fatigue is a very real problem for human performance and, by extension, personal health and life safety risk management and process safety management.

2 Managing Fatigue

Fatigue management is challenging for several reasons. Perhaps the most straightforward is that fatigue is common. Recall the estimate noted earlier that fatigue contributes to approximately 13% of workplace injuries [7]. Research also suggests that significant chronic sleep loss exists in one-third or more of normal adults [16]. Consistent with the idea that fatigue is common in the workplace, a National Safety Council study surveyed over 2,000 American workers, identified nine common fatigue risks, and found the following prevalence for each [17]:

1. Shiftwork – 17% work a non-day shift;
2. High-risk hours – 41% work at high-risk times such as early morning hours;
3. Demanding jobs – 81% have jobs that entail a high risk of fatigue because of requirements for sustained attention or strenuous physical and mental activity;
4. Long shifts – 21% work long shifts of at least 10 hours;
5. Long weeks – 22% work long weeks of at least 50 hours;
6. Sleep loss – 43% get less than seven hours of sleep;
7. No rest breaks – 10% do not get to take a rest break;
8. Quick shift returns – 14% get less than 12 hours off between shifts; and
9. Long commutes – 31% have long commutes.

The commonplaceness of fatigue is compounded by the fact that individuals struggle to accurately assess their own fatigue [18]. This diagnostic problem may result, at least in part, from cognitive biases – departures from rational judgment – inherent and inescapable in human mentality [19].

As mentioned above, fatigue can come from multiple fatigue-hazards in multiple categories: sleep loss, circadian challenge, and difficult task factors. Within these categories, there are potentially many fatigue risks to which the worker may be exposed. The appendix to this paper offers a preliminary fatigue hazard list which contains some hazards for each fatigue hazard category and some that can be found in all three. One point is that this list is sizeable even though it is preliminary. Another is that the sheer number of potential fatigue hazards may make it difficult for organizations with little or no experience in fatigue management to account for, mitigate, and control fatigue risk on a routine basis.

Yet another challenge in managing fatigue arises from the fact that fatigue and its causes can come from outside duty hours. Operators can show up to the facility already fatigued from dealing with long commutes, sleepless children, or any number of other fatigue hazards and risks. Managing fatigue is, therefore, a responsibility shared by individuals and companies.

This variability only adds to that associated with operational contexts that can vary substantially across industries, companies within those industries, business lines within those companies, and even facilities within those business lines. By extension, this operational variability can entail different task demands and schedules. Unique facility and operational features can pose unique fatigue risks – an observation which points up the necessity of evaluating facility- and operations-specific risks. As discussed in the next section, that approach is not what has been traditionally taken to manage fatigue.

2.1 Traditional Approaches

Individuals and companies alike have traditionally met their fatigue management responsibilities by relying on prescriptive approaches where limitations and requirements are imposed across organizational units [20]. Two of the most common prescriptions are hour-of-service restrictions and minimum rest break requirements. These prescriptions are simple and can be helpful in limiting risks associated with some fatigue causes; however, they do not commonly or specifically limit worker exposure to fatigue risks associated with typical human sleep patterns, duty cycles, and non-work-related time such as that spent commuting [20].

Another concern with adherence to the common prescriptive requirements without a thorough evaluation is that such an approach does not itself constitute formal or rigorous management. Formal safety management systems (SMSs) require hazard identification and assessments of risks [21]. Note that such is also the underlying collective theme of OSHA Process Safety Management (PSM) – to help ensure hazard recognition, evaluation, and control.

The foregoing suggests that a one-size-fits-all prescriptive approach to fatigue management can be improved. Organizations in various industries are recognizing this situation and beginning to move away from prescriptions and toward a more customizable risk-based approach [20].

2.2 Fatigue Risk Management Systems

A risk-based approach currently being adopted in many industries is the fatigue risk management system (FRMS). The FRMS can be thought of as a method for monitoring and managing fatigue-related safety risks that relies on scientific principles and operational experience to help ensure personnel work with proper alertness [1].

There currently exist somewhat different FRMS versions across industries. Given that an FRMS is a management-system level administrative control, most comprehensive versions entail the following elements in one form or another:

- Introductory material conveying purpose, objectives, scope, and accountability;
- Policy statement conveying organizational commitment;
- Scientific definitions and information on sleep, fatigue, and fatigue countermeasures;
- Fatigue risk assessment and countermeasure processes, procedures, and guidance;
- Assurance and continuous improvement processes, procedures, and guidance;
- Fatigue awareness and training; and
- Implementation and communication plans.

Within the petrochemical process industries, at least two notable and relevant guidance documents have recently emerged. One is *Assessing Risks from Operator Fatigue* published by the International Association for Oil & Gas Producers (OGP) in 2014 [22], and the other is *Recommended Practice (RP) 755, Fatigue Risk Management Systems of Personnel* published by the American Petroleum Institute (API) in 2010 [23].

The OGP guidance document recommends a two-stage fatigue risk management process by which facilities and operations are first screened and then analyzed for fatigue risks. This guidance document represents a significant step forward in fatigue management for the petrochemical industries. It highlights the importance of a risk-based approach and promotes assessment of fatigue at the level of worker activity.

The OGP document has some limitations, however. It does not present guidance for developing and implementing a comprehensive FRMS. It also presents a method that would not only be relatively novel for most process safety professionals but would place rather significant responsibility on those individuals for assessing specific fatigue-related task issues. Examples include defining action levels for fatigue and predicting levels of fatigue exposure using biomathematical models. These tasks can be quite challenging for process safety professionals not reasonably expected to be well-versed in details of fatigue-related performance. This situation constrains the applicability and practicality of the approach presented by OGP.

Another limitation is that the recommended OGP processes do not offer the opportunity to acknowledge any existing company, facility, or operational features that could be claimed as fatigue countermeasures. This is an important opportunity because using current features of the PSM scheme affords efficiency and promotes implementation of fatigue risk management. It may be much easier, and just as effective, to improve an existing countermeasure than to add a new one. For example, a facility may already have a policy that limits overall overtime hours to help control fatigue but does not say anything about limiting contiguous overtime hours or overtime hours used to extend normal shifts. Adding these additional countermeasure modifications may be easier and more efficient for the company than implementing another countermeasure, such as adding staff to reduce overtime hours altogether.

Countermeasure problems just mentioned are exacerbated by the fact that while the document calls for development of an action plan, it does not offer much in the way of how to construct that action plan. Examples of fatigue countermeasures to consider and implement are not given.

Whereas the OGP document focuses more on assessing fatigue than managing it, API RP 755 sort of does the opposite. In another welcomed step forward, the API document does offer guidance on several components that should be found in a comprehensive FRMS; however, it gives little to no functional guidance on actual fatigue risk assessment and control. A limited set of countermeasures are offered, but they are not connected to more specific features of fatigue that could reasonably be expected of a comprehensive risk-based approach. To help close these gaps, the rest of this paper focuses on a relatable, practical, and straightforward way to conduct a fatigue risk assessment (FRA).

2.3 A Practical Approach to Fatigue Risk Assessment

The fatigue risk assessment (FRA) serves as the core of an FRMS because it consists of processes that enable the company to achieve its fatigue-related safety objectives. The type of FRA presented and advocated below is designed to mirror a hazard and operability (HAZOP) analysis – a more common risk assessment of which process safety professionals involved in PSM activities are typically well aware.

The FRA is conducted in a meeting of relevant stakeholders and is focused on fatigue and the human error that may result. More specifically, the FRA is designed to (a) identify situations where fatigue may pose hazards, (b) assess the risks presented by those fatigue hazards, (c) note existing countermeasures, and (d) decide whether those countermeasures are adequate or whether additional fatigue countermeasures are required to mitigate or control the fatigue risk. These processes therefore offer a method for understanding more specific ways human error can lead to adverse incidents. This focus contrasts more traditional approaches to process and safety risk management approaches in which operator or maintainer error is often noted as a reason that a critical adverse incident can occur, but little is said about the nature of that error or how it can be prevented or mitigated.

Figure 2 summarizes the FRA process sequence.

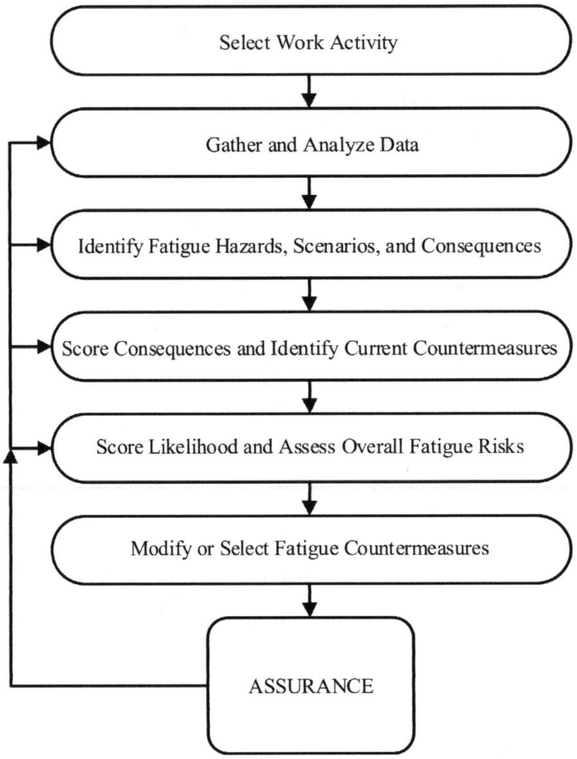

Figure 2. The Fatigue Risk Assessment Process Sequence

The first step is to identify the work activity where fatigue is thought to be a potential risk within the facility. This step should include a prioritization of work activities, and de facto, of the fatigue hazards found therein. This prioritization can be difficult, however, without previously having performed a risk assessment like the one pursued in this process. One way to deal with this issue is to rely on safety criticality of the activities. The company can choose to ensure that all safety-critical work activities are subjected to, or at least carefully considered for, FRAs, or it can rely on a threshold of criticality to decide which tasks get FRAs and when.

Another approach is to focus on whether the activity requires human performance that is sensitive to fatigue. The company can rely on scientific reviews like the ones presented in *Sections 1.2* and *1.3* of this paper to understand which tasks are particularly susceptible, such as those highly demanding of memory, attention, judgment, perception, and fine motor control.

An example work activity that could be selected in the first step of the FRA is starting a boiler. It is best to ensure that the activity is reasonably well defined. For example, it would be helpful to decompose "starting the boiler" into procedural steps, such as (a) put line up vales in startup position, (b) close bottom blowdown valves, and so on. Each procedure step can be subjected to an FRA. Note that it is not required to assess each procedural step in an FRA – this option is just given for illustration.

The second step is to gather and analyze data. An objective of this step is to start understanding what fatigue hazards may exist in the work activity. Continuing with the boiler startup example, analysis may reveal that the boiler is routinely started during the early afternoon hours – a period of the circadian cycle known as a window of circadian low (WOCL) associated with increases in human errors [24]. Or it may reveal that there was a previous incident in which a fatigued operator ran the pressure too high on the boiler with the vent valve closed. These findings would be noted and included in the FRA with the goal of determining specific countermeasures to control the risk associated with the fatigue hazards.

One data source that should be considered entails the results from previous safety studies or other types of risk assessments. Results from these analyses can be useful for identifying significant safety events to which fatigue-related human error may contribute. Another potential data source is any incident investigation in which fatigue was determined to be a contributing factor, as alluded to above. Yet another is any analysis of factors that include fatigue hazards, or otherwise bear on fatigue in a clear way; an example is overtime analysis often done for the sake of cost management that can also have implications for fatigue management.

A final source of data worth mentioning here includes recorded measurements of fatigue itself. Fatigue measurement can be challenging. There is no single fatigue measure that can serve as a "gold standard," mainly because there can be many signs and symptoms of fatigue coming from multiple sources [1]. Two approaches can be used, however: (1) subjective measures, such as fatigue ratings based on worker recall or current impressions, and (2) objective measures, such as performance tests and physical monitoring [1].

One subjective measure is the fatigue report form, which allows individuals to provide feedback on fatigue risks and subsequent fatigue that occur during operations. Another subjective measure is the retrospective survey. This method can be an efficient way to gather current fatigue-related information from workers about levels of fatigue experienced during past operations. Individuals can also give more direct subjective fatigue and sleepiness ratings in the moment.

Objective measures are often used in laboratory studies and focus on specific aspects of performance that may not be exactly representative of actual job task demands [1]. However, they can still be useful in helping to determine whether an individual is fatigued. One objective measure is the psychomotor vigilance test (PVT) which requires an individual to respond as quickly as possible to the presentation of several stimuli in a sequence [1]. The average reaction time can be computed and compared to a database of reaction times to help indicate fatigue. Another option is the actigraph – a small watch-like device worn on the wrist, which contains an accelerometer and a storage drive that continuously record the larger motions of the individual over some period of time [1].

When FRA processes are initially implemented, it is not always possible or necessary to gather new data for this step; it may be possible instead to identify potential fatigue hazards based on previous safety studies, as just mentioned, in addition to operational experience. Over time, the assessments can be expected to mature as more data becomes available through other collection and tracking requirements set forth in the larger FRMS.

The following should be considered when deciding which measures to collect and analyze [25]:

- Fatigue-related impairment can affect multiple skills and can have multiple sources, so there is no single measurement that gives a total picture of fatigue level;
- The level of fatigue-management effort should reasonably correspond to the level of fatigue risk, as assessed in the FRA. All measurements require resources, and a goal should be to use those resources efficiently;
- Balance should be maintained between gathering enough data and the additional demands that data collection can place on individuals;
- A set of measures can be selected for routine monitoring; and
- Additional measures can be used if a fatigue hazard is identified and stakeholders decide that more information is needed.

The third step is to identify the fatigue hazards, scenarios, and consequences. A list of fatigue hazards can be developed or expanded based on the scientific literature, operational experience, and other data sources. The scenario is meant to be a written statement of how the particular work activity and fatigue hazard can lead to a potential consequence. For example, the fatigue hazard of starting the boiler during a WOCL could be associated with a scenario in which the operator forgets to open the vent valve leading to overpressure, catastrophic failure, and potentially multiple fatalities. It is important to note in the scenario to a reasonable degree of specificity the task in which the worker may make an error and the type of error considered.

Step four is to score the consequence severity of the incident using a company risk assessment matrix for safety. Like in the HAZOP method, the scenario to be assessed should be the worst-credible case, and incident severity estimation should *not* consider existing countermeasures. Once the severity of a scenario is scored, any process or facility features that may function as fatigue countermeasures are identified. As mentioned, an example of an existing countermeasure is a policy that limits overtime hours to help control fatigue.

This is a good place to note that the FRA approach does not prohibit or preclude the use of prescriptions. In fact, prescriptions such as hour-of-service limitations and mandatory break requirements can be helpful and should be claimed as countermeasures in the context of the FRA. This point notwithstanding, existing countermeasures should also be evaluated for appropriateness, specificity, and overall adequacy for the particular fatigue activity, hazard, risk, and scenario being evaluated.

General potential countermeasures from which the stakeholder team can select should be developed in the larger FRMS. An example of such a list for sleep loss includes considerations such as:

- Overtime;
- Long work hours and weeks;
- Commutes and travel;
- Health and wellness;
- Personal activities; and
- Company activities.

Step five is to score the likelihood of the scenario using the safety risk assessment matrix and assess the overall fatigue risk. Like in a HAZOP, the likelihood should consider existing countermeasures. The FRA should at all times be focused on hazards, risks, and countermeasures specific to fatigue.

Step six is to determine whether the overall fatigue risk is acceptable. This determination should entail discussion about whether current fatigue countermeasures are adequate. If the team determines they are not, then new fatigue countermeasures should be issued, and action items should be assigned and tracked.

The final step is to develop fatigue-related performance indicators that allow assessment of how well those countermeasures perform in the future. Doing so should be a part of an assurance effort in which the company compares countermeasure performance with fatigue-related safety objectives established in the FRMS.

Recall that in our boiler-startup example we noted that modifications could be made to the facility overtime policy to help reduce fatigue risk. These changes included limiting contiguous or consecutive overtime periods and shift-extension overtime hours. Fatigue-related performance indicators for those countermeasure modifications could be the proportion of (a) overtime versus non-overtime hours worked and (b) employees who have exceeded the maximum number of consecutive working days/hours before taking a rest day. Analyses of these types of indicators for countermeasure performance should be integrated in subsequent FRA reviews to facilitate continuous improvement in fatigue risk management; Figure 2 represents these cyclical integrations as return pathways from Assurance to earlier steps in the FRA process.

3 Integrating FRMS and PSM Schemes

As mentioned earlier, an FRMS can be viewed simply as an SMS focused on fatigue. Given that a PSM scheme can be viewed as an SMS focused on process safety, existing PSM elements can be leveraged to promote the development and implementation of a new FRMS. An initial gap analysis can be performed to identify which existing SMS features can be leveraged to focus on fatigue in an efficient way [1]. The gap analysis should ask how fatigue can be best integrated into four main SMS components: (1) documentation, (2) safety risk management, (3) assurance and continuous improvement, and (4) safety awareness and competency management.

Some specific PSM elements are immediately applicable. Employee participation mechanisms can be utilized to involve relevant personnel in the FRA and to communicate important goals and tasks with those who are expected to perform FRAs and other FRMS-related activities. Training programs can be developed to address fatigue and its symptoms, signs, and countermeasures. The organization may consider developing different tiers of fatigue-related training that range in detail and demand. The most rigorous version could be expected for FRMS and FRA practitioners expected to lead fatigue-related meetings and for supervisory personnel expected to make key decisions about fatigue management. Other less-rigorous versions focused on fatigue awareness and identification can be rolled out to field personnel.

Managing fatigue of contractors can be challenging, but a good place to start is to include in the contract requirements that the contractor abide by company fatigue countermeasures. Specific fatigue countermeasures can be developed for contractors, such as minimum staffing and maximum overtime hours.

It is useful to consider fatigue-related issues in pre-startup safety reviews (PSSRs) for the same reason it is useful to consider how different operations may yield different fatigue risks in an FRMS. Specific fatigue-related hazards here can include the stress and time-pressure associated with performance milestones and turnaround time minimization, high work tempo, task overload, and overtime often used to ensure that new and modified facilities are started up on time.

Related to the incorporation of fatigue into PSSRs is the need to consider fatigue in management of change (MOC) reviews. This need is especially salient when MOC-governed changes entail reduced crew complements, increased intensity or duration of activities, and modified schedules. It is recommended that each MOC review conducted for safety-critical equipment entail an FRA.

Finally, fatigue measurement and symptoms are often salient topics in incident investigations. Thus, special focus should be given in the incident investigation to considering whether fatigue was a contributing factor. There are no simple or objective tests that can conclusively be used to show that fatigue was a contributing factor in an adverse safety incident [1]. Thus, for fatigue to be considered as a contributing factor, at least the following should be evidenced [1]:

- The person(s) took actions or made decisions that were probably causal in the event;
- The person(s) had exposure to fatigue risks at or around the time of the event; and
- Actions and/or decisions were consistent with behavior expected in fatigued states.

Factors that should be considered in determining whether personnel were likely to be exposed to fatigue risks should be developed and incorporated into the PSM scheme. Companies may consider in this update giving more formal, specific, and written guidance criteria that can be used to help determine whether personnel were exposed to fatigue risks that led to a fatigued state at the time of the incident. Examples of such criteria include the minimum number of quality sleep hours expected in previous periods to conclude that the person was rested, whether the incident occurred in a WOCL, and how recent shift rotations and durations were structured.

4 Summary and Conclusions

This paper reviews the adverse safety impacts of human fatigue and discusses some of the core challenges of managing associated safety risks. It draws out that these challenges can dissuade risk managers from formally incorporating fatigue in risk management but also gives practical methods that can be used to assess and control fatigue.

The most comprehensive approach described herein is the fatigue risk management system (FRMS). The FRMS is a modern method that relies on data, operational experience, and scientific principles of sleep and fatigue to identify current fatigue hazards and assess the risk associated with them.

A core element of the FRMS approach is the fatigue risk assessment (FRA). The FRA method presented in this paper capitalizes on existing knowledge of process safety professionals and administrative systems already in place within many petrochemical facilities. Such systems can include OSHA Process Safety Management (PSM) schemes and safety management systems (SMS). In fact, the FRMS can be simply thought of as an SMS focused on fatigue. Taking this perspective allows risk managers to leverage existing PSM and SMS elements to address fatigue-related issues. Specific leverage opportunities include existing hazard identification and safety risk study methods, such as hazard and operability (HAZOP) analyses, pre-startup safety reviews (PSSRs), and management of change (MOC) reviews; existing facility data, such as those from incident investigations and other operational systems; employee participation mechanisms; and any assurance and internal promotion or communication processes.

One of the most critical conclusions of this paper is that the FRA can help fill gaps in relatively recently published fatigue guidelines for the petrochemical industries. The FRA can serve in this augmentative role because it is a feasible-but-thorough means of controlling the very real threat of fatigue-related safety risk. It goes beyond prescriptive limitations that do not commonly or specifically limit worker exposure to a variety of known fatigue risks. The FRA method presented here is also flexible – while it is best-practice to conduct the FRA in the context of a larger FRMS, the method can be applied in and of itself using the steps outlined here. Regardless of the specific characteristics of the FRMS and FRA chosen, the risk-based approach is an effective way to manage fatigue. Companies in several industries and across the globe are recognizing this utility and working to capitalize on it.

5 References

[1] FRMS Implementation Guide for Operators, ICAO / IATA / IFALPA, 1st Ed., 2011.

[2] American Petroleum Institute Recommended Practice 755. Fatigue Risk Management Systems for Personnel in the Refining and Petrochemical Industries. Report No. 2005-04-I-TX, 2010.

[3] U.S. Chemical Safety and Hazard Investigation Board (CSB). Investigation Report, Refinery Explosion and Fire, 2007.

[4] Mitler, M. M., Carskadon, M. A., Czeisier, C. A., Dement, W. C., Dinges, D. F., & Graeber, R. C. Catastrophes, Sleep, and Public Policy: Consensus Report. Sleep, 11(1), 100-109, 1988.

[5] Rogers, W. P., Armstrong, N. A., Acheson, D. C., Covert, E. E., Feynman, R. P., & Hotz, R. B. Report of the Presidential Commission on the Space Shuttle Challenger Accident. Washington, DC, 1986.

[6] Practices that Relate to the Exxon Valdez. Washington, DC: National Transportation and Safety Board, 1990.

[7] Uehli, K., Mehta, A. J., Miedinger, D., Hug, K., Schindler, C., Holsboer-Trachsler, E., ... & Künzli, N. Sleep Problems and Work Injuries: A Systematic Review and Meta-analysis. Sleep medicine reviews, 18(1), 61-73, 2014.

[8] Lombardi, D. A., Folkard, S., Willetts, J. L., & Smith, G. S. Daily Sleep, Weekly Working Hours, and Risk of Work-related Injury: US National Health Interview Survey (2004–2008). Chronobiology International, 27(5), 1013-1030, 2010.

[9] Swaen, G. M. H., Van Amelsvoort, L. G. P. M., Bültmann, U., & Kant, I. J. Fatigue as a Risk Factor for Being Injured in an Occupational Accident: Results from the Maastricht Cohort Study. Occupational and Environmental Medicine, 60 Suppl 1, i88–i92, 2003.

[10] Dembe, A. E., Erickson, J. B., Delbos, R. G., & Banks, S. M. The Impact of Overtime and Long Work Hours on Occupational Injuries and Illnesses: New Evidence from the United States. Occupational and Environmental Medicine, 62(9), 588-597, 2005.

[11] Traffic Safety Facts. Report No. DOT HS 812 446. Washington, DC: US Department of Transportation and National Highway Traffic Safety Administration, 2017.

[12] Pilcher, J. J. & Huffcutt, A. I. (1996). Effects of Sleep Deprivation on Performance: A Meta-analysis. Sleep, 19(4), 318-326.

[13] Krueger, O.P. (1989). Sustained Work, Fatigue, Sleep Loss and Performance: A Review of the Issues. Work Stress, 3, 129-41, 1989, as cited in [14].

[14] Roehrs, T., Burduvali, E., Bonahoom, A., Drake, C., & Roth, T. Ethanol and Sleep Loss: A "Dose" Comparison of Impairing Effects. Sleep, 26(8), 981-985, 2003.

[15] Wickens, C. D., Gordon, S. E., Liu, Y., & Lee, J. An Introduction to Human Factors Engineering. Pearson Prentice Hall: Upper Saddle River, New Jersey, 1998.

[16] Bonnet, M. H. & Arand, D. L. We Are Chronically Sleep Deprived. Sleep, 18, 908-911, 1995.

[17] National Safety Council. Fatigue in the Workplace: Causes & Consequences of Employee Fatigue, 2017.

[18] Lerman, S. E., Eskin, E., Flower, D. J., George, E. C., Gerson, B., Hartenbaum, N., ... & Moore-Ede, M. Fatigue Risk Management in the Workplace. Journal of Occupational and Environmental Medicine, 54(2), 231-258, 2012.

[19] Croskerry, P. The Importance of Cognitive Errors in Diagnosis and Strategies to Minimize Them. Academic Medicine, 78, 775-780, 2003.

[20] Gander, P., Hartley, L., Powell, D., Cabon, P., Hitchcock, E., Mills, A., & Popkin, S. Fatigue Risk Management: Organizational Factors at the Regulatory and Industry/Company Level. Accident Analysis & Prevention, 43(2), 573-590, 2011.

[21] ISO 45001: Occupational Health and Safety Management Systems – Requirements with Guidance for Use, International Standards Organization, 2018.

[22] International Association of Oil & Gas Producers. Assessing Risks from Operator Fatigue: A Guidance Document for the Oil and Gas Industry. OGP Report No. 492, 2014.

[23] American Petroleum Institute Recommended Practice 755. Fatigue Risk Management Systems for Personnel in the Refining and Petrochemical Industries. Report No. 2005-04-I-TX, 2010.

[24] Bjerner B, Holm A, & Swensson A. Diurnal Variation in Mental Performance: A Study of Three-Shift Workers. British Journal of Industrial Medicine, 12, 103–10, 1955.

[25] Manual for the Oversight of Fatigue Management Approaches, ICAO, 2nd Ed., 2016.

6 Appendix

PRELIMINARY FATIGUE HAZARDS LIST		
SLEEP LOSS (SL)	CIRCADIAN CHALLENGE (CC)	TASK FACTORS (TF)
Less than 7-9 hours per night	Work in Window of Circadian Low	Complex
Sleep fragmentation	Shiftwork	Monotonous
Quick returns to work	Time zone changes	Overload
Sleep disorders	Schedule unpredictability	Underload
Medication	Unplanned work extensions	Physical demands
Alcohol	Evening types on morning shifts	Mental demands
Other drugs	Morning types on evening shifts	Diminished light
Medical problems	Early start times	Inappropriate temperature
Lifestyle choices	Later start times	Poor ergonomics
Social commitments	Day-to-night transitions	No or reduced rest breaks
Cumulative sleep debt	On-call status	High work tempo
Work-related worry	Unscheduled maintenance	Equipment resource limitations
GENERAL (G)		
Performance milestones		
Stress		
Poor planning		
Long commutes		
Long shifts		
Overtime		

19 AIChE Spring Meeting
15TH GLOBAL CONGRESS ON PROCESS SAFETY
NEW ORLEANS • 2019

PHA Data Quality – Mining the Hidden Risks!

Susan Barney, EIT
Risk Alive Analytics Inc.
#300, 926 5th Ave SW, Calgary, AB, Canada
sbarney@acm.ca

Prepared for Presentation at
American Institute of Chemical Engineers
2019 Spring Meeting and 15th Global Congress on Process Safety
New Orleans, LA
March 31 – April 3, 2019

PHA Data Quality – Mining the Hidden Risks!

Susan Barney, EIT
Risk Alive Analytics Inc.
#300, 926 5th Ave SW, Calgary, AB, Canada
sbarney@acm.ca

Keywords: Data Mining, Artificial Intelligence, Safety Culture, Data Quality, Comparing Data

Disclaimer:

The information in this paper is general in nature only and should not be relied upon without first obtaining advice from a qualified professional person. The advice and strategies herein may not be suitable for your situation. Any use which a third party makes of this paper, or any reliance on or decisions made based on it, are the responsibility of such third party. Neither the author nor ACM Risk Sciences and Development Inc. shall be responsible for damages, if any, suffered by any third party as a result of decisions made or actions taken based on this paper.

Abstract

There is growing concern among owners/operators regarding the quality and consistency in the Process Hazard Analyses (PHA) of their facilities. When a PHA is conducted, whether through an internal or contracted facilitator, there can be a strong influence on the data from many factors. These factors affect the integrity of the data and, can lead to increased risk exposures in the facility.

While PHAs have been conducted within the industry for many years, there are still distinct differences between the quality and completeness of data generated within each session. Using data analytics and visualizations, the industry can move from inconsistent data quality to a place of quality awareness. This knowledge can reduce risk exposure by helping remove inconsistencies within an organization's PHA data. Additionally, this can increase the percentage of critical scenarios being captured during a PHA session, reducing the likelihood of missing a high-risk scenario and the potential corresponding recommendation required to adequately address the risk. This concept can be adopted for a newly purchased or a newly built facility by having a guideline to start with for the baseline PHA, or for the revalidation of a PHA.

To achieve these visualized analytics, there is a series of transformations while mining the PHA data which will allow it to be in a comparable state. The data must be generalized so specifics such as tag numbers are removed, and comparisons can be drawn. It is then organized into subsections of process safety vulnerabilities, which summarize the threats the facility is exposed to.

From this generalized information, the most critical vulnerabilities across multiple facilities can be extracted to be assessed in the next PHA. Along with the aid of a subject matter expert, it can also be determined if any vulnerabilities were not considered in the PHA, such as prior incidents. This can create a more thorough PHA with the confidence that most scenarios have been analyzed.

1 Introduction

Every organization that processes hazardous materials should have a rigorous Process Safety Management (PSM) Program. Process Hazard Analysis (PHA) is a foundational and key component of a complete PSM program. PHA studies are performance based and an organization must choose what type is used to assess the risk at their facilities. The selection should be appropriate to the complexity of the process and must identify, evaluate, and offer recommendations to mitigate the hazards uncovered in the session [2]. For the purpose of this paper, we will be focusing on Hazard and Operability (HAZOP) as the PHA study for comparing data quality.

HAZOP sessions require the experience, knowledge and technical backgrounds of the team members along with systematic methodologies to comprehensively uncover hazards and risks present at the facility [1]. Risk can be perceived as a function of frequency and consequence severity, therefore, the risk in the scenarios captured within a HAZOP session should be understood by the whole team. Thorough analysis of the hazards during HAZOP sessions allows for necessary information to be available to key stakeholders enabling them to make suitable decisions to manage the risk.

The concern of owners and operators regarding the quality and consistency in the HAZOP has grown in the last few years. Whenever an incident occurs at a processing facility, the HAZOP team will typically return to the documentation to determine if that scenario was captured during the study. This reactive approach can be beneficial to help with future HAZOPs to ensure all previous incidents are captured, but a proactive approach would help in avoiding the incident. The overall culture of process safety should be proactive rather than reactive, as when lives are at risk, personnel safety needs to be treated with the utmost importance.

The most dangerous type of risk to a facility is the unknown risk. This unknown risk exposure can arise from an incomplete, non-concise, poor quality HAZOP session. The risk gaps that are created from these inconsistencies can be improved upon, and the starting point is the data that is already available.

Over decades of conducting thousands of risk assessments for the some of the world's largest operating companies, ACM Facility Safety has accumulated an extensive database of process safety information. Many methods and tools have been developed to help clients get the most out of their HAZOP sessions and help make their facilities safer. One of the latest of these tools is Risk Alive Analytics; a web-based service that uses the PSI data from HAZOP sessions to find, and bring to the surface, information that is typically hidden or not considered. Having this information can help eliminate many risk gaps. One of the analytics within Risk Alive is the Facility Unit Comparison module. Data can be generalized and with the aid of subject matter experts, conclusions can be drawn based on the comparison of similar facilities to find this hidden information. This can help organizations become much more proactive with their knowledge of risk exposure and understanding of potential hazards in their respective industries.

2 HAZOP Introduction and Required Information

HAZOPs are used as a formal and systematic examination of the intentions of new and existing facilities to assess the hazard potential mal-operation or malfunction of individual items of equipment and the consequential effects on the facility as a whole [1]. The HAZOP generates a record and provides proof that a recognized form of hazard analysis has been completed for a project, or a facility. For many operations, a HAZOP study provides all the safety assurance that is necessary to satisfy the regulatory authorities [1]. If an unexpected incident occurs the investigating board will typically use the HAZOP reports to determine if the organization completed its due diligence. If it is found that the organization did not fully investigate the extent of a foreseeable hazard, they could face significant regulatory and financial consequences.

The personnel requirements for a HAZOP session include:

- The HAZOP team typically consists of between five to ten individuals who are familiar with the project and can use critical thinking to provide insight and knowledge on the information being discussed.
- The team members should be competent in three main areas: they are knowledgeable, experienced with the process, and trained on the methodology and how it relates to process safety.
- One or two individuals to facilitate and document the findings of the sessions. They must be familiar and experienced with the HAZOP methodology and able to properly record and report upon the technical discussions.

One of the root issues of HAZOPs is that they are highly subjective and qualitative approach. Subjectively developed reports are often difficult to return to and use as cold eyes documentation as there are many decisions made in session that are critical to scenario analysis that are only in the minds of the attendees. If the suitable members are within the team and the information is recorded properly, the subjective manner should not heavily affect the data. It is also important that a well-defined risk matrix is used to understand the consequences. Valuable insights from experienced members of the team are generated during the session. There are many instances where certain data is not documented as a time saving method for the session, such as the initiating event probability to determine the probability before safeguards or magnitude of likelihood reduction from a safeguard but this lack of explicit information can make it challenging to remember what was discussed during the session. It would be illogical to expect someone to recall a conversation they had months ago based on their memory alone.

The **optimal** minimum amount of required data to complete a PHA session includes:

1. Node
2. Deviation
3. Cause
4. Worst Credible Consequence Description
5. Initiating Event Probability, Worst Credible Consequence Severity and Risk Ranking (Before the consideration of Safeguards)
6. Safeguard Description

7. Magnitude of Likelihood Reductions from the Consideration of Safeguards
8. Frequency Worst Credible Consequence Severity and Risk Ranking (After the consideration Safeguards)
9. Recommendations
10. Magnitude of Likelihood Reductions from the Consideration of Recommendations
11. Responsibility for ensuring the implementation recommendations
12. Proposed Risk Ranking (After Recommendations Are Implemented)

Globally, there is a very large number of facilities that process different types of hazardous materials. Each of these facilities conduct and document a form of PHA. This data comprises of a significant number of hours of sessions where there were multiple subject matter experts within one room discussing potential proposed hazards to a specific facility. Therefore, the data generated within a PHA session is quite valuable because of the experience, insights & expertise that went into generating it.

As methodologies of hazard identification have evolved over time, it is important to ensure the data being compared is in a comparable state. One of the main things that creates difficulty when trying to compare data is the quality of the data itself. As the data generated during a HAZOP session can be influenced by many different factors, there can be large variations within it. Although there are many things that can lead to data variations, the items of focus in this paper are [3]:

- Missing information, some organizations may decide to not fill in certain columns of data to save time within the session, this may include, initiating event probability, or risk ranking, the magnitude of likelihood reduction from a safeguard or recommendation
- The quality of data recorded within the session, for example, brief consequence text, not capturing the full scenario or skipping steps in the sequence, using very generic safeguards that do not capture the proper information required.

Examples below will describe in more detail how the quality and completeness of the data can strongly affect the outcome of the session and the follow up action items.

2.1 The Variation in HAZOP Data Quality

These examples show how the quality of the data recorded within the session does not always capture the proper information required. Figure 1 below shows consequence descriptions from the same type of facility across three different organizations.

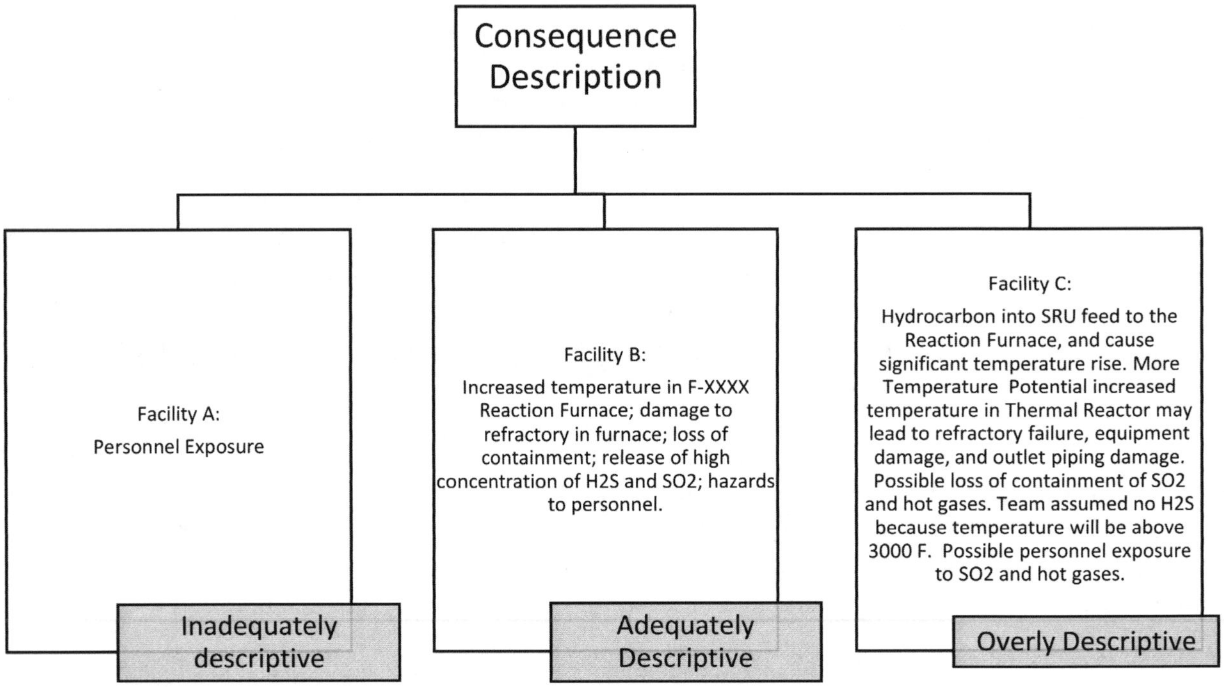

Figure 1: Consequence Variation between HAZOP Files

It would be challenging for someone to look back on the results post-session to understand the sequence of events associated the first consequence or the true risk they were trying to capture. Time would have to be spent looking at drawings, the associated cause, the safeguards and any other applicable information to be able to determine what was being discussed. If the information had been accurately recorded as in the second description, it would be easily understood how the consequence could occur. The third description is quite lengthy and wordy, here the key information can also be difficult to extract and understand as there is a large volume of data to review. Ideally, someone who is familiar with the processing unit should be able to look at a consequence text and determine what it is describing without any major questions.

Figure 2 below shows two examples of safeguards that were used within the same HAZOP, underneath each example, it shows how many times this safeguard was used.

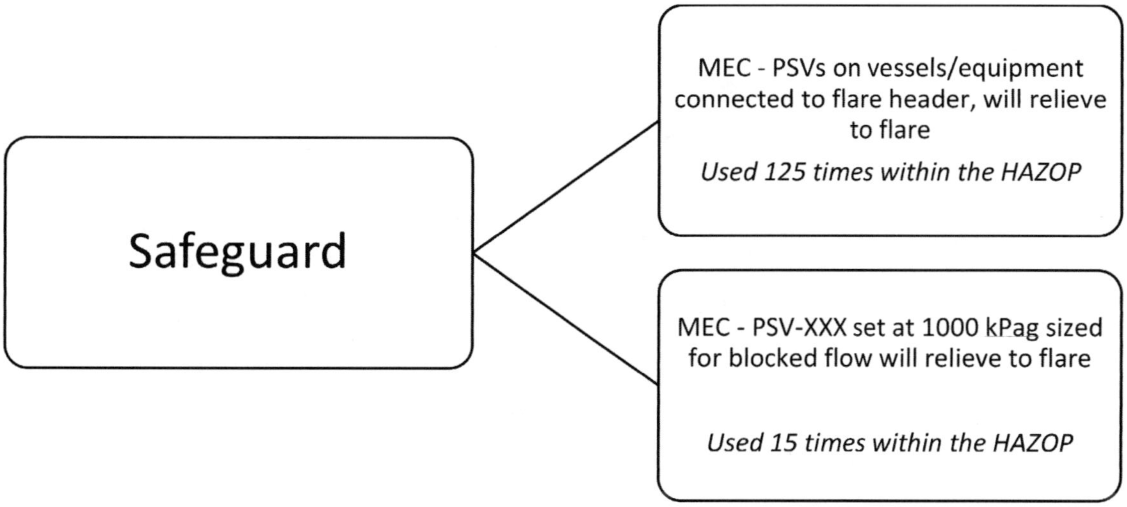

Figure 2: Safeguard Description from Real HAZOP Files

In this example, a generic safeguard such as the top safeguard in Figure 2 does not provide information on where the safeguard is located, the set points, and what type of conditions it protects against. This safeguard was used 125 times within the respective HAZOP. This may raise concern that each time this general safeguard is applied on a consequence scenario, is it truly a risk reducing safeguard? Are the members in the session discussing the set points, locations and functionality of the Pressure Safety Valves? The bottom safeguard in Figure 2 has a description that is more specific. It includes the location, the pressure rating and the condition in which it protects against. Recording safeguards in this manner will create more explicit information regarding the protection layer. To understand why the quality and consistency of a PHA is important, it is essential to understand how the information within the HAZOP report can be used post-session. As the session is focused on identifying hazards and how they create risk within an operating facility, having data where the hazard is not fully understood creates questions regarding the risk at the facility.

Other than the risk within the session, there is also unconscious and unknown risks. An unconscious risk is one where the hazard is not perceived. Unknown risk is one where the full range of consequences is not known or fully understood [1]. Both types of risk, unconscious and unknown can lead to missed hazardous scenarios.

In Figure 3 below, as there is no initiating event probability, as well as no mention of which safeguards are valid for a magnitude of likelihood reduction, there are columns of data left blank. When this report is viewed after the session, the individual would have to try to determine the required likelihood reduction required to ensure this scenario's risk is at a tolerable level. This could pose some difficulty as they may not be aware of the applicability of each of the safeguards listed. To complete this step in the HAZOP, they would first have to determine the probability of the initiating event and then use best practice guidelines to determine which safeguards are valid in the relevant scenario.

Cause	Cause Location	Consequence Description	Consequence Type	Initiating Event Probablity	Conseqeunce Severity	Risk Ranking (Before Safeguards)	Existing Safe Guards			Frequency	Consequence Severity	Risk Ranking (After Safeguards)
							Existing Safe Guard	Existing Safe Guards Category	Order of Magnitude of Likelihood Reduction			
1.1.3. XV-XXXX malfunctions closed.		1.1.3.1. Increased level in Acid Gas K.O. Drum; liquid carryover toReaction Furnace; damage to refractory; damage to furnace shell; leaks; loss of containment of H2S and SO2 when starting up on acid gas; hazards to personnel.	HS				1. SIS - TAHH-XXXX high high temperature will trip SRU including XV-XXXX acid gas isolation valve.			2	6	II
							2. BPCS - TI-XXXX high temperature alarm with operator action (i.e. divert SW acid gas to flare).					
							3. Other - Permanently mounted H2S gas detection in area will alert personnel to follow evacuation procedure.					
							4. PPE - All personnel carry gas detection monitors and know evacuation procedure.					
							5. BPCS - FC-XXXX high flow alarm with operator action.					

Figure 3: Screen shot of example PHA Pro File of HAZOP Session Data

If these columns of data had been recorded within the session when they were discussed, it would create no questions post session when the report is viewed. The inherent safety of the design is important to be understood during the session. If the design is robust and built inherently safer, the initiating event probability can be reduced, because the hazard has been removed at the source. If the design is not built to be inherently safer, you must rely on passive, active or procedural constraints required to compliment the design to control the hazard [4]. They are not required for the process, they are required to take the design to a safe state if the initiating cause occurs, by reducing the likelihood of the consequence. This reflects the earlier point that it is essential to have the proper competent team members within the session that thoroughly understand the design and can critically think about the inherent design of the system. An individual going back to look at this report post session most likely will not have this knowledge to determine the initiating event probability. If this information is intentionally left blank during the session it can create a confusion post session.

3 Introduction to Facility Unit Comparison

A common issue that has been discussed within organizations is "How can we tell if our processing units are being assessed in a complete and consistent way?" To understand this, the HAZOP data must be reviewed to draw comparisons between facilities of similar design. It should now be understood that the quality and completeness of the data recorded within the session is affected by many factors.

The traditional PHA session does not leverage the accumulated knowledge within an organization or across the industry, only that within the room. To resolve this lack of accumulated knowledge, an analytical tool has been developed. This tool is designed to help in two main approaches:

1. This tool can help any organization understand the similarities and differences between their own facilities' PHA or between themselves and other organizations PHAs from similar facilities,
2. It can help prepare for the next scheduled PHA session by having some information that can be reviewed prior to the session or aid with the discussions during the session, such as:

- Sparking conversations on causes or consequences that may not have been considered, leading to conversations to determine if they are exposed to such scenarios.
- Providing reference to see how organizations categorized the severity of a consequence. Although different organizations have different definitions for similar severities, normally health and safety severities have similar overlapping definitions.
- If a group is having difficulty determining a proper risk reducing safeguard, they can use the tool to look at safeguards that have been applied to scenarios with similar consequences.

This method of comparing data can be achieved through visual representation of overlap or uniqueness between similarly designed facilities. Information can be provided to the organization helping them understand which threats/causes have not yet been considered and which safeguards have been deemed risk reducing. Threats that have not yet been considered fit into two groups:

- First are threats that were not considered, but are protected against
- Second are threats that were not considered and are not protected against

It is important to gather this information to ensure a high percentage of critical scenarios are captured. Using this analytical tool and the knowledge it comes with, organizations can begin to create PHAs that are more consistent and complete, ensuring they are leaving no potentially high risk unassessed.

To create this visual, there are a series of transformations while mining the HAZOP data to allow it to be in a comparable state. The data must be generalized so specifics such as tag numbers are removed, and comparisons can be drawn. The method of creating comparable data is completed using a combination of artificial intelligence and machine learning along with the knowledge of a process safety engineer and review from a subject matter expert.

Figure 4 below demonstrates how the visual is displayed and the main three components in the hierarchical format including the Vulnerability, Mode of Failure and the Generalized Cause or Protection Layer.

- *Vulnerability* is the first node on the left-hand side of the visual, which is defined as the final point of failure or damage within a consequence. Vulnerabilities are identified across comparable facilities to determine the similarities and differences.
- The next node in the middle of the visual is the *Mode of Failure*, which describes the general process upset that leads to the vulnerability. Although the Mode of Failure identified often overlaps with the PHA Deviation/Guideword, it is not a consistent pattern, and the consequence description is normally the key source of data.
- The final node on the visual can toggle between two modes, the *Generalized Cause*. The generalized information is used to increase readability and comparativeness, as many organizations only include the tag number on the cause, safeguard or recommendation but not where it is located or what it is protecting against.

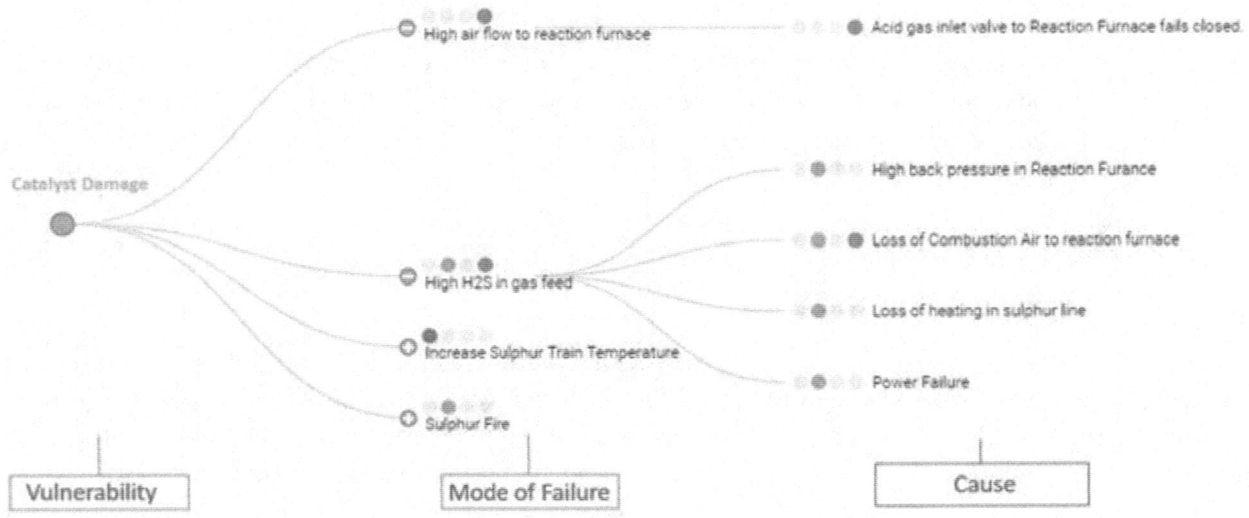

Figure 4: Facility Unit Comparison Visual

When these visuals are completed, a subject matter expert should always review the data to ensure that the "unique" Vulnerabilities and Mode of Failures are comparable to the other facilities. This is important as it is not intended to say that the organization needs to consider every Vulnerability or Mode of Failure listed within the Facility Unit Comparison. What it can do is help create a more complete HAZOP session if this information is reviewed prior to the next session to determine if there is any unknown risk that has not yet been considered. It can also aid in allowing organization to determine where they have inconsistencies within their facilities. Including:

- SIS system requirement comparisons
- Consistency in risk ranking of scenarios
- Completeness of consequence scenarios considered

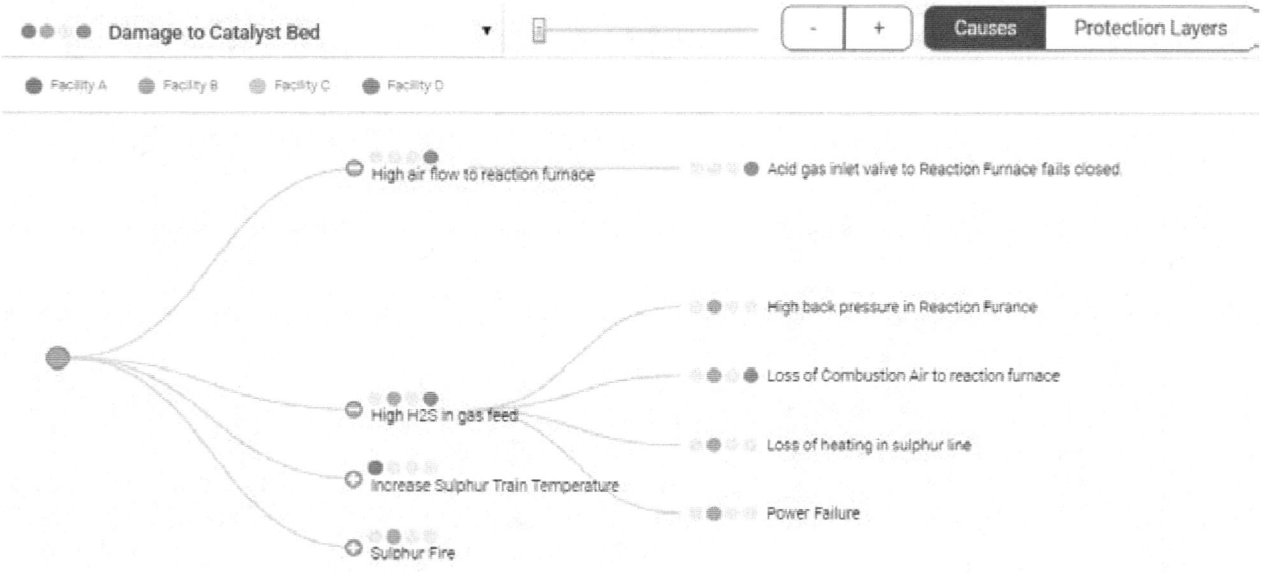

Figure 5: Facility Unit Comparison Sample Visual

The tool has a drop-down list of all the vulnerabilities that have been identified within the facility. In Figure 5, it can be seen which facility considered which vulnerability. Below the drop-down list, the comparable facilities are displayed with a corresponding color dot. In Figure 5, the first vulnerability "Damage to Catalyst Bed" was considered by Facility A, B and D. In the area below, the mode of failures related to damage to catalyst bed can be observed. This demonstrates that "High H2S in Gas Feed" as a mode of failure for damage to catalyst bed was considered by Facilities B and D. The final node shows the generalized causes/protection layers associated with this mode of failure and subsequently the vulnerability.

4 Facility Unit Comparison Case Studies

The concept of Facility Unit Comparison can now be used to demonstrate some case studies on how the data comparison can be utilized and what actions can come from this. Organizations have been able to improve their quality and consistency, while increasing the safety of their facilities and optimizing the number of protection layers a facility requires, potentially resulting in reduced capital and operating costs

4.1 Case Study #1 from Facility Unit Comparison

A Facility Unit Comparison was completed for a client with the goal of comparing HAZOPs from ten (10) similar facilities to determine their overlap and uniqueness. The facilities are in different locations across multiple countries and had different teams conducting each HAZOP session. Figure 6 displays the bold generalized cause of interest on the right-hand side, Valve Failure related to the Amine Regenerator Reflux Pump.

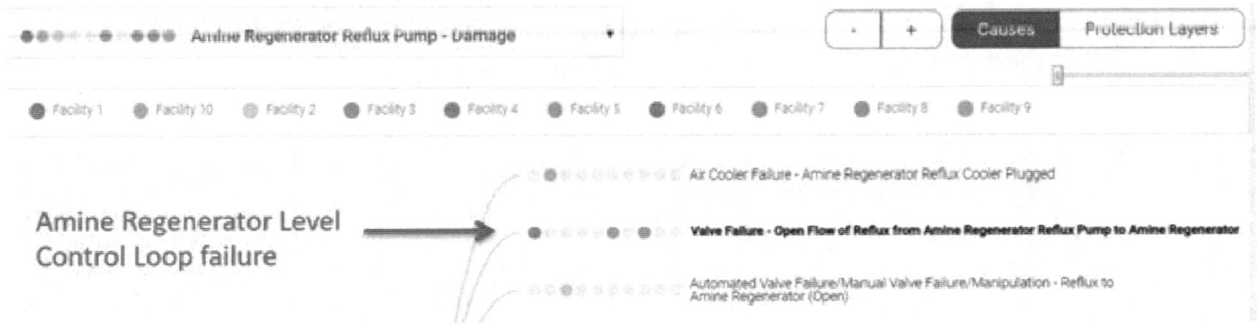

Figure 6: Case Study #1 Facility Unit Comparison

The cause was considered by three of the facilities during their respective HAZOPs. Each consequence related to this cause was classified as a health and safety impact and involved damage to the same piece of equipment, Amine Regenerator Reflux Pump. When the native data associated with this cause was investigated, as seen in Table 1, it was found that the teams had identified conflicting initiating event probability, worst credible consequence severity and in turn, risk rankings for the same cause and consequence description. Although there may be cases where these differences make sense because of design or operational differences, it is important to investigate the differences to see if there is valid justification for the inconsistency.

Table 1: Partial Drill-down Data related to Generalized Cause in Figure 6

	Cause	Consequence Category	Worst Credible Consequence Severity	Initiating Event Probability	Risk Ranking (Before Safeguards)	Recommendations
Facility 1	Valve Failure	Health and Safety	2	4	Medium	No Recommendations
Facility 8	Valve Failure	Health and Safety	6	5	Very High	Three Recommendations
Facility 6	Valve Failure	Health and Safety	4	4	High	One Recommendation

As seen in Table 1, when the same cause and consequence were analyzed to have different risk rankings, it created a different number of recommendations to be applied to each scenario to reach a tolerable level. It needed to be determined which of the initiating event probability, worst credible consequence severity and risk rankings were correct and what number of recommendations are needed based on the existing protection layers that previously existed. This discussion required a detailed consequence analysis with a process safety subject matter experts and operations specialists. The investigation determined the amount of hazardous material that would be released during this consequence would be negligible from this scenario alone, and therefore four (4) recommendations could potentially be eliminated.

This is an example of how conducting a complete HAZOP session with the appropriate individuals in the room, while ensuring the discussions are accurately recorded, can change the outcome of the session. An investigation was completed on how much money the organization saved by removing each of the recommendations. The results are displayed in Table 2.

Table 2: Recommendation estimations over a five-year period

Recommendation	Estimated CapEx cost	Estimated OpEx cost	Total Cost over 5 years
Recommendation 1	$10,000 (assuming software and document updates only)	None expected	**$10,000**
Recommendation 2	$100,000 (assuming new instrument added to SIS)	$10,000 annually for testing and verification (SIS)	**$150,000**
Recommendation 3	$100,000 (assuming cabling required)	$10,000 annually for testing and verification (SIS)	**$150,000**
Recommendation 4	$50,000 (assuming installation of a new horn)	$5,000 annually for testing	**$75,000**
Total Cost	**$260,000**	**$25, 000/year**	**$385,000**

The cost estimate is representative to how much can be saved within an organization, while not changing the risk exposure of the facility. Each of the recommendations were estimated and it was found that a total of $385,000 over a 5-year period could have been eliminated from this one scenario alone. It is believed within this organization and across other organizations that these types of inconsistencies may lead to more over instrumentation cost savings in the future.

5.2 Case Study #2 from Facility Unit Comparison

Another Facility Unit Comparison was completed with major findings related to consistency and completeness within the HAZOP. It was determined that the same generalized safeguard "Alarm – High SO2 in the Incinerator" was being used at the same facility (Unit A) for 1 and 2 magnitudes of likelihood reduction. In the other comparable units, it is seen that the same generalized safeguard consistently has 1 magnitude of likelihood reduction applied to it. This is demonstrated in Figure 7 below.

Figure 7: Facility Unit Comparison Findings #2

This information was reviewed by a subject matter expert and the outcome was best practice for taking safeguard magnitude of likelihood reduction for alarms are as follows:

- If both alarms are ringing into the boardroom at the same time and the operator is expected to take an action for both, then only one magnitude of likelihood reduction should be given
- If one alarm rings in as an early warning signal in advance of the other alarm and there are different actions to be taken due to the progression of the process event, then two magnitude of likelihood reductions may be justifiable

As a result of this facility unit comparison, this organization is now aware of this inconsistency. This will spark some conversation with the associated personnel to ensure they are following best practice on consistently applying magnitude of likelihood reduction for safeguard alarms. If this safeguard when given two magnitude of likelihood reductions is bringing the residual risk to a tolerable level, the scenario should be evaluated to ensure it is adequately protected against.

5.3 Case Study #3 from Facility Unit Comparison

Another Facility Unit Comparison finding related to the quality and consistency in PHA is shown in Figure 8 below. This example highlights the completeness of assessing all scenarios within the PHA study.

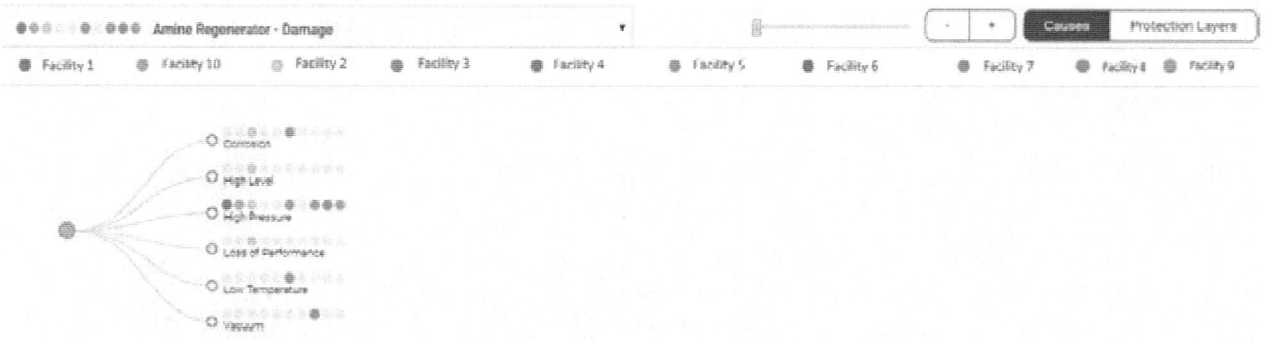

Figure 8: Facility Unit Comparison Findings #3

As seen in Figure 8, seven of the ten facilities considered damage to the amine regenerator, with six different modes of failure. The three facilities that did not consider damage to the amine regenerator did have the equipment within their scope. This began conversations to determine why this had not been considered or recorded within the HAZOP, and if there were any unknown risks related to damage. In addition, by using the information within the facility unit comparison, it can help guide the scenarios that may have been missed.

In addition to these three examples, there are many other ways that Facility Unit Comparison has helped to improve the quality and consistency in HAZOP. It brings together a large amount of data in a valuable way to draw conclusions on the question of "How can we tell if our processing units are being assessed in a complete and consistent way."

5 Conclusions

By using this information, organizations can be much more confident in the quality and consistency of their HAZOPs and reduce the risk of missing scenarios. To ensure a higher quality PHA, organization must ensure proper methodologies are being followed and the appropriate individuals are a part of the discussion. Additionally, ensuring all required information is being recorded accurately will improve the thoroughness of the PHA. The case studies within this paper demonstrate examples of findings while comparing different facilities HAZOP data. This knowledge can help reduce the risk exposure and increase percentage of critical scenarios captured during the session.

This concept of creating a higher quality of data can be adopted by many different types of industries. In a world where the role of artificial intelligence is becoming more prominent; this will aid in simplifying the creation of higher quality data. The HAZOP should no longer feel like a check box exercise but a method of completely understanding all the hazard and risks within a facility. This will help reduce the number of safety related incidents and create a safer working environment for all employees of an organization.

By taking advantage of the data that is available within the industry and using data analytics, artificial intelligence, machine learning and the expertise of subject matter experts, organizations will be able to develop a culture where their approach to safety is much more proactive and the number of process safety incidents are minimized.

6 References

[1] B. Skelton, *Process Safety Analysis an Introduction*, Gulf Publishing Company, Houston, TX, 1997

[2] US Department of Labor, *Process Safety Management,* Available at: https://www.osha.gov/Publications/osha3132.html#pha, Accessed pm January 15, 2019

[3] ACM Facility Safety Inc., *PHA / HAZOP Facilitation Workshop*, ACM Facility Safety Inc., Calgary, AB, Canada, 2005-2018

[4] Center for Chemical Process Safety of the American Institute of Chemical Engineers, *Inherently Safer Chemical Processes A Lifecycle Approach*, John Wiley & Sons, Inc., Hoboken, New Jersey, 2009

Avoiding Quality Pitfalls for HAZOP/LOPA Sessions and Documentation

Steven T. Maher, PE CSP
Risk Management Professionals
Steve.Maher@RMPCorp.com

Carine M. Black
Risk Management Professionals
Carine.Black@RMPCorp.com

Morgan T. McVey
Risk Management Professionals
Morgan.McVey@RMPCorp.com

Prepared for Presentation at
American Institute of Chemical Engineers
2019 Spring Meeting and 15th Global Congress on Process Safety
New Orleans, Louisiana
March 31 – April 3, 2019

Avoiding Quality Pitfalls for HAZOP/LOPA Sessions and Documentation

Steven T. Maher, PE CSP
Carine M. Black
Morgan T. McVey

Risk Management Professionals

Keywords: PSM, RMP, CalARP, SMS, Process Safety, PHA, HAZOP, LOPA, Quality

Abstract

Although Process Safety Management (PSM), requiring a high-quality Process Hazard Analysis (PHA), was promulgated in the United States in 1992, and HAZOP Study techniques date back to the mid-1970's, the HAZOP Methodology (and companion techniques such as LOPA) can still present challenges in consistently applying PHA in a high-quality fashion. Many challenges can exist for the implementation of PHA, and each PHA is unique due to variables such as team composition, the HAZOP/LOPA facilitator, process unit history and past incidents, regulatory environment, safety culture, and available and accurate process safety information. These variables can have a significant impact on the HAZOP/LOPA results; however, the value of HAZOP/LOPA sessions and quality/usability of resultant HAZOP/LOPA documentation is a critical objective. The primary responsibility for achieving this objective must fall on the shoulders of the facilitator to provide streamlined support to navigate through the PHA process while accurately capturing team discussions with the necessary level of detail. Key topics addressed within the paper include:

- Quality objectives, including content and structure conducive to subsequent HAZOP/LOPA usage/updates
- Need for uniform, high-quality application
- Experienced-based tips for driving high-value sessions and for developing high-quality HAZOP/LOPA documentation
- Checklist for assessing the quality of individual HAZOP/LOPA Studies
- Key indicators that a reviewer can use to validate the acceptability of the HAZOP Study
- Structuring HAZOP/LOPA to support other PSM elements and safe plant operation

It is not the intent of this paper to provide basic PHA training, but to focus on key tips to address commonly-observed, and readily-rectified, PHA deficiencies. We have attempted to list the most common quality challenges observed from the work of even experienced HAZOP/LOPA Facilitators, and this can also provide a basis for characteristics to look for by a Quality Assurance Reviewer.

1. Focus on Objectives – Why is a Quality Process Hazard Analysis (PHA) Important?

Most people attempt to avoid undesirable outcomes and organize their actions to minimize risk; however, very few of us apply a structured evaluation of potential undesired outcomes to avoid. Whereas this may be a personal decision for the individual, this is unacceptable for a complex process system like a refinery, where the undesired events can have impacts well-beyond the individuals directly involved. These events may be rare and involve a complex set of initiating events and failure of safety/mitigation systems; thus, it is necessary to perform periodic and thorough evaluations.

FIGURE 1.1 – Tragedies to Avoid

The December 2, 1984 Methyl Isocyanate (MIC) release from the Union Carbide Bhopal Facility is considered a pivotal event in catalyzing the application of Safety Management Systems (SMS) approaches to enhance process safety. The MIC release and the magnitude of the tragedy drew the attention of industry, the public, and the regulatory community to the potential consequences associated with process safety events (Figure 1.1). Industry's response was swift and definitive. The American Institute of Chemical Engineers (AIChE) founded the Center for Chemical Process Safety (CCPS) in 1985, recognizing that the most effective mechanism for addressing process safety was not the application of additional prescriptive mechanisms, or by addressing any specific action, but by effecting fundamental business (i.e., safety culture and management systems). CCPS Guidebooks are currently considered key references in conveying the technologies needed for process safety, and the first CCPS Guidebook ("Guidelines for Technical Management of Chemical Process Safety"[2]), published in 1987, was designed to address this pressing need.

A key part of managing process safety has always been the identification and understanding of potential hazards and their consequences. This need pushed practical techniques for hazard identification, i.e., the Hazard and Operability (HAZOP) Study, developed in the decades prior to the Bhopal tragedy, to the front lines in the effort to manage process safety.

2. Brief History of Key PHA Techniques and Regulatory Requirements

Although there are quite a few tools in the PHA toolkit, the team-oriented, patterned-brainstorming sessions associated with HAZOP Studies[3] are generally considered the workhorse of the industry (Figure 2.1). Layer of Protection Analysis (LOPA)[4], frequently integral with the HAZOP Study, is a complementary tool that provides additional details on failure probabilities and conditional modifiers. HAZOP Studies provide a foundation for scenario identification and high level development, while LOPA provides a consistent and more quantitative framework for further analysis of applicable high-risk scenarios to reveal additional insights, some of which can be directly used for identifying appropriate reliability targets for key safety features. Although HAZOP Studies have been a core part of an acceptable hazard evaluation process referenced in various industry guidelines[2,3], as well as regulatory requirements such as Process Safety Management (PSM)[5] and Risk Management Programs (RMP)[6] for onshore facilities in the United States, LOPA is a relatively new tool that simplifies Quantitative Risk Assessment (QRA) techniques to a manageable level to facilitate usage. LOPA applications are gradually becoming best practice for addressing higher consequence/risk events and are especially useful for an initial assessment of the reliability needed for key safety systems. As well as LOPA being an industry-accepted practice for important applications, for California Refineries, PSM[7] and the California Accidental Release Prevention (CalARP) Program[8] require the development of Safeguard Protection Analysis (SPA), with LOPA as an acceptable tool.

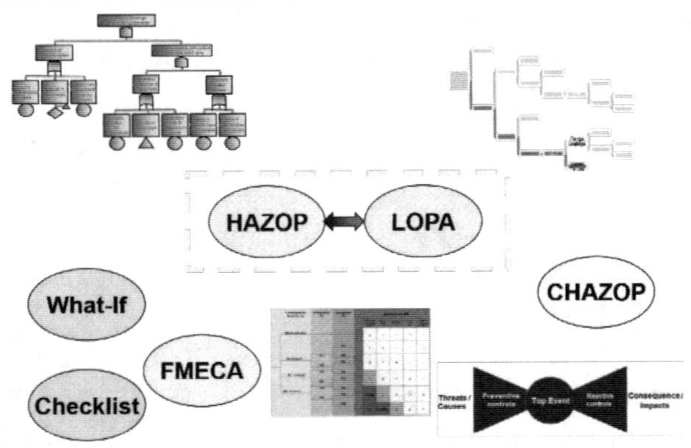

FIGURE 2.1 – HAZOP & LOPA are Core Elements of Hazard Evaluation

Although the core HAZOP has been relatively constant, HAZOP and LOPA applications have been evolving to accommodate higher expectations of both regulators and industry practitioners with respect to quality and level-of-detail. There are fundamental quality requirements that must be met to achieve regulatory compliance, but just as "beauty is in the eye of the beholder," some quality parameters pivot on the objectives of the potential applications anticipated for the HAZOP/LOPA.

Thus, the first step before convening the team and starting on the first scenario is to outline a clear definition of objectives.

3. Defining Objectives

The core objectives of all hazard identification exercises (e.g., HAZOP/LOPA) are to uncover potential weaknesses/vulnerabilities in system design/operation that could result in an undesired outcome (e.g., injury, environmental impact, equipment damage, operational impact, or compromised company reputation). Part of this analysis is to challenge the reliability of applicable safeguards to ensure adequate mitigation for potential process upsets. A quality PHA can address a variety of objectives supportive of successful plant operation:

- Protecting plant personnel, the nearby community, and the environment
- Addressing SMS regulatory requirements

- Providing insight regarding safety/operational/financial advantages/disadvantages of design and operations options
- Facilitating Management-of-Change (MOC) evaluations
- Prioritizing Mechanical Integrity (MI) Program elements and covered equipment
- Formulating Safety Integrity Level (SIL) requirements for Safety Instrumented Functions
- Providing input/verification of operating limits and Operating Procedures with respect to consequences of deviation
- Providing training to inexperienced personnel in the design intentions, expected operation, and potential hazards of the facility
- Supporting safety system design bases and required safety functions
- Providing a strong foundation for future PHA updates
- For capital projects, optimizing the design while minimizing associated costs

The value-add benefits listed above are just some of the ways in which a high quality PHA can streamline independent but related PSM efforts. All too often, the mentality of "checking the box" for a regulatory requirement leads to a poor-quality PHA, which greatly limits its potential future usability, resulting in lost opportunity costs, not to mention, potential regulatory deficiencies.

4. Planning and Preparation Essentials for Supporting a Quality PHA

Success in the development of a quality PHA can be compromised even before the PHA Team Session begins due to poor planning and/or inadequate preparation. Reference 10 can be accessed for additional planning and preparation tips (and timing) that includes the following key elements:

- **Ensure attendance of key personnel during PHA Team Sessions** – Qualified, experienced, knowledgeable, and prepared technical experts (Engineering & Operations) who participate in all phases of the PHA are critical. Ensure Team Member availability for the entire duration of the study for consistency. Table 4.1 identifies the key disciplines useful for any HAZOP/LOPA Team. Don't wait to the last-minute to schedule necessary personnel. Sometimes the best contributors for the PHA are some of the busiest individuals at the plant.
- **Venue logistics conducive to teamwork** – HAZOP/LOPA sessions may be best conducted offsite to facilitate focus. The meeting room layout should be supportive of constructive interaction.
- **Completeness-verification and, as appropriate, field-verification of key design information made available to the PHA Team (e.g., P&IDs, PFDs, Cause & Effect Matrices, Electrical Area Classification Drawings)** – Table 4.2 provides a checklist of key information requirements for a HAZOP/LOPA.
- **Software preparation, including population of key causes, is contributory towards:**
    - PHA Completeness and Organization

- Streamlined Session Time: Causes Prepopulated from P&IDs for Quick Location of Relevant Scenarios During the Session
- Sensible Grouping of Causes for Effective Future Use and Update

Populating key causes can also provide a vehicle for sending a "preview" of key issues to be evaluated during the HAZOP Study to participants, so they can be better prepared to provide definitive feedback.

- **Qualified, experienced, and prepared PHA Facilitator** – Having a competent PHA Facilitator can make or break a HAZOP/LOPA Study. The following characteristics are essential:
  - Understanding engineering first principles and the HAZOP/LOPA Methodology in order to provide effective guidance to the team
  - Understanding when the team has reached the limits of what they can accomplish
  - Managing team dynamics and challenging personalities
  - Motivating the Team Member contribution and promoting creativity
  - Clarity in translation of the phenomenology of process upsets into a HAZOP/LOPA scenario framework

TABLE 4.1[10]
CHECKLIST OF KEY DISCIPLINES USEFUL FOR A HAZOP/LOPA
(R = Required by PSM/RMP)

- ☐ Facilitation/Leadership (R)
- ☐ Team Scribe/Recorder
- ☐ Process/Project Engineering (R)
- ☐ Operations (R)
- ☐ Maintenance (R)
- ☐ Control/Protection Instrumentation Engineer
- ☐ HSE Engineering
- ☐ Rotating Equipment Specialists
- ☐ Specialists to Address Unique Requirements, e.g., Damage Mechanism Review (DMR), Safeguard Protection Analysis (SPA), Hierarchy of Hazard Control Analysis (HCA)

Note that an optimal team size for an Operating Facility HAZOP/LOPA is 6-8 individuals.

TABLE 4.2[10] – GENERAL INFORMATION REQUIREMENTS FOR A HAZOP/LOPA

- ☐ **Process Flow Diagrams**
- ☐ **Piping & Instrumentation Diagrams (with changes identifiable from any previous HAZOP Studies)**
- ☐ **Cause & Effect Diagrams**
- ☐ **Alarm & PSV Setpoints**
- ☐ **Site Layout / Platform Location Drawings**
- ☐ **Accident/Incident History & Reports**
- ☐ **Management of Change (MOC) & Pre-Startup Safety Review (PSSR) Documentation**
- ☐ **Previous HA/PHA Recommendation Status**
- ☐ Equipment Data Books
- ☐ Operating & Emergency Procedures
- ☐ Maintenance Records
- ☐ System Descriptions
- ☐ Previous HA/PHA Reports
- ☐ Toxic, Chemical, and Physical Properties
- ☐ Prevention Program Compliance Audits

Bold items in the list above are generally more important.

- Streamlining progress and maintaining team focus on important issues

- **Qualified, experienced, & prepared PHA Scribe** – The right scribe can:
  - Streamline team time, saving significant cost
  - Allow the Facilitator to more effectively interact with the team, improving team focus and PHA completeness/quality
  - Boost documentation quality and report turnaround time
  - This is an important issue, so Section 4.1 contains key elements from Reference 1.

4.1 Scribe Support

Scribing skills include typing, software usage, and computer interfaces with equipment such as displays. The Facilitator can be tasked with both facilitation and scribing duties depending on various parameters of the HAZOP/LOPA, such as system complexity and team composition. So, for a relatively small, manageable group, a Facilitator might be able to also accommodate scribe duties. The pivotal decision-making parameter is whether the additional cost of a Scribe outweighs the money saved by shortening the HAZOP/LOPA Sessions and the subsequent documentation time.

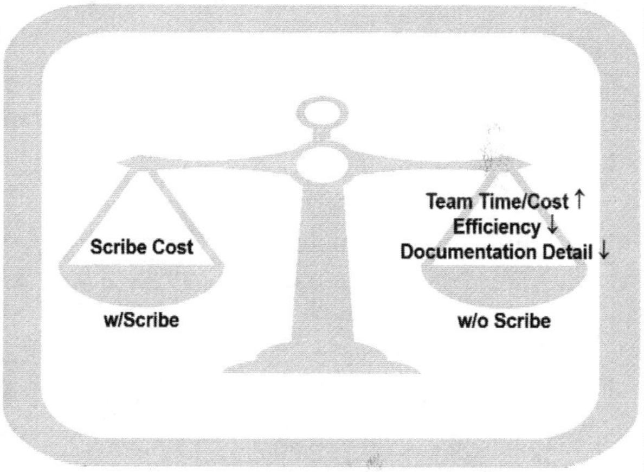

FIGURE 4.1 – Weighing Your Scribe Options

Many professionals, especially in a large company, tend to underestimate the value of their time and that of their peers. The employee time cost can be substantial and shortening the HAZOP/LOPA sessions could have a significant financial benefit. In most cases, this financial threshold is reached once the team size (excluding Facilitator and Scribe) exceeds 3-4 individuals for local projects and possibly 4-5 individuals if travel costs for the Facilitator and the lower-cost Scribe are considered. As another benchmark, Reference 1 recommends to "use a dedicated Scribe, for meetings longer than 4-total hours" and "Use a well-trained scribe to take the documentation load off of the team. This rule can save 30-50% of meeting time and increases brainstorming (because the team is not daydreaming as they wait for Leader to complete the notes)." Along with tangible savings, the Scribe can provide an improvement in the quality of the documentation by capturing more Team discussions and detail than could be accomplished otherwise, and the session itself, by allowing the Facilitator to better focus on engineering issues and maintain an optimal "pace" of the HAZOP/LOPA (too fast, the team gets lost, and too slow, the team tires quickly of watching a Facilitator type and shifts their attention to checking e-mail on their cellphone).

Of course, the decision to involve a Scribe pivots on the Scribe's capabilities. Some key characteristics to seek in a Scribe include:

- Familiarity with HAZOP/LOPA techniques and documentation

- Familiarity with reading engineering drawings and documentation
- Familiarity with the HAZOP/LOPA documentation software and computer skills

For a HAZOP/LOPA, avoid assigning an untrained scribe, or one that is not committed to professional-quality PHAs. Unless properly trained, and motivated, a poor scribe can compromise quality and team progress. Typically, a younger engineer that may be undergoing training to become a Facilitator works best.

5. Tips for Conducting a Quality HAZOP/LOPA

A key variable in the quality and usefulness of a PHA are the session dynamics. Section 4 identifies key personnel and other resource needs, but effective utilization during the PHA is critical. The following are key PHA Session characteristics that are conducive to creating a high-quality and useful outcome. Most of these characteristics must be driven by the PHA Facilitator:

Process Design/Limits and Response to Upset Conditions – It is important for the PHA Team to clearly articulate the process dynamics associated with the upset condition, and for the PHA Scribe to document the maximum unmitigated pressures/temperatures and associated design limits, so that it is clear that the PHA Team's discussions are focused on the same potential hazard scenario, and to ensure that the application of LOPA (if needed) is targeted to the potential hazard scenario.

Overpressure Ratios – Various interpretations of API RP 581[9] are used in industry to characterize the probable worst-case of an event as either a non-leakage event, a flange leakage event, or a credible potential for piping/vessel failure. Although the PHA Team should always be looking at the specific process conditions and equipment, defining a typical consequence for the PHA Team limits the potential for a gross misapplication.

For each overpressure scenario, an overpressure ratio can be calculated and documented allowing the Team to identify a more realistic severity based on whether the scenario is a catastrophic rupture event or a flange leak.

Overpressure Ratio = maximum pressure / MAWP

Where maximum pressure is the highest pressure that can be reached without taking into consideration pressure relief valves.

Consequence Documentation – To help ensure proper convergence by the PHA Team, it is typically helpful to detail the sequence of events when documenting the consequence.

For example: Potential blocked-in discharge of the pump, leading to increased discharge pressure up to 250 psig, leading to no potential to overpressure the 300# discharge piping as it is rated for the pressure. Potential pump deadhead, seal failure, loss of containment of flammables below the autoignition temperature (low H2S), resulting in a pool fire with potential personnel impact/exposure.

This level of detail ensures that all Team members are on the same page and are evaluating the same scenario. Furthermore, it provides sufficient information where anyone can read the scenario and know exactly what was analyzed.

Instrumentation and Setpoints – This may be a little more challenging for a system under design, but in general, specific alarm and protection system setpoints should be discussed (and documented) to ensure that the PHA Team can clearly gauge the ability of a safeguard/Independent Protection Layer (IPL) to be effective, especially considering time for Operator diagnosis and response. Furthermore, documenting setpoints of alarms, PSV's, interlocks, etc. allows the Team to verify that the layers of protection will activate as intended. For example, if an interlock setpoint is higher than a PSV setpoint, this may raise a red flag that warrants further investigation.

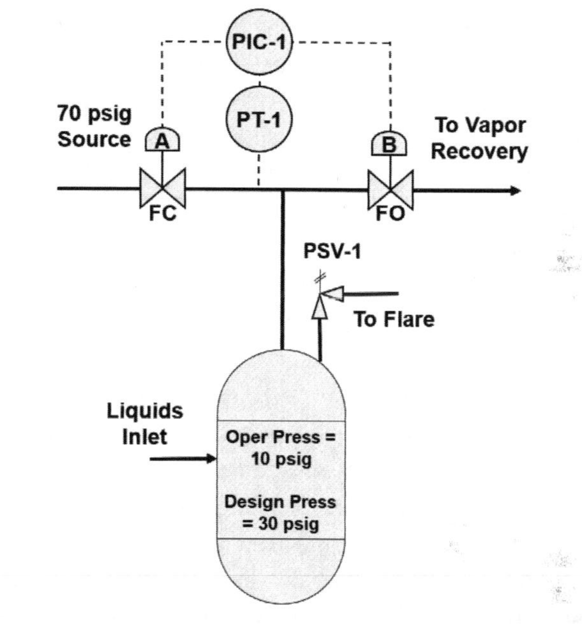

FIGURE 5.1A – Typical "Push-Pull" Blanket Gas Pressure Control System
(control station block and bypass valves removed)

Control and Protection System Actions – Similar to clear discussions on instrumentation & setpoints, clear specification of the action of control & protection system can avoid erroneous credit of an ineffective safeguard.

Importance of Instrumentation Failure Specificity – Transmitter failures, <u>and the failure direction for the causal event</u> (i.e., misreading high or misreading low), should be specifically identified. This is critical for the PHA Team to assess if any common-mode failures (between two credited safeguards or between the causal event and an associated safeguard) exist that could compromise safeguard effectiveness. Transmitters that have a failure mode that could impact multiple end-devices should typically be listed as an independent causal event, with end-devices being controlled clearly specified. Figure 5.1A is a typical "push-pull" blanket gas pressure control system, where a single valve failure may be a simple operational challenge, but the pressure transmitter misreading low could cause both a pressure excursion and disable an overpressurization protection feature. In contrast, Figure 5.1B illustrates a relatively

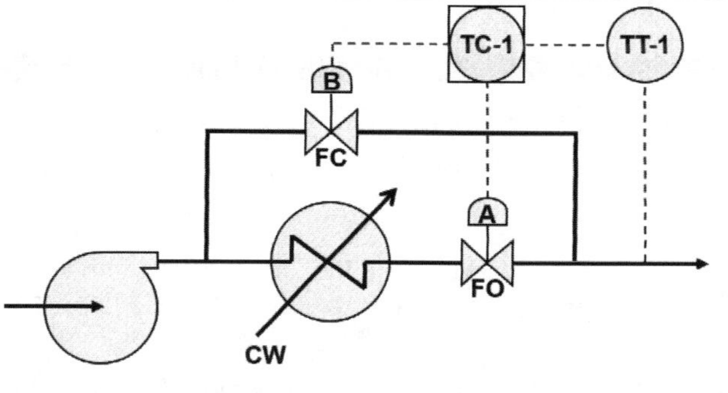

FIGURE 5.1B – Common Temperature Control System
(control station block and bypass valves removed)

common configuration, where a temperature transmitter failure may impact two devices, but which may only have a non-hazardous change in temperature profiles. For the system in Figure 5.1B, a more significant event may occur due to the failure closed of TV-1B, which would initiate a cooling effect that results in the control system commanding TV-1A to also close, resulting a complete blocked discharge potential hazard.

Instrumentation Transmitter vs. Controller Tag Numbers – Although Console Operators often refer to an alarm using the controller tag number (e.g., LC-XXX), to further minimize the potential for the PHA Team failing to identify potential common-mode failures, it is typically a better practice to reference the root transmitter (e.g., LT-XXX) in the causal event and safeguards.

Valve Failure Mode Clarity – As part of the causal event, if the cause clearly, and consistently, states for a fail-closed valve, "fails closed" or "malfunctions open," and for a fail-open valve, "fails open" or "malfunctions closed," the PHA Team is better able to assess the validity of the failure mode. For example, if the risk of a fail-open valve opening is much higher than its closing, it might be appropriate for the PHA Team to reassess the validity of the valve's failure mode. To further support the PHA Team's efforts in properly evaluating failure mode acceptability, incorporate a "Loss of Instrument Air" process deviation where 2-3 minutes are spent before moving onto the next node to list key control valves and their failure modes for each node to validate that the valve failure modes are appropriate.

Crediting Alarms as Safeguards – Does one alarm provide a definitive basis for diagnosis of the upset condition and possible cause to the Console Operator? Does two? Does twenty? Regardless of the number, there is still a common-mode dependency for any number of alarms of the Console Operator to make an effective diagnosis and use their training to identify and implement the appropriate corrective action. To ensure that the most important alarm functions are identified (in case the PHA is used as a basis for the Mechanical Integrity Program), a reasonable practice is to list the one or two alarms that will annunciate "first" and provide the most direct correlation with the process parameters (i.e., flow, pressure, temperature, level) that are driving the process upset. To further emphasize the common-mode dependency of the Console Operator, it may be beneficial to bundle the alarms as a single safeguard (IPL). Periodically, it is often valuable to remind the PHA Team that alarms should only be credited if sufficient time for response exists, given the considerations of:

- Someone present to hear the alarm (local alarms should typically not be credited)
- Alarm prioritization and diagnosis
- Deciding on the corrective action
- Permission for implementing the corrective action
- Initiating and completing the corrective action
- Time needed for the corrective action to mitigate the event (i.e., process safety time)

Failure of a Safeguard as a Causal Event – HAZOP/LOPA are scenario-based analyses. If a safeguard's failure is already implicit in the risk-ranking for a scenario, treating its failure as a separate causal event is unnecessary and potentially confusing. A minimum flow recirculation line is a common example: The action of the minimum flow recirculation valve opening would typically be credited as a safeguard, with the failure to open implicit in the risk-ranking for that scenario. Therefore, evaluating a separate "malfunctions closed" or "fails to open" event for that minimum flow recirculation valve is unnecessary and potentially confusing.

Consistent Grouping of Scenario Consequences and Application of Safeguards Associated with the Scenario – Safeguards should have a clear correlation with the ability to ameliorate the scenario consequences. Characterizing each safeguard as an Independent Protection Layer (IPL) can facilitate segue to LOPA.

Subcomponent Failure Modes – Grouping subcomponent failures should be done with caution. For example, a blocked turbine exhaust may be a cause of a loss of flow from the pump being driven by the turbine; however, a blocked turbine exhaust could also cause an overpressurization event, requiring safety systems to function in order to ameliorate a serious potential personnel hazard.

Operations and Maintenance Modes – Facilitators need to routinely challenge the PHA Team on how the system is operated (i.e., for control systems, always in "auto," or frequently in "manual") and if there have been reliability problems that could affect the initiating event frequency of the causal event or the effectiveness of the safeguards.

PHA Team Training – In addition to PHA Team Training offered at the beginning of a PHA, consider providing a short "daily briefing" to enhance team alignment of key PHA principles.

Consideration of All Salient Perspectives and Input – Everyone on a PHA Team has an important perspective to offer to support the quality of the PHA, and the Facilitator needs to drive participation, involvement, and cooperation.

Session Length Reflecting Process Complexity – Consistent with the importance of Management Commitment previously identified, without adequate time for properly considering whether credible failures have an appropriate level of protection, PHA results will be incomplete or erroneous. Of course, giving the PHA Team too much time results in disengaged team members, which is also a deterrent to PHA quality. It is important to keep the PHA Team engaged.

FIGURE 5.2 – Ensure Drawing Updates are Manageable

Consistent Risk-Ranking – Challenges in the PHA Team's understanding of the process dynamics, scenario consequences, or proper application of the facility's risk-ranking matrix can manifest themselves in inconsistent risk-ranking. The Facilitator needs to be looking for inconsistencies that are indicative of quality problems.

Recommendation Basis – Recommendation significance and priority should be consistently driven by the magnitude of the "gap" to an acceptable risk threshold.

Manageable Drawing Updates – There are many occasions, especially during a design HAZOP/LOPA, where changes to the design information are applied during the course of the study. The Facilitator needs to skip over (and come back to, as appropriate) portions of the process where the design information is insufficient.

Manageable Information Gaps – In the same way that information gaps due to inadequate drawings must be managed, so must gaps in the knowledge-base of the PHA Team participants. PHA Team participants should be encouraged to expeditiously identify when they don't have definitive information, such that other "subject matter experts" can be summoned, as necessary.

Node Completeness Checks – A technique that generally works well for a HAZOP Study is, after each node, take 10-15 minutes to "step through" the equipment, from the beginning of the node to the end, and verify with the HAZOP Study Team that appropriate equipment and failure modes have been addressed. At that time, double-checking piping specification adequacy, appropriate protection of equipment design limits, reviewing the need for heat tracing, identifying potential dead legs that may require demolition or enhanced mechanical integrity inspections, and the appropriateness of the failure positions of instrument-air-controlled valves can be accomplished. The Completion Check allows the Facilitator to verify that key failure modes have been covered and gives the Team a chance to step back and verify that key issues haven't been missed.

PHA Revalidation vs. Re-do – At most, regulatory guidance for "a hazard review [being] revalidated only once between full hazard reviews" only applies to certain RMP Program 2 processes, and not most PSM-covered processes. For situations when past studies cannot be readily revalidated, a new complete hazard review or PHA may be warranted; however, a technical concern with an arbitrary re-do is that if a good Facilitator and Highly-Qualified Team did a high-quality PHA, and the PHA is arbitrarily discarded and replaced with a re-do by a less capable/experienced Facilitator and/or Team, the result can be a decreased thoroughness in the baseline hazard identification process that forms a foundation for the Safety Management System Program. There may also be a liability issue if a past high-quality PHA that properly evaluated a hazard was discarded, the poor-quality replacement didn't properly evaluate a hazard, and an injury accident occurs. Thus, unless the previous PHA is completely unusable, a re-do should take the form of starting with the "PHA of Record" and comprehensively reviewing/updating each-and-every PHA scenario. With this approach, insights from the previous effort would not be arbitrarily overlooked.

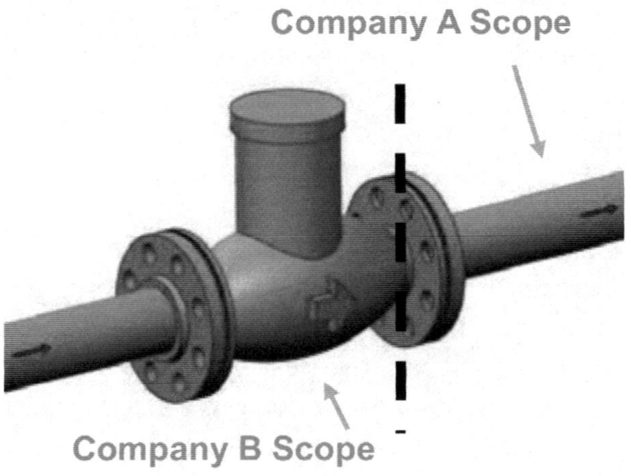

FIGURE 5.3 – Node Boundaries to Avoid

Node Boundaries – Avoid node boundaries in the middle of a flowpath (Figure 5.3). This can originate from the interface from different vendors involved in the plant design, but inserting an arbitrary boundary in the middle of a flowpath can cause arbitrarily inconsistent results.

Avoid Repeating Scenarios – HAZOP Study Teams will often gravitate towards discussing "generalized" scenarios (e.g., no flow and high pressure resulting from the same cause) such that when the Facilitator guides the Team to "drill down" to the specific cause, it is clear that the Team is discussing different direct impacts from the same equipment failure causal event. Generalized scenarios should be avoided, and the Facilitator must continually drive towards causal event specificity. It is much better to do a thorough evaluation of a specific cause once, and thoroughly, to ensure that important nuances of the process dynamics are surfaced.

6. PHA Documentation Tips

Consistent with the objectives defined in Section 3, documentation quality should be tuned to be able to address all of the identified objectives. Typically, the following documentation characteristics support high-quality documentation that can be directly used to address project objectives, as well as providing sufficient information to support quality reviews.

Specific Causes, with Equipment Numbers Identified – Causes should be able to be correlated to a specific equipment failure or credible process conditions associated with valid operating modes. Tag numbers and P&ID references should be used to facilitate reader understanding of the scenario and quickly locate the equipment associated with the initiating event. A good rule-of-thumb is that if it takes more than ~20 seconds for someone who is familiar with the system to locate the equipment on the P&IDs, a P&ID reference is likely necessary. In addition, to avoid misinterpretation, equipment names should exactly match the name on the P&ID and be capitalized for ease of review and specificity.

Prolific Use of Equipment Tag Numbers and P&ID References – All the key PHA objectives identified in Section 3 are enhanced with specificity and the prolific use of equipment tag numbers. Taking just a few moments can greatly enhance the usability of the PHA documentation.

Block Valves Inadvertently Mispositioned – Any references to block valves should be descriptive regarding location. For example, if the block valve is not in the control valve station, there should generally be a reference to a line number and P&ID, or equivalent.

Probable Worst-Case Scenarios should be specific and able to be directly correlated to the causal event associated with the scenario. A qualified reviewer should be able to read the PHA documentation and clearly understand the sequence of events leading to the probable worst-case scenario.

Safeguards Should be Reliable, Active, and Have Sufficient Process Safety Time – Safeguards should also be specific to the cause/consequence scenario being examined. Ordering the safeguards by the progression of the event (i.e., the first safeguard to be encountered documented first) is a good documentation practice.

Recommendations (or gap acceptance) Whenever Clearly-Defined Acceptable Risk Level is Not Achieved – If there is a risk-gap, it should be addressed with an action item or documented as acceptable to the PHA Team.

Risk-Ranking – Consistent and Matched with Scenario – After the PHA, the Facilitator should carefully review risk-rankings for consistency and identify any inconsistencies to the PHA Team to effect a correction.

Level of Detail and Scenario Depth Pivoting on Importance and Complexity – All scenarios should be readily understandable to a qualified reviewer. As appropriate, details may need to be added to clarify process dynamics, especially if follow-up assessments need to be performed.

Recommendation Usability – Recommendations should be understandable, self-standing, logical, and complete. Actions should be clear, include definitive references to P&IDs and equipment numbers, and minimize the need for the assignee to review the HAZOP/LOPA Report (because the recipient of the action item is likely busy and many not have easy access to the HAZOP/LOPA Report).

Scenario Traceability – Scenarios should be logically-developed, complete, and understandable.

Clarifying Comments – Liberal use of clarifying comments should be made to support future usability and expansion of the PHA. The PHA Team's evaluation and basis for conclusions should be readily understood to support future revalidation efforts.

7. Priorities for a Quality Assurance Review

This paper has attempted to list the most common quality challenges observed from the work of even experienced HAZOP/LOPA Facilitators, and this can also provide a basis for characteristics to look for by a Quality Assurance (QA) Reviewer. A QA Reviewer will typically not have the time, resources, or the benefit of having worked with the PHA Team; however, the following represent our top seven characteristics for the QA Reviewer to examine that might reflect quality problems with the PHA or documentation:

- **Completeness Check** – Picking at least one node, have all key causal events been developed as definitive scenarios?
- **Probable Worst-Case Consequences** – Have they been clearly identified and used as the basis for risk-ranking?
- **Safeguards** – Are the identified safeguards clear barriers between the causal event and ultimate consequences? With the additional equipment and transmitter specificity, do IPLs appear independent, with no apparent common-mode failures missed? Are root transmitters for causes and safeguards clearly identified to reflect the PHA Team having clearly evaluated the potential for common-mode failures?
- **Scenarios** – Are they clearly interpretable? Scenarios should present an image of the event.
- **Risk-Ranking** – Consistent application?
- **Actions** – Clear, complete with focused basis, and self-standing?

- **Multiple HAZOP/LOPA Entries for the Same Causes** – Have these been avoided to decrease the potential for confusion and future misapplication?

8. Emphasis Points for Maximizing the Future Usefulness of the PHA

As well as addressing immediate needs, there are a number of ways that the performance of the PHA and resultant documentation can be geared towards maximizing the ability to use and update the PHA.

Documentation Traceability – Application of the above tips on documentation quality can generally improve documentation traceability, also enhancing the PHA's ability to be updated.

Prolific Use of Equipment Tag Numbers and P&ID References – At the risk of being repetitive, this important characteristic is worth repeating for the documentation quality enhancement value achieved for a very small incremental effort.

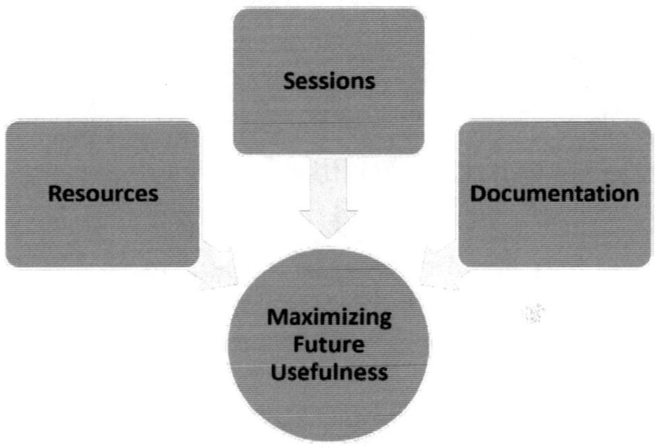

FIGURE 8.1 – Maximizing Future Usefulness

Application of Risk-Ranking – It's tempting to only rank high-consequence scenarios; however, for a small incremental effort, ranking all scenarios reinforces the non-importance of low-consequence scenarios. This can be a significant aid to the design process, as well as MI Program prioritization.

Consistent Scenario Grouping – Consistently placing certain types of scenarios under the same guideword such as "No Flow" or "High Level," etc. can help the reader locate types of events and facilitate future use and update.

Standardized PHA Approach – Although some owner/operators may use a PHA method tuned for their specific needs, standardizing the PHA technique to be consistent with industry best practices can facilitate its ability to be used in the future or updated. Use of a non-standard technique can be even more challenging if the facility changes ownership.

Large Nodes – Although sometimes daunting for a PHA Team, large nodes can allow for a more holistic approach to understanding process dynamics and facilitating scenario location and augmentation. Of course, node size and node complexity must be balanced.

Strive for "Evergreen" Approach – Although a small incremental effort may be required, focus PHA documentation on having an ability to readily update the PHA to address MOC and other future design/operations changes. Consider the long-term use of the PHA.

Software – Software should optimize usage, storage, and retrieval of the PHA. Consider long-term use and the potential for the owner/operator no longer supporting annual licensing fees. Save files in other machine-editable formats, as appropriate.

9. Summary

Implementation of the specific quality tips provided in this paper also require an environment supportive of the performance of high caliber work. Key underlying characteristics of the organization and the individuals involved in the HAZOP/LOPA are critical for supporting a program that emphasizes the importance of the performance of quality Process Hazard Analyses:

- Commitment from Management
- Commitment from the PHA Team
- Dedication of the Facilitator/Scribe in the Performance of Quality Work

These characteristics and the implementation of some of the key tips provided in this paper can provide a platform for the performance and creation of an efficient PHA that has current and future value, and can be used to address many of the objectives identified in Section 3.

10. References

[1] 2009 Global Congress on Process Safety, "Optimizing Qualitative Hazard Evaluations for Maximized Brainstorming (or "How to Complete A PHA/HAZOP Meetings in One-Third the Time Currently Required while Finding More Scenarios")," http://www.process-improvement-institute.com/_downloads/Optimizing_PHA-HAZOP_While_Maximizing_Brainstorming.pdf.

[2] CCPS "Guidelines for Technical Management of Chemical Process Safety," 1987.

[3] CCPS "Guidelines for Hazard Evaluation Procedures," 1985.

[4] "Layer of Protection Analysis: Simplified Process Risk Assessment, " October 2001.

[5] PSM – 29 CFR 1910.119, "Process Safety Management (PSM) of Highly Hazardous Chemicals, Explosives and Blasting Agents," 1992.

[6] RMP – 40 CFR Part 68, "Risk Management Programs (RMP) for Chemical Accidental Release Prevention," 1996.

[7] California Code of Regulations, Section 5189.1, "Process Safety Management for Petroleum Refineries," October 1, 2017.

[8] "California Accidental Release Prevention (CalARP) Program," October 1, 2017.

[9] API Recommended Practice 581, "Risk-Based Inspection Technology," Second Edition, September 2008.

[10] Paper 513595, 2018 Global Congress on Process Safety, "Preparing for a Successful HAZOP/LOPA (Making or Breaking Quality & Efficiency)," Maher, Steven T., Carine M. Black, and John M. Johnson, April 22-26, 2018.

[11] http://www.RMPCorp.com/HAZOP-Study-series-module, HAZOP/LOPA Facilitation Best Practices Webinar Series.

Practical Strategies to Prepare and Lead HAZOP Analysis of Complex Chemical and Petrochemical Processes

Marco-Antonio Medrano-González
RSC. Operation Manager
Zacatecas 70, 309. Col. Roma Norte, Mexico City
marco.medrano@risksc.com

Prepared for Presentation at
American Institute of Chemical Engineers
2019 Spring Meeting and 15th Global Congress on Process Safety
New Orleans, LA
March 31 – April 3, 2019

Practical Strategies to Prepare and Lead HAZOP Analysis of Complex Chemical and Petrochemical Processes

Marco-Antonio Medrano-González
RSC. Operation Manager
Zacatecas 70, 309. Col. Roma Norte, Mexico City
marco.medrano@risksc.com

Keywords: HAZOP Analysis, Process Hazard Analysis, Strategy Implementation, Chemical Process, Petrochemical Process.

Abstract

Hazard and Operability (HAZOP) Analysis was developed by the Imperial Chemical Industries (ICI) in the sixties as a systematized methodology to identify hazards in chemical processes at the design stage. The origin and the development of the main concepts addressed by HAZOP such as deviations and guidewords may have come from the use of What-if? Analysis and Safety Audits in chemical processes. Since then, many HAZOP analysis have been performed around the world in different types of chemical and petrochemical processes with modifications with respect to the original methodology or modifications to address specific issues such as determination of a Safety Integrity Level (SIL) for Safety Instrumented Systems (SIS) using LOPA and/or the identification of hazardous scenarios with potential environmental impact, among others.

The development of new chemical and petrochemical processes, as well as the commercial incorporation of non-conventional unit operations in these processes, has created new hazard conditions that make it difficult in some cases to prepare and lead HAZOP using conventional guidelines. In this paper several cases of HAZOP analyses have been reviewed in applications to complex chemical and petrochemical processes, in order to obtain a set of different practical strategies and guidelines to make it easier and efficient for the preparation and leading of HAZOPs in such processes.

1 Introduction

Hazard and Operability (HAZOP) Analysis was created and developed by the Imperial Chemical Industries (ICI) in the sixties as a systematized methodology to identify hazards in chemical processes at the design stage. HAZOP started to be more widely used within the chemical process industry after the Flixborough disaster in 1974. Flixborough was the most important industrial accident that forced many British chemical companies to revisit their standards, design criteria and engineering processes for new chemical facilities' projects. Since then, HAZOP has been widely adopted as a popular systematized and orderly methodology to hazard analysis in the design stage, being also a complementary tool for reviewing engineering design. Nowadays HAZOP has been extended and adapted to hazard analysis and identification of other types of systems. There are generically four types of HAZOP adaptations applied today.

- Process HAZOP: A traditional application of hazard analysis in chemical processes.
- Human Factors HAZOP: A particular group of specialized HAZOP approaches focused on human error causes rather than technical failures (mechanical, electrical, instrumentation, etc.)
- Procedure HAZOP: a review of operational procedures or sequences for operating process equipment or any type of machinery involving any human interaction.
- Software HAZOP: Identification of potential errors in the development of software applications for controlling automatized operation sequences.
- Control HAZOP (CHAZOP): A review of basic control philosophies or control sequences performed by Distributed Control Systems (DCS).

Each of these HAZOP types involves a different approach to identify hazards and subsequently to apply different strategies. This paper provides a basic review of the Process HAZOP applied in complex chemical and petrochemical processes, in order to obtain a set of different practical strategies and guidelines to make the process of preparation and leading easier and more efficient.

2 HAZOP and Life Cycle Projects

The initial HAZOP Analysis of ICI included guidelines to apply the HAZOP at six levels or stages of a Life Cycle Project [5].

- Project exploration / preliminary project assessment.
- Project definition, a review of immediate, evident or obvious major hazards.
- Detailed design review, an identification of hazards and operability issues.
- Final stages of construction
- Plant commissioning.
- Normal operation, usually after start-up.

In our experience we have found that the scope and the background of the project are two factors that significantly influence the preparation, leading and execution of HAZOP Analysis [3]. The execution of a HAZOP may require some adaptations or basic modifications depending on the stage of life cycle project where this hazard analysis must be performed and the background of the project.

The stage of life cycle of the project defines the scope as well as the objectives of the HAZOP Analysis according to the compliance of local regulations or corporate standards. The background is referred to particular circumstances or situations of the project where a HAZOP Analysis can be a key factor to solve potential design decisions, to define and delineate additional engineering activities and reduce potential future difficulties in the project execution. The background of the project also helps spot which can be the consequences of interest to be identified and assessed by HAZOP. Figure 1 shows a life cycle project and typical scopes and objectives of HAZOP in each stage.

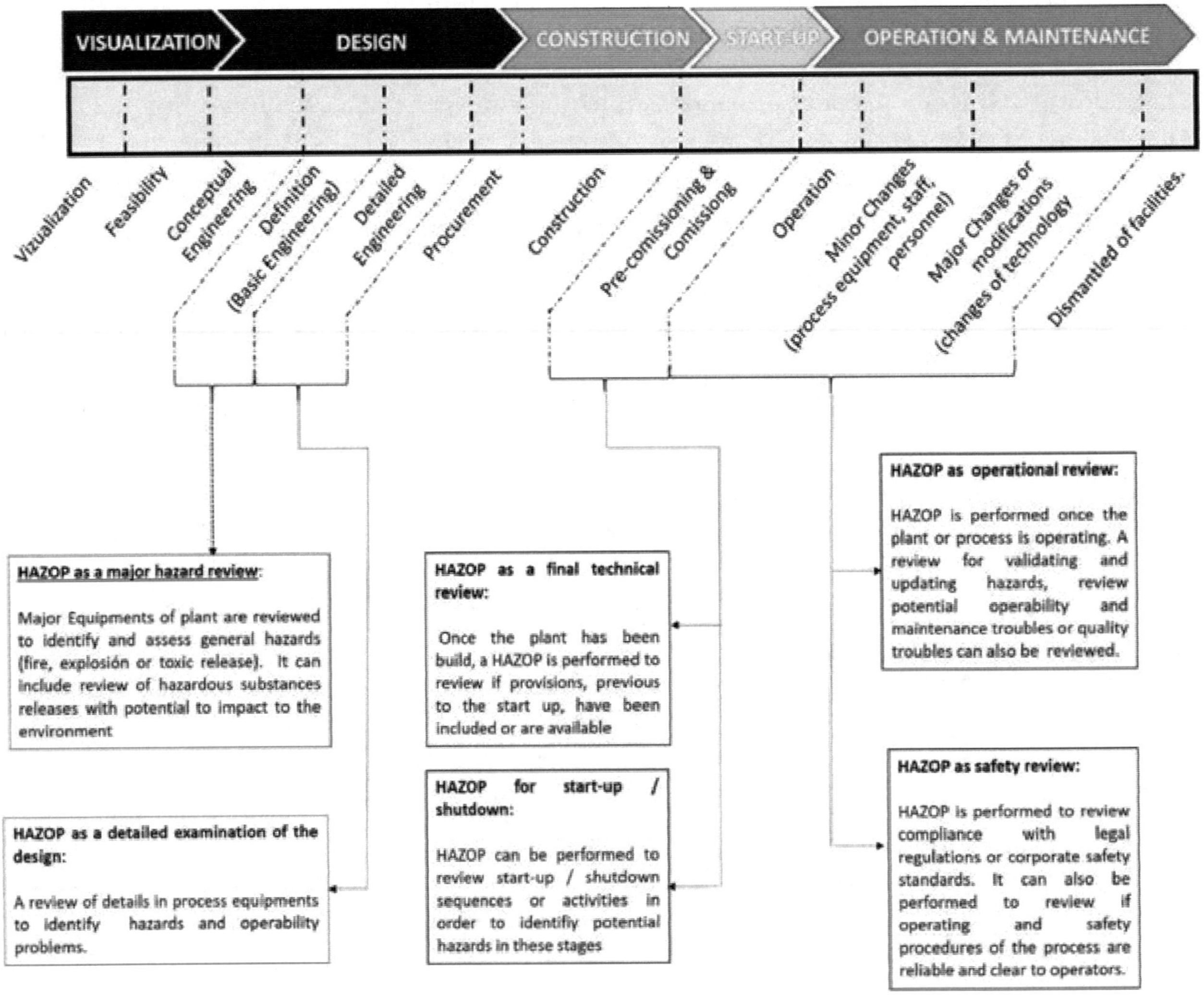

Figure 1. Stages in the life cycle project define the scope and objectives of HAZOP.

Our experience based on understanding the scope and objectives of the HAZOP as well as the background of the project permits to apply several practical strategies and basic guidelines to make the preparation and leading of HAZOP for chemical and petrochemical processes easier and efficient.

2.1 Define scope and objectives of HAZOP

The scope of a HAZOP can be to ensure compliance with local regulations, industry standards or corporate safety guidelines. The main objective of a HAZOP is to identify hazards in a particular chemical or petrochemical process or operation (sequence of activities). This means, HAZOP identifies potential and credible scenarios that would lead to the release of hazardous or flammable material to the environment, thus exposing workers to injury, causing potential damage to the environment, physical damage to assets, damage to civil population near to facilities, financial losses, among others. Other objectives of HAZOP can include a check of the safety of the process design engineering and identify new methods to improve the safety of an existing or modified facility.

There are other topics that a HAZOP can review which be a part of the scope or objectives such as identifying operability and quality issues of a process, problems related to maintenance of process equipment, piping or instrumentation, reliability of operating and safety procedures and potential conditions or situations of environmental impact from loss of containment events. For operability and quality issues, HAZOP is used to identify potential areas for improving efficiency or get economic benefits in the normal production or normal operation of a chemical plant.

2.2 Review the background of the project

Understanding the background of the project in which a HAZOP will be developed, helps to define other topics that must be taken into account to prepare and lead this hazard analysis such as consequence of interest to be that must be reviewed. Consequences of interest can be hazards or dangerous conditions or operating problems, quality of products or raw materials in the process, conditions of potential damage to the environment, among others. The background of the project will also help define the level of detail or depth to which the previously defined consequences of interest should be reviewed during the execution of HAZOP meetings.

HAZOP not only identifies hazards in the normal or continuous operation but also during other modes of operation such as shutdown, start-up, purging, venting, maintenance among others. Each mode of operation has different characteristics and particularities that must also be understood before defining a strategy to perform a HAZOP analysis.

2.3 Identifying consequences of interest

Identifying consequences of interest helps to define the level of detail or depth of the HAZOP [7]. One method to identify consequences of interest is to become knowledgeable about the company's or facility's risk matrix. The risk matrix defines criteria of acceptance or tolerance to risk for each type of consequence of interest to be reviewed. The background of the project establishes limits about the consequences of interest to be reviewed by HAZOP. Example of typical consequences of interest are injuries or fatalities to employees or civil population near to chemical facility, damage to the environment, among others.

Understanding the background of the project involve a review of other topics no related directly to hazard identification such as operability and quality issues. In these cases, the usual approach of plant operators, is to review whether the instrumentation of the process is the most effective option or whether there are other alternatives to improve the existing control of process based on previous experience controlling and operating the process. *In other cases, plant operators decide make HAZOP to ensure trouble-free start-up or shutdown.* HAZOPs for start-up or shutdown review if each sequence or activity for starting-up or shutdown of the process creates new potential hazards conditions or situations that can modify the overall level of the safety of the facility.

3 Applying a strategy for HAZOP Analysis.

Once the scope, objective(s) and consequences of interest have been defined previously, a strategy to prepare and lead the HAZOP comes in. Strategy is defined as the art and science of planning and organizing different types resources to be used in the most efficient way to get a goal in the most effective form [6,8]. A strategy can be made up of different elements. Preparing and leading a HAZOP Analysis requires in many cases, defining a previous strategy for its performance and execution depending on the scope and objectives to be covered by HAZOP as well as knowing the background of the project. A model that comprises a four steps strategy for planning and leading HAZOP is proposed and based on the common five steps used in an HAZOP Analysis: node definition, design intent, process variables, guidewords and deviations. This four-step strategy model is shown in Figure 2.

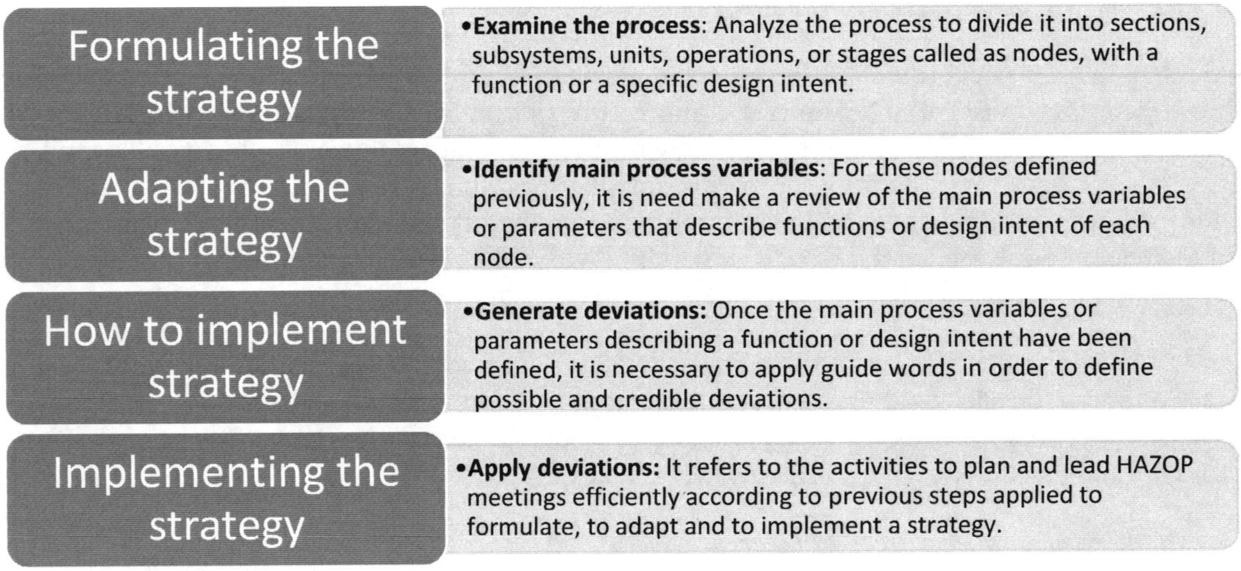

Figure 2. A strategy of four-step to prepare and lead HAZOP Analysis.

3.1 Examining the process, formulating the strategy

Formulate a strategy to prepare and lead a HAZOP requires the facilitator to know some basic characteristics of the process and obtain at least two (2) basic types of documents or engineering drawings:

- Process Flow Diagrams (PFDs).
- Piping and Instrumentation Diagrams (P&IDs).

The basic strategy principle applied for chemical or petrochemical process HAZOP is "divide and conquer". This mean that an overall process or plant is divided into sections called subsystems or nodes in order to make easier the identification of potential and credible hazard conditions involved in each section or node.

Process Flow Diagrams (PFDs) provide a roadmap of a chemical process that describes the order of general flow of materials or chemical substances between process equipment or unit operations involved in this chemical process. A PFD describes operating conditions (usually pressure, temperature and flow) and displays the relationship between the different types of process equipment. It shows basic control loops and does not show other details such as piping diameter, type of control valves and instrumentation involved in the process equipment or its associated piping.

PFDs are used to review points in the process that can present immediate, evident or major hazards. Basic control loops in the PFDs can be used to make an initial identification of potential causes of deviations and spot initial consequences of interest. It can also help to define potential deviations or define boundaries of major nodes. P&IDs describe functional relationship of process equipment linked by piping, instrumentation and valves.

There is not a unique way or a specific rule to sectioning or dividing a chemical process in subsystems or nodes, although there are some basic guidelines to sectioning a chemical process according to the scope, objectives and context of the project using PFDs and P&IDs. For example, the plant or process can be divided into operations, units or systems with a specific operational function in the overall performance of the process. In other instances, sectioning or dividing process can be in response to previous definitions of design intention or the compliance of a particular operational function.

One significant benefit of properly dividing a chemical or petrochemical process into nodes is that it will make leading HAZOP meetings more efficient. A basic rule of thumb to apply in this sectioning process or "nodalization" is that complex subsystems or nodes involving high hazards, must be defined in small sections. For simple subsystems or sections with low hazards, larger sections can be defined [1]. Other basic criteria to "nodalization" are the following:

- Define each major process component or equipment as a section or subsystem.
- Define a line section between each major process component as a section or subsystem.
- Piping and Instrumentation Diagram (P&ID) is defined as a node.

Each of these criteria can take into account anticipated or expected deviations for each section or node. When reviewing PFDs or PIDs, control valves can be used as physical boundaries of a node depending of the function or design intention of the previously defined node.

In other and special circumstances, the definition of subsystems or nodes cannot have any logic relation to the operating process. Instead the nodalization responds to other aspects such as the delimitation of areas of responsibility according to mechanical integrity program, inspection routines or control and supervision activities, among others. The background of the project defines how to deal with these special circumstances.

Other basic options guidelines to nodalize are the following:

- Two major process equipment linked as a node.
- Major equipment and minor equipment classification.
- Defining a preliminary design intent

3.1.1 Two major process equipment pieces linked as a node

This option considers a node as a transfer of materials from a source to a destination. Source and destination can be process equipment such a pressure vessel, a storage tank, distillation column, etc. This case is helpful when it is defined deviations under the concept of "change path" described in section 3.3.1.

3.1.2 Major and minor equipment classifications

In this case, each process equipment and its associated lines are considered as a node [7]. Process variables as temperature, flow or pressure change values in each transition of process equipment, subsequently deviations can be defined in terms of these transitions or modification of their values.

One problem with this method is that many nodes can be generated and unnecessary duplication of nodes can appear. A way to make this "nodalization" efficient is to define a classification of major equipment and minor equipment. Major equipment consists of pressure vessels, reactor, distillation columns or any process equipment that have condition to storage significant inventories of hazardous substances. Heat exchangers, filters or pumps or any other equipment with low residence time of hazardous substances can be considered as minor equipment. In this case, a node will consist of a combination of major equipment and minor equipment as it is shown in Figure 3.

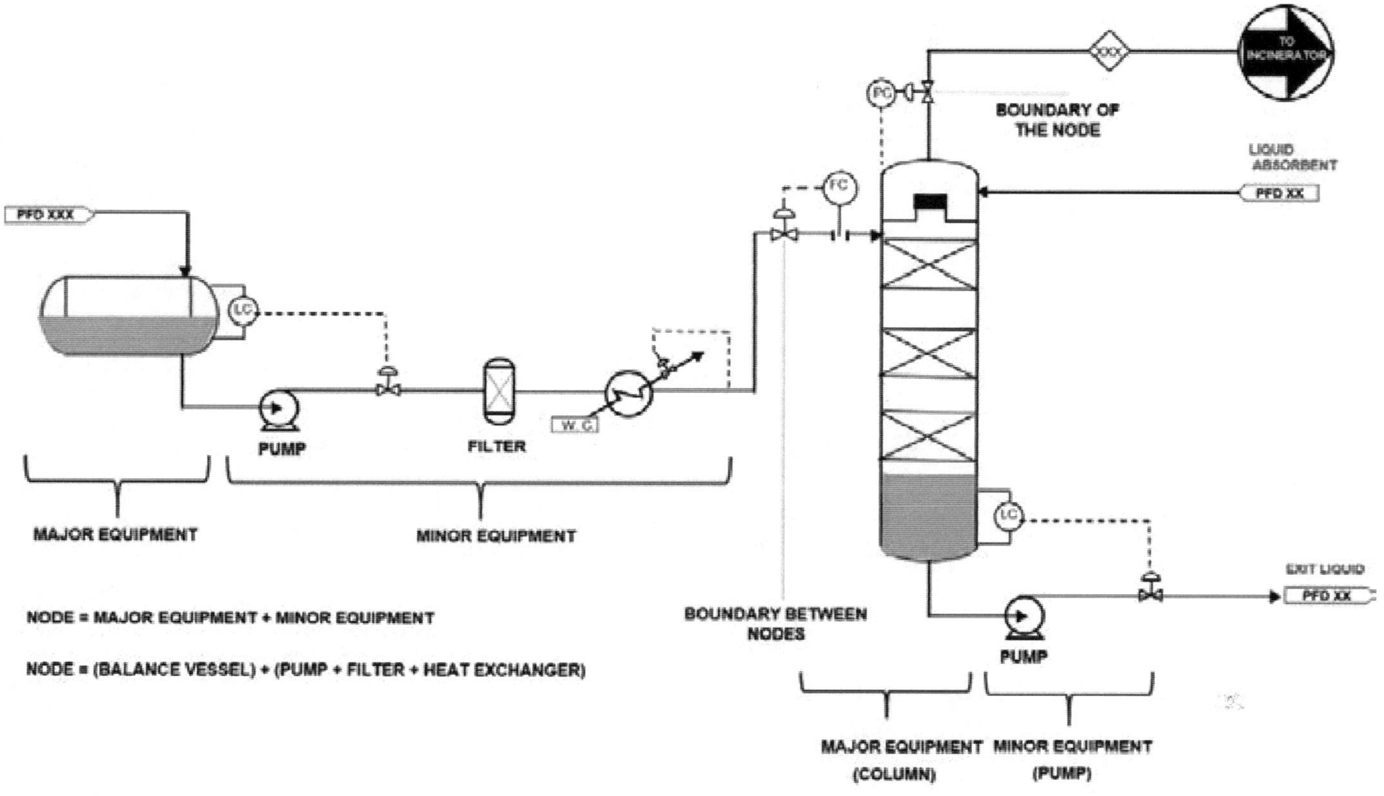

Figure 3. An example of nodalization combining major and minor equipment classification

3.1.3 Defining a preliminary design intention

In this case, each process equipment and its associated lines are defined as a subsystem with a preliminary design intent not necessarily related to a specific operating function in the overall performance of the process. A preliminary design intention describes how a particular section or a set of process equipment is expected to operate or which function must perform in the absence of any deviation.

Under this option, a node can be described with more of one preliminary design intention and this is a hint that the subsystem or node elected can be separated in two or more nodes. An example of the application if this case is shown in the Figure 4.

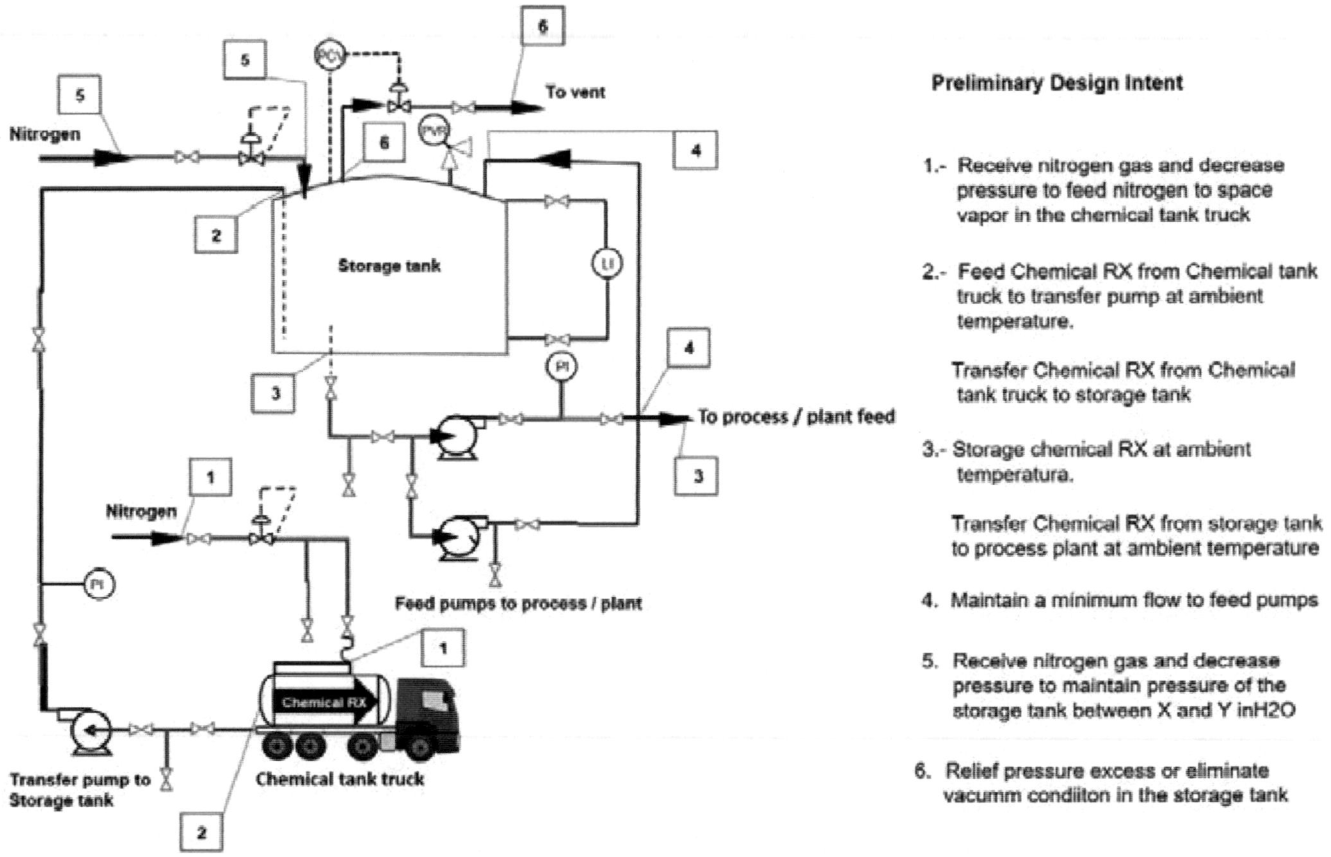

Figure 4. An example of application of preliminary design intention.

3.2 Identifying main process variable, adapting the strategy

Process variables are an important component or element of a node describing how this node must operate and the range of values where these parameters must move under normal operations conditions or safe conditions without any type of induced deviation. Outside of this range of values, is where the effect or consequence of interest of the deviation presents itself.

Most common process parameters applied in HAZOP are pressure, temperature, flow and level. In less proportion, composition and pH are also used. Many petrochemicals, unlike fractions of hydrocarbons, are very reactive to changes in pH or the presence. Figure 4 summarizes process variables commonly applied in HAZOP and other process parameters of interest.

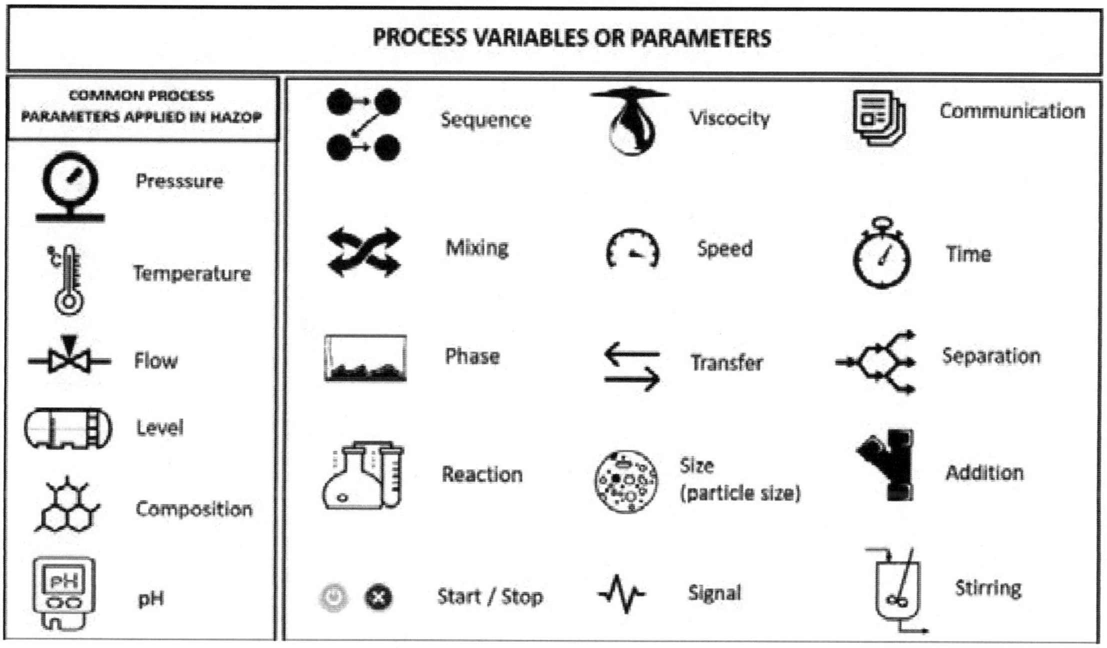

Figure 4. Most common process parameters applied in HAZOP and other less common process variables or parameters applied

In our experience and for continuous operation mode, pressure, temperature, flow and level are the process parameters most commonly applied in the generation of deviations. In several petrochemical processes, composition and pH are also used to generate deviations with consequences of interest, particularly for petrochemical substances made chemically unstable due to changes in pH, presence of contaminants materials or different composition ratios. In these situations, other parameters may be applicable such as mixing or reaction with similar physical meaning of the deviation.

For operating modes such as start-up, purge and venting, among others, the least common operating variables such as phase, speed, communication, time among others can be applied to generate deviations with potential consequences of interest.

3.3 Generating deviations, defining how to implement the strategy

A first rule to generate deviations is that the combination of process parameter with guide word, must have a physical meaning for the operator or members of the HAZOP team.

A guide word is a brief word with a simple meaning that is used or added to a process parameter in order to induce or create an imaginary deviation that may induce a departure to the design intent previously defined. In the practice, there is a set of common guidewords applied for HAZOP according to the Figure 5.

Most common guideword	Meaning
No	Negation of the design intent / Design intention is not achieved
Less	Quantitative Decrease
More	Quantitative Increase
Part of	Qualitative Decrease
As Well As	Qualitative Increase
Reverse	Logical opposite of the design intention or operation function
Other Than	A complete substitution of a design intent / another task or a completely different activity occurs.
Less common guide word	**Meaning**
Where Else	Negation of the design intent / Design intention is not achieved
Before / After	Order of sequence is not achieved
Early / Late	Time to execute a task, activity or stage is different from the design intent
Faster / Slower	A task, activity or a stage (human procedure or machinery) is done faster or slower than the demanded by the operation or process.

Figure 5. Most common and less common guidewords applied in HAZOP and their basic meanings.

The deviation is a combination of a guide word with a process parameter.

Guideword + **Process Parameter** = **Deviation**

This combination generates many deviations, some with a physical meaning and others having no logic or a physical meaning. The deviations without a physical meaning are eliminated and deviations with physical meaning must be reviewed to determine if they are credible, can potentially happen and produce consequences of interest. Other general deviations can also be applied such as: maintenance, shutdown, start-up, sampling, loss of containment, ignition source, service failure, etc.

Once the potential deviations have been defined, they must be applied in the node. One technique to apply deviations is using the "change path" concept [1].

3.3.1 Change path concept

This concept considers that in any process there exists a basic function of transfer of material from a source to a destination. This basic function can also apply when performing an activity on the chemical substance or during the input of a material from a source or to delivery this material to a destination. The concept of change path is easy to recognize because it represents a flow of any material from a source to a destination. A source can be a particular location, equipment, activity or condition and a destination another location, equipment precedent activity or condition implying some type of a physical change in this location, equipment, activity or a change of condition

The operation involved to produce this change (change path) is where deviations from design intent can be applied for the hazard analysis. Figure 6 shows how deviations must be applied under the change path concept. Figure 6 is an adaptation from the original source [1] according to our experience in leading HAZOPs.

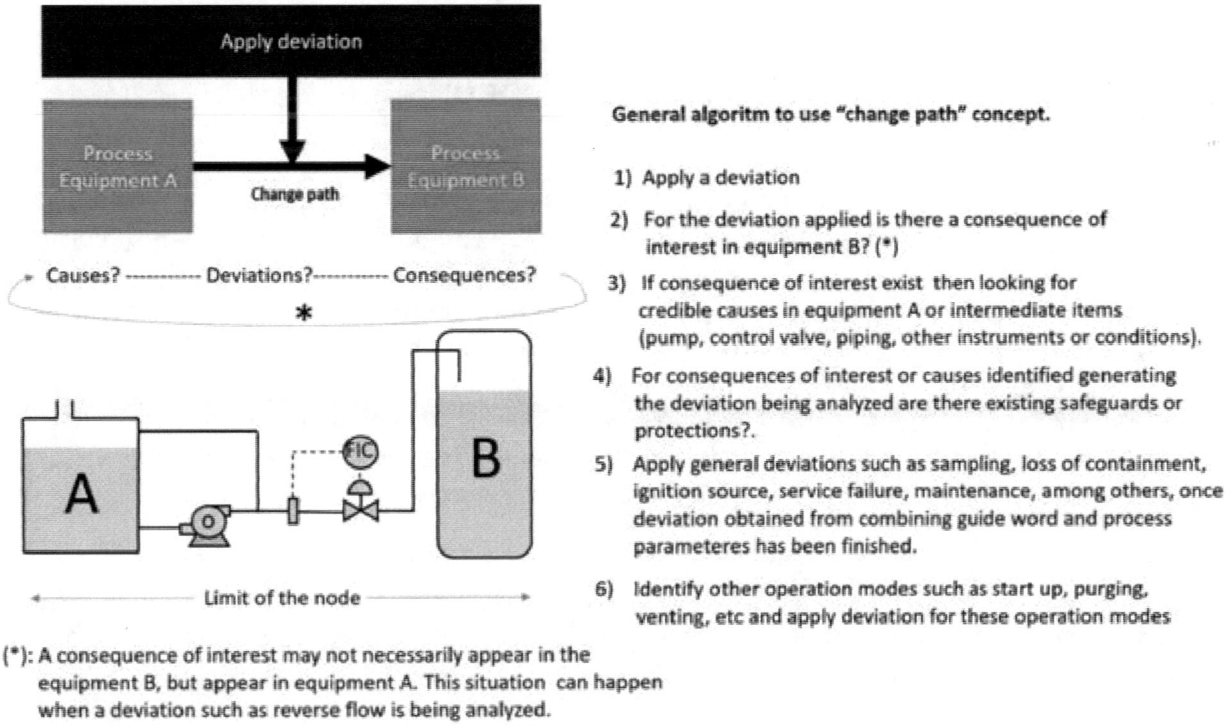

Figure 6. A General Algorithm for the "Change Path" Concept, adapted from [1].

An important part of this concept is to define and identify main or major elements of the node such as main process equipment which serves to identify the essential components or elements of the node as well as the physical boundaries or limits of the node. Under this concept, any element in the node may be able to change or modify process parameters in some way affecting the original conditions of the major process equipment A or B.

At least three (3) elements must be involved in this definition of change path: process items or equipment, the material(s) to move or to transfer and the activity performed in the node.

3.4 Applying deviations, implementing the strategy

Once the nodes of the process have been defined as well as the deviations to apply in the node, then HAZOP meetings with the team begin. Ensure there is a general consensus from the HAZOP team on the consequences of interest to be identified. Start the analysis of each deviation asking first if the deviation is possible happen in the node and second if there are consequences of interest due to that deviation. Other basic rules during the review of consequences of interest, credible and potential causes as well as identifying safeguards and determining recommendations in HAZOP meetings, are described below.

3.4.1 Reviewing consequence of interest in the HAZOP meeting

Sometimes, applying a deviation in a particular node creates a consequence of interest in another part of the process or in a different subsystem or node. For example, a rupture in heat exchanger tubes may result in the transfer of flammable materials into a cooling water system. In this case, the team must decide where the consequence of interest must be included, in the node under analysis or in the other different subsystem.

3.4.2 Reviewing credible and potential causes in the HAZOP meeting

Once the application of deviation creates consequences of interest, the HAZOP leader must ask to team members to look for credible and potential causes makes the deviation occurs. The redaction of causes is a general description of failures such as, misalignment of valves, malfunction of valves, among others. In these cases is convenient to include to these failures, a brief description of the failure mode or the mode in which a system, device or action might fail or how a fail is shown. For example, "failure of a control valve" or "malfunction of a valve or a device" can be a cause of a deviation but does not describe what type of failure mode is involved. *"Valve fails closed" or "valve fails open" describes the failure mode in a clear way for operators.* In other cases, the failure must be described in a particular moment of the operation. For example, "Control valve fails closed or open during unloading / loading / purge / venting, etc".

Describing the failure mode of a system, device or equipment is particularly important in analyzing whether redundancy of process equipment is a real safeguard. For example, process pumps can be electrically-driven or steam turbine-driven pumps and different failures modes can occur for each type of pumps giving the same service.

3.4.3 Reviewing existing safeguards in the HAZOP meeting

Sometimes, a HAZOP member team considers that the failure of a safeguard is cause of a deviation. Before starting HAZOP, advise the hazard analysis team that failure of a device or system defined as a safeguard is not a cause of a deviation. Once consequences of interest and causes have been identified, safeguards related to these causes and consequences must be identified by the team by asking for example, if there are existing safeguards that help to eliminate causes or mitigate consequence effects previously identified.

A review of the PSM program of the company or plant can be helpful to identify other potential safeguards that can be important to eliminate potential causes or mitigate effects of consequences previously defined. Elements of a PSM program as potential safeguard to be reviewed can be training personnel programs, operating discipline, contractors management practices, compliance audits, work permits, mechanical integrity or inspection and maintenance programs among others.

3.4.4 Making recommendations in the HAZOP meeting

When the team arrives at the step to make recommendations, the leader must keep the HAZOP team members focused on developing specific, measurable and realistic recommendations to eliminate causes, mitigate effects of identified consequences of interest identified and/or improve existing safeguards. Some HAZOP team members usually request detailed engineering solutions. It is necessary to remind them that HAZOP is an identification technique not a problem solving methodology.

Some recommendations may involve significant capital investments requiring the execution of new or complementary projects. In these cases, the recommendation might not apply directly as part of the HAZOP, instead these recommendations may be addressed as developing a new or complementary project. The background of the project will define what decision must be taken.

In other cases, for example new plant projects applying RAM analysis (Reliability, Availability and Maintainability Analysis) in the design stage, HAZOP helps to define scenarios with any operability or quality topic that can be used for RAM Analysis purposes.

Figure 7 can be applied to verify whether the HAZOP worksheet developed with the team has identified and included all potential and credible consequence of interest, credible causes, and all the existing safeguards in the node being analyzed. Remember that the scope and background of put a limit on the length of depth of the consequence of interest to review by HAZOP and subsequently in the definition of causes, safeguards and recommendations.

CAUSES	CONSECUENCES	SAFEGUARDS	ACTIONS / RECOMENDATIONS
• Human factors related to operating activities.	• Shutdown of process equipment or control systems. (1)	• Material Specifications / Electric Classification Areas	•Inherently Safety Design Practices
• Human factors / activities related to maintenance of instruments, mechanical, electronic devices or process equipment.	• Delay / breach in receipt, delivery or dispatching programs	• Operating procedures (related to subsystem under analysis, process equipment, operational activity)	•Compliance of maintenance programmes (5)
• Presence of contaminant materials or different compositions	• Out of spec products	• Instrumentation (Monitoring, Measuring and Control involving operational actions).	• Rehabilitate / revamp process equipment, control or safety systems .
• Failure of control valves, instruments, system control	• Partial or full operational shutdown.	• Alarms & Trips (from process interlocks)	Elaborate specific procedures to attend causes or consequences
• Ancillary Services Failure	- Effect modes on process equipment performance (High Vbration, over heatng, excesive cooling, vacum, etc.	• Relief pressure / breaking vacum devices.	Acquire new process equipment, implement new control devices or systems associated with deviation under analysis.
• External Events or Factors	- Mechanical damage or degradation of equipment components	• Drainage / Venting Systems (2)	•Relocate process equipment in Lay Out (as a practice of loss prevention)
• Other potential and credible causes.	• Overpressure / high vacum / overflow conditions in process equipment or subystem	• Detection Systems (3)	• Statics analysis and record of operational variables
	•Loss of Containment / Release of hazardous substance to the environment, build- up or confinement (2)	• Containment Barriers (4)	• Add or configurate alarms in DCS / Safety Systems / Interlocks
	• Flammable or toxic cloud formation and spread out	• Fire water system / mitigation systems / Automatic or manual extinguisment systems	• Implement Alarm Management Practices (6)
	Fire, Flash Fire / Explosion	• Blocking / Isolation Valves	• Improve existing emergency, contingency or operational procedures
	Conditions of Domino Effect	• Safety Instrumented Systems (SIS) or Safety Instrumented Function (SIF´s)	• Apply Safety Systems (detection / mitigation / shutdown)
	• Loss of production	•Work Permit	• Apply security systems or implement security supervisory practices
	• Financial / Economic Losses (business interruption)	•Maintenance programs (corrective, preventive predictive) (5)	• Improve risk comunications protocols to personnel or civil population
	Physical or mechanical damage to other process equipment / other near assets or civil population	• Emergency or contingency procedures (related to loss of containment)	• Develop new or complementary technical and engineering studies
	• Environmental Impact / Contamination (soil, air, water)	• Emergency Response Plan	• Review if some elements of a PSM program applied or require reinforcement
	• Operator / personnel injuries or fatalities	• Personnel Protective Equipment for operating, maintenace activities or emergencies	
		Operator / personnel training programs	

Notes:

(1): In some instances, alarms from control devices are consequence of interest, particulary when the deviation is related to operability or quality issues.

(2): Drainage / venting systems are safeguards only if their design and operation no create new potential hazardous conditions.

(3): As a general category including Fire&Gas Detection Systems, Wind Direction Indicator for toxic substances releases, among other.

(4) Loss of containment event and containment barriers can be related to TIER 1 or 2 according to API 754.

(5): Mechanical Integrity Program can be included as part of a Maintenance Program.

(6): When the amount of process alarms is excessive, an alarm management program should be applied, for example using ANSI/ISA 18.2.

Figure 7. A table to verify if the HAZOP worksheet has identified all potential consequences of interest, credible causes and all existing safeguards.

4 Conclusions

HAZOP studies consume a significant amount of time (man-hours), so in its planning it is necessary to balance the effort, time and human resources. One way to make this balance is under the strategy of "divide and conquer" that involves a review of the process and its respective separation into nodes or subsystems. Formulate a strategy to be used in the most efficient way to get a goal in the most effective form for a HAZOP Analysis requires to know previously, the scope and objectives to be covered by this hazard identifications as well as knowing the background of the project.

Making an orderly and logical separation of the process into nodes, drives the HAZOP team to do a more effective job because clarifies the design intent and what of the node has in the performance of the overall process. The HAZOP team get more awareness on the implications that each deviation may induce in the node or part under study. At the same time, the hazard analysis team members become aware of the importance of their participation in identifying hazards and each member feels comfortable that his or her participation is being well-guided at identifying hazards.

A logic nodalization that is simple to understand will help to the HAZOP team members to revew the process better and to keep better focused on diagnosing safety problems, or even identifying other problems such as operational and quality situations, or reviewing potential impacts from the operation of the plant. Practical strategies to make nodes as simple as possible and easy to understand keeps the group focused on one topic at a time reducing the possibility of presenting other discussions not aligned with the main discussions of the team.

The development of new chemical and petrochemical processes create new hazard conditions making it in some cases difficult to apply the typical or conventional application of HAZOP guidelines. In such cases, the method or technique used to prepare and lead a HAZOP requires a more strategic approach in order to obtain a best hazard identification making the analysis team members feel more comfortable about their participation in the HAZOP and more confident in the final results.

Figure 8 describes some basic differences in key concepts expounded in this paper between the conventional or traditional HAZOP guidelines using practical strategies. The practical strategies described in this paper do not intend to differ from the conventional or traditional HAZOP guidelines. Instead, these practical strategies seek to complement the traditional application of HAZOP to make its use easier when complex chemical or petrochemical processes have to be analyzed, or when the scope or objectives of a HAZOP analysis involves a modification of the original methodology for example, to include Layer of Protection Analysis (LOPA), among others.

Key concept of HAZOP revised		Conventional HAZOP approach	Practical Strategies for HAZOP
Life Cycle Project	Scope and Objectives	Ensure compliance with local regulations, industry standards or corporate safety guidelines	Ensure compliance with local regulations, industry standards or corporate safety guidelines
	Review Background of the Project	No detailed guides on reviewing background of the project where HAZOP Analysis is required.	Experience performing HAZOPs suggest review the background of the project in order to seek other topics that must be taken into account to prepare and lead the hazard analysis.
	Consequences of Interest	General description of the type consequences of interest.	Review background of the project helps to define different consequence of interest to be reviewed with more detail during HAZOP sessions. It also helps to define the level of depth to which consequence of interest should be revised by HAZOP.
Formulate, adapt and implement a practical strategy	Design Intent / Nodalization	A general description how to define a design intent for a node	Provide practical guides to define design intent in several forms according to the type of node or the particular function of a subsystem or node in all the process being analyzed.
	Deviations	A general description how to define and apply deviations in a node	Describe some guides to define deviations that create consequence of interest, for example using the "change path" concept.
	Identifying Causes	A general description of causes and how to be identified.	Give recommendations about redacting causes and the convenience to describe failure modes of credible causes identified by HAZOP.
	Identifying Safeguards	A general description of safeguards and how to be defined in the HAZOP.	Give hints to review existing safeguards specific to eliminate causes or mitigate effect of consequences.
	Determining Recommendations	A general description of recommendations how to be applied.	Give hints about how to manage actions or recommendations defined by hazard analysis team.

Figure 8. Basic differences in key concepts between conventional or traditional HAZOP and practical strategies.

5 References

[1] D. Macdonald. *Practical Hazops, Trips and Alarms.* First Edition. Newness, Burlington, MA, 2004.

[2] M. Pérez-Marín and M. A. Rodriguez-Toral, *HAZOP – Local approach in the Mexican oil & gas industry,* Journal of Loss Prevention in the Process Industries 26 (2013), 936-940.

[3] M. A Medrano-González, *A methodological proposal to support best operational practices and philosophies using risk assessment of transient pipelines.* 4th Latin American Conference on Process Safety, Río de Janeiro, Brazil, 3-5 July 2012.

[4] M. A Medrano-González, *Using process flow diagrams and material & energy balance sheets to.* 9th Global Congress on Process Safety (CCPS), San Antonio, Texas, 28 April -1 May 2013.

[5] T. Kletz. *Hazop & Hazan, Identifying and assessing process industry hazards.* Fourth Edition. Institution of Chemical Engineers, Warwickshire, UK, 1999.

[6] P. Wright, M. Kroll and J. Parnell, *Strategic Management, Concepts and Cases.* Third Edition, Prentice Hall, Englewood Cliffs, NJ. 1996.

[7] The Process Safety Institute, Course 103: *Process Hazard Analysis Leader Training Using the HAZOP and What-if/Checklist Techniques.* ABS Consulting, Houston, October 2004.

[8] http://www.businessdictionary.com/definition/strategy.html (accessed Dic. 21, 2018).

The Effects of Oxygen Enrichment on the Explosibility Properties of Flammable Gases and Vapors

Richard W. Prugh
DEKRA Process Safety
113 Campus Drive
Princeton, NJ 08540
www.dekra-process-safety.com
prughrw@aol.com

Prepared for Presentation at
American Institute of Chemical Engineers
2019 Spring Meeting and 15th Global Congress on Process Safety
New Orleans, LA
March 31 – April 3, 2019

The Effects of Oxygen Enrichment on the Explosibility Properties of Flammable Gases and Vapors

Richard W. Prugh
DEKRA Process Safety
113 Campus Drive
Princeton, NJ 08540

Keywords: Flammables, Oxygen, Explosion

Abstract

In the chemical industry, oxygen-containing chemicals are used in the production of partial-oxidation products, including many in the pharmaceutical industry. The vapor spaces in oxidation reactors can contain high concentrations of oxygen gas, in combination with flammable vapors. Thus, it is important to determine the explosibility properties of organic-compound vapors in atmospheres containing high oxygen concentrations.

1. Introduction

In the chemical industry, "nascent" oxygen is involved in the production of partial-oxidation products, including many in pharmaceutical operations. The oxygen is initially in the form of peroxides, such as hydrogen and sodium peroxides, and also as chlorates and perchlorates, periodates, nitrates, and dichromates [1]. The vapor space in oxidation reactors can contain high concentrations of oxygen gas, in combination with the flammable vapors of organic compounds. To evaluate the possible explosion hazards of such mixtures of fuel and oxidizer, it is important to determine the explosibility properties of organic compounds in atmospheres containing high oxygen concentrations.

Tho basic flammability/explosibility data for this study are from the following references:

Table 1. References for Flammability Data

REFERENCE	FLAMMABILITY/EXPLOSIBILITY PROPERTY					
	Flammable Limits	Auto-Ignition Temperature	Minimum Ignition Energy	Maximum Explosion Pressure	Deflagration Index Kg	Oxygen Conc. Limit
NFPA 49; NFPA 325 [2]	√	√				
NFPA 68 [3]				√	√	
NFPA 69 [4]						√
NFPA 77 [5]	√ *		√			
NFPA 497 [6]	√	√	√			
USBM 503 [7]	√ *					√
USBM 627 [8]	√ *	√ *				
Babrauskas [9]	√	√ *	√ *			
Britton [10]	√ *		√ *			√
Crowl and Louvar [11]	√ *	√	√	√	√	√
Yaws [12]	√	√				

* Limited data for oxygen

Several methods are presented in this paper for estimating explosibility properties for preliminary process design. However, it is prudent to confirm the fuel/oxygen relationships with tests where process safety depends on the reliable control of oxygen concentration.

2. Flammability/Explosibility Properties

The explosibility properties of combustible materials can be divided into Ignition Sensitivity and Explosion Severity [13, 14]. The Lower Flammable Limits, the Autoignition Temperatures, and the Minimum Ignition Energies would be important in evaluating the Ignition Sensitivity. The Maximum Explosion Pressure and the Deflagration Index – a measure of burning rate – would be important in evaluating the Explosion Severity. The following sections describe the relationships of these properties to oxygen concentration.

2.1 Lower Flammable Limit

Data from many sources [15, 16, 17, 18] indicate that there is no significant effect of oxygen concentration on the Lower Flammable Limit [LFL] from the 21 volume-percent oxygen in air to 100% oxygen. At oxygen concentrations greater than the LFL in air, the specific heat of oxygen is about the same as the specific heat of nitrogen; that is, excess oxygen behaves as a diluent at the LFL.

For comparison with the Minimum Explosible Concentrations of combustible dusts and for evaluating the hazards of mixtures of flammable gases/vapors of combustible dusts, the values of LFL can be converted to mass densities by:

$$C = [\, LFL / (100 - LFL) \,]\,[\, MW / 0.024 \,] \quad \text{grams per cubic meter}$$

where 0.024 is the number of cubic meters per gram-molecular weight [at 25°C and one atmosphere pressure (14.7 psia)].

For example, the LFL for propane in air is 2.1 volume-percent and, for the molecular weight of 44.1, the Minimum Explosible Concentration would be given by:

$$C = [\, 2.1 / (100 - 2.1) \,]\,[\, 44.1 / 0.024 \,] = 39 \text{ grams per cubic meter}$$

Since the LFL is not significantly affected by oxygen concentration, oxygen enrichment would not affect the ventilation and dilution or inerting precautions that should normally be taken for a flammable vapor or gas.

2.2 Autoignition Temperature

The Autoignition Temperature [AIT] of a fuel is a kinetic phenomenon; that is, the AIT is low for very-long exposures to a given temperature, and it is quite high for short exposures to a given temperature. This relationship can be shown to follow the Arrhenius "law" [19], as:

$$r = A\, e^{-Ea/RT} \quad \text{fraction per second}$$

where A is the pre-exponential factor (in "per second");
Ea is the activation energy (in calories per gram-mole);
R is a constant (1.987 calories per gram-mole per °K); and
T is the temperature (in °K).

If equal reaction rates are assumed for combustion in air and oxygen, at two different temperatures, a value for the activation energy can be obtained, from:

$$Ea = 1.987\,[\, \ln (100/21) \,] / [\, (1/T_{AIR}) - (1/T_{O2}) \,] = 3.1 / [\, (1/T_{AIR}) - (1/T_{O2}) \,] \quad \text{cal./gmol}$$

The Autoignition Temperatures for 90 chemical compounds in air and in 100% oxygen are available [20]. The average AIT in air was calculated to be about 394°C [about 667°K], and the average AIT in oxygen was about 361°C [about 634°K]. Thus, the differences in AIT for air and oxygen is quite small, at about 5% lower.

The calculated activation energy Ea, as based on the referenced data, would be about 40,000 calories per gram mole, which is a "reasonable" value for an oxidation/ combustion reaction [21].

Thus, the AIT for a given oxygen concentration can be estimated from:

$$AIT_{O2} = AIT_{AIR} / [\ 1 + (\ 1.987\ AIT_{AIR} / 40{,}000\ )\ (\ln (C_{O2}/21)\ ] - 273 \qquad {}^{o}C$$

Oxygen enrichment would not affect the precautions that should normally be taken for a flammable or combustible vapor, such as avoiding exposure to hot surfaces in ovens, in dryers, and on electrical equipment such as motors.

2.3 Minimum Ignition Energy

Previous investigators had found that the Minimum Ignition Energy [MIE] was inversely related to oxygen concentration [22]. The Minimum Ignition Energies for 13 chemical compounds in air and in 100% oxygen are available [23]. The average MIE in air [excluding hydrogen (very low) and trichloroethylene (very high)] was calculated to be about 0.30 millijoule, and the average MIE in oxygen was about 0.0027 millijoule. Thus, oxygen concentration has a very strong effect on MIE, with about a factor of 100 reduction for 100% oxygen. The exponent for oxygen concentration for the 13 chemical compounds [including hydrogen and trichloroethylene] is about 2.9.

Thus, the MIE for a given oxygen concentration can be estimated [+/- 50%] from:

$$MIE_{O2} = MIE_{AIR} / [\ (\ C_{O2} / 21\ )^{2.9} = 6{,}200\ MIE_{AIR} / C_{O2}^{\ 2.9} \qquad \text{millijoule}$$

As indicated above, oxygen enrichment would have a significant effect on the precautions [such as inerting] that should be taken for flammable gases and liquid against very weak electrostatic [and other] ignition sources, including corona discharges [24].

2.4 Maximum Explosion Pressure

The Maximum Explosion Pressure [Pmax] – for combustion in air or in an oxygen-enriched atmosphere – can be calculated rather accurately if the heat of formation of the fuel is known and the degree of dissociation of the carbon dioxide and steam combustion products can be determined. For combustion in an oxygen-enriched atmosphere, the degrees of dissociation are very high; thus, the Pmax in 100% oxygen could be expected to show a rather small increase in pressure as compared to combustion in air.

There is a direct relationship between flame temperature and explosion pressure, as governed by the Ideal Gas Law, as:

$$P_{Max} = 14.7 \, (T_{FLAME} + 273) / (25 + 273) - 14.7 \qquad \text{pounds/sq.in. [psig]}$$

The available flame-temperature explosion-pressure data are as follows [25]:

Table 2. Flame Temperatures and Explosion Pressures for Fuel/Oxygen Mixtures

FUEL	FLAME TEMPERATURES (°C)			MAX. PRESSURES (barg)		RATIO (bara/ bara)	IDEAL-GAS-LAW RATIOS (°K/bara)	
	In Air	In Oxygen	Ratio [°K/°K]	In Air	In Oxygen		In Air	In Oxygen
Methane	1,950	2,705	1.37	7.1	10.4	1.41	274	261
Propane	1,988	2,822	1.37	7.9	10.2	1.26	254	276
Ethylene	2,088	2,902	1.34	8.0	10.7	1.30	262	271
Acetylene	2,262	3,069	1.32	10.6	11.2	1.05	219	274

Thus, approximate values for Pmax in mixtures of oxygen and in air would be given by:

$$P_{Max,O2} = [(T_{FLAME} + 273) / 261] - 1 \qquad \text{barg}$$

Thus, if the flame temperature can be calculated, from heats of formation and specific heats – with consideration of dissociations – the maximum explosion pressure for a hydrocarbon can then be calculated. It is likely that this relationship would also apply to other hydrocarbons and carbohydrates. Also, as indicated in the preceding Table 2, the values of P_{MAX} in 100% oxygen are about 30% greater than the value in air [excluding acetylene].

The data in Table 2 above can be used to estimate the Maximum Explosion Pressure [P_{MAX}] for a given concentration of oxygen as:

$$P_{MAX,C} = 0.035 \, C_{O2} + 0.92 \, P_{MAX,AIR} \qquad \text{bara}$$

The following Table presents the results of calculations, based on the above relationship, for two typical values for P_{MAX} in air [26]:

Table 3. Maximum Explosion Pressures for Fuel/Oxygen Mixtures

$P_{MAX,AIR}$	MAXIMUM EXPLOSION PRESSURES IN FUEL/OXYGEN MIXTURES (barg)							
	30 %O_2	40 %O_2	50 %O_2	60 %O_2	70 %O_2	80 %O_2	90 %O_2	100 %O_2
7 barg	7.4	7.7	8.1	8.4	8.8	9.1	9.5	9.8
8 barg	8.3	8.6	9.0	9.3	9.7	10.0	10.4	10.7
9 barg	9.2	9.6	9.9	10.3	10.6	11.0	11.3	11.7
10 barg	10.1	10.5	10.8	11.2	11.5	11.9	12.2	12.6

Thus, oxygen enrichment would have a significant effect on the extent of damage that could result from ignition of confined flammable gases and vapors, and could affect the design of explosion-protection systems and devices, such as explosion vents.

2.5 Deflagration Index, Kg

The Deflagration Index [Kg for flammable gases and vapors] is a measure of the speed of combustion and, thus, is directly related to the speed of flame through a mixture of fuel and oxidant. In the absence of strong turbulence, the flame speed is related to the fundamental burning velocity [FBV] by the ratio of the densities of the unburned mixture and the density of the combustion gases, by [27]:

$$S_{FLAME} = FBV\ (d_{UNBURNED}/d_{BURNED}) \quad \text{centimeters per second}$$

Typically, the ratio of the flame speed to FBV is about 6.

Previous investigators had observed that there is a relationship between the FBV and the Minimum Ignition Energy [MIE], as given by [28]:

$$FBV = K / MIE^{0.8} \ \text{centimeters per second}$$

Some data are available for both the FBV and the MIE, as indicated in the following Table [29] :

Table 4. The Relationship Between FBV And Oxygen Concentration

GAS/VAPOR	FBV, Air (cm./sec.)	Kg, Air (bar-m/s)	MIE, Air (millijoule)	FBV, O_2 (cm./sec.)	MIE, O_2 (millijoule)	Kg, O_2 (bar-m/s)	Kg Ratio O_2/Air
Hydrogen	312	638	0.016	1400	0.0012	3,000	4.5
Methane	45	55	0.21	450	0.0027	450	4.6
Ethane	47.5	103	0.24	375	0.0019	670	6.5
Propane	46	100	0.25	370	0.0021	720	7.2
Butane	-	92	0.25	-	0.009	-	-
Hexane	-	-	0.24	-	0.006	-	-
Heptane	42.5	-	0.24	-	-	-	-
Averages	-	-	-	-	-	-	5.7

As indicated in the above Table, the Deflagration Index – and the speed of combustion – in oxygen is almost six times greater in 100% oxygen than it is in air [at 21% oxygen]. The data in the preceding Table can be combined to yield two approximate expressions for the Deflagration Index in given concentrations in oxygen, as:

$$Kg_{,O2} = 0.04\ (K_{G,AIR})\ (C^{1.2}) \qquad \text{bar-meters/second}$$

$$Kg_{,O2} = 1.85\ C^{1.1} / MIE_{AIR}^{\ 0.46} \qquad \text{bar-meters/second [excluding } H_2]$$

Thus, if the value of the Deflagration Index in air is known, or if the Minimum Ignition Energy in air is known, an approximate value for the Deflagration Index at a given oxygen concentration can be determined.

3. Combined Relationships

The explosibility/flammability properties of a given flammable gas or vapor can be compared with a standard or typical flammable gas, through the use of Ignition Sensitivity and Explosion Severity. These property combinations were first developed and utilized for the comparisons of the hazards of coal mines, starting in 1940, where combustible dusts presented an explosion hazard [30].

3.1 Ignition Sensitivity

The product of LFL [in terms of grams per cubic meter], AIT, and MIE form the basis for a calculation of Ignition Sensitivity. A high value of this product indicates a low likelihood of ignition by relatively-weak ignition sources, such as electrostatic discharges, friction, impact, and hot surfaces. For propane in a mixture with 40% oxygen, the product is only 39 x 445 x 0.035 = 600. For comparison, the similar product for typical combustible dusts is about 45,000.

To determine the Ignition Sensitivity of a combustible dust, the LFLxAITxMIE product would be divided into the corresponding product for Pittsburgh [bituminous] Coal Dust, at 2,000,000. Thus, the Ignition Sensitivity for typical combustible dusts is about 2,000,000/45,000 = 45. By comparison, the Ignition Sensitivity for propane in a 40% oxygen mixture would be about 2,000,000/600 = 3,300, indicating the much greater sensitivity of this propane mixture to ignition sources.

As another indication of the effects of oxygen enrichment, the Ignition Sensitivity for a stoichiometric [worst-case] mixture of propane and air would be about 450. Thus, the 40% oxygen mixture would be about 7 times more sensitive to ignition [3,300/450].

3.1 Explosion Severity

The product of Deflagration Index Kg and the Maximum Explosion Pressure Pmax [in terms of barg] forms the basis for a calculation of Explosion Severity. A low value of this product indicates a low likelihood of severe damage from explosion. For propane in a mixture with 40% oxygen, the product is about 275 x 9 = 2,500. For comparison, the similar product for typical combustible dusts is about 1,500.

For combustible dusts, the product is divided by the corresponding product for Pittsburgh [bituminous] Coal Dust, at 840. Thus, the Explosion Severity for typical combustible dusts is about 1,500/840 = 1.8. By comparison, the Ignition Sensitivity for propane in a 40% oxygen mixture would be about 2,500/840 = 3.0, indicating the greater likelihood of damage if a confined mixture of propane and 40% oxygen were to be ignited.

The effect of oxygen enrichment can be shown by the Explosion Severity for a stoichiometric [worst-case] mixture of propane and air, at about 0.9. Thus, the 40% oxygen mixture could be about 3 times more damaging than an air mixture [3/0.9].

3.1 Combination of Ignition Sensitivity and Explosion Severity

If the Ignition Sensitivities of fuel/oxidant mixtures are less than 0.2 AND the Explosion Severities are less than 0.5, then such mixtures "are generally not considered to be significant explosion hazards" [31].

For the mixture of propane and 40% oxygen – for which the properties are shown above – the combination of Ignition Sensitivity and Explosion Severity is about 5 orders of Magnitude [a factor of almost 100,000] above the "no significant explosion hazard" criterion. Thus, it is unlikely that a mixture of air with any flammable or combustible vapor could be considered to present "no significant explosion hazard".

3.1 Combination of Ignition Sensitivity and Explosion Severity as Explosibility Index

The product of Ignition Sensitivity and Explosion Severity can be considered to be an Explosibility Index [32 to 37]. This Index has been found to be useful in comparing the hazards of combustible dusts, but had not yet been applied to flammable gases and vapors.

4. Graphical Presentation of Flammability/Explosibility Properties

Figure 1 presents the flammability/explosibility properties of a mixture of flammable fuel, oxygen, and nitrogen, in terms of Ignition Sensitivity [IS], Explosion Severity [ES], and Explosibility Index [EI] at atmospheric pressure and at an initial temperature of 25°C, as compared to the similar explosibility properties of propane at room temperature and atmospheric pressure. The ratios are based on the available data and the results of the calculations that are described above. As could be expected, the explosion hazards increase as the stoichiometric line – where all carbon is oxidized to carbon dioxide, and all hydrogen is oxidized to steam – is approached.

5. Conclusions

To evaluate the explosion hazards of mixtures of fuels with oxygen-enriched atmospheres, the available data were utilized to obtain methods for estimating the values of pertinent explosibility properties. The relationships are as follows:

Lower Flammable Limit	Equal to that of fuel/air mixtures	
Autoignition Temperature	$AIT_{O2}=AIT_{AIR}/[1+(2.0AIT_{AIR}/40,000)(ln(C_{O2}/21)]-273$	°C
Minimum Ignition Energy	$MIE_{O2} = 6,200\ MIE_{AIR} / C_{O2}^{\ 2.9}$	millijoule
Max. Explosion Pressure	$P_{Max,O2} = [[0.035\ C_{O2}) + (0.92\ (P_{Max,AIR} + 1))] - 1$	barg
Deflagration Index	$Kg_{,O2} = 0.04\ (\ K_{G,AIR}\ )\ (\ C^{\ 1.2}\ )$	bar-meters/second
	$Kg_{,O2} = 1.85\ C^{\ 1.1} / MIE_{AIR}^{\ 0.46}$	bar-meters/second

In addition, a method is presented for evaluating the Ignition Sensitivity and the Explosion Severity of fuel/oxygen mixtures, using the above relationships, together with a method for graphically showing the relative hazards of such mixtures.

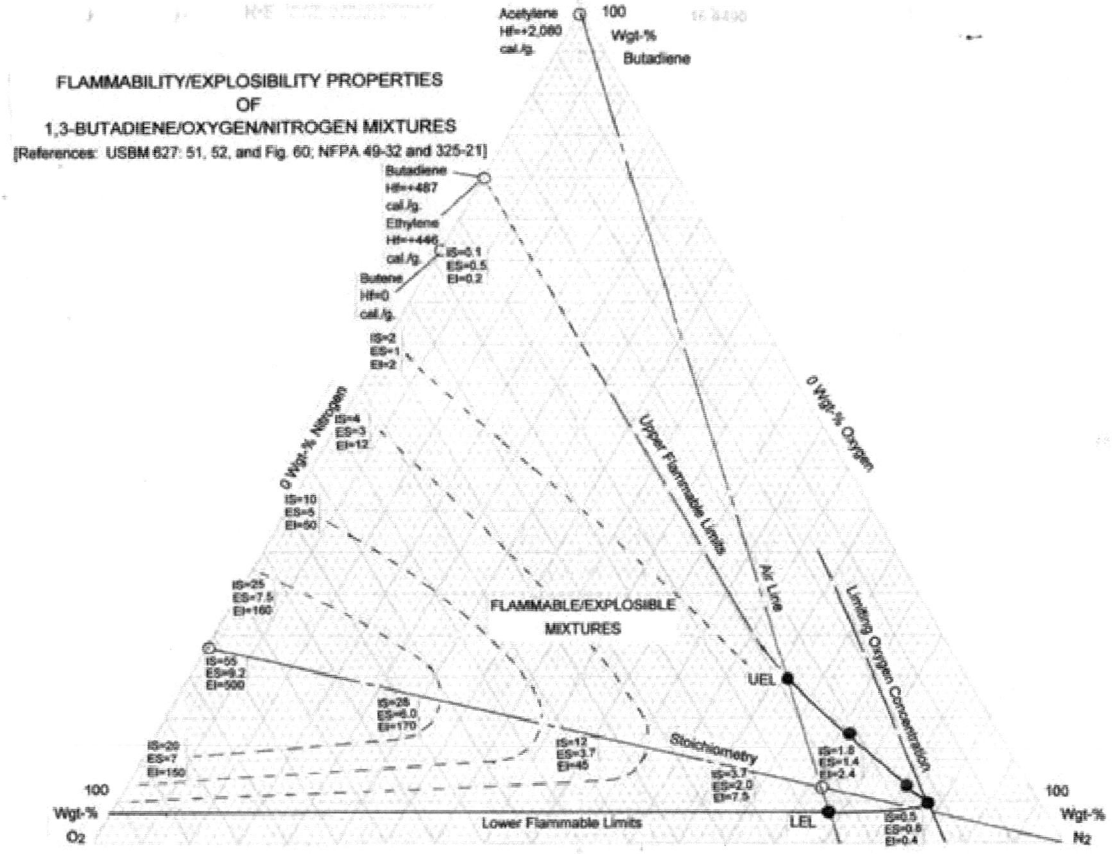

Figure 1 – Weight Concentrations of Gas Mixtures
with
Ignition Sensitivities, Explosion Severities, and Explosion Index
as compared to the properties of Propane

6. References.

1. Urben, P. G. [editor], "Bretherick's Handbook of Reactive Chemical Hazards", 6th Ed., Vol. 2, Page 287 (1999).

2. National Fire Protection Association, "Fire Protection Guide to Hazardous Materials", 13th Ed., including NFPA 49 and NFPA 325 (2002).

3. National Fire Protection Association, "Guide for Venting of Deflagrations", NFPA 68, Table D.1 (2002).

4. National Fire Protection Association, "Standard on Explosion Prevention Systems", NFPA 69, Table C.1(a) (2014).

5. National Fire Protection Association, "Recommended Practice on Static Electricity", NFPA 77, Table B.1 (2014).

6. National Fire Protection Association, "Recommended Practice for the Classification of Flammable Liquids, Gases, or Vapors and of Hazardous (Classified) Locations for Electrical Installations in Chemical Process Areas", NFPA 497, Table 4.4.2 (2017).

7. Coward, H. F., and G. W. Jones [U. S. Bureau of Mines], "Limits of Flammability of Gases and Vapors", Bulletin 503, Table 44 [and many "nose" Figures] (1952).

8. Zabetakis, M. G. [U. S. Bureau of Mines], "Flammability Characteristics of Combustible Gases and Vapors", Bulletin 627, Table A-1 [and many "nose" Figures] (1965).

9. Babrauskas, V., "Ignition Handbook", Table 1A (1992).

10. Britton, L. G., "Avoiding Static Ignition Hazards in Chemical Operations", Page 223 (1999).

11. Crowl, D. A., and J. F. Louvar, "Chemical Process Safety", 3rd Ed., Pages 251, 259, 261, 287, and 654 (2011).

12. Yaws, C. L., "Chemical Properties Handbook", Table 25-1 (1999).

13. National Fire Protection Association, "Recommended Practice for the Classification of Combustible Dusts and of Hazardous (Classified) Locations for Electrical Installations in Chemical Process Areas", NFPA 499, Paragraph A.5.2.3 (2017).

14. Frank, W. L., and S. A. Rodgers, "NFPA Guide to Combustible Dusts", Page 92 (2012).

15. Britton, L. G., "Avoiding Static Ignition Hazards in Chemical Operations", Page 223 (1999).

16. Babrauskas, V., "Ignition Handbook", Page 1023 (1992).

17. Zabetakis, M. G. [U. S. Bureau of Mines], "Flammability Characteristics of Combustible Gases and Vapors", Bulletin 627, Page 9 (1965).

18. Crowl, D. A., and J. F. Louvar, "Chemical Process Safety", 3rd Ed., Pages 263, 265 to 267, and 269 (2011).

19. Crowl, D. A., and J. F. Louvar, "Chemical Process Safety", 3rd Ed., Pages 398 and 400 (2011).

20. Babrauskas, V., "Ignition Handbook", Pages 1023 and 1024 to 1058 (1992).

21. Green, D. W., and M. Z. Southard, "Perry's Chemical Engineers' Handbook", 9th Ed., Page 19-24 (2019).

22. Babrauskas, V., "Ignition Handbook", Pages 75, 76, 203, and 625 (1992).

23. Babrauskas, V., "Ignition Handbook", Pages 1024 to 1055 (1992).

24. Babrauskas, V., "Ignition Handbook", Page 625 (1992).

25. Lewis, B., and G. von Elbe, "Combustion, Flames, and Explosions of Gases", 3rd Ed., Pages 717 and 719 (1987).

26. National Fire Protection Association, "Guide for Venting of Deflagrations", NFPA 68, Table D.1 (2002).

27. Babrauskas, V., “Ignition Handbook”, Page 35 (1992).

28. Babrauskas, V., “Ignition Handbook”, Page 74 (1992).

29. Zabetakis, M. G. [U. S. Bureau of Mines], “Flammability Characteristics of Combustible Gases and Vapors”, Bulletin 627, Page 43 and Figures 44 and 45 (1965).

30. Nagy, J., et al. [U. S. Bureau of Mines], “Explosibility of Carbonaceous Dusts”, Report of Investigations 6597, Page 2 (1965).

31. National Fire Protection Association, “Recommended Practice for the Classification of Combustible Dusts and of Hazardous (Classified) Locations for Electrical Installations in Chemical Process Areas”, NFPA 499, Paragraph A.5.2.3 (2017).

32. Dorsett, H. G., et al. [U. S. Bureau of Mines], Laboratory Equipment and Test Procedures for Evaluating Explosibility of Dusts”, Report of Investigations 5624, Page 2 (1960).

33. Jacobson, M., et al. [U. S. Bureau of Mines], “Explosibility of Dusts used in the Plastics Industry”, Report of Investigations 5971, Tables A-1, A-2, and A-3 (1962).

34. Jacobson, M., et al. [U. S. Bureau of Mines], “Explosibility of Metal Powders”, Report of Investigations 6516, Tables A-1, A-2, and A-3 (1964).

35. Nagy, J., et al. [U. S. Bureau of Mines], “Explosibility of Carbonaceous Dusts”, Report of Investigations 6597, Page 2 (1965).

36. Dorsett, H. G., and J. Nagy [U. S. Bureau of Mines], “Dust Explosibility of Chemicals, Drugs, Dyes, and Pesticides”, Report of Investigations 7132, Tables A-1, A-2, and A-3 (1968).

37. Grossel, S. S., “Safety Considerations in Conveying of Bulk Solids and Powders”, Page 62 in the “Journal of Loss Prevention in the Process Industries”, Butterworths, Vol. 1, No. 2 (April 1988).

7. Legal Disclaimer and Liability

(a) Limitation of Liability. The consulting services conducted by DEKRA Process Safety (the "Company") were performed using generally accepted guidelines, standards, and/or practices, which the Company considers reliable. Although the Company performed its consulting services pursuant to reliable and generally accepted practices in the industry, the Company does not guarantee or provide any representations or warranties with respect to Client's use, interpretation or application of the findings, conclusions, and/or suggestions of the consulting services provided by the Company. Moreover, the findings, conclusions, and the suggestions resulting from the consulting service are based upon certain assumptions, information, documents, and procedures provided by the Customer. AS SUCH, IN NO EVENT AND UNDER NO CIRCUMSTANCE SHALL THE COMPANY BE LIABLE FOR SPECIAL, INDIRECT, PUNITIVE OR CONSEQUENTIAL DAMAGES OF ANY NATURE WHATSOEVER, INCLUDING WITHOUT LIMITATION, ANY LOST REVENUE OR PROFITS OF THE CUSTOMER OR ITS CUSTOMERS, AGENTS AND DISTRIBUTORS, RESULTING FROM, ARISING OUT OF OR IN CONNECTION WITH, THE SERVICES PROVIDED BY THE COMPANY. The Customer agrees that the Company shall have no liability for damages, which may result from Client's use, interpretation or application of the consulting services provided by the Company.

(b) The Company's pricing of the consulting services provided does not contemplate that the Company shall have any liability resulting from its performance of the consulting services, except as otherwise set forth in the Quotation from the Company. Accordingly, the Customer shall indemnify and hold harmless the Company, its shareholders, directors, officers, employees and agents (the "Indemnified Parties") from and against any and all loss, cost, liability and expense, including reasonable attorney's fees and costs, which any of the Indemnified Parties may incur, sustain or be subject to, as a result of any claim, demand, action, investigation or proceeding arising out of or relating to either: (a) the consulting services provided by the Company; or (b) any material, equipment, specifications or safety information (or lack thereof) supplied to the Company (or which should have been supplied to the Company) by Customer and/or any failure of such materials, equipment, specifications and safety information to comply with any federal, state or local law or safety standard.

(c) For additional terms and conditions, which apply with respect to the provision of this report, see the Quotation provided by the Company and executed by Customer. If any terms set forth in the Quotation conflict with the terms set forth herein, the terms set forth herein shall apply.

Bridging Chemical Engineering (Risk Based Process Safety) to the Humanities – 7 Mega Elements' Approach (Protocol) of Accident Investigation in Light of Ethical Dimension, Architecture & Epistemology

Long(Steven) Zhang
No affiliation since Jul 2014 off-hiring from COP
11455 Bissonnet St Apt #514, Houston TX 77099
longz.houston@gmail.com

Prepared for Presentation at
American Institute of Chemical Engineers
2019 Spring Meeting and 15th Global Congress on Process Safety
New Orleans, LA
March 31 – April 3, 2019

Bridging Chemical Engineering (Risk Based Process safety) to the Humanities – 7 Mega Elements' Approach (Protocol) of Accident Investigation in Light of Ethical Dimension, Architecture & Epistemology

Long(Steven) Zhang
No affiliation since Jul 2014 off-hiring from COP
11455 Bissonnet St Apt #514, Houston TX 77099
longz.houston@gmail.com

Keywords: Investigation, Diagnosis, Humanities, *the Art of War, Narrative, Epistemology*

Abstract

In this paper, facing the challenges of Incident Investigation, actually whole CCPS 4th Pillar "Learn from Experience", chemical engineers' (process safety practitioners') brand new vocabulary (7 Mega Elements' narrative on failure diagnosis) would be presented in context of bridging chemical engineering (RBPS) to the Humanities, of which 3 core topics i.e. ethical dimension, architecture and epistemology would be explored and reflected upon so that a new era of RBPS diagnosis (Learn from Experience) predicted.

1 Introduction

In 2007 Center for Chemical Process Safety (CCPS) unprecedentedly presented to process industry Risk Based Process Safety (RBPS) 4 Pillars (and relevant 20 Elements), among which 4th Pillar "Learn from Experience" consists of 4 Elements: Incident Investigation, Measurement & Metrics, Auditing, and Management Review & Continuous Improvement[1].

In latest decades chemical engineers (process safety practitioners) have been facing some big challenges from accident/incident investigations, actually from whole CCPS 4th Pillar (Learn from Experience), for which there might have been many efforts, discussions, and even some 'solutions'. The author argues, however, the challenges comes not only from the needs of high quality and consistency of accident investigation, but, more fundamentally, from building up a brand new vocabulary (protocol) for chemical engineers to competently communicate of, and negotiate on, accident investigations or "why they occur and how ultimately to prevent".

Toward building a new vocabulary/protocol/narrative on accident investigation or, more broadly, on how Experience could be truly Learned (CCPS 4th Pillar), since 9th Global Congress, the author has had a series of efforts for exploration and reflections on the possible way forward, e.g. RBPS Core value (landlord for temple), Process Safety Cultural Communication (landlord's livelihood) & Egg model[2]; 4th Pillar as "espionage" not to enemies but ourselves[3][4]; 7 Mega Elements & 28 infra cultural activities (practices); RBPS Socio-Technical systems[5]; an ethical dimension of Vision 20/20[6] etc.. And in broader context of bridging chemical engineering (Risk Based Process Safety) to the Humanities, the author finally arrived at 7 Mega Elements' vocabulary/ protocol/ narrative/ "cosmology" on RBPS diagnosis for chemical engineers (process safety practitioners)[7].

2 7 Mega Elements' Protocol – Chemical Engineers' Vocabulary

Louis L. Bucciarelli claimed that he had taken seriously the "two cultures" challenge set by C. P. Snow and been working at bridging engineering to the Humanities for 35 years[8]; instrumental rationality does not suffice itself in engineering design, diagnosis and education[9]. And the author's experiences during the humble quest (through relevance of anthropology, intercultural communication, *the Art of War* etc.) for an ultimate solution to catastrophic (process safety) accidents (chemical engineering failure) seems not only resonating with Bucciarelli's bridging journey but a little bit unique, not only toward chemical engineering (Risk Based Process Safety) but also to engineering discipline at large, i.e., through benchmarking RBPS approach against Sun Tzu's *the Art of War,* 7 Mega Elements (vocabulary/ protocol/ narrative) not only were discovered for (chemical) engineers but function as 7 fundamental (chemical) engineering thoughts, which eventually sheds light on those entailed 28 practices of (chemical) engineering discipline.

Unlike Bucciarelli who tried to bridge philosophy and engineering, the author proceeded another way, i.e. first making relevance of anthropology & intercultural communication to chemical engineering (Risk Based Process Safety), then benchmarking Egg model against *the Art of War*, what the author calls 1st main ancient source for bridging (chemical) engineering to the Humanities, compared to 1st main modern one - Bucciarelli's *Engineering Philosophy.*

When guiding all visitors to some biggest lessons ("more original" two cultures) of ancient cultural traditions or civilizations of east and west, the author claimed that there should be an landlord (core value) to live within the temple building (CCPS 4 Pillars) and, furthermore, the landlord should have livelihood (Process Safety Cultural Communication (PSCC)) as well, and that should the landlord died the building seems still beautiful outwardly, however, with a purpose of tomb. In another words, when (chemical) engineers' vocabulary/ protocol/ narrative becomes deficient, or landlord and his livelihood are absent in the vocabulary, catastrophic (process safety) accidents are bound to happen, no matter how technically over competent relevant chemical engineers (process safety practitioners) might be, e.g., the lesson from the case of Bian Que–Duke Cai.

Through anthropology and intercultural communication, **RBPS Core Value** was defined in two ways, an unfolded one: *RBPS Core Value functions as the root of the tree of Process Safety*

Culture. Deficiencies are symptoms or signals for a dying or dead state of a core value, which means that the organization has strategically failed at the highest levels in managing process safety through a risk based approach – even if there has been no occurrence of process safety accidents till that moment. In another words, the RBPS temple is bound to collapse, sooner or later, due to the change or drift from the original design intent of the temple. And RBPS Core Value's simpler definition, through the concept of Power Distance in value studies: *there is no power, authority or governance on process safety except on behalf of process safety expertise (PSE).* (PSE means the divine essence of the highest strategic insights and vision of process safety management; the Tao for process safety culture; and the soul of *The Art of War* against catastrophic process safety accidents)

And process safety culture does not exist except through ***PSCC****, whose ultimate purpose is to convey the message of RBPS Core Value. Without nourishment from RBPS Core Value, the tree of Process Safety Culture or PSCC would get dry and dead.* So the temple building was eventually upgraded to Egg model, as showed within the scheme 1st. And these 2 new universes (or 2 Mega Elements) might be deemed what Bucciarelli called the rhetoric of engineering as the parcel of instrumental explanation[10], and probably 2 most fundamental (chemical) engineering thoughts as well.

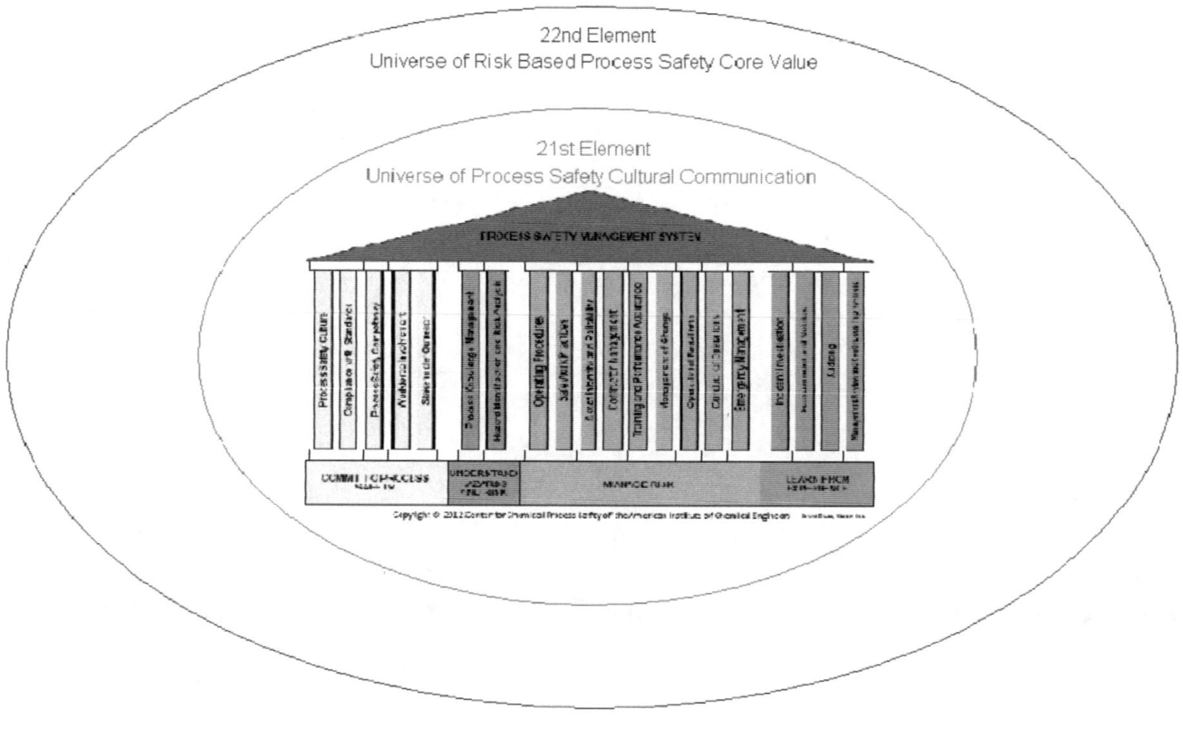

Egg Model (CCPS Temple + 2 universes) - Scheme 1st

The Temple (4 pillars) Figure Source: CCPS, Process Safety Boot Camp, contributed by David Guss, Nexen Inc. Copyright 2010 AIChE and used with permission

Then, in spite of the formidable strengths of most, actually 17, elements of CCPS 4 Pillars, except Process Safety Culture, Process Safety competency & Stakeholder Outreach, the outcome of benchmarking Egg model (4 Pillars + 2 Universes) against *The Art of War,* what the author calls a matured ancient vocabulary on engineering (military) diagnosis, not only shows that those 4 Pillars' (20 elements') narrative, even together with the new 2 Universes, would probably be not so architecturally organized at negotiation and communication on why process safety (chemical engineering) accidents (failure) occur & how to prevent, but makes the author further believe that Sun Tzu's vocabulary (6 Mega Elements, refer to Table 1st) might potentially become a narrative/ protocol paradigm for future (chemical) engineering diagnosis and *The Art of War* an main ancient source of bridging engineering to the Humanities.

For the fact that, however, we could not neglect the historical context of narrative/ protocol of *The Art of War*, the author, in light of modern globalization and global operations, fundamentally redefined Process safety Culture into Process Safety Cultural Communication (PSCC) as the 7th Mega Element, and upgraded another 3 ones i.e. Leadership, Weather/Climate & Espionage as below:

Leadership: a new chain of command, inherently from RBPS Core Value, rather than blindly from superiors;

Weather/Climate: RBPS Core Value enlightened outreach to the stakeholders & community, rather than only tactically maintaining seemingly good relationship with community; &

<table>
<tr><th colspan="2">RBPS Egg</th><th>The Art of War</th><th>Philosophy of Process Safety</th><th></th></tr>
<tr><td colspan="4">RBPS Core Value - Tao - RBPS Worldview</td><td>Worldview</td></tr>
<tr><td rowspan="4">Commit to Process Safety</td><td>Process Safety Competency</td><td rowspan="2">Leadership</td><td rowspan="2">People</td><td rowspan="7">Process Safety Cultural Communication (Process Safety Culture) RBPS Methodology</td></tr>
<tr><td>Workforce Involvement</td></tr>
<tr><td>Compliance with Standards</td><td rowspan="2">Weather/Climate</td><td rowspan="3">Plant</td></tr>
<tr><td>Stakeholder Outreach</td></tr>
<tr><td>Understand Hazards & Risk</td><td></td><td>Terrain</td></tr>
<tr><td>Manage Risk</td><td></td><td>Discipline</td><td>Procedure</td></tr>
<tr><td>Learn from Experience</td><td></td><td>Espionage</td><td>Engineering & Humanities</td></tr>
</table>

Table 1st

Espionage: 4th Pillar *Learn from Experience* (4 elements: *investigation, metrics, audit & review*) plays role "espionage", not to the enemies but to ourselves our organizational "national security"

And 7 Mega Elements might also be deemed as 7 (chemical) engineering thoughts.

3 Ethical Dimension (2 Universes) - What Chemical Engineers Count & How

While declared *there is no power, authority or governance on process safety except on behalf of Process Safety Expertise (PSE)* the author claimed that PSE shall eternally and exclusively chair at the highest strategic levels throughout all the processes of decision making and behavior determining, i.e. chemical engineers (process safety practitioners) need getting inspired from RBPS Core Value and persistently conveying the message of RBPS Core Value, both of which make up the ethical dimension (of 7 Mega elements' vocabulary/ protocol/ narrative), probably what Bucciarelli calls "best left sacred"[11], which ensures, or anthropologically designs, all relevant decision making and behavior determining correct. In this sense, at highest strategical levels, chemical engineers do not design at all but RBPS Core Value does.

From philosophical perspective, Bucciarelli claimed that the tensions at the interfaces between different design participants could not be solved by instrumental means alone [12]. And imitating American anthropologist Edward T. Hall's saying the author ever described process safety culture as a prison, unless we have RBPS Core Value as the key to unlock it and PSCC the brightness of the key. And Bucciarelli pointed out that *designing is a social process of negotiation, of iteration, of rectifying mis-steps, even misconceptions – a process rich in ambiguity and uncertainty* [13]. Similarly, the author in metaphor described PSCC as the photosynthesis of tree of process safety culture, the impulses from the heart (RBPS Core Value) and nourishment (highest designing principles) transportation from the root (RBPS Core Value). What does Bucciarelli mean by that 'social process'? What does he mean by "contribute to a sense of community among the citizenry" [14]? And what does the author mean by "photosynthesis" and "nourishment"? With hesitation & obligations, the author claims that the said social process, what the author puts PSCC, traditionally should be called Constitutional Democracy; and RBPS Core Value a nickname for Universal Value in chemical engineering.

As Universal Value & Constitutional Democracy in (chemical) engineering seem more subtle than those traditional ones the author here might call them Universal Value & Constitutional Democracy 2.0, what Bucciarelli described "old themes but in a new light" [15]. And this 2.0 might be acknowledged as two fundamental (chemical) engineering thoughts and the ethical dimension of (chemical) engineers' vocabulary/ protocol/ narrative on accident investigation and whole (chemical) engineering diagnosis (4th Pillars) as well. And probably only in this way could (chemical) engineers truly acquire what Bucciarelli called "new ways of perceiving and reading the world" [16], eventually leading the industries and communities into a new enlightened era of accident investigation and (chemical) engineering diagnosis. And the author might further states that, as the ethical dimension of (chemical) engineers' vocabulary (protocol), Universal Value & Constitutional Democracy 2.0 should function as the highest communication (designing) principles on accident investigation and (chemical) engineering diagnosis, and on accident prevention (foreknowledge) as well.

4 7 Mega Elements' Architecture – 28 Practices' Diagnosis Map

Inspired from Edward T. Hall's map of culture, the author developed a map of 28 infra cultural activities of Process Safety Culture (refer to 7 ones in the diagonal from upper left to lower right and 21 ones above), what Bucciarelli might like to call another, slightly different, "more complex vision of engineering thoughts and practices" [17]. And each infra cultural activity was defined as the critical statements & behaviors related to the interaction between Mega Element A and Mega Element B, e.g. infra activity 04 means: verbally & behaviorally communicate that the actual death of RBPS Core Value might be one of the root causes of accidents. And those 7 ones, extremely, are in fact of the critical verbal statements & behaviors related to the definition of the Mega Element (or engineering thought) concerned, e.g. infra activity 22: verbally and behaviorally embrace and reflect on Leadership - the engineering thought of new chain of command inherently from RBPS Core Value rather than blindly from superiors.

This matching between verbal statements & behaviors, or the inherent relationship between the message of RBPS Core Value and all relevant decision making & behavior determining, seems some resonance with Wittgenstein's recognition that rules and meanings (what the author calls ***tone***) in talk and texts (what the author calls vocabulary/ protocol/ narrative) were mixed with practice [18], or with what Bucciarelli put another way "the meaning of statements is embedded in practice" [19]. So these 28 infra cultural activities' map, an unfolded version of what & how chemical engineers (process safety practitioners) count, might accommodate as highest strategic encompassing diagnosis protocol (map).

	RBPS C.V. 0	P.S.C.C. 1	Leadership 2	Weather/C. 3	Espionage (Diagnosis) 4	Terrain 5	Discipline 6
RBPS Core Value 0	00	01	02	03	04	05	06
Process Safety Cultural Communication (PSCC) 1	10	11	12	13	14	15	16
Leadership 2	20	21	22 4 dimensions	23	24	25	26
Weather/ Climate 3	30	31	32	33	34	35	36
Espionage (Diagnosis) 4	40	41	42	43	44	45	46
Terrain 5	50	51	52	53	54	55	56
Discipline 6	60	61	62	63	64	65	66

28 Infra Cultural Activities' Diagnosis Map (Table 2nd)

As the Humanities e.g. philosophy, anthropology & intercultural communication, Sun Tzu's *The Art of War* etc. sheds more lights on the old themes (i.e. RBPS Core value, PSCC, Leadership and Weather/Climate), upon this Diagnosis Map, "another" complex vision of engineering thoughts and practices, the author claims that *the yellow grounded, or more exactly red number lettered area, are all about organizational "national security", and that the infra activity 44 (Espionage)'s 2nd rationality, next to the instrumental one, whose protocol area is painted green, SHALL be reactivated so that "espionage" to our organization would shift to a more proper name - "counter-intelligence", on accident investigation & prevention and on engineering diagnosis at large.*

Through maintaining the architecture/protocol of 7 Mega Elements could all these 22.5 infra activities of "national security" area be reliably visible, open and "problematic" over there, together with "traditional" green area (5.5 infra activities), to critique and diagnosis among board of directors, executive management, operation managers, strategy, legal, HR, chemical engineers (process safety practitioners), and even beyond. Pragmatically, from designing perspective, chemical engineers should pursue this more matured architecture/protocol/vocabulary, getting inspired from RBPS Core Value, persistently conveying its message and maintaining alive this complex vision of engineering thoughts & practices i.e. this 28 Practices (infra activities)' Diagnosis Map, otherwise the insightful topic "social construction of failure/ accident" [20] would turn out to be a truism. In another words, either RBPS Core Value highest strategically designs safety (and tranquility) through PSCC, or otherwise (chemical engineering/ process safety) failures/ accidents/ catastrophes are socially constructed.

5 Epistemology – Future (Chemical) Engineers' Knowledge Hierarchy

Walter Vincenti's understanding of *engineering knowledge,* as J. C. Pitt ever explained*, is task-specific and aims at the production of an artifact to serve a predetermined purpose* [21], of which, the author believes, CCPS 4 Pillars (20 Elements), as of (chemical) engineering/RBPS knowledge, might have been satisfactorily reflecting the main characters. Bucciarelli discovered, however, that engineering knowledge, of which he described *what engineers do, the reasons they do what they do, and their justification for believing the latter are good and true,* is *historically and culturally contingent, not fixed by the dictates of science and empirical fact alone.* [22]. The author might explain here, echoing the former statement "the landlord … even to buy a new house with new style of design to live within", that the (chemical) engineering knowledge we have here with CCPS 4 Pillars might not be the single unique design/house for the landlord [23], but the landscape (landlord, livelihood & house) of the contingencies, what Bucciarelli calls "a picture of both possibilities and limitations of engineering thinking" [24], is unique. In another words, both the landlord (RBPS Core Value), deeply rooted in most cultural traditions and civilizations, and his livelihood (PSCC), what the author likes to call Constitutional Democracy, should preside the landscape of the contingencies and chair and guide at the highest strategic levels the whole engineering design process, for surety including on accident investigation and diagnosis, so that the objects of design, which "traditionally" have been confined to artifacts e.g. a mega petrochemical project/ operations, might also need to include all processes of decision making and behavior determining (refer to Table 2nd – 28 Infra Activities' Diagnosis Map).

The author might state that *a complete (chemical) engineering design process, including on accident investigation, diagnosis & prevention, shall consist of 3 components: RBPS Core Value (Universal Value), PSCC (Constitutional Democracy) and the application of (chemical) engineering knowledge (e.g. one with CCPS 4 Pillars).* And at same time, we recognize that *future (chemical) engineers need a new epistemology - a new hierarchy of knowledge which consists of Universal Value Knowledge, Constitutional Democratic Knowledge & the relevant (chemical) engineering knowledge.* Edward T. Hall had an earth-breaking saying: *culture is communication and communication is culture* [25], from which the author got inspired and developed the concept of PSCC. When (chemical) engineers verbally (of text) and behaviorally (of practice) abandon the ethical dimension of 7 Mega Elements' vocabulary/ protocol/ narrative, they lose the whole process safety culture or, we could say, both their knowledge hierarchy is deficient, akin to a genocide, and their engineering design process seriously flawed, like building concentration camps.

6 Conclusion

In context of bridging chemical engineering to the Humanities, on accident investigation, diagnosis and prevention, and on engineering design process at large, the author presents 7 Mega Elements' Vocabulary/ Protocol/ Narrative, 28 Infra Cultural Activities' Diagnosis Map and a newly galvanized old Hierarchy of Knowledge, so that some outlooks might be made: a) as one of the largest communities of engineering professions, AIChE would pursue leading roles at dialogues between engineering & the Humanities; b) some small groups of chemical philosopher/anthropologist-engineers would emerge; c) an enlightened era of (chemical/process safety) accident investigation, diagnosis & learning, and engineering design process at large, would arrive; & d) engineering education's future would get brighter than ever.

7 References

[1] Center for Chemical Process Safety, *Guidelines for Risk Based Process Safety*, John Wiley & Sons, Inc., 2007

[2] Zhang, L., An Upgraded Definition of Risk Based Process Safety Core Value and Its Impact in Light of Anthropology & Intercultural Communication on the Future Evolution of Process Safety Culture & Management, 2013 AIChE Spring Meeting and 9th Global Congress on Process Safety, San Antonio, April 2013

[3] Sun-Tzu and Ames, Roger T., Sun-tzu: *The Art of Warfare*, Ballatime Books, 1993

[4] Zhang, L., Benchmarking Risk Based Process Safety Egg (RBPS Temple+2 Universes) against The Art of War, 1st CCPS Asia-Pacific Conference on Process Safety, Qingdao, Sept 2013

[5] Zhang, L., People Means Leadership, NOT Simply mean People – 4 New Dimensions of Process Safety Competency, 2014 AIChE Spring Meeting and 10th Global Congress on Process Safety, New Orleans, April 2014

[6] Zhang, L., How Committed to Process Safety is Committed Enough? – Doing Process Safety Cultural Communication (PSCC) as the Inevitable Ethical Dimension for Vision 20/20, 2016 AIChE Spring Meeting and 12th Global Congress on Process Safety, Houston, April 2016

[7] Zhang, L. and Wasel, W., Corporate Memory Loss Ongoing – Except CCPS 4th Pillar (Learn from Experience) as Espionage, 2018 AIChE Spring Meeting and 14th Global Congress on Process Safety, Orlando, April 2018

[8] Bucciarelli, L. L., *Engineering Philosophy*, Delft: DUP Satellite, 2003, p.2

[9] Ibid. p.99

[10] Ibid. p.6

[11] Ibid. p.4

[12] Ibid. p.4-5

[13] Ibid. p.7

[14] Ibid. p.5

[15] Ibid. p.2

[16] Ibid. p.7

[17] Ibid. p.101

[18] Ibid. p.6

[19] Ibid. p.6

[20] Ibid. p. 101

[21] Pitt, J. C., "What Engineers Know", *Techne* 5:3, Spring, 2001, p.25

[22] Bucciarelli, *Engineering Philosophy*, p.99

[23] Pitt, "What Engineers Know", p.26

[24] Bucciarelli, *Engineering Philosophy*, p.99

[25] Hall, E. T., *The Silent Language*, Anchor Books, 1990, p.186

548354 Applying PSM in Vaca Muerta Shale Oil & Gas Operations: The Experience

How to develop, implement and sustain Process Safety in greenfield shale oil and gas projects? This has been a challenge to many new ventures.

Although conventional oil operations have been in place for decades, the exploration, drilling and fracking operations in unconventional shale fields are relatively new in the World. Most of the projects have kicked off less than eight years ago and there are some aspects of the operations that are unique to unconventional fields.

The objective of this paper is to share with the Process Safety global community, how a fit for purpose Process Safety program was developed and is being implemented in the Upstream Unconventional shale oil and gas YPF & Chevron joint venture in Vaca Muerta / Argentina. The project started in 2013, has currently 14 D&C rigs, 5 fracking sets, several separation and treatment production facilities, 3,500 employees and is expected to grow even more in the next 5 years. The paper describes how the project team developed the idea, planned the steps, engaged all stakeholders and selected the key elements of the Process Safety Management System that would add more value in each stage of the project, including drilling, fracking and oil and gas production. The paper also explains the key challenges, such as the difficulties faced to instill Operational Discipline and other key cultural and technical aspects related to Contractor field Operation crews. The journey to implement Process Safety in the Unconventional shale oil & gas operations is very dynamic and is in progress, however, there are already several positive aspects as well as pitfalls to take away as a lesson learned.

548596 A Survey of Process Safety Workers' Attitudes, Behaviors, and Outcomes: Moving Beyond Human Error and Toward a Systems Model

In an effort to move beyond the "human error" explanation for safety incidents, we explored issues with procedures from the worker's point of view by using an anonymous, holistic survey developed from interviews with then currently employed operators. The current sample (*N* = 174) included individuals employed in the process safety industry (MKO/Advanced Procedures Initiative) and were primarily from the Oil & Gas and Chemical industry. The survey was deployed over the course of approximately four weeks. Twelve distinct constructs emerged from the survey (e.g., perceptions of procedure quality, procedure deviation, attitudes toward the procedure change process, etc.). Results indicated that perceptions of procedure quality was the focal variable in all analyses including positive relationships with attitudes toward the procedure change process and negative relationships with procedure deviations, and safety incidents and near-misses. We conclude that these elements are important factors to consider when evaluating and developing procedure systems and provide additional information beyond what the industry considered paramount.

546680 A Method of the Evacuation Pathway Optimization Under Toxic Gas Releases in Chemical Plants

Toxic gases release in chemical plant is a key factor contributing to major accidents and may cause tremendous human life and economic losses. A scientific planning evacuation pathway is an important guarantee of reducing the losses of accidents. Many studies regarding evacuation planning in chemical plant have been carried out by simulating the accident scenarios. However, at present, a systemic method of evacuation pathway optimization has not been developed, and the effectiveness of emergency preparedness has been compromised. To address this problem, an approach based on the CFD modelling of hazardous gas release consequences to plan evacuation pathways and appropriate shelters is proposed in this paper. The methods stochastic programming and dose-response model are mainly employed in this paper. The toxic gas concentration can be converted into poisoning fatality probabilities by dose-response model. Then, stochastic programming method is used to select all the optimal pathways from the evacuation point to the safe area. Setting the minimal fatality probability as the optimization goal, the optimized evacuation pathways under all the leakage scenarios are finally determined through solving the stochastic programming model. As for places with limited evacuation opportunities, an approach of setting up a safe room is proposed in this paper and relevant design guidance are put forward in this paper considering the detailed situation of the chemical plant. A case study of the optimization of evacuation pathways and safe room setup in a salt chemical plant that contains 88 leakage scenarios of chlorine leakage is investigated. With the proposed methodology, the optimized evacuation pathway and the appropriate safe room can provide scientific references for decision makers and evacuees on reducing losses when toxic gases release in chemical plants.

19 AIChE Spring Meeting
15TH GLOBAL CONGRESS ON PROCESS SAFETY
NEW ORLEANS • 2019

Four Major Gaps That Are Preventing Most Companies Worldwide from Achieving Excellent Process Safety Performance

William G. Bridges
Process Improvement Institute Inc. (PII)
1321 Waterside Lane, Knoxville, TN 37922 USA
wbridges@piii.com

Jeff Thomas
Process Improvement Institute Inc. (PII)
jthomas@piii.com

PROCESS IMPROVEMENT INSTITUTE

Prepared for Presentation at
American Institute of Chemical Engineers
2019 Spring Meeting and 15th Global Congress on Process Safety
New Orleans, LA
March 31 – April 3, 2019

FOUR MAJOR GAPS THAT ARE PREVENTING MOST COMPANIES WORLDWIDE FROM ACHIEVING EXCELLENT PROCESS SAFETY PERFORMANCE

William G. Bridges
Process Improvement Institute Inc. (PII)
wbridges@piii.com

Jeff Thomas
Process Improvement Institute Inc. (PII)
jthomas@piii.com

Keywords: process safety, PSM, human factors, near misses, PHA, HAZOP, procedures

Abstract

Process safety requires implementing many management systems, specific engineered features, and operating and maintenance practices effectively. Most companies believe they have done just that, and yet major accidents continue to occur. Why is that? What is missing?

This paper looks at the statistics of major accidents, combined with results from audits and assessments from more than 50 chemical, petrochemical, oil/gas, and related processing companies world-wide. The paper illustrates the four major gaps that are common to the companies/sites that keep having major accidents, compared to those companies/sites that do not have such accidents:

- Accurate and clear operating and maintenance work instructions (procedures)
- PHA of all modes of operation, especially startup, shutdown, abnormal operations, and online maintenance modes
- Getting Near Misses reported and investigated
- Addressing human factors that are missing from most management systems

The paper also provides guidance for closing each of these gaps.

Introduction

"Process safety management (PSM) is the application of management principles to the identification, understanding, and control of process hazards to prevent process-related incidents"[1]. PSM entails development and implementation of programs or systems to ensure that the practices and equipment used in hazardous processes are adequate and are maintained appropriately. The primary categories of programs or systems have come to be called elements of PSM. However, the basic elements of PSM have been defined by many groups in a number of ways. Table 1 lists the elements of PSM systems from various industry and government groups. Many of the elements with different names have essentially the same meaning. For instance, "maintenance and inspection of facilities," together with some aspects of "personnel" practices, both under American Chemistry Council's (ACC's, formerly CMA's) Process Safety Code of Responsible Care™[2] are essentially the same as the single element, "mechanical integrity," under 29 CFR 1910.1190) [3].

Table 1. Comparison of PSM systems

OSHA 29 CFR 1910.119 EPA 40 CFR 68	AIChE/CCPS Risk-Based Process Safety (RBPS) Standard	Responsible Care© Process Safety Code
Management System	**Commitment to Process Safety**	**Management Leadership**
Employee Participation	Process Safety Culture*	Commitment*
Process Safety Information	Compliance with Standards	Accountability*
Process Hazard Analysis*	Process Safety Competency*	Performance Measurement
Operating Procedures*	Workforce Involvement*	Incident Investigation
Training*	Stakeholder Outreach	Information Sharing
Contractors	**Understand Hazards and Evaluate Risk**	CAER Integration
Pre-Startup Safety Review	Process Knowledge Management	**Technology**
Mechanical Integrity	Hazard Identification and Risk Analysis	Design Documentation
Hot Work Permit	**Manage Risk**	Process Hazard Information
Management of Change	Operating Procedures*	Process Hazard Analysis
Incident Investigation	Training and Performance*	Management of Change
Emergency Planning and Response	Safe Work Practices *	Facilities
Compliance Audits	Asset Integrity and Reliability	Siting
Trade Secrets	Contractor Management	**Codes and Standards**
	Management of Change	Safety Reviews
	Operational Readiness*	Maintenance and Inspection
	Conduct of Operations*	Multiple Safeguards*
	Emergency Management	Emergency Management
	Learn from Experience	**Personnel**
	Incident Investigation	Job Skills*
	Measures and Metrics	Safe Work Practices*
	Auditing	Initial Training*
	Management Review and Continuous Improvement	Employee Proficiency*
		Fitness for Duty*
		Contractors

* Contains some of the 6 elements of a Human Factors element, but all missing elements not covered completely. Mostly covered in CoC/OD element

Note that the newest definition of process safety is CCPS's *Risk Based Process Safety (RBPS)*[4] replaces their earlier process safety definition. In the older definition from CCPS [1], there was an element on Human Factors, which brought strong focus to this necessary element; in RBPS, the human factors subelements are now spread across 6 different elements, though some human factors were inadvertently weakened in the transition to *RBPS*.

Although nearly the entire industry agrees that implementing PSM is the right thing to do, interpreting and converting the PSM requirements into practices is unique to each company, and even unique to each plant site. Not only can the requirements be interpreted differently for each site based on local needs, but each site also starts from a different point when they begin to implement a system that is consistent with this regulation.

This paper looks at the statistics of major accidents, combined with results from audits and assessments from more than 50 chemical, petrochemical, oil/gas, and related processing companies world-wide. The paper illustrates the four major gaps that are common to the companies/sites that keep having major accidents, compared to those companies/sites that do not have such accidents:

- Accurate and clear operating and maintenance work instructions (procedures)
- PHA of all modes of operation, especially startup, shutdown, abnormal operations, and online maintenance modes
- Getting Near Misses reported and investigated
- Addressing human factors that are missing from most management systems

The paper also provides guidance for closing each of these gaps.

1. Procedures Need to be More Accurate and Clear

Operating and maintenance procedures have always been crucial to the safety, quality, and productivity of process systems. With the advent of new safety and quality standards such as OSHA's process safety management (PSM) regulation, OSHA's personal protective equipment (PPE) regulation, EPA's risk management program (RMP) regulation, and the ISO 9000 quality standard, many companies are facing the daunting task of developing or upgrading their procedures to satisfy varied and sometimes complex and conflicting requirements.

Although there are overlapping characteristics that the various regulations and standards share, each approach the procedure-writing process from a somewhat different perspective (e.g., quality or safety, protection of workers or protection of the public/environment), and the required level of detail for documenting the procedures differs greatly between these regulations and standards. For instance, the ISO 9000 quality standard advocates documenting procedures and work instructions that impact quality and suggests keeping the procedures as simple as possible, while the OSHA PSM regulation requires detailed procedures that address: all operating modes (startup, shutdown, etc.), operating limits, consequences of deviations, means to avoid hazards, safety and health considerations, and safety systems and their functions. How can companies develop and maintain procedures that ensure productivity and simultaneously satisfy the different regulations and standards? The key is to remember the ultimate goal of the regulations and standards: to reduce human errors that can impact quality, productivity, and/or safety.

Procedure-related errors are errors that occur because some characteristic of the procedure caused task performance to fail. **This is currently the most critical human factor at most sites since 90% of accidents have at least one root cause related to mistakes within procedures**. Reducing these procedure deficiencies can reduce human error rates by a factor of 2 to 20 [5].

1.1. Identified problems

Deficient Procedures are the most prevalent problem in process industries since procedures have not traditionally been developed from the perspective of optimizing human factors; instead procedures have been traditionally developed to meet a compliance requirement. Examples of procedure deficiencies (inaccuracies) include:

- Incorrect/incomplete/nonexistent (most procedures we have audited have been only 70-75% accurate – the inaccuracies include missing critical steps, steps as written are not what needs to be done, or the steps are out of sequence) [5]
- No/misplaced/incorrect information in warnings (for example, a warning should **never contain** the action to take; it should instead **emphasize** the action to take) and warnings should be located BEFORE the step applicable to the warning
- Poor format and presentation rules

1.2. Objective

The flowchart (Figure 1) nd the procedure quality checklist (Table 2) describes the basic sequence of steps for writing effective procedures. Note that some basic rules are contained in the flowchart, such as "operators write operating procedures," "maintenance craft-persons write maintenance procedures," and "Lab technicians write lab procedures." Procedures are NOT written by engineers or department superintendents. This is a critical to ensure procedures are written in the user's common language and are written at the right level of detail.

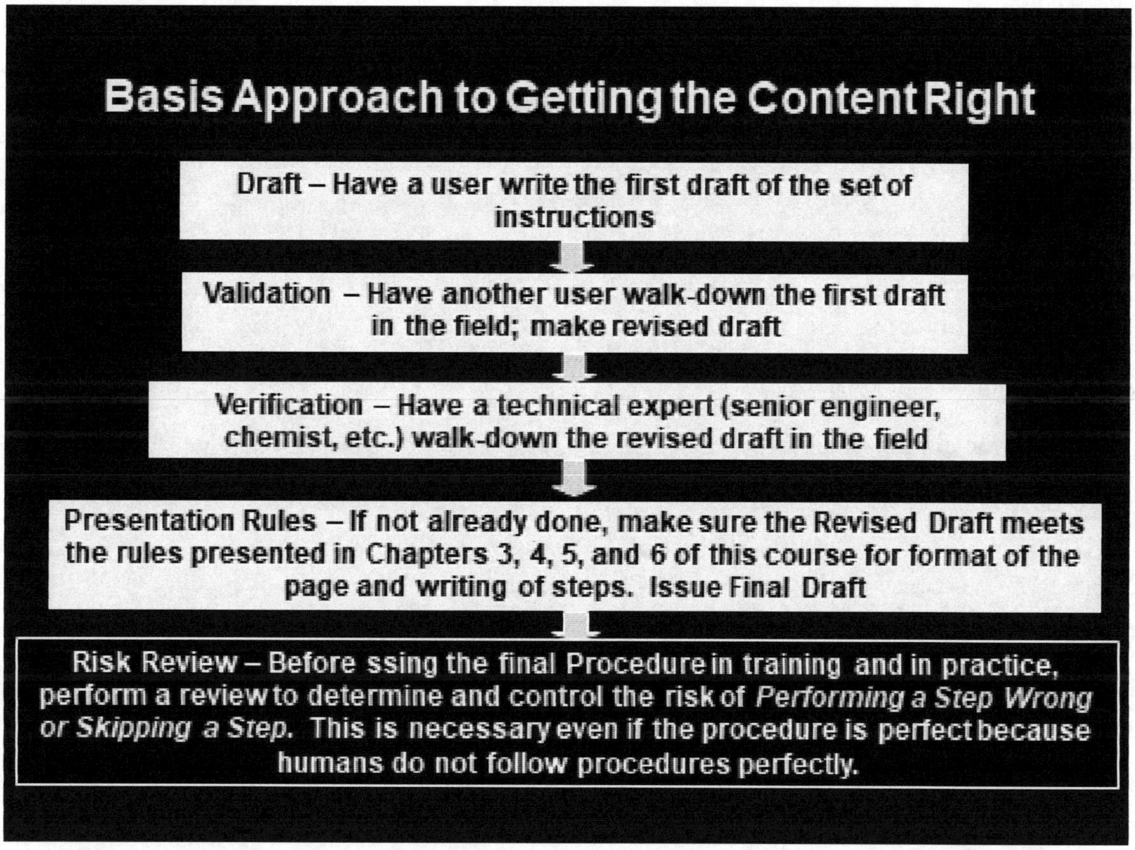

Figure 1. Best Approach to developing written procedures, once the unit has decided which procedures are needed (Extracted from PII Course, Writing Effective Operating and Maintenance Procedures © PII 2003-2017) [6]

The procedure content controls the human error related to "procedure-based" error – if content is wrong, the trainee is very likely to learn the step wrong (trainer uses the procedure as a basis for training or the trainee refers to the procedure step that is wrong when executing the procedure).

Even if the content is correct, not following page formatting best practices can still increase human error rates. If the content of the step is wrong it does not matter if the presentation is clear. If the content is perfect, you can further reduce error rates by 50% to 70% by presenting the information properly.

Rules for developing content[7]

- The procedures should be written at the level of someone who has just completed the basic training for that task. Do not write the procedure for someone just hired or for the 10+ veteran.
- Clearly identify ahead of time what activities need procedures to help ensure error rates are controlled low enough, and identify which activities are considered common "skills" of all of the staff.

> *Example: Starting a pump is typically considered a "skill." This means a procedure step can simply be "Start the benzene recycle pump (P-119)." The procedure does not need to explain how to start the pump. But, starting a pump is a skill that nonetheless must be learned; it requires doing several sub-steps related to positioning of intake and discharge valves, how the pump is throttled at startup, checking local pressure gauge, etc. So, a training module is needed for starting pumps of various types, and the operator needs to learn this necessary skill. Once the skill is learned, it will be applied to so many various cases and so often, that a procedure is no longer needed as a "refresher" on how to "start the pump."*

- The first draft of the procedure must be walked-down in the field by another user. Simply reading through the procedure does not catch enough of the mistakes.
- The revised draft should be reviewed in the field by a technical staff person, such as a process or operations engineer (for operating procedures).
- The procedure needs to be checked to ensure it follow the page format and step writing rules described in the next section. This ensures the accurate steps are clearly presented.

The target is to reach an accuracy of 95% or better[7] (so, no more than one wrong or missing step out of 20 steps). The following have been observed in the field by PII staff by direct data collection at more than 110 sites/plants around the world, based on walk-down review of several to dozens of procedures at each unit/plant:

- At 95% accuracy or better and when the same procedures follow 80% of the rules for procedure clarity (presented next), then most users will follow the written procedures and will try to keep the procedures up-to-date
- At about 85% accuracy or less, about half of the users stop using the procedures
- At about 75% accuracy or less, less than 10% of the users will refer to the procedure or will try to keep it up-to-date. **So, the written procedures are not very effective.**
- Unfortunately, the typical operating procedure walked down by PII staff (accompanied by senior operators, a process engineer, and a shift supervisor) is about 75% accurate (so one step in four is missing or wrong). Usually these inaccurate procedures also follow less than half of the best practices for procedure format, presented next [7].

Procedure Clarity

The importance of the accuracy of the procedure steps were discussed above. The clarity of the procedure steps (how they are written and how the page is formatted) is also important. Following best practices for step and page format reduces human errors by a multiplying factor of 3 to 5. [5], [7]

The best practice rules for writing and validating procedures have been published for many years *(see Bridges & Williams, 1997; Madden & Bridges, 2016, 2017)* [7],[8],[9]. These rules include best practices for formatting of the pages and steps. These have been gradually improved over the past decades and now are incorporated into PII's training materials.[6]

Table 2. Procedure Quality Checklist (courtesy PII, copyright 2003-2017)[6]

#	Issue	Response
	Procedure Content Checklist	
1	Is the procedure drafted by a future user of the written procedure? (Engineers should not author procedures to be used by operators or maintenance staff.)	
2	Is the procedure validated by a walk-down in the field by another future user of the procedures?	
3	Is the procedure reviewed and commented on by technical staff (engineers or vendors)?	
4	Is the procedure checked versus the Page and Step format rules below?	
5	Is a hazard review of step-by-step procedures performed to make sure there are sufficient safeguards (IPLs) against the errors that will occur eventually (when a step is skipped or performed wrong)?	
6	Is the content measured using "newly trained operators" to judge the % of steps that are missing, steps that are confusing or wrong, and steps that are out-of-sequence? (A score of 95% accuracy of content is good; 98% should be the targeted average.)	
	Page Format Checklist	
1	Is the title of the procedure the largest item on the page?	
2	Is the procedure title clear and consistent with other titles, and does it uniquely describes the topic?	
3	Are the document control features the smallest items on the page?	
4	Are temporary procedures clearly identified?	
5	Is white space used effectively? • Is there one blank line between each step? • Does indentation help the user keep their place? • Are the margins large enough to reduce page congestion?	
6	Is type size is 12 pt font or larger?	
7	Is mixed case used for words of steps, with ALL CAPS used only for special cases (such as IF, THEN, AUTO, and WARNING)?	
8	Is the step number very simple (single level of number used)? Only an integer?	
9	Have sections or information not necessary to performing the steps been moved to the back or to another part of the manual or training guide?	
10	Are section titles bold or larger than the text font? Do sections have clear endings?	
11	Is the decision on electronic presentation versus hard copy correct? Are electronic linkages to procedures clear and accurate and easy to use? If paper is chosen, is the color of the paper appropriate?	
12	Is the overall page format (such as Outline format or Two-Column [T-Bar] format) appropriate to the use of the procedure?	
13	Are play script features added for tasks that must be coordinated between two or more users? • Play script is normally used when there are two or more hand-offs of responsibility for steps.	
14	Are rules followed for formatting of Warnings, Cautions, and Notes? (See annotated rules, such as Warnings are for worker safety and Warnings must clearly standout from rest of page.)	
	Step Writing Checklist	
1	Is each step written as a command?	
2	Is the proper level of detail used throughout? This is judged based on: • *Who* will use the procedures • Same level of detail used in similar procedure steps	
3	On average, is there only one implied action per instruction? Best practice is to average 1.2.	
4	Does the procedure indicate when sequence is important? • If sequence matters, each step should be numbered (with an integer or letter)	

	• If sequence does not matter, bullet lists should be used	
5	Are only common words used? Apply "education" level test (5 grade reading level is best)	
6	Do all acronyms, abbreviations, and jargon aid understanding? • Develop a list of such terms for use in procedures *and* communication. • Use terms that users use (within reason)	
7	Is each step specific enough? No room left to guess/interpret: • The meaning of a word or phrase (Check vs. Make sure) • The intent of a step or series of steps • A desired quantity or value • To what equipment the step applies	
8	Is the procedure free of steps that require in-your-head calculations? • Values expressed as ranges rather than targets with error bands • Conversion tables, worksheets, or graphs provided where needed	
9	Are graphics to the user's advantage? • No explanatory paragraphs or lengthy instructions that could be replaced by a picture • No impressive graphics that provide no real advantage	
10	Are references to the user's advantage? • No lengthy explanations or instructions that could be replaced by branching to a reference • No references to a procedure that references still another • No gaps or overlaps between this procedure and a referenced document • If branching, must branch to a procedure, not to a specific step in a procedure	
11	Are rules followed for writing warnings, cautions, and conditional steps? • No more than 2 per page • No actions within a warning or caution (actions must always be numbered steps) • Warnings and Cautions contain descriptions of potential consequences	

For a complete description of best practices, refer to PII's paper *"Best practices for writing operating procedures and trouble-shooting guides"*. [7]

2. PHAs (Process Hazards Analyses) Need to Close Two Major Gaps

Process Hazards is a broad term. The relationship between many components define process hazards. For a PHA team, Process Safety Information (information on the chemicals, technology and equipment) and operating procedures are necessary to identify specific components of and analyze a hazard scenario. These components include: process deviations (parametric and procedural), their causes, the process safety consequence and various safeguards for detecting deviations and causes, preventing causes, detecting and mitigating consequences and intervening safeguards. A thorough review of incident investigations also offers an opportunity to further analyze process hazards.

2.1. Identified problems

Over the last twenty years the CSB has brought needed attention, through its investigations of industry incidences, to two significant gaps in many current PHAs; where the teams did not identify and analyze:

- Process hazards unique to non-routine modes of operations resulting in catastrophic consequences, and
- Process hazards related to Damage Mechanisms.

Though the intent of the US OSHA PSM regulation has always been for PHA teams to analyze the hazards for all modes of operation, OSHA did not give specific requirements for addressing damage mechanisms and only recently have they increased their enforcement attention on PHA of non-routine modes of operation.

Industry has responded by modifying the *Guidelines for Hazard Evaluation Procedures*, 3rd Edition, 2008, CCPS/AIChE[10] to improve the coverage of these two main issues. API in turn issues guidance on addressing damage mechanisms[11]. Industry best practices now exist for addressing both of these weaknesses. Now, with proper approaches, a team with process safety competencies and process knowledge can thoroughly address the issues during a PHA.

Another less significant gap noted by regulators and by the US CSB has been an incomplete consideration, within the PHA, of previous incidents.

This section of the paper lays out the case from the US CSB and US OSHA on these PHA gaps:

- Poor or no coverage of hazards during non-routine modes of operations
- Poor or no coverage of damage mechanisms

2.2. Objectives

2.2.1. Hazard Evaluation of Non-Routine Modes of Operation

Non Routine modes of operation include start-up, shut-down, online maintenance and other abnormal operations. The hazard evaluation of non-routine modes of operation involves reviewing procedures using a HAZOP, simplified HAZOP, or What-if analysis to uncover potential accident scenarios associated with non-routine operations, for continuous or batch operations. Human error is more likely and more critical during non-routine operations. By analyzing procedural steps where human error is more likely, and where human error or component failure could lead to a consequence of interest, risk can be reduced. The hazard evaluation team's objective is to evaluate the risk associated with skipping steps and performing steps wrong.

The purpose of a hazard evaluation of non-routine modes of operation (governed by written procedures) is to make sure an organization has enough safeguards for the inevitable instance when a step is either performed wrong or skipped (inadvertently or due to shortcutting or other reasons).

Industry has found that a HAZOP or what-if analysis, structured to address procedures, can be used effectively for finding the great majority of accident scenarios that can occur during non-routine modes of operation[10],[12],[13],[14]. Experience shows that reviews of non-routine procedures have revealed many more hazards than merely trying to address these modes of operation during the P&ID driven hazard evaluations.

General guidelines for analyzing non-routine modes of operation or batch processes

- Define the assumptions about the system's initial status. "What is assumed to be the starting conditions when the user of the procedure begins with Step 1?"
- Define the complete design intention for each step. "Is the step actually 3 or 5 actions instead of one action? If so, what are the individual actions to accomplish this task?"
- Don't analyze safeguard steps that start with ensure, check, verify, inspect, etc., or where the consequence of skip is "loss of one level of safeguard/protection against…" There is no reason to analyze these steps since they will show up as safeguards of deviations of other steps. This approach is similar to not analyzing a PSV during a HAZOP of continuous mode (i.e., during a parametric deviation analysis); instead the PSV is shown as a safeguard against loss of containment.
- Together with an operator before the review, identify the sections of the procedures that warrant use of:
  - 7-8 Guide Words (extremely large consequences can happen if deviations occur)
  - 2 Guide Words (the system is complex, mistakes are costly, or several consequences could occur)

  - On others, use What-If (no guide words or guide phrases; for use on simpler or lower hazard systems)
- Decompose each written step into a sequence of actions (verbs)
- Apply guide words directly to the intentions of each action

The following preparation steps may also be needed prior to the procedure PHA:
- Walk through procedure in the plant with one or more operators to see the work situation and verify the accuracy of the written procedure.
- Determine if the procedure follows the best practices for "presentation" of the content; the best practices will limit the probability of human error
- Discuss generic issues related to operating procedures, such as:
  - staffing (normal and temporary)
  - human-machine interface
  - worker training, certification, etc.
  - management of change
  - policy enforcement
- Review other related procedures such as lock out/tag out and hot work
- **If the procedures are NOT >90% accurate, then redo procedures before the PHA.**

2.2.2. Damage mechanisms

Recently, more and more regulators, including the US CSB, are making analysis of Damage Mechanisms (DMs) a PHA requirement. This requires that potential process damage mechanisms and their potential consequences of interest be identified. Damage mechanisms can be broken down into three main categories. These categories and examples of each are shown in the following table (Table 3).

Table 3. Types of Damage Mechanisms [15]

Category	Examples
Mechanical	• Mechanical loading failure • Mechanical fatigue • Buckling • Cracking • Embrittlement • Ductile fracture • Brittle fracture

Chemical	• Corrosion
	  - Uniform
	  - Localized
	  - Pitting

Physical	• Thermal failures
	  - Creep
	  - Thermal fatigue
	  - Transformation

The team may use aids such as piping specs, established literature and standards, any applicable MOC documents, etc. to help identify DMs. The PHA leader has several options of how to address damage mechanisms, but PII preferred approach is to discuss during HAZOP under "Loss of Containment" deviation in each node/section.

Damage mechanisms can (and sometimes ***should***) be reviewed prior to the PHA. For example, all MOCs should consider damage mechanisms and their potential consequences of interest before the change is approved. The MOC should also require updates to process safety information which might be relevant to protect against DMs.

Coverage of Damage Mechanism (DM) for each major section [14]

As mentioned above, this is PII's preferred method for thoroughly covering all DM within a PHA/HAZOP of an entire unit. The benefits are that the team can more easily catch changes in DM from section to section and they can more easily identify when unique safeguards, such as Remote Isolation Valves, segregated containments (dike), and unique materials of construction are needed. To facilitate this, a list of generic causes of loss of containment and a generic list of safeguards against loss of containment are covered in the Loss of Containment (LOC) deviation of each section.

For each node (each line, each vessel, each column, etc.), the PHA team should discuss and document each damage mechanism listed in Table 3 (as a cause of loss of containment), consequence of the failure if the damage mechanism occurs, and the safeguards in place to prevent the damage mechanism, detect the mechanism before failure, prevent the release, detect and response to the release, and mitigate or contain the release.

The documentation style varies between PHAs. There are two different styles that have been acceptable in the past.

- One style (Example A, Table 4) uses a reference to a summary table of typical causes (this summary table is not shown but is similar to Table 5 of this paper) instead of listing each individual damage mechanism. In Example A, the same approach is used for safeguards. This approach saves redundant text and some time, but it requires the leader and scribe to be diligent to cover everything in the Typical tables that were referenced in the Cause and Safeguard columns. A modified approach is to relegate some generic causes and safeguards to Typical tables for reference and then to carefully list the specific cause and/or safeguards of interest for LOC for each specific node.

Table 4. Example A – Referencing Generic Tables for Typical Causes and Typical Safeguards [14]

Dev	Deviation	Causes	Consequences	Existing Safeguards	Recommendations
9.10	Loss of containment	Accelerate corrosion External fire High pressure (linked from 9.7) Typical causes of loss of containment (see Table A.1)	Release to atmosphere leading to potential injury of workers and/or community	Pressure Relief Valve's 231A, B, C on Reactor and no valves in line to reactor Most lines / connections are welded construction; only a few flanges Drills conducted each year on evacuation, rescue and isolation Emergency Response personnel are trained at SABIC FTC Generic safeguards protecting against or mitigating process material releases (see Table A.2)	Safety 7. Consider changing the ITPM schedule for managing most PSVs whose inspections are too infrequent based on industry standards and best practice. For instance, the current inspection frequency for PSV-8220 on the Ammonia Receiver (I-2005G) is 9 years, whereas, consensus codes typically recommend testing/inspection every 1-4 years for PSVs in highly toxic services.

- Example B (**Table 5**) does not use or reference a generic or Typical list of causes (damage mechanism) or Typical safeguards, but instead develops a specific listing for the LOC deviation of each node. This style has proven easier to justify to regulators and other outside reviewers but takes more documentation effort.

Table 5. Example B – Specific Listing of Causes and Safeguards [14]

Dev	Deviation	Causes	Consequences	Existing Safeguards	Recommendations
1.9	Loss of containment	Corrosion/erosion External fire and/or flame impingement Gasket, packing, or seal failure Improper maintenance Material defect Operator failing to close or inadvertently opening a valve to the atmosphere (e.g., a valve at a hose connection) Railcar inadvertently derailed Valve leaking to the atmosphere High pressure (linked from 1.5) Acid corrosion caused by high concentration of water (linked from 1.8) High ambient temperature External impact (such as from a mini-engine or another railcar)	Catastrophic release of chlorine from a ruptured railcar Steady release of chlorine from a ruptured connection Steady release of chlorine from a leaky connection High pressure caused by thermal expansion of liquid chlorine if railcar is also over-full	Chlorine repair kit Derailer and warning flag to prevent impact by a mini-engine or another railcar Limited vehicular access to area Maintenance/operator response as required, including isolation if needed Operator periodically monitoring the railcar valves while unloading Personal protective equipment in the area Plugs installed in all chlorine valves to the atmosphere when the valves are not in use Relief valve on each railcar for mitigating releases caused by overpressure Supplier maintenance of railcars (per strictly enforced US DOT requirements) Video monitoring of the unloading area Concrete crossties on rail spur Dike preventing any combustibles spilled nearby from reaching the unloading rack area Concrete railroad ties in chlorine unloading area to prevent fires near railcar	10. Consider installing a chlorine detection system in the unloading and vaporizing area to help detect chlorine releases (especially at likely release points) 11. Verify that periodic maintenance and inspections are being performed in accordance with Chlorine Institute recommendations 12. Review the drainage system for the unloading are, and identify the areas that may be affected by a large chlorine release 13. Consider prohibiting the use of heavy equipment (e.g., cranes) in the unloading and vaporizing area unless special precautions to prevent equipment damage are enacted 33. Consider providing a high pressure alarm for each vaporizer 40. Consider providing a water deluge system in the unloading area to help mitigate chlorine releases from the railcar

Regardless of method, the team leader and scribe must ensure the team rigorously discusses all of the damage mechanisms in Table 5 and adequately documents the results of the damage mechanism review in the PHA/HAZOP analysis tables.

For a complete description of PHA best practices, refer to PII's paper *"Recipe for a Complete Process Hazard Analysis"*. [14]

3. Near Misses Need to be Reported and Enough of Them Investigated

We must learn from accidents and near misses to prevent recurrence. The first step in the learning process is investigation to determine the causes and underlying reasons why accidents and near misses occur. A thorough investigation of root causes will identify the management system weaknesses. Learning which management system weaknesses are leading to near misses and accidents is one of the highest value activities in which a company can invest, and learning from near misses is much cheaper than learning from accidents. Many chemical companies have implemented process safety management systems, and now they are beginning to focus on getting near misses reported and on root cause analysis. This is a very exciting trend. Unfortunately, the chemical industry gets very few near misses reported (the chemical industry is certainly not the only industry with this problem).

To understand more about near misses and getting them reported, it is best to first review the basic definitions.

An incident is either an Accident or a Near Miss.

An Accident is a sequence of unplanned events and conditions that result in harm to people, environment, process, product or image.

A Near Miss is an unplanned sequence of events that could have caused harm if conditions were different or is allowed to progress, but did not in this instance.

Given a consistent understanding of the definition of a near miss, it is possible to estimate how many near misses should be reported for every accident. Studies in several industries indicate that there are between 50 and 100 near misses for every accident. Also, data indicates that there are perhaps 100 erroneous acts or conditions for every near miss. This gives a total population of roughly 10,000 errors for every accident. **Figure 2** illustrates the relationships between accidents, near misses and non-incidents.

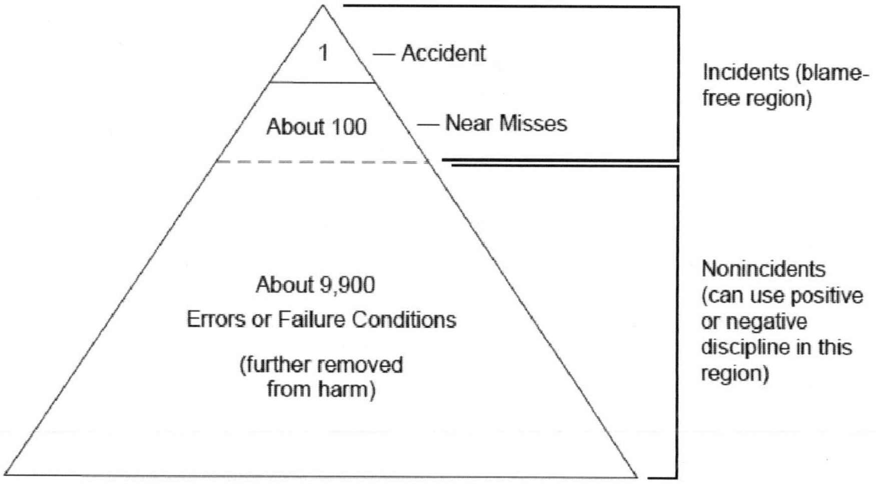

Figure 2. Relationship between Errors and Potential or Actual Impact [16]

Investigating near misses is critical to preventing accidents, because near misses share the causes and root causes of accidents; they are one or two barriers away from the loss/accident. We are very likely preventing many apparently unrelated accidents when we prevent the ones that are obviously related.

3.1. Identified problems

During the past 15 years, we have asked more than 5,000 students of our process safety management (PSM) courses and 3,500 students from our investigator leadership training courses how many near misses they get reported for every accident. The students represented about 400 companies, predominantly in chemical-related process industries (chemical, polymer/plastic, petrochemical, refining, oil and gas exploration, pharmaceutical, pulp and paper, etc.). The answers are quite disturbing. More than 95% said that their ratio of near misses reported to accidents reported falls in the range of 0 to 20. Less than 5% of the individuals said the ratio was greater than 5, and less than 2% said the ratio was higher than 10. Students noted that fear of disciplinary action, lack of management commitment, and lack of understanding of the difference between a near miss and a non-incident were the main reasons why near misses do not get reported. [16]

In conducting about 150 PSM audits and 7000 process hazard analyses (PHAs) during the past 22 years, we have found ratios from 0 to 105. For the first half of the 1990s, more than 90% of the facilities we talked to had ratios in the range of 0 to 0.5, and more than 95% had ratios in the range of 0 to 1. In the last half of the 1990s, more than 90% of the facilities had ratios in the 0 to1 range, and more than 95% had ratios in the 0 to 2 range. In addition, only a few (less than 2% of the facilities) had a ratio higher than 5. By 2005, many facilities had ratios in the 20 to 100 range and each of these had seen great gains (more on this in future articles), but the average across all companies we talked to increased only slightly. By 2011, the ratio of near misses to accidents (any loss event) was about 2 for companies we deal with. But by the end of 2011, many major companies had ratios well above 50! [16]

Our auditors and PHA leaders commented that the primary reasons for lack of reporting were about the same as those found in the survey. The barriers to getting near misses reported, discussed in detail later, are:

- Fear of disciplinary action.
- Fear of teasing by peers (embarrassment).
- Lack of understanding of what constitutes a near miss versus a non-incident.
- Lack of management commitment and lack of follow-through once a near miss is reported.
- An apparently high level of effort is required to report and to investigate near misses compared to low return on this investment.
- There is No Way to investigate the thousands of near misses per month or year!
- Disincentives for reporting near misses (e.g., reporting near misses hurts the department's safety performance).
- Not knowing which accident investigation system to use (or confusing reporting system).
- Company discourages near-miss reporting due to fear of legal liability if these are misused by outsiders.

The good news is that near-miss reporting appears to be improving, and the industry appears to recognize most of the barriers to near-miss reporting. The bad news is that the ratio is still very poor and improvement appears slow!

The number of companies that participated in the informal surveys suggests that the results are a statistically significant representation of the chemical industry.

A formal (written) survey was developed and e-mailed, faxed, and/or mailed to more than 100 companies. Of these, more than 12 replied originally in 1997. Since then, we have reviewed data in detail from more than a dozen major companies. These 25+ companies or affiliates provided data from more than 400 facilities, including more than 150,000 employees in manufacturing and operations. The data were from the chemical industry, polymer industry, refineries, drug/pharmaceutical companies, pulp and paper mills, petrochemical companies, and oil exploration/production.

Table 6 summarizes the results of both the informal and formal surveys.

Table 6. Summary of survey results [16]

Survey Items	Informal surveys		Formal Survey (or actual organization data)
	Classroom polls	Found during Audits and PHAs	
Current Near-miss Reporting Ratio Range (0-5 years prior)	1 to 105	2 to 20	0 to 105
Previous Near-miss Reporting Ratio Range (5-10 years prior)		1 to 5	NA
Current Near-miss Reporting Ratio Average (0-5 years prior)	2-3	3	5
Previous Near-miss Reporting Ratio Average (5-10 years prior)		1	NA
Goal for Near-miss Reporting Ratio (average)	20+	NA	20
Theoretical Upper Limit on Near-miss Reporting Ratio (average)	About 100	NA	100
Number of Participating Companies	400	300+	25+
Number of Participating Facilities	8500+	500+	400+

3.2. Objective

A company should strive to reach a ratio of 50-100 and investigate about 20 near misses per accident. This will provide a statistically significant sample of all incidents (and all important errors) and provide a company with sufficient feedback on which management system weaknesses are causing the errors and component failures. Various companies with different cultures have achieved high ratios with great return on investment.

Table 7 shows a brief description of possible solutions to overcome each one of the barriers identified in the survey.

Table 7. Summary of Survey Results[16]

Barriers	Solutions
Fear of disciplinary action	Implement a policy to NOT punish individuals when their errors lead to accidents and Near Misses.
Fear of teasing by peers (embarrassment)	Ensure that all employees understand the importance of near-miss reporting; demonstrate, through feedback of lessons learned.
Lack of understanding: Near miss vs Non-incident	Develop a list of "in-context" examples that illustrate what you consider to be Near Misses and what you consider to be non-incident
Lack of management commitment and lack of follow-through on reported near misses	Hold management accountable for achieving a Near-Miss reporting ratio
Apparently high level of effort is required to report/investigate Near Misses	Ensure that the data are entered in a database and queried regularly. Share the results with employees so they can see the value of the reported near misses
There is no way to investigate the thousands of Near Misses per month	Let front-line foremen or supervisors decide if a Near Miss or accident needs to be investigated
Disincentives for reporting Near Misses	Ensure that goals and incentives are not tied to lower incident rates (since this discourages reporting), but instead provide incentives for high Near-Miss reporting ratios
Not knowing which accident investigation system to use	Have ONE incident reporting system with ONE approach
Company discourages Near Miss reporting due to fear of legal liability	Involve legal on major Near Misses and accidents to ensure the results are protected as much as possible under attorney/client privilege.

For a complete description of the possible solutions for each barrier, refer to PII's paper *"Gains from Getting Near Misses Reported"*. [16]

4. Key Human Factors are Not being addressed

All accidents (or nearly all, if you consider that there are some natural phenomena that we either cannot guard against or choose not to guard against) **result from human error**. This is because humans govern and accomplish all of the activities necessary to control the risk of accidents. Humans influence other humans in the process – not only do humans cause accidents (unintentionally) by making errors directly related to the process itself, but they also cause errors by creating deficiencies in the design and the implementation of management systems (i.e., we make errors in authorities, accountabilities, procedures, feedback, proof documents and continual improvement provisions). Ultimately, these management systems govern the human error rate by directly causing or indirectly influencing the process. The process-related activities where errors have the most influence include:

- Designing a process
- Engineering a process
- Specifying components

- Receiving and installing equipment
- Commissioning
- Operating a process
- Maintaining, inspecting and repairing a process
- Troubleshooting and shutting down the process
- Managing process, procedure, materials, facility and personnel changes

The 10 human factors categories to be controlled by companies are:
- Available Time (includes staffing Issues) – for responses only
- Stress/Stressors (includes staffing issues)
- Complexity & task design
- Experience/Training
- Procedures
- Human-Machine Interface (includes tools)
- Fitness for Duty
- Work processes & supervision
- Work Environment
- Communication

4.1. Identified problems

Recent major accidents have highlighted the need for increased focus on human factors. The U.S. Chemical Safety Board (CSB) cited (US CSB Video, 2006) human factor deficiencies as one of the main contributors of the catastrophic accident at the BP Texas City Refinery in March 2005[17]. The human factor deficiencies included lack of control of worker fatigue, poor human-system-interface design, poor communication by radio/phone, out-of-date and inaccurate operating procedures, and poor (no) communication between workers at shift handover. The CSB cited similar issues from many other accidents[18] and has urged industry and the U.S. OSHA (the regulator) to pay much more attention to human factors. As a result, the recent U.S. OSHA National Emphasis Program for Refineries included human factors as one of the 12 core elements.

Implementing human factor engineering and policies to prevent accidents is not a new concept. Nearly all (or all, from a more complete perspective) of the causes and root causes of major accidents in the past 30 years have been the result of poor control of human factors. This has been cited in many root causes analysis reports and papers concerning these major accidents.

Process Safety Management (PSM) systems based on OSHA's PSM standard are likely lacking the fundamental human factor elements and implementation guides that, if applied across the applicable PSM elements, would work together to reduce human error. The *Risk Based Process Safety* guideline from the CCPS/AIChE (2007) does contain these human factor requirements.

Many site and company staff do not know there are standards for the control of human factors. However, at last count there were more than 300 non-governmental standards (such as from ANSI, ISO, IEEE, etc.) for the control of human factors, and there are more than 100 government regulations and standards and guidelines. Some of the better government regulations cover aviation, marine operations (shipping) and aerospace. An organization should devote resources to finding and implementing such best practices.

The following is a description of each of the human factors mentioned above and specific ways to control each. Our data from review of more than 2000 incidents reveal that:
- 90% of accidents had at least one root cause related to deficiencies in written work instructions.
- 70% of accidents had at least one root cause related to miscommunication between workers or between workers and supervisors.

- 40% of accidents had at least one root cause related to excessive fatigue of the worker. [5]

4.2. Objective

Not all errors can be prevented. Since the beginning of time, humans have tried to control error rate with more or less success. Human errors have been measured for hundreds of years. Psychologists have studied why humans make mistakes and have gradually put a science around human error probability. Today, the best models for control of human error in the workplace are generally agreed to be related to control of human factors; in turn, these have grown out of what was previously denoted performance shaping factors. What can be done to control each Human factor is discussed in greater detail next.

For a complete description of each human factor, its importance and how to control them; refer to PII's paper *"Human factors and their optimization"* [5].

Time available for the task

Task design is directly linked with this category. Process design, procedure development and worker scheduling need to take into consideration not only the time it takes to complete each task but also the time it takes to manage usual as well as potentially unusual or unexpected situations. True; we do not create norms for exceptions, but exceptions sometimes are more usual than realized.

Stress/Stressors

A well-designed, organized and managed work process helps maintain and promote individual health and well-being, but stresses will arise through no fault of the organization. Supervisors should be trained to recognize stress in workers and, in severe cases, reassign them and help them get assistance to deal with the stress and in other cases, reduce their job pressures for a day by shifting assignments. A wellness program may also be necessary to help cope with stress. Simple approaches, such as training workers to recognize and deal with stress, can help most individuals. Approaches to deal with stress include positive self-talk and practicing relaxation techniques (i.e., breathing control).

Complexity & task design

If roles and responsibilities for accomplishing tasks are not clearly defined, then there will be a risk of serious errors. It is also important that employees and engineers/designers understand the characteristics of the work elements involved, how each element passes information to the other, how each person involved communicates, and that each person has guidelines to adhere to during task design and procedure development.

For optimal performance, task designers often must integrate human and automated equipment in their task analysis.

It is crucial for task designers also to have an understanding of the appropriate allocation of roles and responsibilities to the various participants in a task. Understanding the characteristics and limitations of all of the humans and automated systems involved in the task (particularly critical tasks) will allow for establishing additional controls where needed.

The task analysis process includes *Task definition* which is the initial operational description of the operator tasks required for execution of a given system function; *Performance analysis* of all tasks with respect to human performance requirements; *Error and workload analysis* to be included in the design process, particularly when the likelihood of unacceptably operator error is high; and *Evaluation* since evaluating the design of a new or complex task is important to ensure that the task can be effectively performed as designed. For very high consequence scenarios involved complex human interaction, it is likely prudent to invest in a human reliability analysis (HRA).

Experience/Training

Training to master the Knowledge, Skills, and Abilities (KSAs) required for a job may be obtained from a variety of sources. Companies provide the training or it is obtained from other sources (e.g., trade schools, universities, contractor organizations, etc.) to reach the skills and knowledge established in job descriptions. There are several factors that may result in personnel not mastering the required KSAs for a job. These include course design and delivery methods, course completion, practical skills demonstrations or simulation, and training frequency. Course design begins when the learning objectives have been identified. The design process consists of determining the delivery methods (simulator, on-the-job training, etc.), number of hours required to cover the materials, instructor qualifications, etc. Although some methods, materials and instructors may be more effective or efficient than others, the important issue is that the course content is complete and addresses all of the relevant KSAs, so that the learning objectives are met and the KSAs mastered.

Another determinant of KSA mastery is course completion. Although this factor appears obvious, there are often competing demands on personnel that may pull them out of training at times. As a result, they may miss the instruction related to specific KSAs. Testing may not identify the KSA deficiency because it is impossible to test mastery of all KSAs. Sampling techniques are used to generate examinations. If attendance and participation are not controlled, some personnel may miss training on specific KSAs and testing may not identify the deficiencies before an error is committed.

Another factor affecting KSA mastery is forgetting. An individual's ability to perform a task will degrade over time unless the relevant KSAs are refreshed. Proficiency training (i.e., refresher) will be required for some tasks to maintain the level of mastery that was demonstrated following initial training. One function of training programs is to identify those tasks that require proficiency training.

If certain tasks are performed frequently, proficiency training may be unnecessary. By performing a task, personnel practice the task and obtain feedback on where they have weaknesses. Task performance refreshes the KSAs and successful task performance verifies that proficiency has been maintained. Furthermore, since humans learn by comparison to similar activities, as a worker practices one task, they are learning about similar tasks. Care must be taken here to avoid the practice drifting away from what the procedure requires. Also, if new operators are trained by experienced operators, you must ensure that the new employee is trained according to the procedure, not whatever practice is actually being used. (Of course, ideally, the practice and procedure will be identical.)

Operating and maintenance procedures

Many procedures do not follow best practices for controlling human error, and so the written procedure actually "contributes" to increased error rates. Additionally, many organizations do not have guides on how to troubleshoot (what to do when process deviations occur). The best practice rules for writing and validating procedures have been published for many years. [6], [7], [8], [9]

Procedures have not traditionally been developed from the perspective of optimizing human factors; instead, procedures have been traditionally developed to meet a compliance requirement to have written procedures. Examples of procedure deficiencies (inaccuracies) include:

- Incorrect/incomplete/nonexistent (most procedures we have audited have been only 70-85% accurate – the inaccuracies include missing critical steps, steps as written are not what needs to be done, or the steps are out of sequence)
- No/misplaced warnings or misuse of warnings (for example, a warning should never **state** the action to take; it should instead **emphasize** the action to take an the statement of the action should be in a numbered step)
- Poor format and presentation rules

Human-machine interface

Design for operability refers to designing the Human-System Interface (HSI) to be consistent with the abilities and limitations of the personnel who will be operating it. Weaknesses in the design processes can result in an HSI that is not well suited to the tasks that personnel must perform to ensure plant safety, resulting in increased workload, decreased performance by personnel, and an increased likelihood of errors.

Design for maintainability refers to designing the HSI and associated plant equipment to ensure that personnel are able to perform necessary maintenance activities efficiently. Weaknesses in the design process can result in systems that impose excessive demands on personnel for maintenance and, therefore, are prone to maintenance errors or problems with reliability and availability. *If it is hard to reach, workers will make more errors, including the error of deciding it is not worth it.*

Design for flexibility refers to the way that changes, such as upgrades to the HSI, are planned and put into use. A new HSI component may require the user to perform functions and tasks in new ways. Skills that the user developed for managing workload when using the former design, such as ways for scanning information or executing control actions may no longer be compatible with the new design. The new HSIs must also be compatible with the remaining HSIs so that operators can use them together with limited possibilities for human error. Also, HSI modifications may not be installed or put into service all at one time, causing the user to adapt to temporary configurations that are different from both the original and final configurations. Weaknesses in HSI implementation can increase operator workload and the likelihood of errors.

Fitness for duty

Implement a company fitness-for-duty program with a primary responsibility for detecting and preventing impaired personnel from performing tasks that may affect productivity and safety. Medical evaluations of personnel, behavioral observation programs, employee assistance programs, and drug and alcohol testing are used to detect impairment. Weaknesses in this program may allow impaired personnel to have access to vital areas in a plant where they could commit errors.

Overtime Policies and Practices – Most companies establish limits for work hours to reduce on-the-job fatigue and the potential consequences for poor task performance. Routine authorization for work hours in excess of those recommended may result in fatigued workers. Furthermore, a practice of excluding training or meetings that occur outside of an individual's normal work schedule from work-hour limitations will also contribute to fatigue. But in many cases these have weak enforcement, especially during shutdowns when the company is under stress to re-start soonest.

Shift Scheduling – Shift scheduling may also affect the likelihood that personnel will show performance decrements due to fatigue. A change in the assigned shift or a rotating shift schedule will disrupt circadian rhythms and may increase the likelihood of errors. So, a company must choose the proper shift rotation to allow adjustments to sleep patterns.

Work processes & Supervision

Written plant policies and procedures are meaningful only when they are enforced; otherwise, worker practices are the policy. Once discrepancies are tolerated, individual workers have to use their own judgment to decide what tasks are necessary and/or acceptable. Supervisors and managers should vigorously enforce plant policies/procedures and ensure that plant policies/procedures and practices are revised as necessary to be consistent. To enforce well, a supervisor must (1) lead by example, (2) watch workers often to ensure they are following company policies and procedures, and (3) determine the correct course of action (including recommending changes to policy and procedures) when workers are found deviating from procedures.

Work environment

Programmatic causes of task environment errors are typically found in the company's processes for designing human-system interfaces or in managing maintenance activities. Other programs may also be implicated. Common programmatic causes of task environment errors include:

Industrial Hygiene and Radiation Protection – These programs are responsible for ensuring that task environments have been evaluated to identify hazards and that needed controls are implemented to minimize exposures. Weaknesses in these programs may result in personnel working in task environments that are conducive to errors.

Work Planning and Control – Weaknesses in the work planning and control system may allow work to be planned without consideration of adverse environmental conditions and performed without the necessary compensatory measures. For example, communication devices may not be provided in noisy environments to support task performance. For tasks that involve unusual physical positions or cramped workspace, additional time to complete the task may not be scheduled. Rest breaks for hot and cold environments may not be planned into the work, or additional temporary lighting may not be provided if the work site is not adequately lighted.

Procedures – Weaknesses in the company's procedure development process may result in the design of procedures that are inappropriate for the conditions in which they will be used. For example, procedures that may be used at night, outside and in the rain should be laminated and the type size should be larger to ensure the procedure can be read. Procedures that will be used in vibration conditions may also require larger type size than procedures read in the stationary environment of the control room.

Human Factors Engineering – Weaknesses in the human factors engineering program may result in the installation of new equipment or systems without consideration of task environment characteristics. For example, the impact of control room lighting on the visibility of digital displays or effects of vibration on the legibility of dials or gauges at local control stations should be considered before installation.

Communication

Successful communication requires several steps. The sender first develops the intention to communicate either verbally or in writing. The sender then composes a message that presents the meaning as clearly as possible. The receiver must pay attention to the message and then interpret its meaning. If the communication is successful, the receiver interprets the message consistently with the sender's intended meaning. Table 8 summarizes the rules for successful verbal communication.

The similarity of the meanings given to the message by the sender and receiver can be verified through feedback. An example of feedback verification in verbal communication is when the receiver "repeats back" the message and the sender either agrees with the receiver's repeat back or corrects it. Verification feedback serves an important error-checking function in the communication process. It also allows supervisory oversight of communications to catch errors before they have consequences.

A sender and receiver must both be active for communication to be effective. The sender and receiver share responsibility for ensuring successful communication. However, when companies analyze the causes of events, errors in sending messages are more often identified than errors in receiving. The reasons for the difference are unclear. A company's investigation should consider sending and receiving errors and corrective actions should address both to be effective.

Shift handover is another critical aspect of communication. The status of equipment, repairs that occurred, problems faced, upcoming chores, etc., are all key information the oncoming worker will need to know. One good guide on shift turnover is U.S. DOE 1038-93. [19]

Table 8. Rules for effective verbal (face-to-face, radio, phone) Communication (© PII, 2012)

#	Communication rules
1	Senders MUST require repeat back by the receiver; ask questions to make sure of understanding.
2	Receivers of messages MUST repeat a message they receive.
3	Communicate one task at a time.
4	Sender & receiver must use the standard name and component number (and building name and number) and/or for related reference material (drawing, procedure name/number) in each message.
5	Sender must require receiver to give feedback of foreseeable conflicts. Sender (if message is to subordinate) should follow-up in the field if possible, to make sure message was understood.
6	Receiver must request clarification anytime they believe the message is confusing.
7	Do NOT give verbal instructions to workers who do not have the demonstrated skills to correctly understand and perform the task assignment (unless you plan to supervise/coach the task yourself).
8	Communicate to the most senior member on the work crew.
9	Know your audience & change message accordingly to compensate for lack of knowledge (when appropriate; see other related rules).
10	Use units (and use approved units, such as SI) for process parameters.
11	Always count from left to right when giving instructions related to multiple choices of equipment.
12	Use approved jargon only.
13	Use the words **check**, **make sure**, and **actions** consistently (see procedure writing rules).
14	When possible, write out the task (use a special procedure or work order) rather than communicating verbally. Use formalized templates for consistency and make sure the receiver is provided all information (such as work orders and drawings) necessary for the task.
15	Perform pre-job meeting with work crew (including with contractors or construction company).
16	Perform walkthrough at the location where the work will be completed.
17	Have a backup communication method. Don't rely on one mode of communication. In emergency situations have hand signals as backup to loss of verbal communication.
18	Talk to the operator or technician who is doing the work (no delegation of work); but see Rule 8.
19	Receiver (worker or group) must report back when work is complete.
20	If confusion exists in implementing a task, the shift supervisor (and higher, if supervisor is unavailable) must be contacted to make sure he/she understands the problem.
21	Do NOT use a PA (public address system) for process instructions, since this method of communication does not allow repeat-back.
22	Use Open questions and non-confrontational questioning methods when requesting clarification.
23	When communicating remotely (by phone or radio), if the message is not understood on the second attempt at clarification the Sender must find the Receiver and communicate face-to-face.
24	For large or complicated jobs, in addition to repeat back, the sender must ask the receiver for an assessment of the pre-job briefing to ensure the workers are not confused.
25	Workers and supervisors must keep a shift log to aid in turnover between shifts. Workers and supervisors and support staff (if necessary) must have a minimum or 15 minutes overlap with a relieving shift. *See shift turnover standard for more details.*

Conclusion

Closing the gaps in these 4 areas has a greater than 100:1 payback!

Acronyms and Abbreviations

ACC - American Chemistry Council
AIChE – American Institute of Chemical Engineers
API – American Petroleum Institute
CCPS – Center for Chemical Process Safety (of AIChE)
CMA - Chemical Manufacturers Association
CSB – Chemical Safety Board
DM – Damage Mechanisms
FMEA – Failure Mode and Effects Analysis
GCPS -- Global Congress of Process safety
HAZOP – Hazard and Operability; as in HAZOP Analysis or HAZOP Study
HRA – Human Reliability Analysis
HSI – Human-System Interface
KSA – Knowledge, skills and abilities
LOC – Loss of containment
LOPA – Layer of Protection Analysis
MOC – Management of Change
OSHA - US Occupational and Health Administration
PHA – Process Hazard Analysis
PSM – Process Safety Management
RBPS – Risk Based Process Safety
SOP – Standard Operating Procedure

References

[1] *Guidelines for Technical Management of Chemical Process Safety,* CCPS/AIChE, 1985.

[2] "The Cost and Benefits of Process Safety Management: Industry Survey Results," W. Bridges, *Process Safety Progress*, AIChE, January 1994.

[3] "Necessity of Performing Hazard Evaluations (PHAs) of Non-normal Modes of Operation (Startup, Shutdown, & Online Maintenance)", W. Bridges and Mike Marshall (US OSHA), *18th Annual International Symposium, Mary Kay-O-Connor Process Safety Center, College Station, TX, October 2015.*

[4] *Guidelines for Risk Based Process Safety*, CCPS/AIChE, 2007.

[5] "Human factors and their optimization" W. Bridges and G. Collazo-Ramos, *8th Global Congress on Process Safety*, Houston, AIChE, April 2012.

[6] PII Course 5, *Writing effective operating and maintenance procedures*, © 2003-current

[7] "Best practices for writing operating procedures and trouble-shooting guides", W. Bridges and R. Tew, *13th Global Congress on Process Safety*, San Antonio, AIChE, March 2017.

[8] "Effective Procedure Writing – A Two-Part Series (Part 1: Reducing Risk of Worker Errors; Part 2: Addressing Process Safety Management Requirements for Operating Procedures)," *Chemical Engineering Progress*, 1997, W. Bridges with T. R. Williams (both of JBFA).

[9] "Best Practices for Writing Operating Procedures and Trouble-Shooting Guides", L. Madden and W. Bridges, *12th Global Congress on Process Safety*, Houston, TX, AIChE, March 2016.

[10] *Guidelines for Hazard Evaluation Procedures*, 3rd Edition, 2008, CCPS/AIChE.

[11] "API RP 571 - Damage Mechanisms Affecting Fixed Equipment in the Refining Industry", API, 2nd edition, 2011.

[12] Bridges, W.G., et. al., "Addressing Human Error During Process Hazard Analyses," *Chemical Engineering Progress*, May 1994.

[13] "Necessity of Performing Hazard Evaluations (PHAs) of Non-normal Modes of Operation (Startup, Shutdown, & Online Maintenance)", W. Bridges and Mike Marshall (US OSHA), *18th Annual International Symposium, Mary Kay-O-Connor Process Safety Center, College Station, TX, October 2015 and 12th Global Congress on Process Safety, April, 2016.*

[14] "Recipe for a Complete Process Hazard Analysis – Especially Addressing the Key Demands from US CSB", R. Tew and W. Bridges, *13th Global Congress on Process Safety*, San Antonio, AIChE, March 2017.

[15] PII Course 8, *PHA/HAZOP Leadership training*, © 2003-current.

[16] "Gains from Getting Near Misses Reported", W. Bridges, *8th Global Congress on Process Safety*, Houston, AIChE, April 2012.

[17] US Chemical Safety and Hazard Investigation Board (CSB), *Anatomy of a Disaster: Explosion at BP Texas City Refinery,* published 2006.

[18] US Chemical Safety and Hazard Investigation Board (CSB), *"Pesticide Chemical Runaway Reaction Pressure Vessel Explosion – Bayer CropScience, LP",* published 2008.

[19] "DOE-STD-1038: Guide to Good Practices for Operations Turnover", US DOE, 1998.

546668 Recycle Line from NGL Spheres to Residual Gas Header

Abstract:

HNGL plant is received 1.6 BSCFD of Haradh gas flows into the DGA Gas treating units where CO2 is removed and the produced sweet gas is directed to the NGL recovery units .Hawiyah sales gas is pretreated at HGP for CO2 removal ; hence the produced 2.4 BSCFD flows directly to the NGL units where it is mixed with the Haradh gas. The mixed gas then is processed in the NGL units producing C2+ NGL and sales gas stream. The NGL is pumped to storage spheres and subsequently shipped to NGL piping network through Shedgum Gas Plant. The Residue Gas flows to the sales gas compression area which increase the pressure allowing routing back the gas to the sales gas grid.

The purpose of this project is to install vent line from the two NGL Spheres to Sales Gas (Residue Gas) common header at Hawiyah NGL Plant. The aim of this project is to minimize the continues flaring from the NGL spheres , safe plant resources and protect the environment

Sphere PVs were venting the NGL vapor to the cryogenic burn pit when the pressure of the sphere exceeds 350 psig. In order to lower sphere pressure without losing NGL production, newly installed PVs are set at lower pressure set point of 305 PSIG to recycle the spheres vapors to NGL trains residue gas line, downstream of the brake-compressors..

During summer months July and August (if associated with maximum plant production), spheres are venting about 4.0 MMSCFD of Methane. Based on Methane price of 4.4 $/MMBTU then total saving is 1.05 MM $/year.

547776 Practical Notes on Ragagep Compliance

Process Safety Management (PSM) and Risk Management Plan (RMP) regulations have long required facilities to ensure that they comply with Recognized and Generally Accepted Good Engineering Practice (RAGAGEP). In recent years, authorities having jurisdiction have placed increased emphasis on RAGAGEP compliance. Specifically, emphasis has been placed on documentation and compliance with the industry codes and standards to which a facility is designed and maintained. This poses a unique challenge to older facilities, which have been designed to codes and standards long out of date and especially to owners and operators that do not have access to engineering documents used in the original design of their facility. For facilities using outdated codes, OSHA requires that the operating company document that "equipment is designed, maintained, inspected, tested, and operating in a safe manner."

Over time, the gradual evolution of industry codes and standards may cause facilities to fall out of compliance with the RAGAGEP requirements of the regulations. This paper provides example cases based on authors' experience with various owners and operators of older facilities. Practical methods are discussed for achieving and maintaining RAGAGEP compliance. It includes examples of comparing systems designed to outdated codes and standards against the latest codes, to ensure that they are able to maintain compliance with RAGAGEP.

548482 Propane Safety: Investigation Findings and Lessons Learned in the 2014 Philidelphia FOOD Truck Explosion

On July 1, 2014, a propane cylinder catastrophically failed and exploded on the back of a food truck in Philadelphia, PA. The explosion led to a sudden release of pressurized propane vapor and superheated propane liquid, which rapidly evaporated resulting in a large white cloud that entered and engulfed the rear of the food truck. This cloud ultimately found an ignition source within the food truck and resulted in a significant fireball burning both inside and outside of the food truck. The explosion and ensuing fire fatally injured the food truck owner and her daughter, and also caused injuries to 11 other individuals. This paper will first present the results of our investigation into the cause and origin of the catastrophic failure and explosion event, which included: (1) analyzing video footage of the explosion and ensuing fire; (2) applying propane phase diagrams to determine the expansion and resulting pressures within the cylinder prior to failure; and (3) applying blast techniques to determine the post catastrophic blast and ignition during the event. These techniques included advanced computational fluid dynamic modeling.

The paper will also discuss safety critical design features and processes in the propane industry that must be in place to ensure the safe use and filling of propane cylinders, which include: cylinder requalification; having a fixed maximum liquid level gauge; having a pressure relief valve; and filling cylinders by weight or volume.

548391 Using the Risk Analysis Screening Tool (RAST) in Process Design to Manage Risk

The Risk Analysis Screening Tool (RAST), available through the Center for Chemical Process Safety (CCPS), courtesy of the Dow Chemical Company can enable users to identify hazards and risk scenarios to better manage process safety risks. RAST is a spreadsheet based tool that uses simplified algorithms to identify and evaluate risk scenarios of a unit operation. The risk scenario identification is based on generic HAZOP deviations applied to the unit operation. An engineer can use RAST to identify risks to proposed process designs or plant modifications in the design phase of a project, before P&IDs are developed and a full process hazard analysis study is done. This can save time in process design by weeding out higher risk process alternatives as the process flow diagram is being developed.

The utility of RAST is demonstrated by application to actual case studies. RAST is applied to the runaway reaction at T2 Laboratories in 2007 and to the chlorine release at DPC Enterprises in 2003 showing how the hazardous events and consequences could have been analyzed during the design phase of the projects.

Drug Product vs. Process Safety: Integrating Both into the Design Process from Concept Development to Start-up

Jeffrey Heil, P.E.
Lead Process Engineer
DPS Group Global
959 Concord Street, Suite 100
Framingham, MA 01701
Jeffrey.Heil@dpsgroupglobal.com

Prepared for Presentation at
American Institute of Chemical Engineers
2019 Spring Meeting and 15th Global Congress on Process Safety
New Orleans, LA
March 31 – April 3, 2019

Drug Product vs. Process Safety: Integrating Both into the Design Process from Concept Development to Start-up

Jeffrey Heil, P.E.
Lead Process Engineer
DPS Group Global
959 Concord Street, Suite 100
Framingham, MA 01701
Jeffrey.Heil@dpsgroupglobal.com

Keywords: Process Safety, Process Design, Biotechnology, Inherently Safe Design

Abstract

Process safety has been around for over 350 years but may easily be overlooked or treated as an afterthought in the design process, often leaving hazard reviews and safety valve considerations to the end of detailed design. In the chemical industry, emphasis on safety is generally in respect to process safety, but in the growing biotechnology industry the emphasis is often shifted to product and patient safety. Due to this subtle yet significant difference in focus, it is often challenging for the design team on these projects to secure the necessary project resources to ensure both robust product and process safety in the facility being designed, often on an accelerated schedule. Leveraging experience from working in the oil and petrochemical industry for over 10 years, an in-depth understanding of the project design cycle, and by asking some simple questions along the way; process safety in design can be assured for even the most challenging projects, without a significant impact on project resources.

1 Introduction

1.1 Brief History

The first safety valve, developed in 1681 by Frenchmen Denis Papin, protected the first ever pressure cooker, one of which was used to make meat jelly for England's King Charles II. [1] Papin first invented the steam pressure cooker in 1676 to make the jelly and other food products but then realized no one would be willing to operate the cooker given the risks. Having identified the hazard, he set out to invent a way to mitigate that hazard while ensuring the King got his jellied meat pies.

Following development of the first weighted lever style safety valve invented by Papin, little progress was made until the early 1800's when locomotive steam engines gained notoriety both as a superior means of transportation and a hazardous machine often operated beyond its limits in the name of more power and speed. Safety valves were revised throughout the early 1800's to make these locomotives safer and formally modernized in 1856 with the invention of the first spring actuated pressure relief device that could not be modified to a different pressure, while in service [2].

Modern process safety, as we can still see it in existence today, dates to the early 1900's with the development of modern oil refineries and all the hazards that they include. With these refineries and their complex chemical processes came a wealth of experimentation, safety system design, and industry supported safety initiatives. The goal was the same as it is today, ensuring safety to the employees, the environment, and the facility. The problem with many of these early safety initiatives is that different companies had different ideas of what safe meant and relied on historical data to predict the future. The notion that "it has been running like this for 20 years and will continue for another 20 years" lead to a raising number of deadly process safety incidents, particularly throughout the 1970's to 1980's (i.e. Bhopal, India Methyl Isocyanate release and Piper Alpha explosion in the North Sea.). Plants were operating to the status quo, not improving their practices new technology became available. In parallel, urban sprawl forced industrial plant operation closer to populated areas, increasing risk to the community. The OSHA process safety management (PSM) initiative was introduced into law in the United States to address this very problem [3]. The PSM standard was performance-based meaning that as new technology was developed and safer systems could be designed, those safer designs had to be implemented; i.e., the status quo was no longer good enough. The PSM standard was ratified in 1992: hence, even though the oil and petrochemical industries in the United States have been around for about 100 years, they have only been operating under this standard for 26 years.

Contrast this with today's biotechnology companies. Gentech, one of the earliest, received FDA approval for the insulin drug Humulun in 1982. Like any new technology, investment and growth in biotechnology was slow to start, not taking off until early 2000's. There are many metrics for tracking growth of the Biotechnology industry such as venture capitalist investment and the number of companies in operation, but the high number of mergers and acquisitions over the past 30 years makes such comparisons incomplete. The Nasdaq Biotechnology Index (NBI) was established in 1993 with a baseline index value of 200 and provides a good metric for tracking industry growth over time. After holding steady in the low 200-300s throughout the 1990's, the NBI surged to 1,300 in 2000 as the Biotechnology bubble was forming. The index has steadily risen to a peak of 4,003 on 01-July-2015 and is hovering currently around 3,500 (Q1 2019), exemplifying the continued growth of the industry over the past 20 years. [4]

Pharmaceutical companies (e.g., Aspirin, invented by Bayer in 1899) have been around almost as long the oil companies have and are sometimes confused with biotech. Biotechnology companies are adolescent at best, having typical company life of less than 20 years with little process safety oversight beyond building code requirements as they do not meet the hazard requirements of the OSHA PSM Standard. Not meeting the hazard requirements of the PSM standard does not exempt these companies from a responsibility to ensure a safe working environment. If anything, their proximity to high density population areas and minimal operating experience places them at a higher risk with respect to process safety, but in many cases the operators and designers have only worked in these types of facilities and may not fully appreciate the safety concerns present. Coming from a background where process safety is paramount and the product quality, while important, takes second place, the author has made it his personal mission to increase awareness throughout the Biotechnology industry one project at a time.

1.2 Process vs. Product Safety

In the chemical industry, emphasis on safety is always in respect to process safety; ensuring that the public, employees, environment, and facility (in that order) are sufficiently protected from inherent hazards of a process. "Safety" refers to containment of the process and protection of personnel from harm. If a hazardous material cannot escape the process; it cannot be breathed in, released to the environment, set on fire, or result in damage to equipment that may cause injury to nearby personnel. The previously described process safety management standard formalized the set of tools used by these industries to ensure current best practices are used when it comes to process safety in hazardous environments. The goal of the PSM standard and implications of it being "performance based" is summarized as follows:

> *"... Recognizing the dangers associated with highly hazardous chemicals, Congress mandated (by the Clean Air Act Amendments of 1990) that OSHA set requirements for the management of highly hazardous substances to prevent and mitigate hazards associated with catastrophic releases of flammable, explosive, reactive, and toxic chemicals that may endanger workers. As a result, OSHA developed the Process Safety Management (PSM) standard (issued in 1992), which covers the manufacturing of explosives and processes involving threshold quantities of flammable liquids and flammable gasses (10,000 lbs.), as well as 137 listed highly hazardous chemicals. The PSM standard (29 CFR 1910.119) requires employers to implement safety programs that identify, evaluate, and control these hazards. Unlike some of OSHA's standards, which prescribe precisely what employers must do to comply, PSM is a "performance-based" standard-it outlines key features of safety programs for controlling highly hazardous chemicals, and employers have the flexibility to tailor their safety programs to the unique conditions at their facilities..." [4]*

The biotechnology industry on the other hand often has a more focused emphasis on product safety. Product safety ensures that all current good manufacturing practices (cGMP) are being followed with the goal of protecting the end-user or patient from hazards of a contaminated biological drug product. Product safety was important for the pharmaceutical industry because the medicines being produced were intended for use in the human body and would chemically alter the bodies function. Product cross contamination could result in serious health side effects, especially with highly potent or specialized substances. Biotechnology processes work by growing viruses and manipulating this viral material to deliver highly specialized medicines to the patient. Poor manufacturing conditions could allow unintended bacterial growth alongside the desired product, leaving the future patient with a potentially deadly virus or bacterial infection instead of a cure. The US Food and Drug Administration (FDA) passed the first cGMP regulations in 1969 [6], making them older and with many respects better understood in the industry than the PSM standard is in the oil and petrochemical industry. The practiced engineer will note several similarities in the quoted statements describing the PSM standard and the cGMP standards.

> *"The CGMP requirements were established to be flexible in order to allow each manufacturer to decide individually how to best implement the necessary controls by using scientifically sound design, processing methods, and testing procedures. The flexibility in these regulations allows companies to use modern technologies and innovative approaches to achieve higher quality through continual improvement. Accordingly, the "C" in CGMP stands for "current," requiring companies to use technologies and systems that are up-to-date in order to comply with the regulations. Systems and equipment that may have been "top-of-the-line" to prevent contamination, mix-ups, and errors 10 or 20 years ago may be less than adequate by today's standards." [7]*

Product safety requires process containment too, but in many cases the consequences are less likely to cause bodily harm so long as product quality control procedures ensure quarantine of the affected drug product. Note that this statement applies to the biotechnology industry, the consequences of a loss of containment in the pharmaceutical industry can be as deadly as in a chemical plant, depending on the substance released. Fifty years ago, when the cGMP's were adopted, pharmaceutical operators already knew about process safety and adopted it as readily as oil refineries and industrial chemical plants did. Twenty years ago, biotechnology was in its infancy and there were few large-scale biological processes in operation. Ten years ago, most biological processes ran at atmospheric pressure using water-based chemistry. The risk associated with potential process safety events was minimal in comparison. Today, many biotech processes use advanced solvent-based chemistries needing large volumes of hazardous flammable solvents, a key element in the severity of an incident when a loss of containment occurs. Add to this unique process chemistries such as dichloroacetic acid, methyl chloride, hydrochloric acid, and sodium hydroxide; all of which can easily corrode stainless steel, a common metal alloy used in biological processing, and unique hazards become a reality even at small scale.

2 The Design Process

Designing processes to produce goods takes time, time that the future producer of those goods doesn't always like to part with because "time is money". Each month spent designing or building the facility to manufacturer a product is a month of revenue missed or often in biotech, a month of trials that could be running. A typical design process includes several independent stages or phases. An ideal project would separate each of these phases with well-defined checks creating structure, an opportunity for detailed reviews, and adding time to the project. Initial concept development, feasibility, and design of large-scale chemical or industrial facilities often involves acres of real estate, spans years, and costs millions of dollars. By contrast small-scale bioprocessing facilities, often less than 40,000 ft^2, are fully designed in terms of weeks or months and leverage parallel work paths where possible.

The impact of such a short design schedule is that time for staggered activities, separated by well-defined stage-gates, is often sacrificed in the name of the project schedule. It is possible to crash a project schedule, completing the design early, but doing so limits the amount of design work that can be completed based on previously completed design elements. If a project schedule allots for 6-8 weeks between establishing the basis of design for a facility and documents that are to be issued for vendor bid, scheduling one week for a hazard analysis and another week to implement the resulting design changes (25% of the schedule) is a challenge. By the time the hazard analysis is complete, many aspects of the design have been progressed and resulting changes will result in unbudgeted rework. The same is true for the design of key safety systems such as relief devices or process interlocks that require a complete system design to be finalized. The engineer can anticipate credible overpressure scenarios and safe operating limits for a process but cannot complete these design elements until piping to and from the relief device is complete and the hazard analysis is approved, accepting the level of protection assigned to each process step. It is for these reasons that it is not uncommon to see relief devices for a process being sized and planned for as one of the final project tasks. Explaining to an architect that a storage tank must be moved to meet a flammable liquid set-back identified during a hazard review or that another roof penetration is required for a relief device tail pipe two weeks before the completion of a design are just two examples of real-world challenges that project teams may be faced with.

Understanding the different design phases or stage-gates, the tasks that occur during them, and their impact on the ultimate facility design is important if one is to create a plan for addressing process safety on such a short schedule. Table 1 summarizes one example of a design process, key stage-gates, and design elements to be considered in each phase. Note that jumping a phase and for example combining schematic design and design development (the most common violation of the design process) into one phase can confuse the separation of tasks between these phases and eliminate checks important to the design process.

Table 1 – Key Aspects of the Design Process

Phase	Process Design	Process Safety
Feasibility Study	• Identify product to be manufactured and major unit operations • Establish block flow diagram of process • Identify facility requirements • Confirm business case for proposed facility	• Identify hazardous materials used in the process • Identify hazardous operating conditions (i.e. high pressure) • Establish acceptable risk criteria for operation of the facility

Phase	Process Design	Process Safety
Concept Design	• Breakdown individual processing steps to the equipment level • Develop process flow diagrams for each major process step • Develop preliminary heat and material balance • Develop equipment list and identify long lead equipment (to prioritize design development tasks) • Detailed layout of process equipment • Identify utility requirements and size utility systems • Size and layout major process headers	• Develop a comprehensive hazardous and flammable materials inventory / determine if additional design precautions are required • Identify potential material compatibility concerns and material specifications for each process step • Complete a code review in collaboration with the full design team to ensure all building, fire, and design codes are being followed • Ensure facility citing requirements are met in the layout
Design Development	• Develop initial P&IDs for the process • Identify and document key control loops and automation requirements • Finalize design of all process equipment (size, pressure rating, operating conditions) • Complete detailed sizing and specification of major process headers and utility lines • Finalize process heat and material balance • Prepare equipment and piping specifications and datasheets	• Confirm inherently safe design where possible (pressure profile versus equipment design) • Identify safe operating limits for each unit operation, add control points as required • Review P&IDs for relief devices and confirm their discharge locations • Complete scenario analysis and preliminary sizing for all relief devices. Based on preliminary sizing, define relief device inlet/outlet piping • Confirm piping specifications versus services • Confirm vessel sizing versus hazardous material inventory and update as required • Complete a preliminary hazard analysis on the full process at the end of the design development stage

Phase	Process Design	Process Safety
Detail Design	• Incorporate all recommendations from the design development preliminary hazard analysis • Finalize all equipment and line sizing • Fully specify all process components including equipment, lines, valves and instruments	• Ensure that all actions from the preliminary hazard analysis are incorporated into the design • Review and finalize relief device sizing • Complete sizing and design of effluent handling system for relief device discharges • Double check the set points of key process variables versus the operation's safe operating limits, previously defined. • Review final equipment specifications against material compatibility matrix and equipment specifications

3 Safety Throughout the Design Process

Process safety should be considered in the earliest stages of design. One principle of process safety design is that inherently safe design, creating a system that physically cannot fail, is the only way to prevent an incident from occurring. The easiest example of an inherently safe system is designing a pressure vessel to withstand maximum system pressure, even if all process controls were to fail. While an easy goal on paper, inherently safe design is not always possible and often never fully achievable. A vessel can be designed to the proper pressure, but an unforeseen corrosion issue could reduce the design pressure making the vessel susceptible to premature failure. In anticipation of this and to protect against it, engineers design-in multiple layers of protection to a process starting with inherently safe design at the concept level and ending with proper emergency response procedures developed during commissioning and simulated during plant operation. Continuing with the vessel example, minimizing the quantity of material stored and having an inventory management plan to support smaller on-hand inventories is another way to reduce hazards associated with the process. Additional design principles or layers of protection used are summarized in Figure 1. Different layers of protection against a process safety event are established during concept development and built-up throughout the design process.

Figure 1. Typical Layers of Protection Applied to a Design

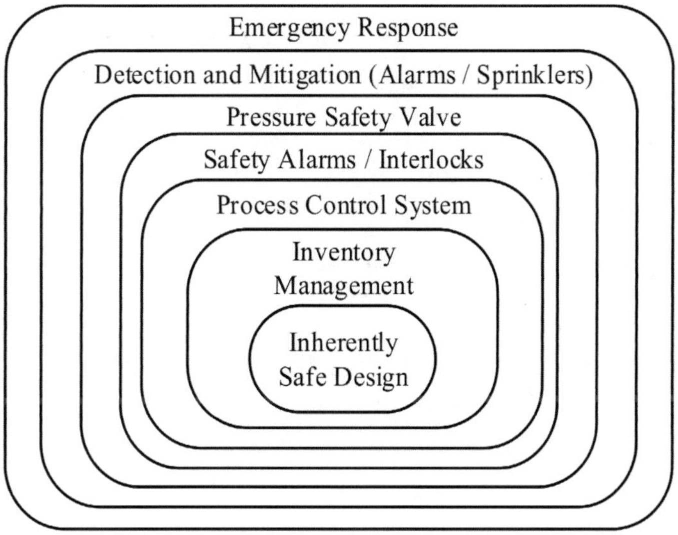

3.1 Concept Design

Concept design is one of the earliest stages of the design process. During concept design types of process equipment are first selected, their location mapped out, and consideration given to how materials will be heated up, cooled down, transferred, and stored when not in use. By asking a few questions as these fundamental design decisions are made, the design predisposes its self to being inherently safer.

- What is the worst-case operating condition, can the system be mechanically designed to handle those conditions?
- Where will hazardous materials be present and can that equipment be installed away from high concentrations of personnel?
- What hazards do the materials pose (toxic, flammable, corrosive)?
- How much of a potentially hazardous raw material must be on-hand? Weekly vs. monthly delivery may reduce stockpiles by 75%.
- Is there a safer way to produce the desired product?
- Is there a safer way to complete this unit operation?

3.2 Detail Design

Concept design develops enough detail to prove that the processing goal is obtainable and business case can be made. Concepts developed such as system location, connectivity, and required unit operations are further developed into a constructible design that can ultimately be operated to produce a finished product. In detailed design, detailed questions are needed as more layers of protection are added. These questions are to be asked informally throughout design and will be formalized in a process hazard analysis (PHA), completed during the detailed design phase.

- Where are the areas of high pressure, temperature, and hazardous material inventories; can they be reduced?
- Are there any specific code requirements or industry best practices for this system?
- What are the expected operating parameters, how can they be exceeded, identified ahead of time, and mitigated?
- Are all systems protected by a mechanical pressure relief device (if needed), do those devices discharge to safe locations?
- How and where could a loss of containment occur, how can it be quickly identified (leak or fire detection)?
- How will the consequences of an event be mitigated once started; how can the event be stopped (process shutdown or fire protection)?
- How can the public, employees, and environment be protected from the consequences of an event if it were to occur (emergency management)?

3.3 Start-Up

Following detailed design and construction, a process must be started, commissioned, qualified, and transitioned to manufacturing. Throughout this process, design assumptions must be confirmed, formal procedures verified, and changes to the design noted so that their impact on the process can be determined. Process safety considerations during start-up must include the full team from engineering to operations and maintenance.

- How does the as-built design compare to the final design, what are the impact of any changes?
- Were changes made to the operating envelope for any of the systems (pressure, temperature, level, concentration)?
- Has a pre-start-up safety review (PSSR) and as-built PHA been completed?
- Do operating procedures match the actual operation of the process?
- Have emergency drills been completed, are there lessons learned from these?

Addressing process safety throughout the design process, versus at the end of design when final design reviews are completed; minimizes late stage design changes, results in a better designed facility, and ensures safe facility operation for years to come.

4 Conclusion

Process safety is a well-established element in the design of oil and petrochemical facilities, taking a front row seat in the design process and being an area of focus once operations begin. Product safety holds a similar role and has been common place for a similar period in pharmaceutical and recently bioprocessing facilities. While the gap between these two elements of design has been closing in recent years as more pharmaceutical companies enter the bioprocessing arena and bioprocessing operations grow, a considerable gap remains. Based on the authors experience designing processes and facilities in both of areas, a few causes for the gaps are easily identifiable:

- Historically, bioprocessing facilities have not contained hazardous processes. Areas of pressure were limited to gas delivery systems (i.e. nitrogen and compressed air) and storage tanks contained aqueous based chemistries (i.e. bioreactors) minimizing the hazards with a loss of containment of these products.
- The shortened design timeline for bioprocess facilities challenges the ability to incorporate process safety into the design at distinct stage gates. The schedule does not allow for a discrete process safety design effort while other design work is on hold.
- There is a general lack of knowledge regarding safety system design in the biotechnology industry. This goes back to the opening statement that in many cases engineers responsible for these systems don't know what they don't know.

Product and patient safety are well defined requirements for biomanufacturing companies and are considered in the earliest stages of the drug discovery process. Applying similar principles to process safety throughout the design lifecycle, and not at distinct stage gates, will ensure a smooth transition from process development to design, validation, and ultimately commercial manufacturing.

5 References

[1] F.M. Editors of Encyclopedia Britannica. "Denis Papin." *Encyclopedia Britannica*, Encyclopedia Britannica, Inc., 18 Aug. 2018, www.britannica.com/biography/Denis-Papin.

[2] Ramsbottom, J. "On an Improved Safety-Valve." *Process Institute of Mechanical Engineers*, vol. 7, 1856, pp. 37–47.

[3] US Department of Labor Final Rule, "Final Rule on Process Safety Management of Highly Hazardous Chemicals; Explosives and Blasting Agents", 1910.109, *Fed Register #: 57.6356*, 24 Fen. 1992.

[4] NASDAQ Biotechnology Index, NBI, https://finance.yahoo.com/quote/%5Enbi/.

[5] *Process Safety Management Introduction.* Occupational Safety and Health Administration, www.osha.gov/dsg/psm/index.html.

[6] US Food and Drug Administration, *Good Manufacturing Practices (GMPs) for the 21st Century - Food Processing*, "GMPs Section One: Current Food Good Manufacturing Practices", 09 Aug. 2004, www.fda.gov/food/guidanceregulation/cgmp/ucm110907.htm.

[7] US Food and Drug Administration, "*Facts About the Current Good Manufacturing Practices (CGMPs)*", 25 Jun. 2018, www.fda.gov/food/guidanceregulation/cgmp/ucm110907.htm.

546706 LOPA and Sil Risk Analysis As Critical Tools for PSM Culture and Accidents Prevention

Industrial facilities that operate with dangerous products, whether chronic or acute, are increasingly common to be present and part of society at large, often allowing an aggregate of increasingly rich and economically beneficial professions and specialties to the community, be it directly or indirectly. Accidents in these types of units, especially those operating with hazardous products of acute consequence (those that have the capacity to cause fatalities when released to the environment), such as flammable, toxic, explosive products, due to the possibility of occurrence of undesirable events and the behavior of the effects resulting from releases of these types of product, could cause serious damage to the people and / or the installations that are in its vicinity. The various types of accident studies are known as EAR (Risk Analysis Studies), some of which are used more widely and others are specific to specific response needs.

According to BERNSTEIN (1997), the word "risk" derives from the ancient Italian risicare, which means "to dare". While DUARTE (2002) describes that the meaning of the word risk is an expression of thought that has a lot to do with the social, scientific and technological evolution of man. As BILLINTON and ALLAN (1992) note, society has great difficulty distinguishing between a hazard, which can be prioritized in terms of its severity, but does not take into account its probability, and risk, which considers not only the dangerous event, but also their probability of occurrence. KAPLAN (1997) explains that when a person asks "What is the risk?", 3 (three) questions are actually being questioned: What can happen? How can this happen? And if this does happen, what are the consequences? The answers define the scenario, frequency and consequence.

Within the agencies that tend to manage corporate risk, it is noticeable that the risk associated with a particular accident scenario is calculated according to its frequency of occurrence and its severity. Frequency is related to the temporal chance of occurrence of the initiating event, which leads to the release of dangerous product, such as LPP (Liquefied Petroleum Gas) emanation by ball rupture through overpressure of this equipment or, as a second example, the intrinsic rupture of a stretch of line. The consequence, also called severity, considering personal safety, is the capacity that the physical effects, due to the release of dangerous product, associated to the analyzed initiating event, can cause fatality (s), related to the potential of damage related to each effect analysis. The risk, qualitatively or quantitatively observed, and defined in a simplified way, is evaluated as the multiplication of frequency of occurrence and severity. The total social risk related to a particular facility, which can be defined as the risk that a community outside the facility is subject to by its presence in its vicinity, can be calculated as the sum of the risks of all the initiating events arising of the same. In order to do this, for each accident scenario, not only the generating causes, known as Initiating Events, but also the likelihood of ignition, the probability of affecting people, and several other parameters are considered, being one of them very important and permissible of analyzes, which are the failures

safeguards. These, if they act effectively when demanded, can prevent the accident from occurring, since the Initiating Event happens.

Hazardous product risk studies, such as qualitative (Hazard Operational Preliminary Analysis (APPP), Hazard Operability (HAZOP, among others) or those known as quantitative analyzes (Vulnerability Analysis and Quantitative Risk Analysis per se), in addition to being increasingly requested by regulatory agencies, have often been seen as an internal requirement of companies, representing one of the pillars of industrial safety management, in order to keep their risks at low or acceptable levels. However, these analyzes are usually performed in stages

until the final definition of the risk, comparison with acceptability criteria, and later mitigation, if necessary, each of these stages having a minimum time required for its execution, which may make the analysis complete rather time-consuming. Situations similar to these are detrimental to Risk Management when, in fact, what is desired is not a complete risk analysis, but only the achievement of a specific response, such as if my identified risk scenario is effective and properly protected.

The extent of possible damage related to each identified accident is delimited by the intensity of the physical effect causing the damage, and the relationship between the intensity of the physical effect and the corresponding damage is established through the vulnerability models.

On the other hand, semiquantitative risk analyzes such The extent of possible damage related to each identified accident is delimited by the intensity of the physical effect causing the damage, and the relationship between the intensity of the physical effect and the corresponding damage is established through the vulnerability models.

On the other hand, semiquantitative risk analyzes such as LOPA (Layer of Protection Analysis) and SIL (Safety Integrity Level), even though they are less demanded by chemical and petrochemical companies than qualitative or quantitative steps can have beneficial and effective results in terms of knowledge of all the characteristics of the analysis scenarios and of their protection barriers and, in a practical way, much faster to be developed. In addition, they can provide results that do not make up the analysis range of qualitative and / or quantitative studies in depth, such as knowledge and prior review of safeguards, many of which are known as Independent Protection Layers (CIP or IPL - Independent Protection Layers).

Therefore, with the anticipation of an increase in the number of future installations operating with hazardous products and the current installations that already do so, the semi-quantitative studies of LOPA represent an enormous value for the observance of the existing protections in each scenario of process accidents, also serving as the basis for the determination of new Independent Protection Layers or even the reduction of their number, tending to simplify the steps of determining the frequency of the scenario., even though they are less demanded by chemical and petrochemical companies than qualitative or quantitative steps can have beneficial and effective results in terms of knowledge of all the characteristics of the analysis scenarios and of their protection barriers and, in a practical way, much faster to be developed. In addition, they can provide results that do not make up the analysis range of qualitative and / or quantitative studies in depth, such as knowledge and prior review of safeguards, many of which are known as Independent Protection Layers (CIP or IPL - Independent Protection Layers).

Therefore, with the anticipation of an increase in the number of future installations operating with hazardous products and the current installations that already do so, the semi-quantitative studies of LOPA represent an enormous value for the observance of the existing protections in each scenario of process accidents, also serving as the basis for the determination of new Independent Protection Layers or even the reduction of their number, tending to simplify the steps of determining the frequency of the scenario.

546375 Controlling Static Discharge Using Antistatic Additives

Controlling Static Discharge using Antistatic Additives.

Abstract

The presentation will discuss controlling static discharge in non-fuel, non-conductive/low-flash point fluids. The scope of the presentation will concentrate on industrial producers, distributors, consumers of these fluids. Furthermore,furthermore, the presentation covers several industries where these fluids are most commonly found. Common Static-Accumulating Flammable Liquids That May Form Ignitable Vapor-Air Mixtures:

- VM&P naphtha
- Cyclohexane
- n-Heptane
- Benzene
- Toluene
- n-Hexane
- Xylene
- Ethyl benzene
- Styrene

Static discharge has caused catastrophic explosions and fires (U.S. Chemical Safety Board, Static Spark Ignites Explosion Inside Flammable Liquid Storage Tank No. 2007-06-I-KS).

Issues

- Nonconductive flammable liquids can accumulate static electricity during transfer and storage.
- Static sparks can readily ignite flammable vapor-air mixtures inside storage tanks.
- Material Safety Data Sheets (MSDSs) often do not adequately communicate hazard data and precautions.

Key findings:

- Safety publications omit antistat options. (see Process Safety Beacon, August, 2016)
- US affiliates of multinational corporations have different guidelines for antistatic use than their parent companies.
    - NFPA®77 and the American Coatings Association's, *Generation and Control of Static Electricity in Coatings Operation* has guidelines that are different than EU regulations ATEX- Appareils destinés à être utilisés en ATmosphères EXplosibles and ESIG's *Flamability: Safe Working With Industrial Solvents.*
    - The presentation will highlight these differences and how they contribute to the disparity in antistatic additive applications.
- Normal Bonding and Grounding May Not Be Enough! Companies that handle, transfer, and store flammable liquids should contact manufacturers to determine if these liquids can accumulate dangerous levels of static electricity, and if they can form explosive vapor-air mixtures inside storage tanks. If so, *extra precautions*—beyond normal bonding and grounding—may be necessary. Antistatic Additive is an extra precaution.

Recommendations: Knowledge members can use

- Who adds the Antistat?
- How much Antistat is added?
- How is the Antistat added?

546677 Led Your Process Safety Culture, Lead Beyond Hierarchies

The Oil & Gas industry is trying to advance at the next crossroads: "How to achieve that leaders without skills and without training in the management of culture, transform their own culture?"

Our research presents three key ideas that leaders from all over the world

can apply immediately to lead risk management with effectiveness:

frequency, consistency, meaning.

When specific leadership competencies to manage risk have not yet been developed,

Leaders can use this three concepts of easy adoption.

Increasing frequency of interactions with people will allow us to generate safer operations. In full attention economy and contexts of increasing complexity, for risk management, more communication is better communication. But it is not enough to improve the frequency of interactions. That frequency must be accompanied by consistency. The challenge is to achieve possible consistency between what we say and what We do. Leading by example is one of the central keys of risk management. The lack of consistency provides symbolic permissions to normalize deviations. We must maintain a high frequency of face-to-face interactions, but in addition, we must be consistent at every opportunity. And finally, to give meaning to our collaborators. It is necessary to ensure the understanding of risks. Each collaborator need to build his or her own sense of vulnerability and Operational Discipline.

We need workers who operate according to procedures and not adjusted to what the bosses say. This is what meaning allows.

548192 The Importance of Using Color Scale in Process Safety

Everyone knows the importance of color. From physics to fashion going through the automotive industry she will always be present. There is no world without color. Often, they come up on the scale form or not, but the variation in the scenarios are always important. It is so on the traffic sign and the ph tape. From the Diamond of Hommel to across the street. Colors, colors and colors. In addition to not necessarily needing ididomas, such as color of the sky and molten metal have a variety of shades. The main idea of the work is to choose a suitable color to be used in communication within a work area, choosing which and how many elements can evaluated suggest a color scale from the chosen one and present its contribution in the process safety.

550372 Flammable Gas Release Modeling for Real-Time Analysis Using Adversarial Variational Bayes

In this study, we have suggested a method to calculate high-accuracy damage range from flammable material in petrochemical complex. There are many flammable and toxic substances in chemical processes. Computational Fluid Dynamics(CFD) is one of the best way to calculate and simulate scenarios of release range of various fluids or gases. However, it is not used to calculate damage range, because of a lot of time to simulate one scenario. Thus, lots of researchers have enthusiastically studied the surrogate model or reduced order model to get the results of release range in real-time [1-3]. Recently, deep learning based generative model has receive attention because of its superior performance of non-linear manifold estimation [4]. Herein, we propose surrogate model based on adversarial varational Bayes (AVB) [5] for real-time analysis of probability of death. The proposed surrogate model can extract the latent space of high-dimensional flammable and toxic substances dispersion contours. Furthermore, it can generate the estimated contour image from variables of interest much faster and more precise than CFD data memorization with linear interpolation because it does not memorize the data but train the non-linear manifold regarding exact posterior distribution on given observation which basis is predefined in standard variational autoencoder (VAE) [4, 6].

The acrylonitrile was studied about risk assessment for health damage like cancer hazard.[7] But the component has flammable and toxicity, so we should calculate range of damage. The main scenario in this study is the acrylonitrile released from underground model, and the result is shown different damage following hole size, mass flow rate, wind velocity, and wind direction. The superior performance of the proposed model was exemplified by comparing with other surrogate models.

[1] Chen, T., Hadinoto, K., Yan, W., & Ma, Y. (2011). Efficient meta-modelling of complex process simulations with time–space-dependent outputs. *Computers & chemical engineering*, *35*(3), 502-509.

[2] Wang, K., Chen, T., Kwa, S. T., Ma, Y., & Lau, R. (2014). Meta-modelling for fast analysis of CFD-simulated vapour cloud dispersion processes. *Computers & Chemical Engineering*, *69*, 89-97.

[3] Long, K. J., Zajaczkowski, F. J., Haupt, S. E., & Peltier, L. J. (2009). Modeling a Hypothetical Chlorine Release on a College Campus. *JCP*, *4*(9), 881-890.

[4] Na, J., Jeon, K., & Lee, W. B. (2018). Toxic gas release modeling for real-time analysis using variational autoencoder with convolutional neural networks. *Chemical Engineering Science*, *181*, 68-78.

[5] Mescheder, L., Nowozin, S., & Geiger, A. (2017, August). Adversarial Variational Bayes: Unifying Variational Autoencoders and Generative Adversarial Networks. In *International Conference on Machine Learning (ICML)* (pp. 2391-2400). PMLR.

[6] Kingma, D. P., & Welling, M. (2013). Auto-encoding variational bayes. *arXiv preprint arXiv:1312.6114*.

[7] Haber, L. T., & Patterson, J. (2005). Report of an independent peer review of an acrylonitrile risk assessment. Hunam & Experimental Toxicology, 24, 487-527.

546615 Modeling and Analysis of Flammable Gas Dispersion and Deflagration from Offshore Platform Blowout

The operational safety of offshore platform is challenged by a series of adverse factors including harsh environmental conditions and process accidents. Blowout is a type of major accident that may result in catastrophic consequences. This paper uses a computational fluid dynamics (CFD) model to simulate the blowout accident on an offshore drilling platform. The simulation includes the dispersion of flammable gas and the resulting deflagration based on dispersion results. Through the modeling, the blowout gas dispersion processes under different wind speeds and directions are analyzed, and a deflagration scenario due to the ignition of flammable gas is predicted in which the impacts of jet fire, overpressure and heat radiation are assessed. The simulations indicate that wind speed and direction have a great impact on the dispersion behavior of blowout gas. During the deflagration, the maximum overpressure is 280 KPa, the jet fire mainly distributes in the derrick, and the maximum radiation flux of thermal radiation field is 2300 W/m^2. Essentially, the outcomes from simulations could help to conduct a prior risk assessment and make an emergency response plan. Additionally, it could give a reference for the inherent safety design of offshore platform to mitigate the unexpected consequences due to blowout accidents.

548682 Application of Inherently Safer Design Tool (*i*-SDT) for Hazards Assessment and Safer Operation

Inherently ***S***afer ***D***esign ***T***ool (*i*-SDT) is a newly developed semi-quantitative metrics for hazards identification and characterization[1]. It is built on property platform, probabilistic analysis of historical accidents information and characteristic equations of safety parameters (i.e., flammability, toxicity, explosiveness, reactivity, etc.) which are developed in-house to assess potential hazards under various operating conditions. Designers can identify dangerous components and vulnerable equipment of the investigated industrial process and can take necessary corrective actions by applying the inherent safety principles[2] (e.g., minimization, substitution, moderation and simplification). The final output of this *i*-SDT tool is a cluster safety parameter score (C_{SP}) which provides insights regarding the development of safe design and operational parameters to manage the hazards using a very limited amount of process information and safety incidents information.

In this work, the application of the proposed *i*-SDT tool is explained for assessing different alternative routes of an industrial process supply chain. The proposed *i*-SDT safety metric identifies the chemical components with highest risk and the major accident-prone operational units. The detail analysis using *i*-SDT tools shows how different operating conditions can affect the safety performance of the given process. The ultimate object is to find the safer route while considering the techno-economic and environmental performance. It provides Pareto curve to compare alternative routes and gives the freedom to the end-users choosing their safer operating route/technology and process conditions. Therefore, safety can be considered as an important selection criteria for choosing best alternative from the beginning.

Final cluster safety parameter scores (C_{SP}) obtained from the analysis enables designers to compare between available technologies and provides information on various hazardous situations. It can also provide in depth knowledge regarding the cause behind the hazardous behavior. The proposed *i*-SDT metrics allows designer to do the safety analysis simultaneously while doing the initial process synthesis which will reduce the future modification cost of any add-in safety features. This work shows how the unique features of *i*-SDT tool can be applied in a real life scenario while doing inherent safer design during early stage of design.

1. Kazi, M.–K, Eljack F. T., & AlNouss, A. Application of a property-based inherent safety quantification framework for integrating risk assessment into process safety life cycle. *AIChE Spring Meeting, Orlando, Florida, USA, 2018.* (156c), p. 2260.

2. Kletz, T. A., & Amyotte, P. (2010). Process Plants: A Handbook for Inherently Safer Design, Second Edition: Taylor & Francis.

546631 An Approach to Understanding Worker Interactions with Procedures Under Normal Working Conditions

Incidents in high-risk industrial settings are often attributed to the way in which procedures are used. Despite the identification of procedures misuse, little research has been conducted on how workers have interacted with procedures in normal conditions. To understand these interactions a behavior assessment technique has been developed which can describe how a worker interacts with a procedure to complete tasks in everyday work situations. interactions under normal conditions. Using the concept of work-as-imagined/work-as-done, the techniques describes the interactions at a macro whole-of-task level as well as an individual step level. At the macros level, assessment considers if the task was completed correctly and safely (Tier 0). At the micro step-level, assessment considers if the step was completed as it was imagined (Tier 1) or if the step was done in an alternate way (Tier 2). For the alternate approaches, assessment considers if the worker struggled or required assistance, completed the steps out of order or with iterations or if the step was not completed as intended.

Two studies, one involving industrial tasks Second Life® (Study 1), the other at Shells Robert, LA training facility (Study 2), were used to test and develop the behavior assessment technique. In Study 1, the behavior of 13 participants from offshore drilling conducting tasks in either day or night conditions was assessed. In Study 2 the behavior assessment technique was used to test if experience accounted for behavior difference in 25 participants from high-risk industries. Results demonstrate the success of applying the assessment techniques through identifying how the behavior of the participants changes according to their experience and familiarity with the tasks the were performing. Further analysis identified recommendations for design and work flow changes of both written procedures and specific equipment. In conclusion, the initial results suggest the assessment technique is an important contribution to the methods use for understanding early interventions to improve safety in the process industry.

PROCESS HAZARD ANALYSIS (PHA) PREPARATION: A ROLE BASED OPTIMIZATION

Susan S. Frizzell, PE
Adley Services, LLC
Lafayette, LA
Susan@AdleyServices.com

Biography:

Susan Frizzell, PE is the President of Adley Services. Shortly after high school, Susan began her career alongside her father on a drilling rig. Upon graduating from the University of Louisiana with a BS in Mechanical Engineering, she obtained her MBA and began working for a firm performing engineering and process safety reviews on refineries and other land facilities in the New Orleans area. After Susan relocated to Lafayette, LA, she began engineering offshore facilities, complimented with process hazard analysis facilitation. During her almost 20-year career, she has worked upstream, midstream, downstream, and all points in between. She is a licensed professional engineer and an executive member of various organizations.

Abstract

A PHA meeting consumes large resources and can place a strain on departmental manpower. Teams are often faced with the want to be more efficient in PHA conflicting with the need to be thorough and accurate. The key to solving this paradox is efficient preparation. By examining the lessons learned from past PHA's, a role by role analysis of best practices with regards to preparation will allow a PHA team to reduce the costly time spent in the meeting and yet still stay true to the methods and deliver a great, safe result.

When it comes to an optimal PHA, preparation is everything. For example, in a 6-member team PHA, 60 minutes of misused time equates to 6 hours of misused manhours. Whether it is managerial, operations, engineering, or even facilitation, there are lessons learned from previous PHA's that when shared, will not only make the PHA more efficient and the facility safer, but the industry as whole better. The goal being that those reading this be able to take back to their facility these practical preparation tips to share with their present and future PHA team members so that the resources of the facility can be better spent.

1. PHA Preparation Optimized

To fully optimize the Process Hazard Analysis (PHA) preparation, the team needs a clear vision of the end product: "To identify, evaluate, and control the hazards" [1, 2] of an operation or process to minimize the likelihood of occurrence and the severity of consequences of release of a dangerous substance [3]. With that in mind, the team can focus time and energy preparing for the PHA on factors within their control to meet that goal. An efficient, cost-effective PHA begins with including the different roles in the preparation. The PHA planning team should be comprised of management, engineering, operations, the scribe, and the facilitator.

An active commitment to reduce risk begins by assigning tasks to each role prior to the PHA. From the start, management should provide support and direction for the PHA team. They set the tone for obligation and importance. Second, engineering involvement in the preparation ensures accurate and complete process safety information (PSI) is available and accessible. Without correct information, the hazards may not be fully identified. Next, operations should walkdown the facility and identify areas of concern. They have the most intimate knowledge of the workings of the process, which is extremely valuable to the PHA when properly communicated to the team for hazard recognition. The scribe should have the technical skills to ensure equipment is functional and to properly and efficiently document the team's findings. Lastly, the facilitator needs to coach the team before and during the PHA to ensure the methodology is followed.

Successful PHA preparation can improve risk identification and limit cost through appropriate management direction, current and detailed documentation, and clear expectations of the timeline leading up to the PHA. We identify ways that companies can assist in preparation in the months leading up to a PHA and the benefits that preparation has on hazard recognition, team morale, and, most importantly, the bottom line.

2. Management – "*Set the Tone*"

It is important for management to set the tone of the PHA. More specifically, management needs to set the tone before, during, and after. In a PHA, the attitude, the posture, and the dedication of the team is usually a direct reflection of management's attitude, posture, and dedication. In our experience of both facilitating and participating in PHAs, we've seen attitudes of management in a variety of flavors. The following are observations we've made in the past regarding a successful PHA and one not as successful.

The corporate culture and support infrastructure set by management begins well before the first day of the PHA. A typical PHA is done every 5 years [1]. Given that schedule, it is imperative that management starts out on the right foot and begin planning well in advance. A management component that is very serious and very organized will typically induce an attitude of seriousness and organization to the team members. For example, a management team that waits until the last moment and rushes preparation will typically have PHA members that delay and do little preparation, if any at all, for the PHA. PHAs are very expensive and very time consuming. Lack of adequate and efficient preparation will be very damaging to an efficient use of PHA budget and resources. When management in advance sets the tone that a PHA needs to be done correctly and efficiently, the PHA team members tend to take preparation seriously. Just as it is well documented that an effective safety culture of a company starts at the top with management, effective preparation for a PHA also starts at the top with management [4]. The management teams we've seen in the past that do this most effectively set the tone of preparing and doing a PHA as part of your job, not as additional work on top of your daily duties. In other words, management sets the tone that PHAs are in your job description and that we are all PHA team members. For example, a PHA is not additional work given to engineers, it is in an engineer's job description.

The other things we have observed that effective management does by setting the tone is setting up a support team for the PHA team members. These PHA team members will have to transition out of their daily duties, be present at the PHA, and transition back into their regular duties. Often, their project managers and/or functional managers may give some resistance to that. It is imperative for upper management to set the tone with the managers of the team members that they need to have a plan for them to transition into the PHA and transition out of it with regards to their regular duties. Most of these facilities run around the clock all year long and the work never stops. Therefore, it is imperative for management to have a plan for these employees to be present at the PHA while it is going on. A PHA which is constantly stopped for employees dealing with issues outside of the PHA is an unacceptable waste of resources. For example, if a PHA consists of ten team members and one team member must stop the PHA to deal with outside problems unrelated to the PHA, the nine other team members are wasting company resources. When management sets this tone early on that we are dealing with real human lives and significant company resources, the employees and their managers tend to put greater weight on a PHA. And sometimes, the employees just need the help of upper management to persuade lower and mid-level management that their presence, uninterrupted, is necessary to conduct a PHA.

During the PHA, management's role is typically only at the very beginning. It is very effective during the opening remarks of a PHA for a member of management to also make an opening remark. During this address, management will set the tone on several things. It is not uncommon for management to begin opening remarks with a thank you for the team members for their hard work in preparing and committing of their time. Some members of management stress employees' safety; others may stress the importance of safety to the community or environment. Opening remarks is a great time for management to state known worst-case consequences initiated within the limits of the PHA scope or define the assumptions regarding severity. This guides the team towards consistent company-wide risk rankings and avoids the team paranoia over risk ranking too high or too low. It's not management's position to force upon the team a subjective belief; it is management's position to reaffirm the team uses best judgement and not be hesitant of ranking the risks as they believe they should be. Finally, management can give the PHA team members feedback from previous PHAs as to the wording, substance, and ranking of the recommendations to assist with review and completion post-PHA.

Once the PHA begins, management's role is very limited. We've seen effective managers from time to time pop-in, review recommendations, and discuss with team members their thoughts on such. Effective managers will pick a recommendation, look at the supporting data, and ask specific questions of how the team arrived at its conclusion. This serves the purpose of correcting any doubts that the PHA team may be having and serves the purpose of reaffirming to the PHA team what they are doing is correct. Management can also review recommendations with only the facilitator at specific progress intervals of the PHA. The facilitator can then bring suggestions back to the team without the direct watchful eye of management. This minimizes the intimidation factor but gives the team the opportunity to clarify wording or details.

We find in PHAs that the PHA team members want to be very cooperative with management. That is why it's imperative for management to not only set the expectations and tone, but also provide a supportive means and network for a high-quality PHA to be conducted. By management providing a support team and stressing the importance of a PHA during the preparation phase, the PHA team members are motivated to attend the PHA prepared with their full attention. It goes back to the old saying, garbage in, garbage out. Anyone who has sat in a PHA will attest to the opinion that PHAs are grueling and a marathon of being attentive. The PHA team members tend to be very technically sound and understand the PHA process and goal. The biggest hurdle is usually morale and minimizing outside distractions. A great opening statement by management and periodic check-ins to keep morale and encourage the team members goes a long way. Nothing lifts the moral of battle wary troops like a visit to the battlefield by the general. After the PHA is over, feedback and encouragement go a long way in instilling confidence for the PHA team members that will carry over to their next PHA. Although hardly any of this is required by the statutes, these are observations we have noticed with regards to effective management participation in PHAs.

3. Engineering – "The Devil is in the Details"

As an engineer, it is understood that your job is to ask the questions that have not been answered, review the material that has not been presented, and observe that which is not obvious. PHAs are no different. One of the duties of the engineer during a PHA is to gather all the process safety information (PSI). However, we've observed the more effective engineers will take this list, check how current it is, and review the design pressures and temperatures, such as pump discharge pressures. This review helps determine if the PSI is accurate. It also serves the dual purpose of observing any abnormalities that will need to be discussed during the PHA. Looking at this before a PHA is very effective when the engineer's eyes are fresh, and his mind is clear. Once the PHA is ongoing, this task becomes a little more complicated with lower morale and PHA fatigue. This holds true to the majority of the engineer's tasks.

Drawings, drawings, drawings. One of the largest time and money wasters we observe in PHAs is showing up to a PHA with either outdated drawings or needing drawings that are missing. It is imperative for the effective engineer to not only review the drawings to make sure they are up to date and accurately reflect what is present in the field, but it is equally important to review which drawings have been gathered and submitted to ensure that there are no surprises during the PHA. Anyone who has sat in a PHA and heard the words "these are not the most up to date drawings" or "where are all the other drawings at", will attest that it is going to be a long day. The PHA facilitator, the scribe, and pretty much everyone else in the room relies on the drawings being both accurate and complete in their preparation. If the drawings submitted for the PHA are either outdated, incomplete, or just wrong; it will have a ripple effect for the rest of the team both before and during the PHA. If the reader of this paper takes nothing away but one thing, please note that accurate and complete drawings are imperative for an effective PHA team.

PHA team members include those who are present in the room full time and other team members who may be called upon as needed. Effective engineers anticipate ahead of time what possible other resources will be needed during the PHA. For example, engineers for adjacent facilities need to be made aware of an upcoming PHA and given a heads-up as to be ready to answer questions regarding upset conditions and their effects. Likewise, engineers for skid packages or other equipment requiring expertise of vendor engineers need to also be made aware of an upcoming PHA and request their availability if called upon. Similarly, engineers of other disciplines may have better knowledge of failure mechanisms or a specific design review already conducted which is beneficial to the PHA team. It is not uncommon in a PHA to have a question that needs to be answered by a vendor engineer, subject matter experts, or an adjacent facility engineer. Examples of information needed include possible flow rates or pressure to/from receiving/feeding units, adequacy of safe guards in packages provided by vendors, or electrical area classifications. Oftentimes, a simple phone call to these engineers asking if they would be available by phone during the PHA is enough. Other times, effective engineers will require that other personnel be available to come in as requested. PHAs involve a lot of moving parts with each part having its own calendar or schedule. Giving the other possibly necessary personnel a heads up will ensure that should their presence become necessary, the PHA team will not have to recess until they are available.

Effective engineers will typically review MOCs and make phone calls or e-mails to update their current status. The adage that change is constant is very applicable to MOCs. MOCs may have been implemented, postponed, in progress, or completed. It is important to update the status of an MOC. One that has been canceled may require different attention than an MOC that is in progress or completed.

Effective engineers provide the technical expertise of the specific process under review [1]. Before a PHA, effective engineers will take a closer look to evaluate operating trends and upset conditions looking for a pattern that is not readably apparent. Engineers should discuss the facility and how it runs day-to-day with operations prior to the PHA. They should review procedures to ensure they are current and understood. It can be an awkward moment during a PHA when the engineer says, "we've never had a problem with that", and the operator rebuts with "we have a problem with that every day". Engineers know that engineered controls are the strongest defense against hazards. They look for ways to solve problems without relying on administrative controls.

Finally, effective engineers are present during a PHA. This does not mean they are sitting in a chair, it means they are focused, engaged, and completely dedicated to the PHA while they are in that room. Engineering is a life of dealing with problems. To be an effective PHA team member, it is important to make sure the engineer has support to deal with problems while he/she is in the PHA. Effective engineers can have a clear mind because they know that if normal, everyday problems occur, someone else is taking care of them. An engineer that is constantly worrying about what great surprises are awaiting them when they check their email at the end of the day, are not present during a PHA.

4. Operations – "The Ambassador of the field"

The operator is the ambassador of the conditions in the field. The operator, not engineering, management, or the facilitator, is the eyes and ears of the field. As an ambassador, it is important that the operator speak to his constituents (i.e. opposite crews), learn how to effectively communicate, and be present for the meeting of the PHA team. Effective operators speak to their coworkers and get a history of problems, concerns, or other issues they may be having that might not be well documented. Effective operators also walk around and observe conditions of the facility and document such. Documentation being key. Effective operators also review previous incidents not only in their facility, but other similar facilities or neighboring facilities to bring up any concerns that may not be readily apparent when reviewing documentation during a PHA. A great ambassador becomes the eyes and ears of his constituents by listening to what the other operators have to say and physically observing the condition of the facility.

Other things effective operators will do is review operating procedures and note deficiencies. Sometimes operating procedures are written in a way that no one understands them making them ineffective as a safeguard. Other times, there is no written procedure, forcing operations to react to conditions "the way we always do it." If operators are not following the procedures because they are vague, ambiguous, confusing, or out of date, it is up to the operator during the PHA to bring that to the attention of the team members. Frankly, effective operators understand that management and engineers assume that everything they write is perfect, clear, and easy to follow. It is up to the operator to explain why that may not always be the case. Some of the most effective operators we have encountered in a PHA not only describe what the problems are but are very specific about why these are problems.

The next characteristic of the operator as an ambassador for the field conditions is communication. For an operator to communicate with a PHA team, they must speak the same language. Effective operators know that to communicate with a PHA team, they must become a master at reading P&IDs. Being able to understand how the elements of a facility and its attributes are represented on a P&ID are extremely important for an operator. Effective operators can look at a P&ID with little or no guidance and interpret it to see if it is an accurate reflection of what is going on in the field. When the engineer asks a question regarding the P&ID, effective operators are on the same page. If the reader wanted to take away one important thing from this section on operators, please take away that an operator being a master of a P&ID is extremely important to be an effective operator in a PHA.

5. Scribe – "The Show Must Go On"

In American football, a great offensive lineman only receives attention when things do not go as planned. The same can be said about a scribe in a PHA. An effective scribe will give the outward appearance of very little responsibility or having very few complicated tasks. Nothing can be further from the truth. With the goal of quickly and accurately recording information [4], effective scribes understand that a lot of thought and preparation must go into making sure that nothing goes wrong, and if something does go wrong, the PHA suffers minimal interruption. Also, effective scribes are great at being very efficient in the tasks they complete.

At the end of the day, a scribe's responsibility is to ensure that the audio/video equipment works as planned. Although some facilities may be tasked with providing projectors and so forth, the scribe cannot scribe unless all this equipment works. Therefore, an effective scribe will oversee the equipment being used to make sure the parties who are responsible for providing it are communicating and that all the items are compatible. The scribe will ensure that both the scribe's computer and the display have the proper connectors and cables needed. An effective scribe not only brings the connectors and cables needed, but also bring the connectors and cables that were not anticipated as being necessary. An example is the conference room being used for the PHA has HDMI and ethernet and Wi-Fi...or at least that what was told to the PHA facilitator. In the past, we have shown up to conference rooms to find that the HDMI port is in fact a VGA port, or that the HDMI cable is three feet long and the scribe's position is ten feet away. Very effective scribes keep with them a toolbox of supplies such as VGA to HDMI converters, extra batteries for mice and Bluetooth keyboards, and extra cables. Sometimes, the right connectors and cables are present but are malfunctioning. The truly great, effective scribes seamlessly work through these problems and the show continues without interruption. Along those lines, a great practice for scribes to have is to not only back up the PHA file regularly, but also have a written procedure that has been reviewed regarding the backup. We always joke that we PHA the PHA. In other words, what are the risks for file corruption or data loss. Which risk are we willing to accept, and which risk are we going to mitigate? Knock on wood, so far, it's been very effective.

A lesser problem, but a problem none the less, is having to wait on a scribe to catch up. Although not as catastrophic as the PHA having to be stopped completely while technical issues are worked out, a scribe who is not able to keep up with the PHA team stifles the free flow of thought and can leave the door open for other interruptions such as taking phone calls and checking email. Very effective scribes will have experience and practice with the software, know the shortcut keys, and have a vocabulary list prepared for technical terms and names. A typical vocabulary list will be something as simple as a word document open where the scribe has already checked the spelling of names and technical equipment and as the scribe goes along, they can copy and paste the names and equipment labels very quickly without having to correct for spelling. A truly exceptional scribe will have already created templates with standard verbiage for safeguards, recommendations, and comments. For example, when the team decides to make a recommendation, the scribe can copy and paste it into the software and the team can edit the few words that need to be changed to modify the recommendation for the specific application. What this does is it allows the team to make a recommendation and move on to the next node without getting bogged down in the weeds of language choice. We find it is always easier to give a group of people a starting point for a recommendation as opposed to generating the language of a recommendation by consensus of the group. Along those lines, effective scribes review the drawings, MOCs, and other information to become familiar with the vocabulary in the drawings and have an inventory of what PSI they have and what node it corresponds to. Therefore, if a team member erroneously refers to a drawing that does not exist or refers to a drawing or MOC by incorrect name, the scribe will be quick to point out that the item does not exist in the inventory.

Effective scribes will also begin opening vital documents for a specific node as soon as the review of that node begins. Therefore, when the team would like to reference a specific document such as a drawing, MOC, or incident report, the document will be readily available. A common, effective means of doing this is having a second monitor only seen by the scribe, in addition to the screen shared with the projector. The scribe can work on his/her monitor opening documents or correcting language. As the team is ready to view specific information, he/she can drag the window over to the extended screen that is being displayed overhead. This allows the scribe to work without being micromanaged by the team and minimizes the distraction of the team from the scribe flipping through windows or searching for files. Again, the overall theme for an effective scribe is to keep the free flow of thoughts and the team moving. Anytime the team must pause or take a break because of the scribe performing its duties, the team's momentum and productivity can be lost.

Finally, truly effective scribes are great at covering the intangibles. This can include filling all the supplies or doing things to the environment in the room to lift morale. For example, effective scribes will typically have a box of random office supplies, whether it be pens, hi-liters, or post-it notes. At the beginning of each day and after each break, he/she should refill the table or the workspace of the PHA team with extras. So, if a team member would require a specific color pen, they would simply grab it and not have to pause the PHA. This gets back to the overall theme that an effective scribe keeps the evaluation moving forward without interruption. Another intangible that we often see effective scribes will do is they will place motivational posters around the room, such as large check-off list of nodes to visually see progress, to hang inspirational quotes, and so forth. PHAs are really an endurance test of morale and motivation and these little intangibles that the scribes do are of very limited cost, but of great benefit. In all honesty, we often do not realize how great and effective our scribe is until we do a PHA with an ineffective scribe. When you do a PHA with an ineffective scribe, you feel like you're stuck in the mud and going nowhere fast. Although realistically, the delay may add just a few minutes a day, with regards to morale and motivation, it feels like an eternity.

6. Facilitator – "What's Your Plan, Coach?"

Like a good coach, an effective facilitator enables great people to do great things. Effective facilitators setup the team members to be successful and, during the PHA, guide them to produce a high-quality product. Although a facilitator's duty is to ensure the methods are adhered to and the regulations concerning a PHA are complied with [1], a great PHA facilitator is so much more.

A great PHA facilitator, like a coach, ensures that the team is prepared come game day. Effective PHA facilitators will do such things as set deadlines for collection and development of information and suggest further actions to better prepare for game day. Not only do they set the deadlines, an effective facilitator will check in with the team members prior to those deadlines to ensure that progress is proceeding as planned and that the deadline will likely be met. Effective facilitators are great at helping team members overcome obstacles and hurdles in collecting this information. A team being well prepared on game day when they take their chairs and listen to management's opening remarks is no accident. Also, during the preparation, an effective PHA facilitator will ask the hard questions of the team members, such as: Do you have a transition plan for someone to fulfill your role during the PHA? Or, are there any allergies, physical, or dietary restrictions a team member may have so that these can be addressed well ahead of the PHA? To be an effective PHA facilitator, it really comes down to having the team prepare for the work they anticipate doing and addressing as many possible obstacles as one can prior to the PHA commencing. The litmus test for a PHA facilitator is simple - once the PHA commences, no surprises.

In addition to ensuring the rest of the team is preparing for the PHA, the effective facilitator puts in just as much effort as the rest of the team into preparing the PHA file, determining the node segregation, and sharpening his skills as a technical guide. On the day of the PHA, the PHA file should be prepopulated with node descriptions, drawings, common safeguards, etc. The facilitator should have reviewed the node list and noded drawings with other member(s) of the team to ensure the segregations make sense and all parts of the process are included. The effective facilitator should also review recent changes to regulations or process conditions that might affect the hazard identification. He should gain a basic understanding of the process under review along with the potential hazards specific to the facility. Effective facilitators have a list of questions to pose to the team during the appropriate node to ensure all hazards have been addressed.

Again, following the theme of the PHA facilitator being a coach, once the PHA commences, it is game day. It is the PHA facilitator's job to adhere to the methods and regulations regarding PHA facilitation. However, great, effective PHA facilitators understand that the true challenge in a PHA is morale, motivation, and focus. Effective PHA facilitators will typically start off the PHA with a "rules of the road" presentation. They do not assume that everyone in a room regularly participates in PHAs and is familiar with the process. They review the PHA process, the end goal of risk mitigation, the timeline, the daily schedule, acceptable and unacceptable behaviors, and expectations of participation. An effective PHA facilitator will go over these rules of the road and get each team member to commit to their role, to fully participate, and to strive towards the goal. It creates a sense of responsibility and obligation among the members. Once the PHA commences, the effective PHA facilitator will have a sense of when progress is slowing down and when to step in to move the team along. The effective PHA facilitator knows that the team must be given opportunity to discuss causes and consequences, but also has goals and objectives that need to be met within the time allotted.

As the team leader, the facilitator should plan on how to react to differences among team members. The great PHA facilitators have a way of getting conflicts resolved without upsetting or demeaning anyone in the room. Typically, in an argument, somebody must be wrong and somebody must be right. However, the great PHA facilitators have a way of allowing both parties to express their beliefs, guiding them to understand the beliefs of others, and concluding with a common solution. A great way to kill the free flow of ideas in a PHA and stifle productivity is to tell someone, "you're wrong, please be quiet". Great PHA facilitators know how to do that in a way that allows the person to maintain dignity and continue to perform in a PHA.

Another attribute of great PHA facilitators is that they are systematic and thorough. They have a plan of attack for each node. This usually includes the node description and limits review, a review of the appropriate drawings, written list of deviations to encourage the team, and review of MOCs, previous recommendations, and incidents related to the node. The effective facilitator also has a plan in place to limit idle time when critical information necessary to properly review a node is found to be missing. The action is documented as a follow-up for a specific team member to complete prior to the next day allowing the PHA to move forward. Any outstanding to-dos are documented and presented to the team members at the end of each day prior to dismissal. The PHA facilitator is responsible for meticulously documenting the team's findings, which cannot be done without adequate preparation of a game plan.

Finally, great PHA facilitators understand they have an obligation to future PHA facilitator to send their team off with a sense of accomplishment and pride in PHA participation. If the team members of the PHA leave with a bad taste in their mouth or feel discouraged, their participation in the next PHA may not be to the level it could be. Great PHA facilitators compliment their team members and make sure they leave the PHA on a positive note. Therefore, when the team members approach their next PHA, the new facilitator will not be dealing with baggage leftover from the previous PHA. It is a courtesy that great facilitators give to and expect from other facilitators. PHA facilitators are often asked by team members during a PHA, "how can you do this every day?" The response we often use is, "what can we do better to make you volunteer for the next PHA?" Soliciting feedback from team members besides a multiple-choice survey is imperative to be an effective facilitator. The feedback needs to be of substance that can be acted upon and improved upon. We truly love doing PHAs and at the end of every PHA, we express that to the team and solicit their feedback to what we can do better to make them truly love doing it too.

7. Summary

In conclusion, when it comes to an optimal PHA, preparation is everything. Each team member plays an important and specific role. Management sets the tone with their level of commitment to preparation, resources, and action. Engineering ensures the quality of information available to the team and contributes a deep understanding of the process design. Operations represents the field conditions, the procedures, and the most intimate knowledge of the process. The scribe provides accurate documentation through technical skill, proficiency with hardware and software, and team morale support. The facilitator leads the team to identify, evaluate, and control the hazards utilizing a systematic approach to preparation. It has been proven that a small amount of time spent preparing for each role will result in improved hazard identification, higher team morale, and reduced PHA cost.

8. References

1. Occupational Safety and Health Standards, Toxic and Hazardous Substances, Hazard Communication, Available at https://www.osha.gov/laws-regs/regulations/standardnumber/1910/1910.1200 Accessed on December 12, 2018.
2. Bureau of Safety and Environmental Enforcement, Department of the Interior, Offshore, Oil and Gas and Sulphur Operations in the Outer Continental Shelf, Available at https://www.ecfr.gov/cgi-bin/text-idx?tpl=/ecfrbrowse/Title30/30cfr250_main_02.tpl Accessed on December 12, 2018.
3. American Petroleum Institute Recommended Practice 750: Management of Process Hazards, First Edition, American Petroleum Institute, Washington D.C., 1995.
4. CCPS. *Guidelines for Hazard Evaluation Procedures*. Center for Chemical Process Safety, American Institute of Chemical Engineers, New York, NY, 2008.

551204 Building Process Safety Intelligence: What Should Every Graduating Chemical Engineer Know?

Process safety intelligence is derived from derived from a vast corpus of sources - including textbooks, handbooks, journals, conferences, incident reports, safety datasheets, and available materials and thermophysical property data. Dedicated decision-support tools enabling practitioners to harness this intelligence in their daily work, such as Knovel are an important component of chemical engineering practice today. However, many students don't realize the breadth and depth of technical knowledge needed to support process design and safety – they tend to overrely on non-specialist search, and miss pedigreed information available on specialized knowledge platforms, such as Knovel.

The undergraduate engineering space offers an excellent opportunity to plant the foundational information seeking behaviors in future process safety professionals and leaders. Knovel focuses on applied a preview of the types of insight they will be forced to leverage in industry careers, without a direct cost to students and faculty.

Common Process Safety Management Gaps Based on Occupational Safety and Health Administration Process Safety Management of Highly Hazardous Chemicals Violation Trends

Janet Persechino
Engineering Planning and Management, Inc.
959 Concord Street
Framingham, MA 01701
jap@epm-inc.com

Ryan Welch
Engineering Planning and Management, Inc.
959 Concord Street
Framingham, MA 01701
rdw@epm-inc.com

Prepared for Presentation at
American Institute of Chemical Engineers
2019 Spring Meeting and 15th Global Congress on Process Safety
New Orleans, LA
March 31-April 3, 2019

Janet Persechino
Engineering Planning and Management, Inc.
959 Concord Street
Framingham, MA 01701
jap@epm-inc.com
Keywords: OSHA, violations, regulations, process safety

Abstract

On July 17, 1990 the U.S. Occupational Safety and Health Administration (OSHA) published in the Federal Register the proposed standard *Process Safety Management (PSM) of Highly Hazardous Chemicals (HHCs)*. Approximately four months later it was enacted into law via Clean Air Act Amendments (CAAA). These amendments require that the U.S. Secretary of Labor, in coordination with the U.S. Administrator of the Environmental Protection Agency, promulgate, pursuant to the Occupational Safety and Health Act of 1970, a chemical process safety standard to prevent accidental releases of chemicals that could pose a threat to employees [1]. The full text of the final standard was issued in the Federal Register on February 24, 1992.

The OSHA standard is listed under Title 29 (Department of Labor) of the Code of Federal Regulations (CFR) as 1910.119. This standard leans towards a performance-based guidance document that outlines methods and programs to decrease the likelihood of flammable, toxic, reactive and explosive chemical releases. The intent is to reduce the risk to employees, visitors and contractors that may enter or work at chemical processing facilities.

Facilities are subject to the PSM regulation if they hold a threshold quantity of any HHC that is listed in Appendix A of the standard, with some exceptions. The violations that OSHA issues for the 1910.119 standard will cite the specific element and its relevant sub-section. Published data allows for review of the violations, and based on that information an indication of common PSM program gaps can be assessed.

A ten-year span of OSHA PSM violation data was examined (2007-2017). The data shows the top four OSHA cited PSM sections (often referred to as elements) are (highest to lowest):

1) Process Safety Information
2) Mechanical Integrity
3) Operating Procedures
4) Process Hazard Analysis

Each of these elements have known solutions, and specific corrective actions can be easily implemented given the lessons learned, the current level of expertise that exists in the process safety industry and the cost-effective software tools that are now available.

Regulatory compliance is not the only goal of adequate PSM programs, nor is it the goal of this review. The purpose of reviewing the OSHA violation information is not meant to help facilities avoid OSHA citations, but to uncover gaps found by OSHA and bring awareness and attention to some common issues, hopefully leading to focused solutions for safer, more reliable facilities.

1.0 Introduction

There are approximately 24,000 sites that are subject to the OSHA PSM standard nationwide [2]. Facilities are subject to the standard if they process any of the 130+ chemicals listed in Appendix A (of the standard) in excess of the associated threshold quantities, also listed in Appendix A, on-site in one location. They are also subject to the standard if they process flammable liquids and/or gases in quantities equal to or greater than 10,000 pounds. OSHA defines "on site in one location" as areas under the control of an employer, or affiliated employers, that have control over any group of vessels that are interconnected, or in separate vessels that are located in close enough proximity that a HHC could be involved in a catastrophic release.

Exceptions to the applicability include [3]:

- ✓ Retail facilities
- ✓ Oil/gas well drilling or servicing operations
- ✓ Normally unoccupied remote facilities; the standard defines "remote" as a facility that is operated, maintained or serviced by employees who visit the facility periodically to check its operation, no permanent employees are stationed at the facility, and is geographically remote from all other buildings and persons.
- ✓ Hydrocarbon fuel used solely for workplace consumption as a fuel if such fuels are not part of a process containing another HHC covered by the standard
- ✓ Flammable liquids with a flashpoint below 100°F stored in atmospheric storage tanks or transferred, which are kept below their normal boiling point without benefit of chilling or refrigeration

The above information outlines specific threshold quantities that trigger, and exclude, the PSM regulatory requirement. It is strongly recommended that some form of PSM be practiced at plants that may not be subject to PSM, especially those that fall into the following categories:

- Plants that store quantities of HHCs close to the threshold quantities listed in Appendix A of the standard
- Plants that utilize hazardous chemicals not listed in Appendix A of the standard
- Processes exempt from the standard, for example, those listed as exceptions in the standard, plants that have runaway reaction potential, etc.

An overview of "PSM Lite", and related fire code requirements, can be found in these conference proceedings in the paper titled *Fire Code Chemical Process Safety Requirements*. Implementation of an abbreviated form of PSM should be evaluated on a case-by-case basis which will take into account various factors that would clarify the actual risk and the residual risk that would remain after programs are incorporated into the operations.

2.0 Violation and Corrective Action Evaluation

The PSM elements are listed in the 1910.119 standard in the following order/sections:

(c) Employee Participation
(d) Process Safety Information
(e) Process Hazard Analysis
(f) Operating Procedures
(g) Training
(h) Contractors
(i) Pre-startup Safety Review
(j) Mechanical Integrity
(k) Hot Work Permit
(l) Management of Change
(m) Incident Investigation
(n) Emergency Planning and Response
(o) Compliance Audits
(p) Trade Secrets

OSHA has been administering violations for the 1910.119 standard since 1992. For purposes of relevancy, the last 10 years of violation data was examined (2007-2017). These violations were issued when OSHA completed an inspection and/or accident investigation visit to a particular facility.

Each one of the elements listed above has subsections pertaining to the requirements for that element. For example, when OSHA administers a PSM violation during a visit it is listed as: 1910119 E03 I. The nomenclature correlates to a violation for section (e) Process Hazard Analysis (PHA), part 3(i), or (3) the process hazard analysis shall address (i) the hazards of the process. This particular example indicates that the PHA did not properly address any or all of the hazards of the process. The OSHA inspection information also includes other data such as penalty dollar amounts, dates and type of inspection. The information provided in this summary reviews a few segments of this data from years 2007-2017 [4].

3.0 Violation Data Interpretation

All OSHA PSM violations were tallied over the 10-year period (2007-2017), this included:

- ✓ All citation classifications types (serious, repeat, willful etc.)
- ✓ All types of inspections/visits (accident, planned etc.)
- ✓ All types of facilities

This review does not distinguish whether violations were given during the same visit or during separate visits.

The data shows that there are 4 elements, or 4 deficiencies, that clearly stand out as the most prevalent (see Figure 1):

<u>Most Commonly Cited PSM Elements (highest to lowest):</u>

1) Process Safety Information (PSI)
2) Mechanical Integrity (MI)
3) Operating Procedures (OP)
4) Process Hazard Analysis (PHA)

<u>Total Monetary Amount (highest to lowest):</u>

1) Mechanical Integrity
2) Process Safety Information
3) Operating Procedures
4) Process Hazard Analysis

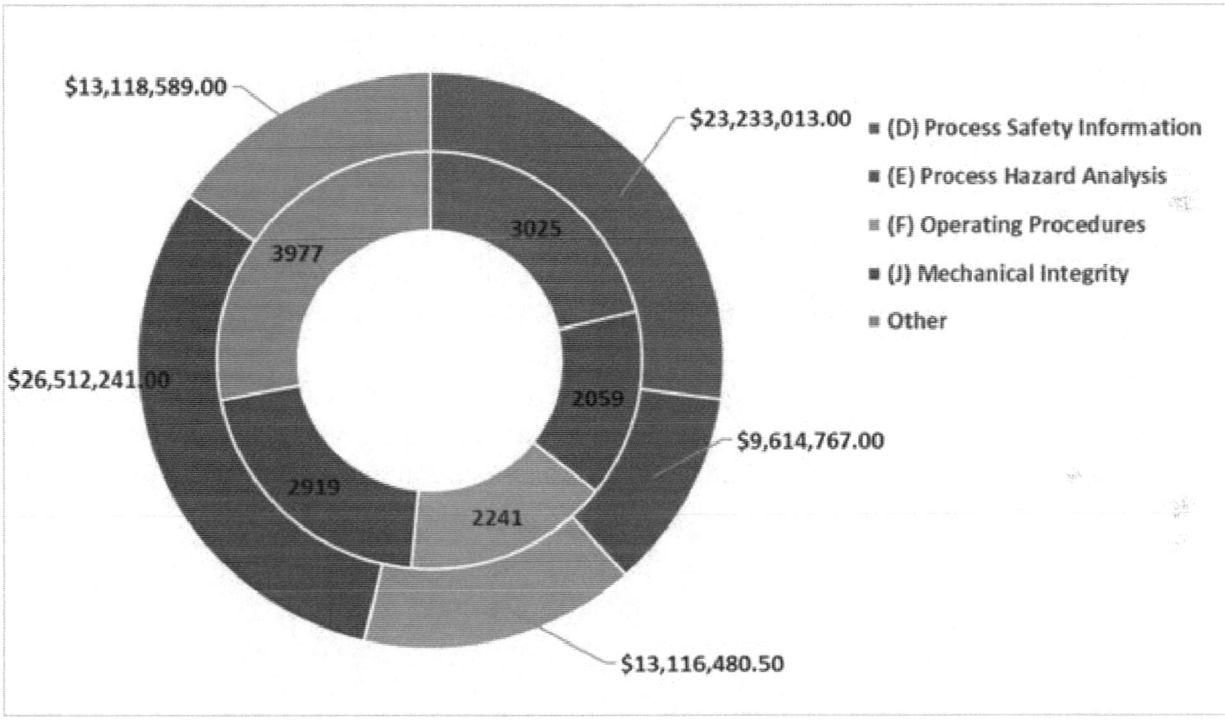

<u>Figure 1:</u> Total Number of OSHA PSM Violations and Total Citation Dollar Amounts by Element: 2007-2017

The top two elements (PSI and MI) have close to the same number of violations and the same goes for the third and fourth most common violations, giving the impression that the struggle to execute PSI and MI is about equal and OP and PHA both come in as close second.

It is alarming that PSI is the most cited element. Having correct and adequate PSI is the foundation for implementing, not only all PSM elements correctly, but any and all chemical safety precautions. PSI is crucial in completing an adequate PHA, the analysis that examines the hazards of the process and ensures that the proper safeguards are in place. The lack of PSI could lead to process hazards being missed, or incorrectly evaluated, during a PHA. Incomplete PSI can also lead to having inappropriate MI programs since some of these programs can be based on the hazards of the process, for example, more frequent inspections than the manufacturer's recommended practices based on the characteristics of the material being processed.

MI is the second most cited element but has the largest monetary amount, this could point to more instances of MI being noted within a single violation. OSHA will increase a fine for a single violation based on the number of "instances", or how many times they see a violation for the same element during the same visit. This review did not take into account the number of instances per violation.

MI can be difficult to implement, but new software solutions and training programs exist to help with proper execution. There are MI software products on the market that include capabilities to capture PSI, PHA and OPs, addressing all of the top-cited elements within one program.

It is helpful to utilize PSM experts for implementation and execution of PSM programs, as they are aware of the depth of information that is needed to meet the intent of the standard and for overall safety. They can often recommend products depending on your particular operation. Individuals dedicated to PSM will have access to resources that might not be available to others who are not embedded in the PSM industry. For instance, access to fee-based databases with previous incident data, one item that must be documented as part of a PHA. The expenditure for bringing in this type of expertise can provide a beneficial return on the investment when examining the average cost per violation and average number of violations per visit (Figures 2 and 3).

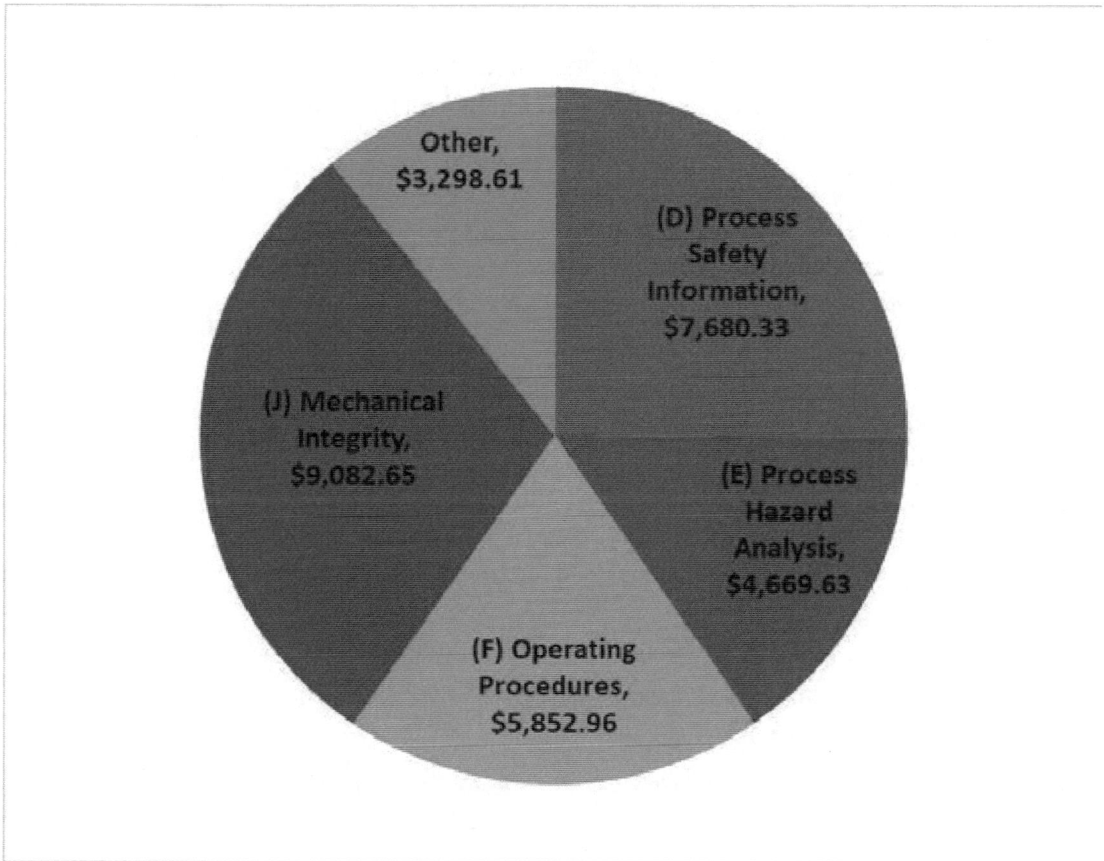

Figure 2: Average Cost per Violation by Element 2007-2017

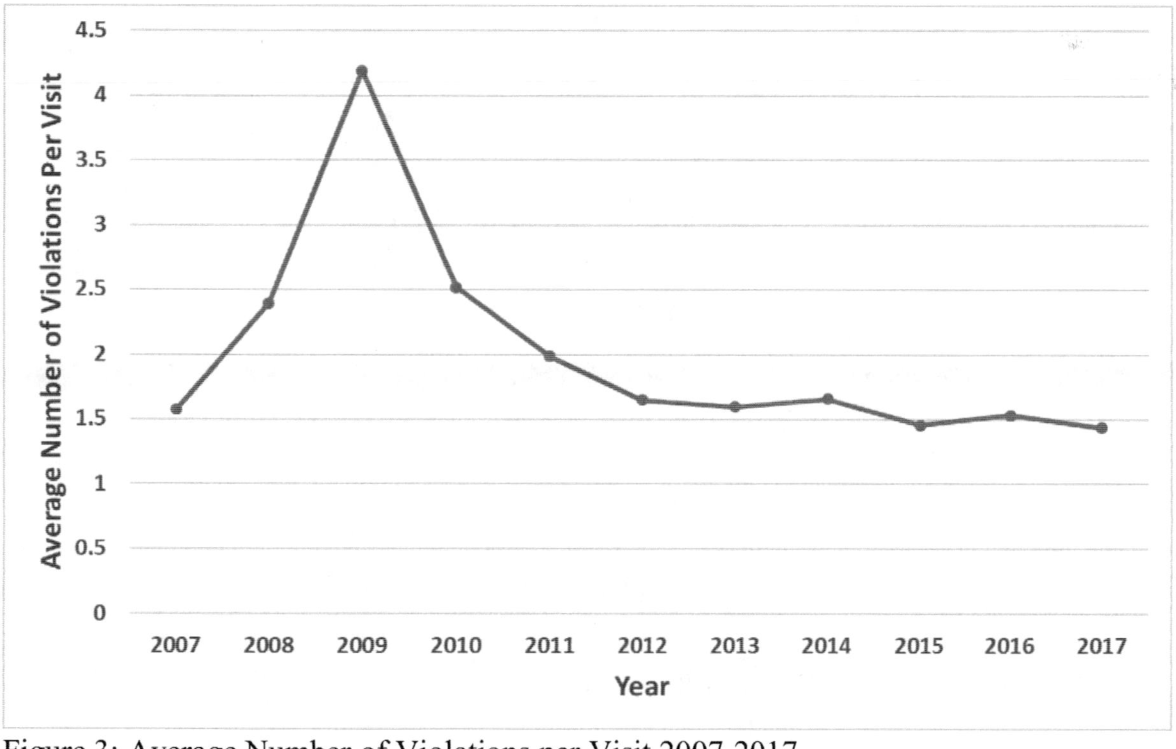

Figure 3: Average Number of Violations per Visit 2007-2017

(To note about Figure 3: the spike in the number of violations in 2009 could be due to the National Emphasis Program (NEP) in which OSHA extended to include a pilot on PSM-covered facilities from the 2007 NEP which only included refineries. The NEP went on from 2011 to only include chemical facilities and not petroleum refineries)

The citation costs are only one aspect to consider when reviewing the return on the investment. Other benefits that can be quickly realized once the programs have been established include but may not be limited to:

- ✓ Improvement in operating efficiencies
- ✓ Increased employee morale
- ✓ Lower turnover rates
- ✓ Safer working conditions
- ✓ Uptime gains
- ✓ Insurance benefits
- ✓ More consistent product quality
- ✓ Decreased likelihood of an incident

A deeper dive into the subsections of the most cited elements yielded the results in Figure 4.

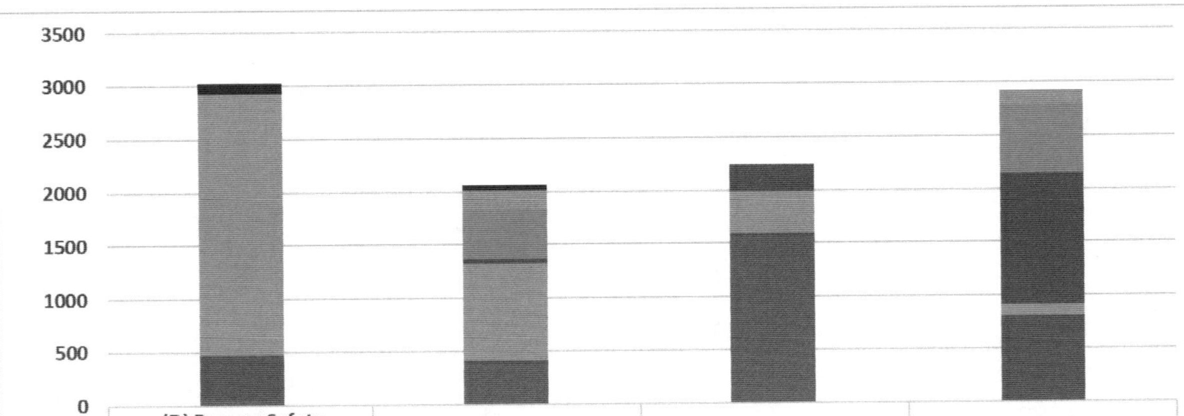

	(D) Process Safety Information	(E) Process Hazard Analysis	(F) Operating Procedures	(J) Mechanical Integrity
■ Blank	101			
■ 07		48		
■ 06		186		135
■ 05		463		640
■ 04		37	255	1235
■ 03	2452	918	391	105
■ 02	418	26	23	798
■ 01	54	381	1572	6

Figure 4: Sub-section Cited in the Top 4 OSHA Most Cited PSM Elements 2007-2017

Figure 4 shows that the biggest gaps over the last 10 years include the following sub-sections of these elements:

1. Process Safety Information (d)
   - Equipment Information (section 3)
2. Mechanical Integrity (j)
   - Written Maintenance Procedures (section 4)
   - Inspection and Testing (section 2)
   - Correcting Equipment Deficiencies (section 5)
3. Process Hazard Analysis (e)
   - Inadequate coverage of the hazards of the process (section 3)
4. Operating Procedures (f)
   - Develop written operating procedures (section 1)

There is a common theme among these four (4) subsection violations. While PSI was the most cited element, the most cited PSI sub-section is Equipment Information, which ties into the MI portion of PSM. Based on this, deficiencies in equipment related information and MI ranks very high on the list. MI related programs help to ensure that chemicals remain inside the vessels, pipes, etc. A review of the OSHA data shows that the industry may be struggling with proper execution of this critical piece of process safety.

As noted above, MI programs are complex and can be difficult to establish, but there are current solutions on the market to help correct these deficiencies, such as economical software solutions that improve compliance by organizing maintenance tasks. The right solution can alert staff to overdue maintenance, and store the maintenance history for each particular piece of equipment. Some tools can also track inventory of spare parts and link Management of Change documents related to the process and/or the equipment. Consultants versed in the regulations and MI techniques can help establish appropriate programs and find solutions useful to your specific operation.

Development of operating procedures for both process and maintenance activities has a high number of citations in the MI and OP elements. It is essential that these procedures are comprehensive, accessible, updated and easy to interpret. Mistakes in operating steps, or failure to recognize abnormal conditions, can quickly lead to catastrophic releases.

There are various writing and auditing techniques that can be used to ensure that these procedures include the correct information and address all stages of the process (start-up, normal conditions, abnormal conditions and shut down). Software tools currently on the market can also assist in organizing this critical information. This approach, while time consuming up-front, is time well spent in providing a safer and more efficient workplace.

Once authored, it is easier to audit the documents for accuracy and audit the operators using a behavior-based system to determine that procedures are being followed correctly.

Vast benefits can be gained using this approach, and many PSM consultants offer services to help author these documents and provide suggestions for document control and review processes.

4.0 Resources

Despite the intent for OSHA PSM to decrease the likelihood and severity of accidents, the OSHA data is showing that PSM gaps still exist in the most fundamental PSM elements. These gaps present an unacceptable risk. This is evidenced by the accidents that continue to occur. The various incident investigations performed by the U.S. Chemical Safety and Hazard Investigation Board often attribute inadequate PSM programs as the root cause.

There are many resources available for assistance with PSM implementation and improvement for plants of all types and sizes. There has been an increase in consulting organizations providing guidance and software solutions for PSM. The list of resources readily available to even the smallest of plants are much more prevalent than they were 10 years ago.

OSHA has published documents with guidelines for PSM for storage facilities and for small businesses. They also have a PSM resource page: https://www.osha.gov/SLTC/processsafetymanagement/. The Center for Chemical Process Safety has many publications, classes and tools to help with all aspects of PSM. Some of these resources are free, or provided at reasonable cost, and CCPS/AIChE (American Institute of Chemical Engineers) membership increases those benefits.

Small businesses often find it difficult to execute PSM. In 2016-2017, in almost the same number of inspections, OSHA administered 57 PSM related violations to companies with under 100 employees and only 18 violations to companies with over 100 employees. OSHA also points to a study that estimates "*employers with 1-25 employees are 47 times more likely to have a release and 17 times more likely to suffer an injury, per employee, than facilities with 1500 or more employees.*" [5]

There are a few PSM consulting firms that target small to medium companies, the companies that cannot dedicate resources to PSM, nor have the expertise in-house to meet the intent of the standard. Many of these consulting firms not only focus on OSHA compliance but becoming a partner to help embed PSM concepts into the plant's overall safety culture. Bringing in these third parties is worth the cost because they are embedded in the PSM realm and can bring along expertise from other industries, the latest information and years of experience.

There are cost effective software solutions, some aimed at small businesses from a price point and feature standpoint. These solutions can improve PSM while improving the bottom line through improved efficiency, production uptime gains, insurance benefits and violation and incident avoidance. The smaller chemical processing companies are yet to realize that these solutions exist at very reasonable prices, and the small upfront investments have very beneficial returns that easily justify the expenditure.

It is ideal to use the same consulting firm for various locations to standardize PSM solutions and audits across multiple sites that may have different risk. This type of PSM implementation helps to normalize the oversight and tracking to give a better picture of risk at the corporate level.

5.0 Conclusion

The research objective of this paper was to spot the potential for common gaps and issues in implementing PSM across the United States at all types of chemical processing industries using OSHA violation data, and to use that information to identify opportunities for solutions and preventative/corrective actions. The OSHA violation picture is only one piece of the gap analysis puzzle, but it is one place that collects enough data to increase awareness of PSM weaknesses nationwide regardless of the type of operation. This information can aid in creating education/awareness programs and campaigns aimed at improving PSM execution.

The top four cited violations are PSI, MI, PHA and OPs. Solutions to the deficiencies revealed by the OSHA data are well-known and PSM experts are aware of the many practical and feasible methods that exist to close these deficiency gaps. The return on investment on items outside of the safety aspect are usually realized soon after the solutions are put in place, including: more efficient and consistent work procedures (which can lead to improved product quality), increased up-time and insurance benefits to name only a few.

A review of the OSHA data can help quantify the costs that can be expected from OSHA PSM violations and citations. However, other impacts of violations and/or PSM incidents that are not so easily quantified include the business, legal, liability, company reputation and the impact that the injuries have on families. Sometimes plants are not able to rebuild in the same location or regain the customer share they may have lost due to an incident. Additionally, the OSHA violations are public record and can be accessed by lawyers, local authorities having jurisdiction, insurance companies, activists, etc. Companies may also find it difficult to obtain insurance due to OSHA violation or incident history. While these impacts are difficult to quantify, they must be considered in the overall cost to the business.

PSM implementation is a continuous improvement journey with new lessons learned, new technologies and new resources being created every year. Using the plethora of tools available can help close the deficiency gap.

References
[1] U.S Department of Labor OSHA 3132 Publication *Process Safety Management* 2000
[2] Washington State Department of Labor & Industries July 2010
[3] U.S Department of Labor OSHA 3133 *Publication Process Safety Management Guidelines for Compliance*, 1994
[4] OSHA Violation Data 2007-2017
[5] OSHA *Process Safety Management for Small Businesses* 2017

546725 Safety Performance Evaluation: A Benchmark of Practices across Various High-Risk Systems

The concept of performance has invaded all management spheres and imposed itself as a strong driver of decision making. Following the funding work of Kaplan and Norton (1996), performance appears as a multidimensional concept which evaluation and use require adapted tools and well calibrated decision processes.

Safety performance does not escape these requirements as it englobes various dimensions (workplace safety, process safety, environmental safety), can be approached through various evaluation paradigms (auto evaluation, external assessment through audits and KPIs) and may be used by several stakeholders (risk owner, regional and national authorities, public perceptions). Indeed, different combinations of these three dimensions allow for a large variety of performance governance paradigms carrying various understandings and practices with regard to the way performance should be assessed and used.

By comparing safety performance management in three different risk systems (process, railway and health) in Europe, this works aims at providing safety management actors with a description of the richness and variety of safety performance understandings and practices. Insights gained from this work should enhance safety managers' abilities to grasp the strengths and weaknesses of their approaches and explore the possibilities of a beneficial benchmarking.

19 AIChE Spring Meeting
15TH GLOBAL CONGRESS ON PROCESS SAFETY
NEW ORLEANS • 2019

Lessons learned from analyzing an explosion accident at Shanghai Secco chemical plant

Hanhai Dong
China University of Petroleum (East China)
No. 66, West Changjiang Road, Huangdao District, Qingdao, China
donghanhaiupc@126.com

Yi Liu
China University of Petroleum (East China)
No. 66, West Changjiang Road, Huangdao District, Qingdao, China
liuyi@upc.edu.cn

Dongfeng Zhao
China University of Petroleum (East China)
No. 66, West Changjiang Road, Huangdao District, Qingdao, China
zhaodf@vip.sina.com

Jian Chen
China University of Petroleum (East China)
No. 66, West Changjiang Road, Huangdao District, Qingdao, China
836771528@qq.com

Yanchun Wang
China University of Petroleum (East China)
No. 66, West Changjiang Road, Huangdao District, Qingdao, China
Wychun25@163.com

Wei Wang
China University of Petroleum (East China)
No. 66, West Changjiang Road, Huangdao District, Qingdao, China
15621097585@163.com

Prepared for Presentation at
American Institute of Chemical Engineers
2019 Spring Meeting and 15th Global Congress on Process Safety
New Orleans, LA
March 31 – April 3, 2019

Lessons learned from analyzing an explosion accident at Shanghai Secco chemical plant

Hanhai Dong
China University of Petroleum (East China)
No. 66, West Changjiang Road, Huangdao District, Qingdao, China
donghanhaiupc@126.com

Yi Liu
China University of Petroleum (East China)

Dongfeng Zhao
China University of Petroleum (East China)

Jian Chen
China University of Petroleum (East China)

Yanchun Wang
China University of Petroleum (East China)

Wei Wang
China University of Petroleum (East China)

Keywords: event sequence diagram; Multilayer Bow-tie Analysis; FLACS

Abstract

On May 12, 2018, six workers from Shanghai Aijinke Engineering Construction Services Co., Ltd. died on the spot of an explosion when they were maintaining a benzene tank in the tank farm at Shanghai Secco Petrochemical Co., Ltd. The accident was characterized as a disastrous accident caused by hot work in a confined space, which resulted in massive casualties and a direct economic loss of 11.66 million yuan. This paper aims to analyze the accident occurrence process in a chronological order firstly by using the event sequence diagram (ESD). Multilayer Bow-tie Analysis was adopted to explore and classify the causes of the explosion of the benzene tank, along with the corresponding consequences, and the preventive measures. The analyses show that the primary inducements were the unsafe behavior of on-site workers, hazardous conditions of in-tank substances, risky factors in the construction environment, and deficiencies in management. Casualties, fire and explosion, equipment damage, and property losses were all included as consequences. The FLACS software was used to simulate the variation of explosion pressure and explosion temperature varying with time when benzene vapor was ignited at the tank bottom during the accident, and a flow field diagram for the explosion was achieved correspondingly. The results obtained are of direct significance for investigation of similar accidents in the future and are instructive to equipment safety evaluation as well. Based on the qualitative and quantitative analyses, we emphasize three considerable reasons leading to the accident, which includes inapposite flammable gases detection points before the hot work, failing to test the flammable gases during the hot work, and deficiency of the Management of Change (MOC). Finally, the paper concludes with a summary of lessons learned from the accident as well as suggestions for accident prevention and safety improvement.

1 Introduction

The term confined space refers to closed or semi-closed facilities and places with limited entrance and exit, and poor ventilation, which may lack oxygen or contain flammable, explosive, toxic, or harmful substances. The main causes of accidents in a confined space include fire, explosion, poisoning, asphyxiation, and physical injury [1]. According to the statistics of the Ministry of Emergency Management of the People's Republic of China, from January 2012 to April 2018, a total of 64 accidents involving three or more deaths in a confined space occurred in China, resulting in 225 deaths and 82 injuries [2]. Among them, on April 8, 2014, three people were killed and two injured by an explosion during the repair and welding of the desulfurization tank at a factory in Wuhai, China. On March 18, 2015, the operators of a chemical company in Binzhou, China, suffered an explosion accident during an operation in a confined space, resulting in four deaths and two injuries. Hot work in a confined space is the common cause of the two accidents above.

Due to the poor ventilation, a large amount of flammable gas may be accumulated in a confined space. When construction workers perform hot work inside, they will face potential safety risks and problems, such as difficulty in escaping, delayed personnel rescue, and inconvenient communication. Hot work in a confined space can easily lead to fire, explosion, and personal injury accidents, which threaten not only the personal health of operators but also the property of enterprises.

Researchers have done a lot of risk analysis and exploration for the particular operating environment in a confined space. Selman et al. [3] analyzed the mortality rate of confined space operations in similarly industrialized countries from 0.05 to 0.08 per 100,000 workers and pointed out that the main risks are toxic atmospheric hazards and physical hazards. Selman et al. [4] divided the emergency rescue of confined space accidents into three levels: individual self-rescue, non-entry rescue, and progressive rescue. Botti et al. [5] applied the Internet of Things (IoT) technology and proposed a holistic approach to analyzing the procedure of confined space operations. Meanwhile, many studies adopted experimental and numerical simulation methods to observe the evolution of characteristic parameters during the diffusion and detonation of flammable gas in a confined space. The results obtained are helpful to the quantitative analysis of confined space accidents. Bang et al. [6] established a theoretical model to predict the pressure rise caused by the deflagration in a confined space. He et al. [7] presented an LES turbulence model to simulate the diffusion process of LPG in the underground confined space.

This paper mainly analyzes a disastrous accident caused by hot work in a confined space at Shanghai Secco Petrochemical Co., Ltd (Secco). On May 12, 2018, the workers from Shanghai Aijinke Engineering Construction Services Co., Ltd. (Aijinke) triggered the explosion of flammable gas when they were performing inspection and maintenance inside a benzene tank. The serious accident resulted in the immediate death of six on-site workers and direct economic loss of 11.66 million yuan. Through a series of qualitative and quantitative analysis methods, we reveal the root causes of the accident, simulate the development process of the explosion, and propose corresponding rectification suggestions. Here we want to emphasize that the reason why we write this paper is not to pursue responsibility and punishment but to reflect on how to avoid the recurrence of similar accidents.

2 The occurrence of the explosion accident

On the morning of May 12, 2018, Aijinke workers arrived at the utility tank farm of Secco to prepare for maintaining TK-75-0201 benzene tank (0201 benzene tank). Before the start of the operation, a Secco field operator used a hand-held gas detector to measure oxygen and flammable gas at the manhole outside the tank. The on-site supervisor of Aijinke, the on-site supervisor of Secco, and the on-duty monitor of Secco utility tank farm signed the work ticket successively without verifying the measurement data and checking the personal protective equipment (PPE) and explosion-proof tools of the on-site workers according to the requirements specified in the operating permit.

At 1:15 p.m., six workers of Aijinke continued to dismantle the floating pontoon in the tank and two workers worked outside for monitoring. At the scene, another Secco field operator was in charge of regular oxygen and flammable gas measurement. At 3:25 p.m., a sudden explosion occurred in the 0201 benzene tank.

After the explosion, Secco immediately launched their emergency response plan and reported the accident information to the response center of Shanghai Chemical Industry Zone Management Committee (SCIZMC), the fire brigade, the medical center, and the production operations monitoring center of Sinopec. At 3:50 p.m., the fire brigade put out the fire by spraying water. At 5:50 p.m., the on-site rescue work ended and a total of six Aijinke workers in the tank died on the spot. Figure 1a shows the scene of the 0201 benzene tank after the explosion.

Figure 1. The 0201 benzene tank and floating pontoons

2.1 *Description of the timeline*

The occurrence of accidents is a complex system problem, and it gives useful information on how accidents arise in practice [8]. In March 2018, Secco found that the breather valve of the 0201 benzene tank leaked slightly, which caused the VOC concentration to exceed the standard. After inspection, it was judged that the floating pontoon leaked. In order to find the root causes, we describe in detail the fundamental behaviors of the primary responsible party and the contractor prior to the explosion.

- On April 16, Secco established a maintenance order and a work list in the SAP system.
- On April 19, Secco began to empty the 0201 benzene tank, performing four days of steam dilution and eight days of nitrogen replacement.
- On April 24, Secco informed Aijinke to formulate maintenance plan.
- On April 26, Secco carried out the Job Hazard Analysis (JHA) around the maintenance work.
- From May 1 to 7, Secco opened the manhole of the tank for natural ventilation and inspection, and found that more than half of the floating pontoons had effusion.
- On May 8, the relevant departments of Secco held a special meeting and decided to change the original plan to replace all the floating pontoons.
- On May 9, Secco asked Aijinke workers to enter the tank and destructively dismantle all floating pontoons at the regular OSBL equipment maintenance meeting.
- From May 9 to 11, Aijinke workers entered the tank and dismantled the floating pontoons. A total of 38 floating pontoons were dismantled, as shown in Figure 1b.

2.2 Analysis by event sequence diagram

Event Sequence Diagram (ESD) is an accident scenario-oriented simulation method, which treats events as the basic research object, and takes into account the domino effect and risk transfer effect [9]. ESD is also an intuitive reflection of the accident scenario. It provides a more literal representation of the system state, indicating the behavior of crucial process variables, and even operator and hardware state changes [10].

In order to further illustrate and point out the critical links of the accident, this paper makes an intuitive description of events by using ESD. As shown in Figure 2, the basic events of accidents are divided into initial events, judge events, action state events, and termination events. At the same time, the path indicated by the dashed line represents the potential events that may occur in this accident, and helps to make a comprehensive and logical analysis of the whole process. The graphic symbols in the ESD are explained in the Appendix.

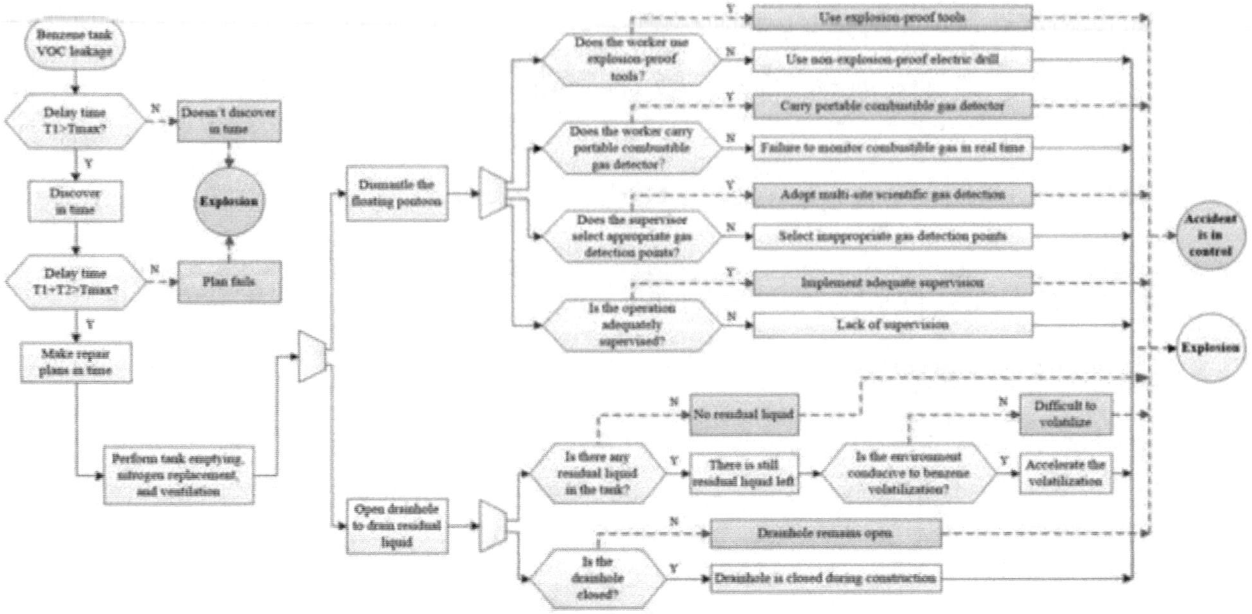

Figure 2. The ESD of the accident

As can be seen from the ESD, there are nine pivotal judgment nodes on the occurrence of the event chain, which can determine the direction of the incident. Among them, whether to discover leakage in time and whether to make a repair plan in time are the nodes at the time aspect. When the detection instrument and worker cannot find leakage and formulate the response plan in the maximum allowable delay time, they will directly lead to the explosion. Enterprises need to enhance equipment reliability and inspection frequency, so as to reduce emergency response time. Whether there are residues in the tank and whether environmental conditions are conducive to benzene volatilization are two nodes at the environmental aspect, which are also the leading causes of accidents. They require the workers to make a detailed hazard identification of the construction environment, and improve scientific detection. The critical nodes regarding human factors include whether the drainhole is closed, whether the on-site workers use explosion-proof tools, whether they carry portable combustible gas detector, whether they select appropriate detection points, and whether the supervisors take adequate monitoring [11]. Strengthening safety training before the operation, improving individual risk identification ability, perfecting risk control measures, and enhancing safety supervision level are essential means to avoid accidents caused by human factors.

All of these factors, organizational and operational, contributed to the accident, with a hierarchy of casualties among them. Through the comprehensive description, ESD can not only find out the nodes where the problem lies, but also guide the prevention of similar accidents. Therefore, enterprises should pay attention to the basic events expressed by nine judgment nodes and take appropriate interventions to block the development path of dangerous accidents.

3 Multilayer bow-tie analysis

3.1 The process of MBTA

Among many techniques available, bow-tie analysis is prevalent because it organically combines the two sophisticated risk analysis methods of Event Tree Analysis (ETA) and Fault Tree Analysis (FTA) [12], systematically summarizes the causes and consequences of accidents, and further integrates the prevention and mitigation of accidents or risks [13]. Bow-tie analysis has been proven to be a reliable and efficient tool, mainly used for process safety analysis, risk management, and accident risk analysis. It helps to understand which significant event combinations will lead to the top events, and which security barrier failures would escalate the top event to a certain consequence.

Traditional bow-tie analysis mainly includes top event, causes, consequences, preventive barriers, and mitigation barriers, which shows a lot of information. However, the single-level structure lacks logic and is not conducive to people's deep understanding. Thus, this paper establishes a Multilayer Bow-tie Analysis (MBTA), which resolves the top-event-related ETA and FTA individually into hierarchical structures, to deeply analyze the root causes, specify the severe consequences, and propose the corresponding safety barriers and scientific disposal methods. MBTA uses graphics to visually display the relationship between different levels and avoid repeated analysis effectively. Figure 3 shows the MBTA model, in which the code explanations are listed in Table 1.

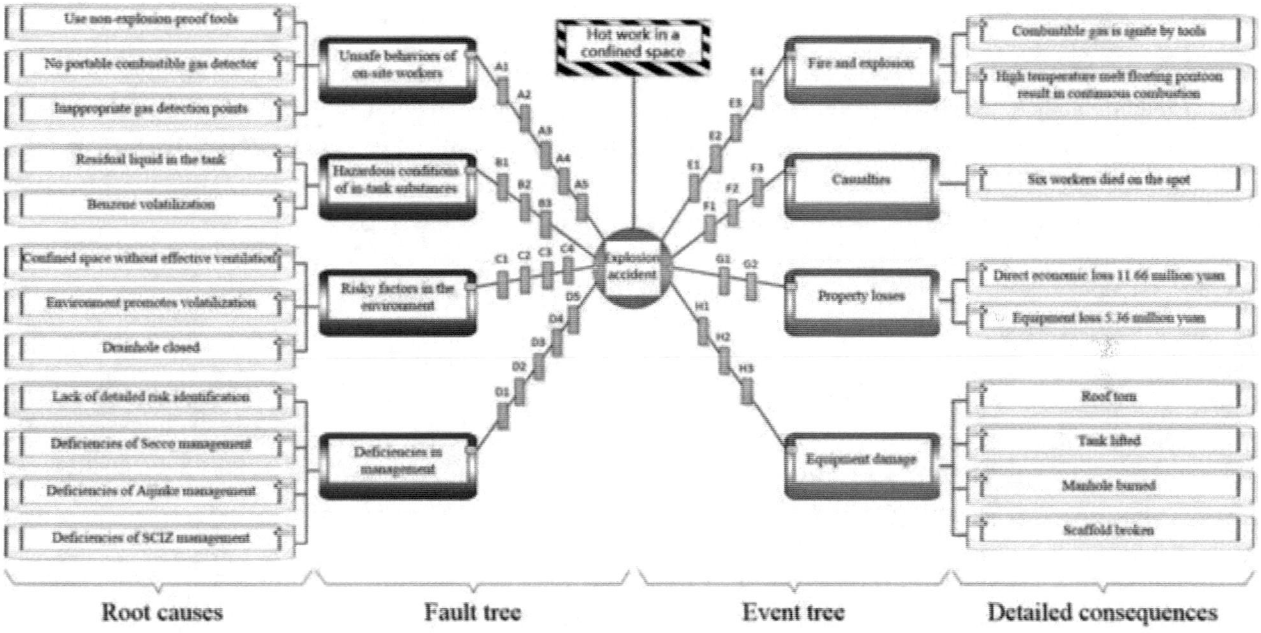

Figure 3. The MBTA model

Table 1. The explanations of MBTA

Code	Safety barrier	Code	Safety barrier
A1	Use explosion-proof tools	D4	Implement the main responsibility
A2	Carry portable combustible gas detector	D5	Improve contractor management
A3	Strengthen operation training	E1	Call the fire brigade in time

A4	Enhance supervision and assessment	E2	Spray water to extinguish fire
A5	Set up warning signs	E3	Fire isolation
B1	Remove the residual liquid	E4	Launch the emergency plan
B2	Hazardous material isolation	F1	Call for medical treatment
B3	Combustible gas detection in the tank	F2	Search and rescue the wounded
C1	Forced ventilation	F3	Launch the emergency plan
C2	Open drainhole	G1	Insurance claim
C3	Continuous detection during operation	G2	Launch the emergency plan
C4	Take environmental risk assessment	H1	Dangerous equipment isolation
D1	Compliance with rules and regulations	H2	Equipment repair
D2	Increase the intensity of punishment	H3	Equipment replacement
D3	Strengthen process safety management		

In the above MBTA model, we adopt the analytic hierarchy process (AHP) to extract the common characteristics of 12 causes of accidents and classify the causes into four categories: unsafe behaviors of on-site workers, hazardous conditions of in-tank substances, risk factors in the construction environment, and deficiencies in management. Moreover, we could conclude the direct and indirect reasons for the accident.

3.1.1 Direct reasons

Benzene is a colorless and transparent liquid with a strong aromatic odor. It has a boiling point of 80.1 ℃, a saturated vapor pressure of 10 kPa at 20 ℃, a flash point of -11 ℃, an explosion limit range of 1.2%-8.0%, and minimum ignition energy of only 0.20 mJ. Benzene is a highly flammable substance. The explosive mixture of benzene vapor and air can burn and explode when exposed to ignition source or high temperature.

At the construction site, the temperature is 24.7 ℃, and the pressure is 100.68 kPa. Under these conditions, benzene is easy to volatilize and form a gas explosion environment in a confined space. During the process of dismantling, residual benzene in the floating pontoons leaked onto the bottom of the tank and was not cleaned up in time. Because there was no effective ventilation in the tank, benzene vapor mixed with air to form an explosive environment and benzene concentration reached the explosion limit. The non-explosion-proof tools used by the operators produced ignition energy, which brought about the deflagration of the flammable gas. The high temperatures generated by the combustion melt the other aluminum alloy pontoons, resulting in sustained combustion for a short period.

3.1.2 Indirect reasons

Aijinke failed to abide by the relevant safety production rules and regulations. Managers did not fulfill their duties of safety management and supervision. Aijinke failed to give safety technical information to workers following the operating rules before the construction. Without implementing and supervising PPE, carrying portable gas detector, and using explosion-proof tools, operators were arranged to work in a confined space. When significant changes occurred in the job conditions, Aijinke did not promptly request workers to stop working and implement the Management of Change (MOC).

Secco had some omissions in safety risk management, professional management, and special operation management. In addition, Secco did not follow the safety production rules and regulations. On-site inspectors failed to carry out gas detection in accordance with specifications. After the change of operation, the manager did not ask the contractor to stop the process in time and revise the construction plan according to the regulations. Secco failed to report the maintenance plan to SCIZMC as required.

SCIZMC failed to implement the safety management system fully and did not discover that Secco performed maintenance operations without permission.

3.2 Lessons learned from MBTA

This paper puts forward a series of preventive measures, such as strengthening operation training, removing residual benzene in the tank, improving risk assessment of the environment, and enhancing process safety management. It is noticed that the environment in a confined space is harsh and there are many potential unsafe factors. Furthermore, the contractor's maintenance work cycle spans the statutory weekend. Based on the two reasons above, the enterprise should upgrade the level of hot work and adopt a broader system of operation approval and management. On-site workers need to stop working immediately when safety problems are found in the construction process and report to their superiors by relevant regulations.

Through refining the consequences of accidents, we summarized that detailed safety emergency plans and energy isolation methods are universally applicable in various measures to control and mitigate the impacts of accidents. They require Secco and Aijinke to formulate a comprehensive emergency response plan before, including fire extinguishing, personnel search and rescue, equipment maintenance, property damage remediation and so on. When an accident occurs, workers have to stay away from dangerous substances and places immediately.

As with many disasters, what ended up in a sequence of equipment damage and human errors started mostly as a management problem. Many unsafe practices were tolerated and even encouraged. The risk management of confined space operations should be divided into ten steps according to relevant regulations, namely training, pre-operation planning, job approval, hazardous energy isolation, ventilation, gas detection, personal protection and tool safety, warning signs, monitoring, and emergency response, so as to strengthen process safety management comprehensively. It is necessary to enhance the issuance of management of work tickets for hazardous operations such as hot work and confined space operations. Enterprises must strictly inspect illegal operations and supervise safety precaution measures to ensure that the operation process is safe and controllable. For changes occurring in the construction process, enterprises ought to strictly implement the MOC, analyze the potential risk and possible accidents, formulate corresponding safety measures, and carry out safety education for all operators.

MBTA provided a detailed analysis of the accident and established the corresponding safety barriers. However, there are still some certain subjective problems, and its content should also be flexible. Therefore, we encourage employees, accident investigators, case learners, and other relevant personnel to improve and develop the content of MBTA continually, thereby effectively expanding the application scope and effectiveness of MBTA.

4 Quantitative simulation of the explosion

The analysis of the accident should include not only qualitative analysis but also quantitative analysis [14]. We use FLACS software to simulate and analyze the occurrence of the benzene explosion. According to the relevant information, the 0201 benzene tank is an internal floating roof tank with a volume of 10,000 m^3. Its diameter is 30 m and height is 19 m. The design temperature range and pressure range are -10 – 65 ℃ and -0.49 –1.96 kPa, respectively. The tank adopts box-type aluminum alloy internal floating pontoons, and a manhole with a diameter of 0.6 m is set at the position of 0.75 m above the ground. Figure 4 shows the 0201 benzene tank model.

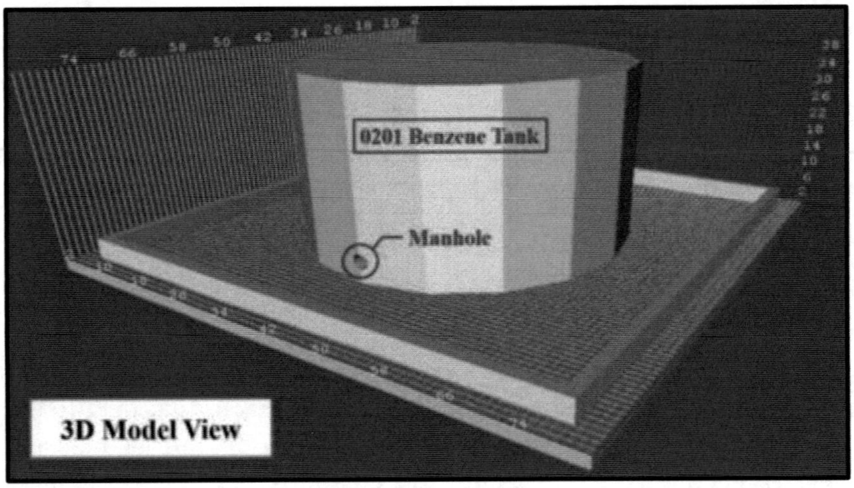

Figure 4. The 0201 benzene tank model

The simulated environmental conditions are fixed at the temperature of 24.7 ℃ and the pressure of 100.68 kPa, which are the same as the construction environment. We make reasonable assumptions: (a) The residual benzene in the tank evaporates continuously into the confined space and forms a uniform mixture of flammable gas. When the volatilization concentration of benzene reaches the lower explosion limit (LEL), the ignition source that persists during the construction operation detonates the flammable gas and results in an explosion. (b) Six on-site workers dismantled the float pontoons at the bottom of the tank, so the ignition source is set at the height of 0.75 meters from the geometric plane center, as shown in Figure 5 point I. There are several detection points of pressure and other characteristic parameters in the model of benzene tank, as shown in Figure 5, points M1-M5. (c) The benzene tank is windless, and the flammable gas accords with the real gas state equation. (d) The gas explosion process satisfies a series of fundamental equations, such as the continuity equation, momentum equation, and energy equation.

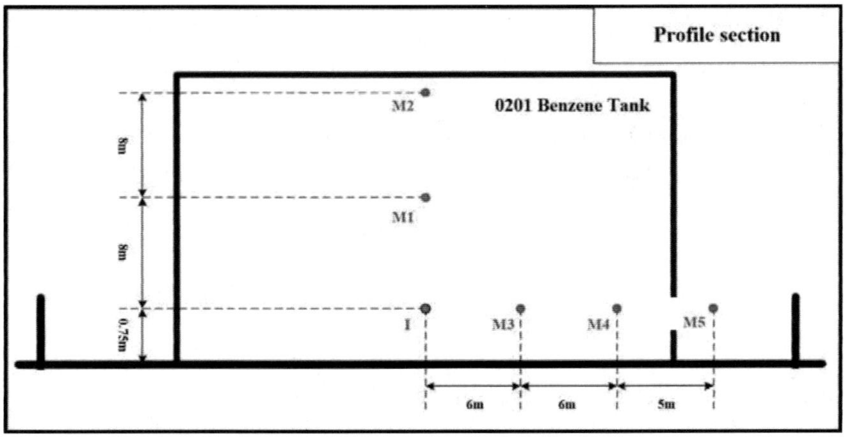

Figure 5. The profile section of the model

We selected the maximum profile section of the tank as the research object and intercepted the temperature field and pressure field at four different moments in the explosion process, as shown in Figure 6 and 7. The simulation results intuitively describe the development process of the explosion. When the flammable gas in the tank was ignited, the flame diffused around in the form of a spherical wave and reached the bottom of the tank at 0.35 s, as shown in Figure 6a and 7a. Since the radial distance of the tank is less than the vertical axial distance, the flame propagated horizontally to the wall, as shown in Figure 6b and 7b. At 1.52 s, the flame reached the manhole and sprayed out to the outside of the tank in the form of jet flow. During this period, the pressure and temperature increased continuously, which are the leading causes of the manhole being burned, as shown in Figure 8a. When the flame propagated to the top of the tank at 2.11 s, the uncombusted gas was recirculated by the viscous effect and blocking effect of the wall, which further caused the recirculation vortex. The recirculating vortex enlarged the contact area of the flame front, resulting in flame acceleration and pressure increase, as shown in Figure 6c and 7c. Figure 6d and 7d illustrate the flammable gas burned out at 3.24 s. Simultaneously, the explosion temperature and pressure reached the peak value contributing to the tearing of the roof and the lifting of the tank, as shown in Figure 8b and 7c.

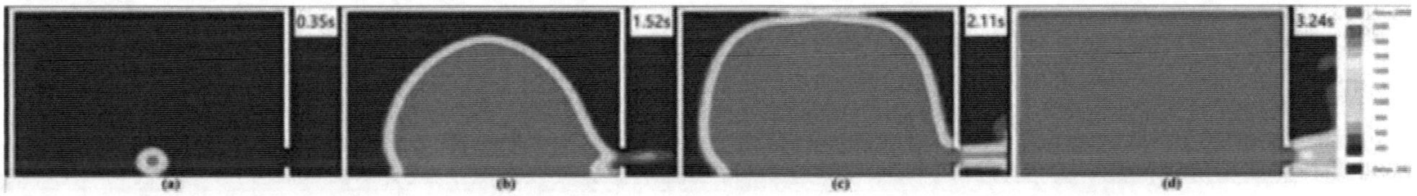

Figure 6. Temperature simulation of the explosion

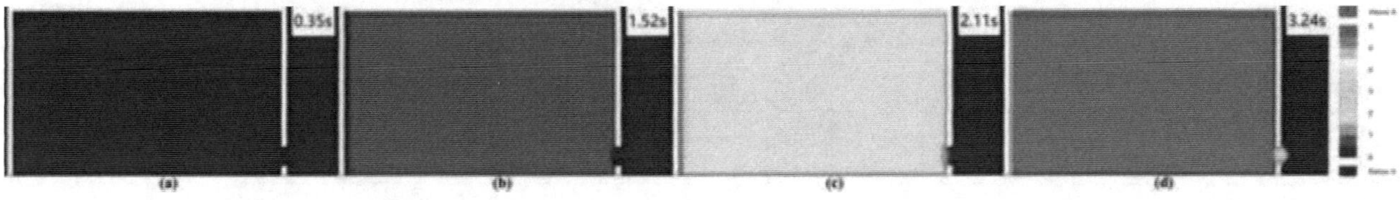

Figure 7. Pressure simulation of the explosion

Figure 8. The results of the accident

Figure 9 reveals the change of temperature and pressure recorded by the detection points M1-M5 during the explosion. Table 2 describes the data of the detection points, including maximum and minimum explosion pressures, maximum and minimum explosion temperatures, peak data arrival times and so on. Combining the data of detection points and the pressure field diagram in Figure 7, it is shown that the pressure in a confined space will be uniform immediately for a certain time and further proves the accuracy of the analysis of the flame propagation process. In particular, the external detection point M5 was affected by the propagation of explosion flame in the tank at 1.6 s, resulting in a peak explosion overpressure of 0.031 bar and a peak temperature of 1895 ℃. Subsequently, due to the pressure rise in the tank, the pressure at the detection point M5 dropped to negative pressure until the flammable gas burned out.

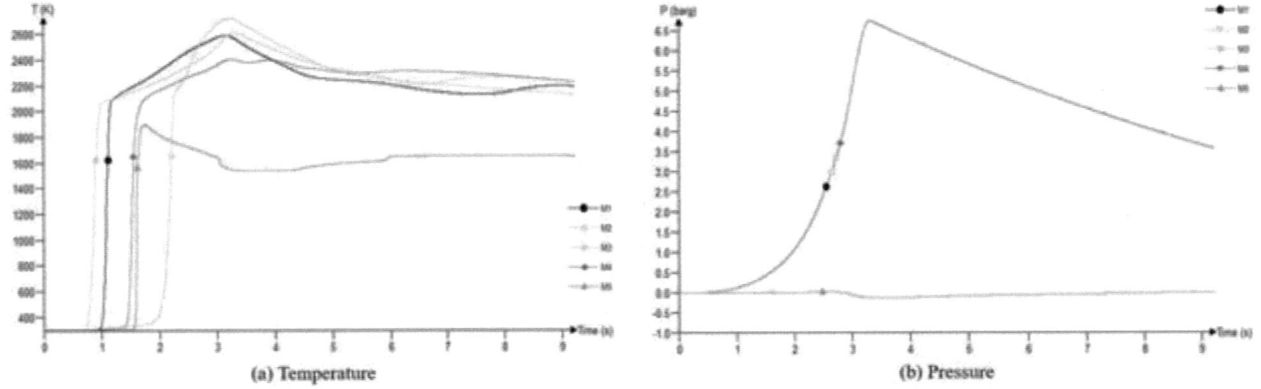

Figure 9. Parameters change of the detection points

Table 2. Data from the detection points

No.	Temperature				Pressure			
	Tmax/K	Time/s	Tmin/K	Time/s	Pmax/bar	Time/s	Pmin/bar	Time/s
M1	2589.4	3.11	298.1	0.04	6.75	3.24	0.00	0.04
M2	2732.2	3.24	298.1	0.04	6.76	3.24	0.00	0,04
M3	2609.3	3.28	298.1	0.00	6.76	3.24	0.00	0.00
M4	2306.2	3.25	298.1	0.00	6.75	3.24	0.00	0.00

M5	1895.5	1.68	298.0	0.69	0.05	1.65	-0.15	3.29

Although there is only a small amount of residual benzene, the benzene volatilization forms a flammable gas with the air inside the tank, which generates a substantial destructive power when ignited. The simulation results show that the flame propagation velocity of the flammable gas increases constantly, with the peak value reaching up to 41 m/s. Since the flame is propagated in the form of subsonic velocity, it can be judged that there are mainly the phenomena of deflagration occurred in the process of explosion [15]. Soloviova-Sokolova et al. [16] calculated the flame burning rate of the benzene vapor was 0.6 m/s at 298 K, 0.1 MPa and 0.6 eq., which clearly illustrated that the peak propagation velocity of the benzene vapor in a confined space would be several times higher. Meanwhile, the explosion pressure in a confined space is usually greater than that in an open area due to lack of heat loss and volume expansion.

When workers are engaged in hot work in a confined space, they should carry out comprehensive risk identification, select several representative flammable gas detection points, and detect the flammable gas in real time. Workers ought to stop work immediately and withdraw in an orderly manner once the flammable gas exceeds the standard. Through the analysis of the explosion process, the flame propagates to the wall in a very short time. Thermal radiation and explosive shock wave are the primary causes of casualties in the tank. Owing to the special working environment in a confined space, the phenomenon of mass casualties and injuries often occurs under the effect of high temperature and high pressure, which needs people's sufficient attention.

5 The subsequent event

The explosion accident of Secco has aroused widespread concern, and 22 responsible persons have also been punished. But unfortunately, lessons from the accident did not prevent a subsequent similar event. On November 26, 2018, three contractor workers of the Ningbo Engineering Company (NEC) and one worker of Secco inspected the discharge tank of cracking gas compressor in the olefin separation unit. Without ascertaining the concentration of oxygen in the tank, a contractor worker of NEC entered the tank for construction and fainted unexpectedly. A Secco worker immediately went into the tank to rescue him, but he also fell into the tank. The accident resulted in the death of two workers who had been rescued after asphyxia.

It is the second confined space operation accident in the same company within six months. The main causes of the accident are still the deficiency of contractor management and risk identification in a confined space. While expressing sadness, the public should profoundly reflect on why accidents continue to occur, what is the role of punishment and accountability, and how relevant regulations and reform measures can be truly implemented. Drawing lessons from the previous accidents is necessary. Enterprises ought to raise the safety awareness of their employees and use strict supervision to ensure safe production of chemical plants.

6 Conclusions

This paper provides an introduction of an accident caused by hot work in a confined space, together with the detailed ESD-based description of nine pivotal judgment nodes leading to this disastrous accident. It concludes that enterprises should take suitable intervention measures to block the pathway of the accident. The MBTA, covering the causes and consequences of the accident, offers proactive and reactive measures based on an intuitive analysis. Meanwhile, it emphasizes the importance in selection of appropriate detection points, real-time detection of in-tank operation, and implementation of MOC. Also, we apply the FLACS software into quantitative simulation of the explosion process in the tank. The results contribute to the guidance of accident investigation and safety measures selection. Finally, it suggests that enterprises adopt technical remedies and better safety management practices, since they are the essential basics for effective implementation of risk control measures.

Appendix

Symbol	Annotation
	Initial event: Beginning event of the ESD.
	Judge event: The event has two possible outcomes that determine whether subsequent events occur.
	Action state event: Describe the action state of the event.
	Termination event: Ending event of the ESD.
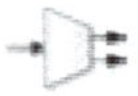	Output AND gate: Input event lead to multiple output events occurring.

7 References

[1] D.Burlet-Vienney, Y.Chinniah, A.Bahloul and B.Roberge, Design and application of a 5 step risk assessment tool for confined space entries, Safety science 80 (2015), 144-155.

[2] The Office of the State Council Security Committee Notified 8 Cases of Blindly Rescuing in a Comfined Space Leading to Casualties Enlargement, Available at http://www.chinasafety.gov.cn/default.htm Accessed on June 8, 2018.

[3] J.Selman, J.Spickett, J.Jansz and B.Mullins, Safety science 109 (2018), 333-343.

[4] J.Selman, J.Spickett, J.Jansz and B.Mullins, Safety science 113 (2019), 78-90.

[5] L.Botti, V.Duraccio,M.G.Gnoni and C.Mora, A framework for preventing and managing risks in confined spaces through IOT technologies, ESREL (2015), 3209-3217.

[6] B.Bang, C.Ahn, J.Lee, Y.Kim, M.Lee, B.Horn, D.Malik, K.Thomas, S.James, A.Yarin and S.Yoon, Theoretical, numerical, and experimental investigation of pressure rise due to deflagration in confined spaces, International Journal of Thermal Sciences 120 (2017), 469-480.

[7] B.He, X.Jiang, G.Yang and J.Xu, A numerical simulation study on the formation and dispersion of flammable vapor cloud in underground confined space. Process Safety and Environmental Protection 107 (2017), 1-11.

[8] I.M.Shaluf, F.Ahmadun and A.M.Said, Fire incident at a refinery in West Malaysia: the causes and lessons learned, Journal of Loss prevention in the Process Industries 16.4 (2003), 297-303.

[9] S.Swaminathan and C.Smidts, The event sequence diagram framework for dynamic probabilistic risk assessment, Reliability Engineering & System Safety 63.1 (1999), 73-90.

[10] J.Zhou, G.Reniers and N.Khakzad, Application of event sequence diagram to evaluate emergency response actions during fire-induced domino effects, Reliability Engineering & System Safety 150 (2016), 202-209.

[11] E.Zarei, M.Yazdi, R.Abbassi and F.Khan, A hybrid model for human factor analysis in process accidents: FBN-HFACS, Journal of Loss Prevention in the Process Industries 57 (2019), 142-155.

[12] N.Khakzad, F.Khan, P.Amyotte, Dynamic risk analysis using bow-tie approach, Reliability Engineering & System Safety 104 (2012), 36-44.

[13] M.B.Mulcahy, C.Boylan, S.Sigmann and R.Stuart, Using bowtie methodology to support laboratory hazard identification, risk management, and incident analysis, Journal of Chemical Health and Safety 24.3 (2017), 14-20.

[14] C.Zhu, J.Zhu, L.Wang and M.S.Mannan, Lessons learned from analyzing a VCE accident at a chemical plant, Journal of Loss Prevention in the Process Industries 50 (2017), 397-402.

[15] I.C.Tolias and A.G.Venetsanos, An improved CFD model for vented deflagration simulations–Analysis of a medium-scale hydrogen experiment, International Journal of Hydrogen Energy 43.52 (2018), 23568-23584.

[16] J.V.Soloviov-Sokolova, V.A.Alekseev, S.S.Matveev, I.V.Chechet, S.G.Matveev and A.A.Konnov, Fuel 175 (2016), 302-309.

Strengthening Process Safety Culture throughout Learning from Each Other´s Vulnerabilities

Luciana Poli SIlva
PETROBRAS
Av. Republica do Chile, 65, 4th floor, aisle 402.
lucianapoli@petrobras.com.br

Prepared for Presentation at
American Institute of Chemical Engineers
2019 Spring Meeting and 15th Global Congress on Process Safety
New Orleans, LA
March 31 – April 3, 2019

Strengthening Process Safety Culture throughout Learning from Each Other´s Vulnerabilities

Luciana Poli Silva
PETROBRAS
Av. República do Chile, 65, 4th floor, aisle 402.
lucianapoli@petrobras.com.br

Keywords: andragogy, dialogue, leadership, safety leadership culture, trust, vulnerabilities.

Abstract

A fundamental approach to any program that targets the implementation of a process safety culture is to work within reality, going beyond procedures and compliance. To facilitate the emergence of the unspoken is essential to stimulate the dialogue among leaders that brings about their vulnerabilities and opens new perspectives to address them effectively. This paper presents a pilot program designed to increase the Process Safety Leadership Culture in a Brazilian Energy Company. Based on a benchmark made within the industry and in andragogic principles, the program created a trust ambiance among peers and inspired leaders to recognize and share their vulnerabilities, as well as to take actions towards the strengthen the role of being a safety culture leader.

1 Introduction

A lot has been said and written about Leadership Process Safety Culture, but there is a gap between what the compliance procedures dictate and what managers really do. When it comes to the Brazilian Energy Industry, the pressure leaders are under in order to achieve the results they are expected to can be considered significant. How to balance the production challenges with a safety culture that must be among the core values of the operation? Are these factors compatible?

The pilot program entitled '*Strengthening Process Safety Culture*' dealt with reality, using andragogic approaches to bring about the complex factors forty managers face in their routine, such as the existence of several levels separating top managers from the operational team, privileging safety over production goals, and being an effective representative of the Company´s policies. All these factors can jeopardize the safety culture from becoming stronger every day.

This paper describes the implementation of a methodology designed to create a safe environment, in order to build trust among participants and at the same time, make them reflect and dialogue about how they can evolve from compliance to commitment.

In the coming sections, we will present the central concepts used to develop the program – Culture, Safety Culture and Leadership - exploring its implication on an actual work environment.

In the sequence, we will talk about what it takes to make the unspoken, unwritten rules emerge. From there, how we stimulated the increase in the self-conscious level as well as in the collective one, building a matrix entitled 'Personnel Influence on Safety' to diagnose how one really sees himself concerning his leadership in safety culture.

We will then present the method used to address the issues managers pointed out, in order to transform the discussions into actions to strengthen the safety process leadership culture.

In the conclusion, we will present our findings and discuss possible future approaches to amplify the safety leadership culture further.

2 Safety Culture and Leadership: definitions and implications

There are many possible definitions for culture. Taylor (2015) says culture has to do with the behaviour patterns that are encouraged, discouraged or tolerated over time in an organization. Schein (2015) says that culture is "the total sum of everything an organization has learned in its history in dealing with the external problems – which would be goals, strategy, how we do things – and how it organizes itself internally."

Since we humans are gregarious social beings, we are naturally wired to fit in with our community. We use our perceptions to read what the tribe accepts and we adapt by behaving accordingly, because we want to fit in and be rewarded. Therefore, we reinforce the existing norms. This process is perpetuated over time; established behavior influences the behavior of new employees and so on.

Most cultural sources are non-verbal. Taylor (2015) says it is a myth to believe that culture reflects the values statements that are declared in the company's annual reports. Most commonly, the statement will be the organization´s intents, not the culture as it actually is.

If we want to change culture, it is necessary to have a clear direction and a focused, effort-driven leadership to lead people towards the goal culture. Culture has to do with what is encouraged and allowed. For Taylor (2015), three major sources send cultural messages: behaviours - of other people, especially important, recognized ones; symbols - artefacts, events and decisions that are meaningful to the community; and systems - mechanisms that manage people and tasks.

In this sense, culture has to do with the manifestation of what is valued in the organization.

According to Lafraia (2014), culture shapes reality. Cultural lenses structure the perception of the people involved in the organization.

When it comes to safety culture, its meaning may vary from one organization to another, or even in different departments of the same company. It is perfectly possible to have diverse lenses that translate the meaning and the importance of safety practices in diverse ways.

OGP, in its 452 Report (2013), says that a positive Safety Culture has to do with a culture in which safety is a core value for everyone who works in the organization, and it is pursued in an everyday basis. In order to build a Safety Culture, there are factors that must be considered, such as "leadership commitment to safety; employee involvement and motivation; responsibilities and accountability; production and bottom line pressures versus quality issues; and actions, or lack of action, to correct unsafe behaviours and unsafe conditions"; among others.

In order to build a consistent safety culture, leaders must educate their teams by demonstrating, by their practices, that some policies are fundamental. Safety culture should use this lens, rather than a command and control opaque one. Leaders must educate people to understand why safety is a core value that concerns the most precious gift: one´s life.

Safety culture is not about only following the rules, but making the team understand that the rules serve a greater principle. This approach can influence values and beliefs and change the individual and team´s behavior. The challenge leaders face is to conciliate the educational role with the policeman role, guaranteeing that compliance is required and everyone in the organization, starting by himself, values that as an unnegotiable principle.

In order to develop a strong safety culture, some elements must be considered (OGP, 2013):

Elements of a Strong Safety Culture

- an informed culture: the company has quality data concerning safety and uses it well;
- a reporting culture: people feel they can report safety concerns without worrying about being punished;
- a learning culture: the organisation actually learns from its unsafe conditions and changes them, banning any equipment, practice or whatever takes place that hazards safety;
- a flexible culture: the company changes its command chain if necessary to adapt;
- a just culture: people understand what is acceptable and unacceptable. What is not tolerated is dealt fairly.

For Peter Drucker (Cohen, 2018), leadership can be defined as the lifting of a man's vision to higher sights. Throughout leadership, people perform in higher standards, they are inspired and evolve in multiple levels.

When it comes to safety leadership, we can quote Ram Charan (2007) when he says that "a leader who does not produce other leaders is not a great leader." A safety leader should act in a way to enhance in everybody´s minds and actions a safety leadership commitment.

In order to make it possible, it is crucial that the leader himself is aware of his beliefs and constantly monitoring his behaviours. It takes time and a lot of effort to build a strong safety culture. Despite that, a safety culture is fragile and easily threatened by wrong messages. (OGP, 2013)

In the next section we explore the method used to address self-consciousness about safety.

3 Creating a trust environment that favors Safety Process Leadership Culture

According to Malcon Knowles, who develop the paradigm of andragogy, adults have a learning style characterized, among other factors, by changing one´s concepts according to new experiences one lives; the pursuit for autonomy; increasing one´s own motivational level to match the social role assumed; and acquiring knowledge to immediately solve practical issues (Muresan, 2013).

Considering this practical approach, we began the program to strengthen safety leadership culture by making each participant reflect on its daily practices concerning management, as well as the safety implications involved in routine decisions.

To start this reflection, we did a benchmark and decided to invite a senior top manager from another company in the Energy Industry to give a testimony. He made an honest statement about his team perception of his management practices. Concerning safety process, he mentioned that he realized that, in order to influence people, he must first change himself. He then shared with an amazed group of participants some practical examples of safety interventions he changed in his auditing procedures, for example, in order to be coherent with the message he wanted to pass forward.

The testimony had an impressive impact on the managers. A remarkable phrase that was repeated by managers was: "*Simplify. Be the Powepoint. Less Presentations, More You.*"

After the testimony, there was a presentation on the invited company's safety philosophy, about its journey into safety leadership. The practical steps mentioned were followed by videos with managers from different parts of the world talking about the way they see safety culture benefits.

With that background environment set, we forwarded to a self-perception diagnosis. The goal was to stimulate the increase in the individual conscious level as well as in the collective one. We used a matrix entitled 'Personnel Influence on Safety' that was obtained in the benchmark process with the above mentioned company.

It aimed to diagnose how one really sees himself concerning its leadership in safety culture. It involved two dimensions: the personal influence towards oneself; and the personal influence concerning other people´s safety.

The figure below illustrates the matrix:

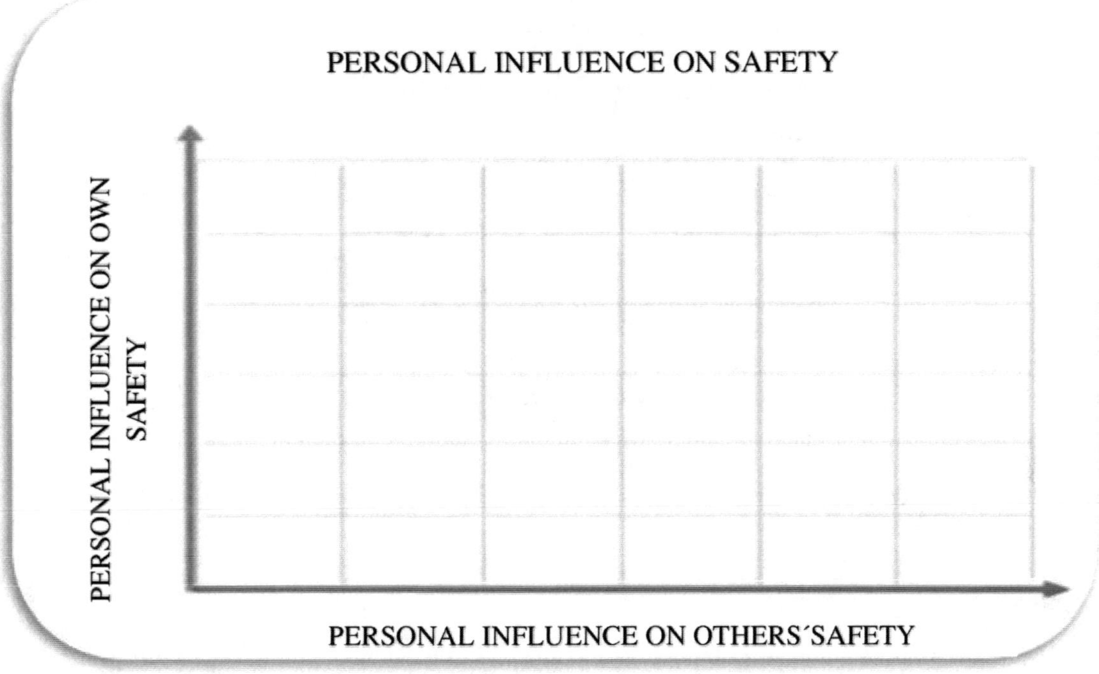

Figure 1. Personal Influence on Safety Matrix

We used the Self-perception as a tool to promote reflection and consciousness. We asked the participants to fill in the matrix in two different papers: one in his scrapbook (explained below) and another one on a plain sheet to hand over to the program facilitator. All the matrices were compiled and the result is shown in figure 2. As indicated, most participants initially evaluated themselves as influencing more themselves than others.

The coordination staff evaluated the matrix result as requiring attention from the participants. The staff observed the fact that some managers attending the program were obese, others worked extra hours, and most of them had reported having trouble balancing personal and professional life. Therefore, to promote deeper self-reflection, the consultant who was facilitating the program posed some questions to the participants regarding personal care. The remarks made were an invitation to honestly look at oneself and feel if the level of personal influence corresponds to what one would recommend to a loved person. Quoting Brené Brown (2015), the consultant said "Treat yourself like someone you love. Do you do it on a regular basis? Do you put yourself first?" Those remarks made some participants change their self-evaluation.

The intention was to call leaders ´attention to the fact that genuine influence on safety starts with their own safety, and goes beyond process or operational safety. To be legit and truly be a safety leader, one should take good care of oneself.

Figure 2. Compiled Personal Influence on Safety Matrix with the results from all the participants

We also developed a scrapbook that was given to each participant as a job-aid material to be used both during the program and after. The content of the scrapbook followed the program plan and was an invitation to self-analysis. The participants had special pages to do some journaling about what they had just listened to or dialogued about. There was also a special page dedicated to register insights. Figure 3 illustrates the scrapbook.

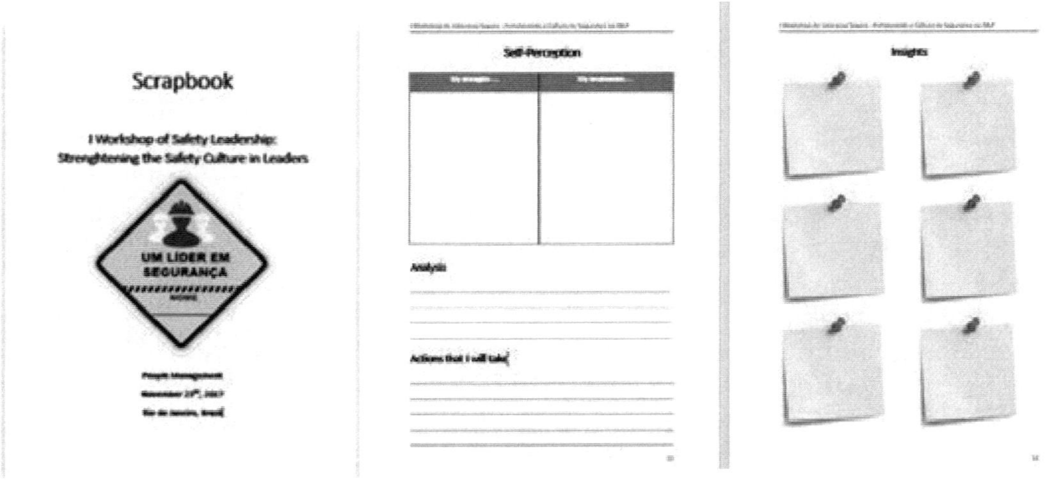

Figure 3. Pages from the Scrapbook used to stimulate self-consciousness and to plan actions

4 Exposing and Addressing Vulnerabilities

4.1 Control Room

We will now present the method used to address the issues managers pointed out, in order to transform the discussions into actions to strengthen the process safety leadership culture.

We prepared a special control room, where we designed a pathway decorated with images of accidents from the company and from the industry, and the number of fatalities. We also put an alarm sound playing during the route and videos showing accidents. These resources were intended to create an emotional ambiance to take the analysis to another level.

Figure 4. Picture of the Control Room and example of poster from accident used to decorate it.

Besides the planned journey passing several industry accidents, the Control Room was decorated with red material, to create an atmosphere of urgency. After going through the posters and watching the videos, the participants were invited to give their testimony of accidents they´ve been through.

The group tone got emotional as some of them shared unpleasant stories they had lived. The goal of inducing the reflection to an in-depth level was successfully achieved.

After that point, the participants went back to the regular room and dialogued about some questions that were in their scrapbook. The questions were related to the *Elements of a Strong Safety Culture* preconized by OGP (2013), mentioned in section 2 of this paper.

The methodology privileged social interaction among participants and the active listening of each other's considerations. Below is the list of the questions addressed to the participants.

Questions to Reflect and Dialogue:

1. Am I able to influence my operational team?
2. How many hierarchical levels are there between the operational team and me? To what extent does my speech actually makes sense to them?
3. How much of what is said by the operational team do I believe genuinely applies to me? Do I believe I receive good quality feedback?
4. Of the security actions implemented in my management, do I receive direct feedback on the effectiveness or otherwise of its implementation?
5. What degree of tolerance do I accept as a leader, regarding violations of safety standards?
6. Do I, as a leader, make it mandatory that the urgent inspection recommendations are executed immediately?
7. Do I, as a leader, take every opportunity to motivate, mentor and develop my team to enhance individual safety awareness and safety leadership?
8. Do I praise when I see a safe behavior?
9. Do I perform Behaviour Audits in a protocolled way, to meet the Scorecard, or do I do it because I believe in the effectiveness of what can arise from it?
10. Is Safety a value or is it a Priority for me as a leader?
11. Do I have the courage to step in and stop an activity when I see someone at risk? Do I always intervene?
12. Do my behaviours as a leader have safety as a foundation?
13. When I find that the environment is unsafe, do I privilege safety over other goals?
14. Safety comes first in my management?
15. How do I, as a leader, ensure that safety in my management is dynamic enough to ensure continuous improvement in operational procedures?
16. Thinking Safety as a chain, how can I, as a leader, be an effective link between operations and the company policy?

4.2 Results

In order to capture the precious discussions and share insights among the participants, in the final part of the workshop there was a collective presentation based on what each small group discussed about the questions above. Each group had to have at least one delegate, and others could participate if they wanted to.

Some of the findings reported were, for example: "*I have to go to the platform more often*"; "*Behaviour audits can be very useful, I should profit more from them*"; "*There are too many levels between my operational team and I. I have to get together with them on a regular basis.*"

After sharing insights, all managers filled in the Personal Influence on Safety Matrix again. The group reported that, this time, the self-perception was lower, since many managers realized they need to invest more efforts and more quality time in Safety, in both dimensions – with oneself and towards others.

After that, each participant revised and complemented his individual action plan (figure 3). They also wrote a personal commitment as a Safety Process Culture Leader.

At the very end of the program, we asked everyone to do a check-out by sharing with the entire group one action of the action plan. This intended to create a deeper commitment, as well as possible insights to others. Also, it was a practice that could be used with the managers team.

According to the group evaluation, presented in the word cloud below, the program succeeded in strengthening the safety culture in leaders, who mentioned as the top benefits from the program: (*Safety as a*) Value; Communication; Life; Commitment; Example; Integration; Consciousness; and Synergy.

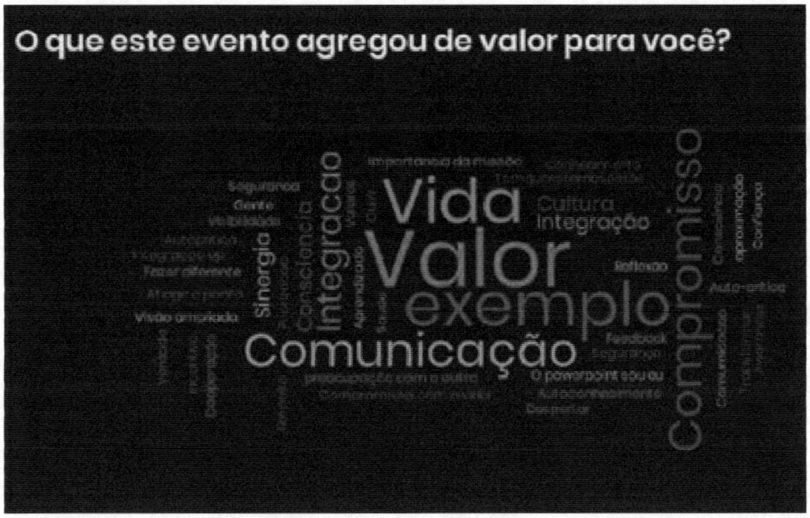

Figure 5. Evaluation: Word cloud containing the managers´ perception about the contribution brought by the program to participants.

5 Conclusions

The feedback given by the program´s participants was encouraging. It indicated that testimonies from partners, combined with relevant questions based on the daily problems managers face concerning safety, created a trust ambiance among managers.

The activities planned based on andragogic approaches favored dialogue and in-depth self-analysis. The use of the same instrument (Personal Influence on Safety Matrix) twice, as well as the individual scrapbook, facilitated the self-consciousness process and lead managers to review their initial assessment and to elaborate an action plan.

Leaders play a decisive role in strengthening the safety culture in the organizational world. Bringing about their vulnerabilities and having honest conversations with peers definitely helps managers to find ways to evolve from compliance to commitment.

Strengthening the safety leadership culture is a process that does not come to an end. It is an on-going process that must be continuously taken care of by leaders who value life and put safety first. Future projects that aim to implement effective actions in Safety Culture should take into consideration the company´s culture, as well as daily operational problems and safety as a core value to preserve life and business.

6 References

B. Brown, "*Daring Greatly: How the Courage to Be Vulnerable Transforms the Way We Live, Love, Parent, and Lead*", Penguin Random House, NY, 2015.

C. Migueles, J.R.B. Lafraia, G. C. Souza, "Criando o Hábito da Excelência: Compreendendo a Força da Cultura na Formação da Excelência em SMS", Quality Mark, Rio de Janeiro, 2014.

C. Taylor "*Walking the Talk: Building a Culture for Success*" Random House Business Books, London, 2015.

M. Muresan, "*Using Cybergogy and Andragogy Paradigms in Lifelong Learning*", 5th World Conference on Educational Sciences – WCES. Romania, 2013

OGP, *Shaping Safety Culture through Safety Leadership.* Report N° 452. International Association of Oil & gas Producers. October, 2013.

R. Charan, *Ram Charan on Leadership*,

Available at https://hbr.org/ideacast/2007/01/harvard-business-ideacast-27-r.html. Jan, 2007 (accessed Dec 4, 2018)

T. Kuppler, *Edgar Schein on Culture*,
Available at https://www.leadershipandchangemagazine.com/edgar-schein-on-culture/ Accessed on January 2, 2019.

W. Cohen, *Drucker's Top Tips to Successful Leadership*
Available at https://www.managementmattersnetwork.com/strategic-leadership/columns/druckers-top-tips-to-successful-leadership. Jun, 2018 (accessed Dec. 4, 2018)

7 Acknowledgments

I would like to thank Petrobras who fully supported this project, as well as Shell Brasil, for a very profitable partnership. A special thanks to my Human Resources, HSE and E&P work colleagues, who had a decisive role in making this project a reality.

548424 Development of Human Reliability Analysis Method for Oil and Gas Operations

The oil industry has grown in recent decades in terms of quantity of facilities and process complexity. However, human and material losses still occur due to major accidents at the facility. The analysis of these accidents reveals that many involve human failures that, if prevented, could avoid such accidents. These failures, in turn, can be identified, modeled and quantified through Human Reliability Analysis (HRA), which forms a basis for prioritization and development of safeguards for preventing or reducing the frequency of accidents. The most advanced and reliable HRA methods have been developed and applied in nuclear power plant operations, while the petroleum industry has usually applied Quantitative Risk Analysis (QRA) focusing on process safety in terms of technical aspects of the operation and equipment. A joint project from the B. John Garrick Institute for the Risk Sciences and a big company from the oil and gas industry is focusing on developing a HRA method focused on this industry operations, to bridge the existing gap between HRA and the Petroleum industry. The Phoenix HRA methodology was used as a basis, which has three layers i) a crew response tree (CRT), which models the interaction between the crew and the plant; ii) a human response model, modeled through fault trees, that identifies the possible crew failures modes (CFMs); and (iii) "contextual factors" known as performance influencing factors (PIFs), modeled through Bayesian networks. In addition to building on such a structure, the method development will rely on interviews with key personnel, visitations to installations, and analysis of past incidents data.

547327 Studying the Dependence of Operator Performance on Step Level Complexity

In the chemical, oil, and power industry, procedures are used by operators to perform their work correctly without compromising safety. Unfortunately, many procedures fail in adequately conveying information to their users. Heuristics tell us that technical jargon, vague language, and operator training all play a role in determining the effectiveness of procedures. To what extent these variables play a role is still not fully understood. Knowing this information will allow plants to more effectively write procedures. To counter this, researchers at Texas A&M's School of Public Health and Mary Kay O'Connor Process Safety Center are investigating the underlying factors that cause operators to deviate from procedures.

One possible factor may be that complex steps hinder an operator's ability to perform a task. This work aims to understand why certain steps are more complex than others. To do this, a set of procedures on how to operate a gas chromatography mass spectrometry (GCMS) machine is written. A study is performed in which participants use a set of procedures to analyze samples through gas chromatography. Using a scoring rubric, the complexity of each step is determined. The procedures have been designed to obtain an even distribution of complexity. Participants for this study are chemical engineering students with little to no prior experience in handling a GCMS machine. Thus, the experience, education, expertise, and training of the participants is controlled. The performance of each participant is reviewed and is grouped into one of four quadrants. For this study, a decrease in operator performance is expected as the complexity of each step increases.

548400 Prevent Them All or Stop Them Small: Tripod Best Practices in Reducing Chemical Accidents and Enhancing Response Preparedness

Prevent Them All or Stop Them Small: Tripod Best Practices in Reducing Chemical Accidents and Enhancing Response Preparedness

The connection between industry, government, and public safety, which we call the Tripod, is vital when implementing the 3-P's - Prevention, Protection, and Preparation. As identified by AICHE's Vision 2020 program, there are significant opportunities for organizations that address industrial process safety to work together. Responsible Collaboration between industry, regulatory and investigatory authorities, labor organizations, communities, and technical organizations promotes mutual understanding of risks, effects process safety and can reduce chemical accident risk in both small and large communities. Community based safety partnerships and leadership are fundamental to the success of such efforts.

For the past five years, a series of geographically focused Chemical and Ammonia Safety Days have cultivated and showcased chemical emergency prevention and planning efforts between Industry, NGOs such as the Ammonia Safety Training Institute (ASTI), and a myriad of public agencies including fire departments, health departments, LEPCs, SERCs, local regulators, and state and federal agencies such as EPA, CSB, DHS and OSHA. Within the context of Chemical Accidents and Releases, the focus of these efforts has been to Prevent Them All or Stop Them Small.

These Safety Day programs are designed for and with local community stakeholders from industry and first response agencies with a goal of seeding a permanent, engaged geographic collaboration for accident prevention and response readiness ownership and partnership. Agenda and curriculum elements include technical training focused on local risk, first responder field practicums and case studies, accident prevention program prioritization, practice and accountability, bowtie analyses of event propagation and management, and emergency response planning centered on the first 30 minutes and one-plan approach. The programs often include field demonstrations and exercises for fire departments and facility responders.

As a best practices demonstration, the presenters for this session will share the developed programs, recruitment methodologies and the demonstrated results of these efforts. In addition to Responsible Collaboration, other Vision 2020 principles embraced and promoted by the Safety Day Approach include Disciplined Adherence to Standards, Intentional Competency Development, Enhanced Application & Sharing of Lessons Learned, and Enhanced Stakeholder Knowledge. As a bonus, the societal theme of Harmonization of Standards is enhanced by the community of practice element of engaging federal, state and local regulators as well as industry stakeholders and process safety professionals.

548394 Risk Analysis Screening Tool – CCPS Project 280

A spreadsheet based Risk Analysis Screening Tool (RAST) is available through the Center for Chemical Process Safety (CCPS), courtesy of the Dow Chemical Company. The tool allows users to identify hazards and understand risks including causes, consequences and probabilities. RAST can be used for existing or proposed installations to facilitate MOC reviews, assess compliance with risk criteria, or evaluate proposed designs.

The users can input the type of equipment being used along with the equipment characteristics, chemical data, process conditions, and the plant layout. Using these inputs, the tool is used to help identify hazard scenarios and evaluate the consequences of representative loss of containment scenarios. RAST also allows the users to specify their own hazard scenarios based on their own knowledge of the process. Users can also input their own risk ranking matrix. With the hazard scenarios and the risk matrix, the user can select those scenarios of most concern and use the tool to create a Layer of Protection Analysis (LOPA) spreadsheet for the process.

As an example a comparison of RAST calculated scenario identification and consequence results to actual case studies from the CSB will be presented. Courses on how to use RAST, and how to adjust the pre-loaded default parameters are available.

550863 Impacts of Climate Change on Process Safety Management in Oil and Gas Operations

Impacts of Climate Change on Process Safety Management in Oil and Gas Operations

Chizaram Nwankwo

Coventry University

Project objectives and goals

The aim of this study is to analyse the impacts of climate change on process safety management in oil and gas operations. The objectives for this study include the following: -

- To evaluate past, present and future trends of climate change around various global regions
- To establish the relationship between climate change and process safety management in oil and gas operations
- To proffer adaptation and mitigation measures for climate change impacts on process safety management in oil and gas operations

In the last two decades, climate change has emerged as one of the most controversial topics that has been argued on several fronts (Hulme 2009). The idea of global warming and the negative impact of human-driven activity on the global weather has been fiercely debated (Cook *et al.* 2016). However, Field *et al.* (2012) argue that the facts and figures tend to reiterate the continuous occurrence and proliferation of climate change, as well as its increasing impacts in different industrial sectors. Indeed, the warming of the climate is unequivocal (Panteli and Mancarella 2015). According to IIGCC *et al.* (2017), 70% of global greenhouse gas (GHG) emissions are attributed to energy-related activities. The oil and gas industry contributes to 60% of these energy-related emissions due to production of fossil fuels as their main operational activity (Puig *et al.* 2015). Evidence from NASA (2018) suggests that there has been 1^{0}C increase in temperature which is close to the set mark in accordance to the Paris Agreement.

There are several case studies such as the flood in San Jacinto River, Houston Texas in 1984 and Coffeyville Refinery, Kansas in 2007. In the case of the San Jacinto flooding event in 1984, 8 pipelines were ruptured and 29 were undermined, 35,000 barrels of oil were spilled and ignited, while 547 people were injured and $23 million was incurred in damage costs (Matek 2017).

Hence, adaptive measures need to be applied to manage the severe consequences of climate change on oil and gas operations (Smith 2013). Such adaptive measures include more robust design specifications for oil and gas assets exposed to harsh climatic conditions (Cruz 2010). Ramasamy and Yusof (2015) suggest that this could be vital in offsetting the impacts of climate change on assets such as offshore rig platforms and equipment, transport pipelines, storage facilities etc. However, there is still little attention accorded to design and adaptation of oil and gas assets to offset the effects of climate change on them (Skjaerseth and Skodvin 2010). Therefore, this study seeks to analyse the impacts of climate change on deterioration of oil and gas assets.

Description of methods and results

Firstly, this study adopts a pragmatic paradigm since it explores the consequences of actions, is problem-centred, pluralistic, real-world practice oriented and allows for use of mixed models (Creswell and Creswell 2018). The area of investigation for this study is focused on impacts of climate change on oil and gas operations. The predominant approach taken for this study is a correlational research approach which tries to establish a relationship between climate change and oil and gas operations. The selected approach is due to the continuous occurrence of accidents in the oil and gas industry due to asset deterioration.

After thorough literature review of climate change trends and asset integrity issues in the oil and gas industry, this study identified a gap in knowledge that the oil and gas industry has not duly considered climate change during design of assets; hence posing a risk to oil and gas operations conducted under harsh climatic conditions. Therefore, the data collection for this study employed "concurrent triangulation" method of quantitative and qualitative data to ensure robust analysis of key variables such as climate change and oil and gas operations. Consequently, data collection tools for this study include questionnaires, interviews, archive data and documentary analysis. The data collection for this study was conducted over a period of 3 months.

Firstly, it involved piloting the questionnaire among academic and industry professionals to identify any errors, as well as testing and refining the data collection instrument. Next, the study was approved as a medium-risk project by Coventry University Ethics Committee, as primary and secondary data were collected from archive data, documents, questionnaires and interviews. The population sample for the study includes oil and gas industry professionals such as Mechanical Engineers, Facility Managers, Production and Well Engineers, Process Engineers, Health and Safety Officers, as well as Safety and Reliability Engineers. The questionnaires used for this study were designed using Bristol Online Survey (BOS) tool and covered questions around climate change and oil and gas operations. These data were predominantly collated and analysed using various software such as Zotero, Microsoft Excel and SPSS software. Statistical analysis conducted in this study include trend analysis, descriptive statistics, Pearson's correlation, Cronbach's alpha reliability analysis and Relative Importance Index (RII). Lastly, the findings of this study were benchmarked and justified with peer-reviewed academic literature.

Potential for application of results

This research could aid in curbing several accidents caused by asset deterioration due to increasing number of climate change events. It will help the oil and gas industry to avoid the enormous financial, legal and moral costs associated with accidents. Also, this study will provide a deeper insight on current climate change statistics and proffer adaptation and mitigation measures on how to deal with these problems. Similarly, it could broaden the scope of researchers to start incorporating impacts of climate change on various industrial sectors and subsectors asides the oil and gas industry.

References

Cook, J., Oreskes, N., Doran, P.T., Anderegg, W.R., Verheggen, B., Maibach, E.W., Carlton, J.S., Lewandowsky, S., Skuce, A.G., and Green, S.A. (2016) 'Consensus on Consensus: A Synthesis of Consensus Estimates on Human-Caused Global Warming'. *Environmental Research Letters* 11 (4), 048002

Creswell, J.W. and Creswell, D.J. (2018) *Research Design: Qualitative, Quantitative, and Mixed Methods Approaches*. 5th Edition (International Student Edition). Sage Publications

Cruz, A.M. (2010) *Vulnerability of Oil and Gas Infrastructure to Climate Change and Extreme Weather Events* [online] available from <http://indico.ictp.it/event/a09141/session/35/contribution/25/material/0/0.pdf> [5 March 2018]

Field, C.B., Barros, V., Stocker, T.F., and Dahe, Q. (2012) *Managing the Risks of Extreme Events and Disasters to Advance Climate Change Adaptation: Special Report of the Intergovernmental Panel on Climate Change*. Cambridge University Press

Hulme, M. (2009) *Why We Disagree about Climate Change: Understanding Controversy, Inaction and Opportunity*. Cambridge University Press

IIGCC, CERES, and IGCC (2017) *Global Climate Disclosure Framework for Oil & Gas Companies* [online] available from <http://www.iigcc.org/files/publication-files/Global_Climate_OG_DisclosureFramework.pdf> [14 November 2017]

Matek, B. (2017) *CHARACTERISTICS OF OIL AND GAS PIPELINE ACCIDENTS CAUSED BY CLIMATE CHANGE INTENSIFIED HURRICANES* [online] Baltimore, Maryland: Johns Hopkins University. available from <https://jscholarship.library.jhu.edu/bitstream/handle/1774.2/46088/Benjamin%20Matek.pdf?sequence=1&isAllowed=y> [26 June 2018]

NASA (2018) *Global Climate Change* [online] available from <https://climate.nasa.gov/> [9 May 2018]

Panteli, M. and Mancarella, P. (2015) 'Influence of Extreme Weather and Climate Change on the Resilience of Power Systems: Impacts and Possible Mitigation Strategies'. *Electric Power Systems Research* 127, 259–270

Puig, D., Haselip, J., and Naswa, P. (2015) *Adaptation to Climate Change in Colombia's Oil and Gas Industry: Recommendations to Promote Risk Management* [online] UDP-CRD-OLADE-01(EN). Denmark: UNEP DTU Partnership. available from <www.unepdtu.org> [19 September 2018]

Ramasamy, J. and Yusof, S.M. (2015) 'A Literature Review of Subsea Asset Integrity Framework for Project Execution Phase'. *Procedia Manufacturing* [online] 4, 79–88. available from <http://linkinghub.elsevier.com/retrieve/pii/S2351978915011336> [14 May 2017]

Skjaerseth, J. and Skodvin, T. (2010) *Climate Change and the Oil Industry: Common Problem, Different Strategies*. Manchester University Press

Smith, M.H. (2013) *Assessing Climate Change Risks and Opportunities for Investors: Oil and Gas Sector* [online] Analysis of Climate Change Risks and Opportunities. Technical Report. ANU and Investor Group on Climate Change. available here. [5 April 2018]

Further Information

Chizaram Nwankwo (E: nwankwo4@uni.coventry.ac.uk)

546246 Study of the Effects of Wind Speed and Surface Roughness Under Stable Atmospheric Condition Using Aloha Model and CFD Model

Study of the effects of wind speed and surface roughness under stable atmospheric condition using ALOHA model and CFD model

Many substances released in chemical processes are toxic, flammable, or even both. When these substances accidentally release to the atmosphere, human health and the environment could be threatened. As results, a reliable quantitative risk analysis tool is crucial to provide decision makers useful information on downwind concentration and hazards. ALOHA consequence model and Computational Fluid Dynamics (CFD) numerical method are widely used in chemical industries. Validation of these models under different scenarios is a necessary preliminary step before applying them to risk analysis. CFD models generally have acceptable concentration predictions on dense gas dispersion. ALOHA model is verified with DEGADIS model, but there is lack of study on effects of wind speed and surface roughness on dense gas dispersion. This paper compared field data from the Kit Fox experiments, which were CO_2 gas releases under stable atmospheric conditions with increased roughness, with CFD and ALOHA models' predictions. Also, statistical performance indicators were utilized to evaluate the accuracy of ALOHA model. Statistical performance indicators showed that ALOHA had systematic errors for dense gas dispersion with increased roughness under stable atmospheric conditions. Comparing effects of wind velocity profile and surface roughness values in CFD model and ALOHA model, this paper provided a better understanding of effects of key parameters in both ALOHA and CFD models. Furthermore, engineers would have better quantitative risk assessments under stable atmospheric conditions when they are using ALOHA consequence model and CFD model.

548222 Consequence Assessments and Collaboration to Streamline Project Siting

Collaboration is often sought after across disciplines, but all too often disregarded in favour of linear project development. For siting and consequence, as with other areas where design interacts with environment, health and safety constraints, this often leads to more costly design due to increased project iterations.

Stantec recently completed a consequence assessment for a compressor station to assess flammable extents during an emergency blowdown. At the same time, the client was experiencing siting issues due to sound emissions from the same process. The facility was grappling with competing factors: blowdown the facility in time to avoid catastrophic failure of their infrastructure, avoid excessive flammability extents from the blowdown stack location, and avoid extraneous noise exposure to nearby receptors. Using transient consequence modelling tools to model the blowdown sequence, Stantec was able to simultaneously and efficiently assess the flammable extents and sound emissions from the blowdown operations.

Consequence modelling for flammability was completed using a transient pipeline blowdown modelling tool to simulate the mass, energy, and momentum throughout the pipeline system. The model first simulated normal operating conditions, and then simulated the blowdown to estimate mass flow rate, liquid mass fraction, and release temperature from the blowdown stack. Several sizes of reducers upstream of the blowdown stack were introduced to estimate the pressure drops in the piping that would be required to appropriately reduce noise levels but also achieve full depressurization of the facility in time to meet safety criteria.

Consequence modelling of flammability was completed using SLAB, while the predicted change in sound pressure level from the blowdown stack was completed using Cadna acoustic modelling software. By combining these two analyses, the assessment costs streamlined for both required studies. In addition, the acoustics study benefited from additional insight provided by the consequence modelling tool.

548348 CFD Simulation of the Effect of Wind Speed and Release Elevation on Dispersion of Flammable Fluids at LNG Export Facilities

Facility siting studies for LNG export terminals are required by federal regulations to assess the potential hazard from accidental releases of flammable fluids. Federal regulations require under 49 CFR 193.2059 the evaluation of flammable vapor-gas dispersion distances from releases from flammable fluids in the facility for a range of representative wind conditions based on historical weather data expected to produce the maximum dispersion distances. A parametric CFD study will be conducted to evaluate not only the effect of wind speed but also the effect of the release elevation on the dispersion distance from flashing and jetting fluids. Hazard distances will be compared between the 2D software PHAST and the 3D software FLACS for a set of exemplar fluids found at LNG export facilities. The ability of the FLACS software to handle detailed 3D geometry provides for the evaluation of the effect of obstructions, congestion, confinement and mitigation measures on the dispersion distances. Release scenarios typical of LNG export facilities will be presented, with a focus on jetting and flashing releases of LNG, ethylene, propane, mixed refrigerants and heavy hydrocarbons.

546616 A Comparative Study of the Optimization Models for the Gas Detector Placement in Process Facilities

Gas detector network is an important layer of protection in process facilities for prevention gas release accidents of turning into fire and explosion disasters. But traditional standards just provide basic principles of the installation of gas detectors. Recently mathematical programming gradually became an important method to solve the problem. When using the mathematical programming method to optimize the detector layout, different optimization strategies will result in different optimization target models and detector layouts. In this study, three models have been developed and contrasted, namely minimal detection time P-Median model (MITP), minimal leakage concentration P-Median model (MICP), and minimal individual risk P-median model (MIRP), for the optimization of the gas detector placement based on stochastic programming. And as a basis, possible release scenarios were identified. In order to achieve the goal of using finite discrete scenarios to represent all scenarios effectively and quantitatively, clustering analysis was used to filter similar scenarios and select representative leak scenarios. The input data of these models was obtained from computational fluid dynamics. Furthermore, an adaptive particle swarm optimization (APSO) algorithm was applied to solve these models. The gas detector network optimization in a diesel hydrogenation unit was carried out as a case study. Results demonstrate that the MICP model is superior to the MITP model and MIRP model in describing the integrity of hazardous gas leakage and diffusion distribution. But the cumulative detection time of its optimized layout is slightly longer than that of MITP. Moreover, the control effect of MIRP model on personal risk caused by hazardous gas leakage was increased by 24.2%, although the minimum number of detectors was slightly higher than that of MICP model.

549411 Reducing Industrial Risk in a Converging It/OT World: 20 Experts Share Their Strategies

Rapid adoption of Industrial IoT requires tight integration between IT and OT. CISOs and corporate OT leaders are now scrambling to develop strategies, secure budget, and implement solutions to secure their mission-critical industrial control systems against burgeoning cyber threats.

Effective integration of IT and OT cybersecurity strategies is essential to address risks successfully. To secure industrial facilities and ensure safe, reliable production, IT and OT security teams – traditionally two separate disciplines with different priorities – must collaborate and share cybersecurity and risk management best practices.

In a series of interviews in late 2017 and early 2018, more than 20 different experts on the front lines of OT cybersecurity risk mitigation shared practical advice for making control systems more secure. These experts represented a diverse range of industries including oil and gas, chemicals and refining, and power generation.

In their interviews, they shared their insights into how to leverage IT/OT convergence and develop an effective OT cybersecurity program that meets challenges in real-world ICS environments. Technology plays an important role, but as these experts also point out, more technology isn't always the solution. Just as important is the need to establish effective communication and trust between cybersecurity professionals and OT engineers.

Security project teams must make a concerted effort to create alignment between business management, the cybersecurity community, and OT professionals. Without alignment, security initiatives will fail to attain the support required to ensure success.

In this presentation, attendees will learn:

- Practical, real-world recommendations from OT cybersecurity experts
- Best practices for breaking down barriers and building trust between cybersecurity professionals and OT engineers
- How to align business management and plant operations on cybersecurity strategies

549417 The Intersection of OT Security and Process Safety

The rise of cyberattacks on critical infrastructure creates troubling concerns for security, economic stability, and personnel and public safety. However, many companies are still relatively unprepared to successfully defend their facilities from cyberattacks launched by increasingly sophisticated threat actors. Safe production is at risk.

Safety focuses on incident prevention and measures risks to personnel. Cybersecurity focuses on protecting systems from threat actors and measures risk based on the threat landscape. The increased network connectivity that exists within industrial facilities today means that cybersecurity is now closely linked to the safe operation of industrial facilities and their processes. Successful defense of cyber-physical systems from nation-state attacks requires close coordination and collaboration between cybersecurity and process safety. Every safety initiative must now consider cyber risk. Every cybersecurity initiative must now consider safety.

Considerable overlap exists between risk analysis in safety and cybersecurity. Safety and cybersecurity disciplines can learn much from each other. However, safety and cybersecurity standards and corporate organizational structures remain siloed in far too many companies. This session focuses on the importance of executing a cohesive strategy for both safety and cybersecurity to ensure safe and reliable operations.

In this session, we will:

- Examine the existing cyber threat landscape
- Discuss how established process safety best practices can be applied to OT cybersecurity
- Provide practical guidance for improving cyber defenses and reducing safety and environmental cybersecurity risks

551385 Process Safety at Offshore Aerodromes (Helidecks)New Application for Process Safety

Offshore air transport is an importance activate of the oil & gas industry. In this way, the management of air operations is essential for maintaining the safety and reliability. The present work has the objectives of divulging the methods used by
Petrobras to audit maritime units with helideck and to demonstrate the contribution of such methods to the safety of process in the air operations. To achieve these objectives, the paper was based on the Center for Chemical Process Safety (CCPS) publication "Guidelines for the Safety of Risk-Based Processes", and aims to share the Bow tie methodology for the qualitative analyzes of risks inherent in Air Operations, focusing on the barriers typically faced by oil and gas industry, in order to contribute to the process by providing a graphical and structured approach to the required barriers, both to avoid catastrophic consequences and to prevent events recognized as precursors.

Thermodynamics and Computational Methods for Relief Valves

By Umesh Mathur, P.E.
umesh.mathur@aggreko.com , mathuru@gmail.com
Aggreko Process Services
3732 Magnolia Street, Pearland, Texas 77584, USA

15th Global Congress on Process Safety (AIChE, March, 2019, New Orleans)

1. INTRODUCTION

The computation of flow through a relief valve is one of the most important aspects of safety system design. This paper discusses the fundamental question of choosing the proper thermodynamic basis for such calculations. We also discuss the discrepancies we have observed in the literature on this crucial question and attempt to dispel some of the misconceptions we have found.

Our concern with this issue arose from a study of the relevant literature on this subject, which shows that the flow of a compressed gas (or gas-liquid mixture) through a rigidly held valve or hole in a pipeline must be treated as an irreversible, isenthalpic process. This reasoning flows simply from the First Law of Thermodynamics as applied to a flow process.

For example, as early as 1947, Hougen and Watson[1] discussed "Free Expansion" of gases (p. 551) and noted as follows:

"The unrestrained expansion of a gas is known as free expansion. Under conditions of no restraint no work is done, and under adiabatic conditions no heat is added. Free expansion under flow conditions is commonly known as throttling or as the Joule-Thompson effect referred to in Chapter XII.

The Joule-Thompson effect is measured experimentally by expanding the gas slowly and in steady flow through a well insulated porous plug; in this way, potential work is lost, and no heat is allowed to enter or leave the system through the walls. The fluid flows reversibly into and out of the process, but the expansion step is completely irreversible. Since no heat is added or lost, the process takes place under conditions of constant enthalpy, according to Equation (VII-11), page 207. However a change in internal energy results if any change in flow energy pV occurs."

Similarly, in a discussion of valves and throttles, Elliott and Lira[2] observe as follows:

"A throttling device is used to reduce the pressure of a flowing fluid without extracting any shaft work and with negligible fluid acceleration. Throttling is also known as ***Joule-Thompson expansion*** *in honor of the scientists who originally studied the thermodynamics. An example of a throttle is the kitchen faucet. Industrial valves are modeled as throttles....Changes in kinetic and potential energy are small relative to changes in enthalpy as we just discussed. When in doubt, the impacts of changes in velocity can be evaluated as described in Example 2.9. The amount of heat transfer is negligible in a throttle. The boundaries are not expanding, and there is also no mechanical device for transfer of work, so the work terms vanish. Therefore, a throttle is isenthalpic."*

Finally, in a discussion of nozzles, Klein and Nellis[3] observe that:

"...The potential energy terms of the inlet and outlet flows are not usually significant. Nozzles neither require nor produce power (W_{in} = W_{out} = 0) and are very nearly adiabatic. Note that the kinetic energy terms associated with the flows, particularly the outlet flow, cannot be neglected as the purpose of the nozzle is to increase the kinetic energy of the fluid. With these simplifications, an energy balance on the nozzle becomes:

$$H_{in} + \tfrac{1}{2}V_{in}^2 = H_{out} + \tfrac{1}{2}V_{out}^2 \qquad (4\text{-}54)"$$

In the light of these assertions by these noted authorities, and others[6][10], it is hard to understand how the flow through a relief valve, or a hole in a vessel that is open to the atmosphere, can be treated as isentropic (***reversible*** adiabatic) in nature, as opposed to the isenthalpic (***irreversible*** adiabatic) basis that is consistent with free expansion. This is because an isentropic expansion process would necessarily show an outlet enthalpy for the gas from the nozzle or valve (or an orifice in a punctured vessel) to have a lower enthalpy than the inlet, the difference in such enthalpies being equal to the work done, for example on a rocket or airplane. However, a relief valve or vessel remains stationary and absolutely no work is done during a free expansion by the gas flowing through.

Nevertheless, several standard references have used isentropic formulas for free expansion[4],[5] in stationary valves, or nozzles, from the inlet to the throat. We believe that this methodology is flawed as it violates the steady flow energy equation (i.e., the first law of thermodynamics written for an open system). This procedure would result in an erroneous prediction of outlet gas temperature when entropy is held constant. This, in turn, would predict an erroneous outlet gas density and thus, for a desired gas relief rate, the required cross-sectional area for the valve throat and also any conduit or pipe downstream of the valve.

The practical consequence of this anomaly in predicting outlet gas properties (especially the gas density at the throat) would be that the relief valve sizing could be incorrect. Therefore, it is important that the proper thermodynamic basis (isenthalpic expansion) be used in all such computations.

Conversely, for a given relief valve, errors in throat density based on an isentropic path would also affect the relief rate. Both types of errors could lead to potentially serious process safety concerns.

For these reasons, when calculating the release rate for gas at a high pressure through a relief valve, it is important to be able to use the proper basis for gas enthalpy (and therefore accurate computations of fluid thermodynamic properties). The same considerations apply to the gas release rate calculations resulting from a rupture in a pipeline or vessel containing a gaseous mixture under pressure.

To help clarify the proper methodology for a relief valve - or a hole in a pressurized vessel - the computations that should be performed are described below. We also show a few simple examples. Note that the possibility of sonic velocity at the throat, or choking, of the valve (or at the orifice in a ruptured vessel) is explicitly considered. In both such cases, the procedure described below is subject to the following assumptions:

- The opening can be represented by a convergent nozzle of exit area equal to the area of the throat of the relief valve, or hole in the case of pipeline rupture

- The compressed gas pipeline, or leaking container, is a large reservoir in which upstream pressure and temperature are maintained constant
- The compressed gas undergoes an irreversible, adiabatic process of expansion through the nozzle.

The flow through such a device from either (a) the inlet to the throat, or (b) from the throat to the outlet follows the steady flow energy balance equation (Equation 9.21, Himmelblau & Riggs[6]):

$$\Delta E = Q + W - \Delta(H + KE + PE) \qquad (1)$$

Where:
ΔE = energy accumulation (zero at steady state)
Q = net heat transfer into the system (very close to zero for a relief valve)
W = net work done by the system (zero for a relief valve)
ΔH = change in enthalpy of the system
ΔKE = change in kinetic energy of the system
ΔPE = change in potential energy of the system (zero for a relief valve)

Clearly, for steady flow through an orifice, valve (or relief valve), ΔE, Q, W, and ΔPE are all zero. Therefore, we are left with:

$\Delta(H + KE) = 0$, or

$\Delta H = - \Delta KE$, or

$$H_{out} = H_{in} - \Delta KE \qquad (2)$$

In other words, the outlet enthalpy is equal to the inlet enthalpy minus the change in kinetic energy in going through the throat or orifice. (It should be noted that the same logic would apply for discharge through an ordinary valve, a relief valve, or an orifice plate).

If there is no change in velocity, the outlet enthalpy will be the same as that of the inlet gas. This phenomenon is generally referred to as a Joule-Thompson (J-T) expansion. For an ideal gas, enthalpy is a function of temperature only; therefore, a J-T expansion would result in no change in temperature. However, for most real gases, a J-T expansion results in a decrease in temperature. There are a few notable exceptions, however, such as hydrogen or helium gas for which the J-T coefficient has the opposite sign[10]. On the other hand, if the gas speeds up, its kinetic energy will increase and, therefore, its enthalpy will decrease in accordance with equation (2) above. This kinetic energy increase would result in a decrease in outlet gas enthalpy, and therefore temperature, even for an ideal gas.

For ideal gases, the computation of enthalpy for ideal gases is straightforward as it is a function of temperature only. For real gases, however, enthalpy is also a function of pressure, temperature, and composition and use of an equation of state (EOS) becomes necessary[2] to calculate the enthalpy departure from the ideal, as discussed below.

COMPRESSIBILITY FACTOR AND ENTHALPY FOR REAL GASES

For real gases, enthalpy is found using a suitable equation of state (EOS), such as the Soave-Redlich-Kwong (SRK), Peng-Robinson (PR), or Lee-Kesler (LK). This procedure may be summarized as follows:

At the given upstream pressure (P), temperature (T), and composition, solve the EOS for the compressibility factor (z). This requires solving a cubic polynomial in z when using any cubic EOS, such as the SRK or PR, using either an explicit analytical or numerical trial and error method. For the LK, however, a trial and error procedure is always required as the LK EOS requires solving the 8-constant Benedict-Webb-Rubin[2] (BWR) equation, which is not a simple cubic in z, once for the simple fluid and again for the reference fluid.

Using the appropriate thermodynamic expressions for the EOS, use this value of z to calculate ΔH_d, the enthalpy departure from ideal.

Calculate the ideal gas enthalpy, H_0 (function of temperature T_0 only).

Find the real gas enthalpy using:

$$H_r = H_i - \Delta H_d \qquad (3)$$

Where:

H_r = real gas enthalpy
H_i = ideal gas enthalpy
ΔH_d = enthalpy departure from ideal

This procedure must be followed for calculating enthalpy at both upstream and downstream conditions.

The gas velocity at the throat may be sonic or sub-sonic. Also, as the gas speeds up in passing through the valve throat or hole (since the cross-sectional area is lower than that of the pipe), its kinetic energy increases and this results in a reduction of the enthalpy (Elliott & Lira[2], Eq. 2.53):

$$H_t = H_0 - \Delta KE \qquad (4)$$

Where:
H_0 = upstream gas enthalpy[J/kg]
H_t = exit gas enthalpy at the throat or hole[J/kg], and
ΔKE = change in kinetic energy[J/kg]

ΔKE is found as follows (Elliott & Lira's Eq. 2.53):

$$\Delta KE = (u_t^2 - u_0^2)/2 \qquad (5)$$

Where:

ut = throat (or hole) velocity[m/s], and

u_0 = upstream velocity[m/s].

Clearly, the outlet and inlet enthalpies are equal only if the change in velocity is negligible.

RELEASE RATE CALCULATION

The computation of the release rate from a pipe rupture should be done subject to the following equations:

Bernoulli's Equation

This equation relates upstream and downstream conditions for an arbitrary flow geometry, as follows:

$$P_0 + \rho_0.u_0^2/2 + \rho_0.g.z_0 = P_e + \rho_e.u_t^2/2 + \rho_t.g.z_t + \Delta P_{friction} \quad (6)$$

Where the subscripts o and e refer to the inlet and exit planes, respectively, and

P	= pressure[Pa]
ρ	= density[kg/m^3]
u	= velocity[m/s]
g	= gravitational constant[m/s^2]
z	= elevation above datum[m]
$\Delta P_{friction}$	= friction losses[Pa]

The friction losses have been ignored for simplicity in this presentation.

Obviously, for a hole, or throttle (or relief) valve, the elevations z_0 and z_t are the same.

Sonic Velocity

The sonic velocity in any medium[2] may be expressed as:

$$u_s = (\gamma \,.\, z \,.\, R \,.\, T / M)^{0.5} \quad (7)$$

Where:

u_s	= sonic velocity[m/s]
γ	= gas C_p/C_v ratio[]
z	= gas compressibility factor[]
R	= gas constant = 8314.46[J/(kgmol.K)]
T	= temperature[K]
M	= gas molecular weight[kg/kgmol]

In Equation (7), properties at the throat must be used for choked flow calculations. For ideal gases, the compressibility factor z is equal to 1. For real gases, however, the compressibility factor z is generally not equal to 1 and can be found only using an appropriate EOS.

C_p, C_v, and $\gamma=C_p/C_v$ ratio

Similarly, for ideal gases, the value of γ is found from the value of C_p, which is a function of temperature only, and the value of C_v is found from the following:

$$C_v = C_p - R \quad (8)$$

For real gases, however, the value of γ again requires use of an equation of state to calculate the ΔC_{pd} and ΔC_{vd} departures from ideality. The real gas C_p and C_v values are found using the ideal values and these calculated departures from ideality, by subtraction, in a manner similar to that described above for enthalpy:

$$C_{p1} = C_{p0} - \Delta C_{pd} \quad (9)$$

Where

C_{p1} = real gas C_p[J/(kg.K)]

C_{p0} = ideal gas C_p[J/(kg.K)]

ΔC_{pd} = departure in C_p from ideal gas value[J/(kg.K)]

Similarly, $C_{v1} = C_{v0} - \Delta C_{vd}$ (10)

In Equation (10),

C_{v1} = real gas C_v[J/(kg.K)]

C_{v0} = ideal gas C_v[J/(kg.K)]

ΔC_{vd} = departure in C_v from ideal gas value[J/(kg.K)]

In a conduit, orifice, or valve, the actual gas velocity, u_e, can never exceed the sonic limit, u_s. Note that u_s, in general, is different at the upstream conditions compared to the value at throat conditions. The upstream and throat temperatures and pressures of the gas may be different owing to the kinetic energy effect, Equation (4).

At a sufficiently low upstream pipeline or vessel pressure, the hole velocity would be lower than the sonic velocity limit. Above some critical limit for the upstream pressure, however, the maximum discharge flow is limited by the sonic velocity criterion at the hole.

Density

Gas density is computed using:

$\rho = P \cdot M / (z \cdot R \cdot T)$ (11)

Where:

ρ = density[kg/m^3]

P = pressure[Pa]

M = gas molecular weight[kg/kgmol]

z = gas compressibility factor[]

R = gas constant = 8314.46[J/(kgmol.K]

T = temperature[K]

The gas compressibility factor is 1 for ideal gases; it is calculated from the EOS for real gases.

1.2 Velocity

The velocity in a pipe or hole (or valve throat) is found using:

$u = W / (\rho \cdot A)$ (12)

Where:

u = velocity[m/s]

W = mass flow rate[kg/s]

ρ = gas density[kg/m^3]

A = cross sectional area[m^2]

For a circular pipe, the cross-sectional area is defined as:

$A = \pi.D^2/4,$ (13)

Where:
D = inner diameter[m]

For non-circular cross sections, the equivalent diameter D_e is defined using the concept of the hydraulic radius which is defined as:

D_e = 4 (Cross-sectional-area/Wetted-perimeter) (14)

It can readily be seen that this definition holds for a circular pipe also: $4 . \pi . D^2 / 4 / (\pi . D) = D$

As noted earlier, for ideal gases, the compressibility factor z is equal to 1. For real gases, however, finding the gas density ρ requires first calculating z from the EOS. This is done using the procedure as described above.

It should be noted that conditions at the throat of a valve (or at the hole in a pipe) are not known. Therefore, the computation of the gas flow rate must be done using an iterative procedure. When using Microsoft Excel, this is best performed using the Solver Add-in. Solver is a non-linear optimization technique that can simultaneously adjust the unknown quantities and solve the problem by enforcing the constraints imposed by the Bernoulli Equation (6) for throat velocity, and the Steady-Flow Energy Equation (1) for hole enthalpy. Also, the calculated throat velocity from the Bernoulli equation can never exceed the sonic velocity limit, Equation (7), and this constraint must be added explicitly when the upstream pressure exceeds a threshold value.

When solving for an unknown gas flow rate, if the upstream pressure is sufficiently low, hole velocity could remain sub-sonic. However, if the upstream pressure is sufficiently high, the pressure at the hole and also the gas mass flow rate would vary in a manner that obeys the sonic velocity limit at the hole.

Example

CASE (A): Methane is flowing through a 300[mm] diameter, above-ground pipeline at an absolute pressure of 46[bar] and a temperature of 300[K]. The pipeline develops a hole with an equivalent diameter of 50[mm]. Calculate the mass rate of leakage to atmosphere and also the velocity of the gas jet emerging from the hole. The pressure inside the pipeline may be assumed to be constant owing to continuous supply from the source.

CASE (B): Rework Case (A) assuming that the pressure in the pipeline is 1.5[bar] instead of 46[bar].

Case (A) P_o = 46[bar]

PROBLEM: Solve for release rate from a hole in a pipeline - sonic velocity at throat.

WE SHOW THE FOLLOWING DETAILED SOLUTION FOR THE IDEAL GAS CASE ONLY. Note that the solution of the real gas case requires use of complex computer programs to solve the chosen equation of state (e.g., Lee-Kesler) iteratively for the required thermodynamic quantities. Since these routines are imbedded in an outer loop for the unknowns (T and P at the throat, and the mass flow rate), the computational burden is too great for hand calculations.

The POLYMATH software[7] was used for solving this problem. All variables are defined in Table 1. The problem statement for Case (A) for the ideal gas case is as follows:

 # Initial guesses

$T_t(0) = 300$ # Initial guess for T_t[K]
$W(0) = 20$ # Initial guess for W[Kg/s]
$P_t(0) = 46$ # Initial guess for P_t[bar]

 # Aly-Lee ideal gas enthalpy constants:

A = 33298 # Ideal gas enthalpy constant A
B = 79933 # Ideal gas enthalpy constant B
C = 2086.9 # Ideal gas enthalpy constant C
D = 41602 # Ideal gas enthalpy constant D
E = 991.96 # Ideal gas enthalpy constant E
F = -120280000 # Ideal gas enthalpy constant F

 # The following three functions must be made equal to zero by adjusting the three unknown variables W[kg/s], T_t[K], and P_t[bar]:

$f(T_t) = h_t - h_{t2}$
$f(W) = u_t - u_{t2}$
$f(P_t) = u_t - u_s$

 # Initial guess ranges for non-linear equations

W(min) = 1
W(max) = 50
$T_t(min) = 200$
$T_t(max) = 400$
$P_t(min) = 10$
$P_t(max) = 46$

This problem requires solving the three simultaneous, non-linear algebraic equations (defined above) by varying T_t (the hole temperature), W (the mass flow rate), and P_t (the hole pressure):

$f(T_t) = 0$
$f(W) = 0$
$f(P_t) = 0$

In this case, the upstream pressure is high enough to cause the velocity at the hole to reach the sonic limit. This problem requires solving three simultaneous non-linear algebraic equations. The POLYMATH solution shown below for Example A was obtained using a software package called POLYMATH[7]. Note that any other similar package can also be used. We verified this numerical solution using the MATHCAD[8] and EXCEL[9] software packages also.

NOTE: The numerical solutions for Cases (A) and (B) are shown in Table 1, following the problem statement for Case (B).

Case (B) $P_0 = 1.5$[bar]

PROBLEM: Solve for release rate from a hole in a pipeline. Assume that the throat velocity is sub-sonic. This means that the outlet pressure is fixed at 1[atm] = 1.01325[bar].

When the outlet pressure is fixed, we are left with two non-linear equations. Also, for consistency with the methodology in (A) above, we use a very large upstream diameter to make the inlet velocity very small.

NOTE: If the calculated outlet velocity exceeds the sonic limit, we would revert to the problem formulation under (A) above. (This would add an equation requiring outlet velocity to equal sonic velocity, and thus again require solving three simultaneous non-linear equations.)

As noted earlier, the solution of the real gas case requires use of complex computer programs to solve the chosen equation of state (Lee-Kesler, in these examples) for the required thermodynamic quantities.

In this case, the upstream pressure is low enough so that the velocity at the orifice does not reach the sonic limit. Accordingly, this problem requires solving only two simultaneous non-linear algebraic equations. The solution shown below was obtained using a software package called POLYMATH. Note that any other similar package can also be used. As with Case (A), we verified this numerical solution using the MATHCAD and EXCEL software packages also.

The POLYMATH problem statement for Case (B) is as follows:

$f(T_t) = h_t - h_{t2}$

$f(W) = u_t - u_{t2}$

 # Initial guesses

$T_t(0) = 300$ # Initial guess for T_t[K]

W(0) = 20 # Initial guess for W[Kg/s]

Using POLYMATH, the solution for Case (A) shows that the gas release rate, W, for an ideal gas is 17.16[kg/s]. As noted above, identical results were obtained using the Mathcad and Excel software packages.

Umesh Mathur, P.E., Aggreko Process Services, Houston, TX

TABLE 1: SONIC OR SUB-SONIC FLOW THROUGH A VALVE OR HOLE

VARIABLES AND EQUATIONS

Variable	Description & Units	Explicit equations
A	Ideal gas enthalpy constant A	A = 33298.
B	Ideal gas enthalpy constant B	B = 79933.
C	Ideal gas enthalpy constant C	C = 2086.9
D	Ideal gas enthalpy constant D	D = 41602.
E	Ideal gas enthalpy constant E	E = 991.96
F	Ideal gas enthalpy constant F	F = -120280000.
P_t	Throat pressure [bar]	$P_t = 1.01325$
kcal_j	Conversion: 1[kcal] = 4184[J]	kcal_j = 4184
M	Gas molecular weight[kg/kgmol]	M = 16.0
T_0	Inlet gas temperature[K]	$T_0 = 300$
P_0	Inlet gas pressure[bar]	$P_0 = 1.5$
z_0	Inlet gas compressibility factor[]	$z_0 = f(T_0,P_0,x)$
$(c_p/c_v)_0$	Inlet gas C_p/C_v ratio[]	$(c_p/c_v)_0 = 1.32$
R	Gas constant[J/(kgmol.K)]	R = 8314.46
D_0	Diameter of pipe[m]	$D_0 = 0.3$
D_t	Diameter of Throat[m]	$D_t = 0.05$
h_0	Inlet enthalpy[kcal/kg]	$h_0 = (A*T_0 + B*C/\tanh(C/T_0) - D*E/\tanh(E/T_0) + F)/M/kcal\_j$
rho_0	Inlet density[kg/m^3]	$rho_0 = P_0*M*1.e5/R/T_0$
A_0	Pipeline cross-sectional area[m^2]	$A_0 = 3.14159/4*D_0^2$
A_t	Throat flow area[m^2]	$A_t = 3.14159/4*D_t^2$
h_t	Throat enthalpy[kcal/kg]	$h_t = (A*T_t + B*C/\tanh(C/T_t) - D*E/\tanh(E/T_t) + F)/M/kcal\_j$
u_0	Upstream velocity[m/s]	$u_0 = W/rho_0/A_0$
rho_t	Throat density[kg/m^3]	$rho_t = P_t*M*1.e5/R/T_t$
u_t	Throat velocity[m/s]	$u_t = W/rho_t/A_t$
u_s	Sonic velocity at throat[m/s]	$u_s = (z_0*(c_p/c_v)_0*R*T_t/M)^{0.5}$
u_{t2}	Throat velocity from Bernoulli Eq.[m/s]	$u_{t2} = (2/rho_t*(P_0*1.e5 + rho_0*u_0^2/2 - P_t*1.e5))^{0.5}$
dh_{KE}	ΔKinetic Energy = Δ(u^2)/2[kcal/kg]	$dh_{KE} = (u_{t2}^2 - u_0^2)/2/kcal\_j$
h_{t2}	Calc. enthalpy at throat = inlet enthalpy - Δ(u^2)/2[kcal/kg]	$h_{t2} = h_0 - dh_{KE}$
T_t	Temperature at throat[K]	$f(T_t) = 0$
W	Mass flow rate[kg/s]	$f(W) = 0$
P_t	Pressure at throat[bar]	$f(P_t) = 0$

TABLE 1: SONIC OR SUB-SONIC FLOW THROUGH A VALVE OR HOLE (CONTINUED)

CASE (A) AND CASE (B) SOLUTIONS (USING POLYMATH)

Variable	CASE (A) Value (SONIC)	CASE (B) Value (SUB-SONIC)
A	33300.0	33300.0
B	79930.0	79930.0
C	2086.9	2086.9
D	41600.0	41600.0
E	991.96	991.96
F	-120300000.0	-120300000.0
P_t	27.71686	1.01325
kcal_j	4184.0	4184.0
M	16.0	16.0
T_0	300.0	300.0
P_0	46.00	1.50
z_0	1.00	1.00
$(c_p/c_v)_0$	1.32	1.32
R	8314.46	8314.46
D_0	0.30	0.30
D_t	0.05	0.05
h_0	226.2098	226.2098
rho_0	29.50683	0.96218
A_0	0.07069	0.07069
A_t	0.00196	0.00196
h_t	205.2641	210.3263
u_0	8.22527	7.71172
rho_t	20.86568	0.73254
u_t	418.7383	364.6543
u_s	418.7383	427.2981
u_{t2}	418.7383	364.6543
dh_{KE}	20.94576	15.88352
h_{t2}	205.2641	210.3263
T_t	255.6215	266.1833
W	17.15555	0.52433
P_t	27.71686	1.01325

CASE (A) RESULTS

This problem was also solved using the Lee-Kesler equation of state to calculate non-ideal gas properties, using a suitable computer program. This solution is provided below, with the ideal gas results quoted alongside for comparison:

CASE (A)

	IDEAL GAS	REAL GAS
Inlet pressure[bar]	46.00	46.00
Inlet temperature[K]	300.00	300.00
Calculated mass flow[kg/s]	17.28	18.25
Inlet velocity[m/s]	8.27	8.09
Throat velocity[m/s]	424.65	421.48
Sonic velocity at throat[m/s]	424.65	421.48
Inlet Enthalpy[kcal/kg]	225.61	214.87
Delta Kin. Energy[kcal/kg]	21.54	21.22
Outlet Enthalpy[kcal/kg]	204.07	193.64
Throat pressure[bar]	27.32	25.73
Throat temperature[K]	254.27	250.83

It should be noted that these results are quite similar for this case because the behavior of the gas (methane) is close to ideal at the conditions specified. However, these results would very likely differ significantly if, for example, the inlet temperature was considerably lower and the pressure much higher in which case the gas behavior would be markedly non-ideal.

CASE (B) RESUTS

Case (B) was also solved using the Lee-Kesler equation of state to calculate non-ideal gas properties. This solution is provided below, with the ideal gas results quoted alongside for comparison:

CASE (B)

	IDEAL GAS	REAL GAS
Inlet pressure[bar]	1.5	1.5
Inlet temperature[K]	300.0	300.0
Calculated mass flow[kg/s]	0.524	0.525
Inlet velocity[m/s]	7.71	7.70
Throat velocity[m/s]	364.7	364.2
Sonic velocity at throat[m/s]	435.06	435.3
Inlet enthalpy[kcal/kg]	226.21	226.21
Delta Kin. Energy[kcal/kg]	15.88	15.85
Outlet Enthalpy[kcal/kg]	210.33	210.36
Throat pressure[bar]	1.01325	1.01325
Throat temperature[K]	266.18	266.25

As might be expected, the gas mass flow rate is far lower for Case (B) compared to Case (A). Also, for Case (B), it was found that the conditions at the throat are virtually identical for the ideal versus the non-ideal gas.

2. ADDITIONAL EXAMPLES

To illustrate the importance of the issues discussed in this paper, we show a few additional examples for the differences in our numerical results compared to others.

2.1 Gas Release From A Hole In A Vessel

A 0.1[in] hole forms in a tank containing nitrogen at 200[psig] and 80[°F]. Determine the mass flow rate through this leak (Crowl[4], Example 4-4, p. 143).

The solution for this problem using the Lee-Kesler EOS was 4.26e-2[lb/s], compared to 3.86e-2[lb/s] in the reference (gas velocity at the throat was sonic). This difference is over 10% and not negligible.

2.2 Flow Through A Nozzle

Air at p_0=7.0e5[Pa] and T_0=293[K] discharges to the atmosphere. Compute the discharge mass flux, G, the pressure, Mach number, and velocity at the exit[5].

Using the Lee-Kesler EOS, the results are as follows:

Quantity	Lee-Kesler (this article)	Ideal gas (isentropic)[5]	Difference
Mass flux[kg/(m^2.s)	1589	1650	-3.68%
Exit temperature[K]	238.5	244	-5.5 [K]
Exit pressure[Pa]	212041	370000	-42.7%
Mach number	1.0	1.0	-
Sonic velocity[m/s]	338	313	+8%
Velocity[m/s]	338	313	+8%

Note that the nozzle is not predicted to be choked when the downstream pressure is above 2.12[bar] (kinetic energy effects must be included), as opposed to 3.7[bar] in the original reference.

If a relief valve is to be designed for this scenario, the size would be under-predicted somewhat using the isentropic formulas used in the original reference. However, this discrepancy could be significantly larger in cases where an isentropic expansion incorrectly predicts partial condensation of the gas at the throat. In that situation, the density of the two-phase mixture at the throat could be far lower than that of the uncondensed gas (for the isenthalpic case) and the valve could be under-sized significantly.

3. CONCLUDING REMARKS

It has been shown that calculation of the release rate from a relief valve (or a hole in a vessel) should be done using an isenthalpic (i.e., irreversible adiabatic) path for the flow process. Using an isentropic thermodynamic path for this problem violates the First Law of Thermodynamics, as expressed for steady flow in a stationary relief valve or a hole in a pipeline. This could lead to an incorrect design both when sizing relief valves and estimating release rate from an existing valve or hole in a vessel.

In addition, it is important to use the proper equation of state to estimate real gas properties reliably when compared to the ideal gas assumption, especially when gas conditions are far from ideal. The Bernoulli equation (for throat pressure), and the sonic velocity limit (when flow is choked at the throat) must also be included for a complete definition of the problem. If the upstream pressure is low enough and the throat is un-choked, the isenthalpic methodology - as described in this paper - should again be followed to ensure a theoretically sound and consistent basis. Iterative computations for non-ideal cases (real gases) generally are too intensive for hand calculations and require use of computerized chemical engineering software.

REFERENCES

1. Hougen, O.A., K. M. Watson: "Chemical Process Principles" (John Wiley & Sons, New York, and Chapman and Hall, London, 1st Edition, Part II, p. 551, 1947, and 2nd Edition with R.A. Ragatz, Vol. II, p. 653, 1959)
2. Elliott, J.R., and C. T. Lira: "Introductory Chemical Engineering Thermodynamics" (Pearson Education, Inc. publishing as Prentice Hall, pp. 68-69, 2nd Edition, 2012)
3. Klein, S. and G. Nellis: "Thermodynamics" (Cambridge Univ. Press, 1st Edition, p. 168, 2011)
4. Crowl, D. A., and Louvar, J.F.: "Chemical Process Safety" (Pearson Education, 3rd Edition, pp. 130-131, 2011)
5. Green, D. W., and R.H. Perry: "Perry's Chemical Engineers' Handbook" (McGraw-Hill, 8th Edition, Section 6, pp. 23-24, 2008)
6. Himmelblau, D.M., and J.B. Riggs: "Basic Principles and Calculations in Chemical Engineering" (Pearson Education, 8th Edition, Fig. 9.18, Item 3, p. 525, 2012)
7. Cutlip, M.B., and M. Shacham: "Problem Solving in Chemical and Biochemical Engineering with POLYMATH, Excel, and MATLAB" (Prentice Hall, 2nd Edition, 2007)
8. Mathcad 2001 Professional: "Mathcad User's Guide With Reference manual", (Mathsoft Engineering and Education, Inc., 2001)
9. Freund, S.M., M. Jones, and J.L. Starks: "Microsoft Excel 2013: Complete (Shelly Cashman Series)", (Course Technology, 1st Edition, 2013)
10. Lewis, G.N., and M. Randall: "Thermodynamics", (McGraw-Hill, 2nd Edition, Revised by K.E. Pitzer and L. Brewer, p. 48, 1961)

548434 Engineered Solutions for Effective Flare System Analysis

Relief and flare system documentation is an important part of Occupational Safety and Health Administration (OSHA) Process Safety Management (PSM) 29 CFR 1910.119. Flare system analysis (sizing or rating) can be performed using commercially available software. Effective flare system analysis requires reliable Process Safety Information (PSI) such as process and instrumentation diagrams (P&IDs), relief system evaluation, Heat and Material Balance (H&MB), relief valve data sheets, isometrics, and disposal system loads. Good engineering methodologies can be implemented to effectively design header systems and rate existing header for any change. Some methodologies include use of dynamic simulation to predict realistic relief loads, reboiler pinch analysis for applicable systems, and selection of global relief scenarios. This paper will describes some methodologies and a combination of instrumented systems for effective flare system analysis. Effective flare analysis helps save capital expenditure spent on flare headers and possible flare load reduction yields reduced radiation footprint on a facility.

548364 Flares Are the End of the Line; How to Permit Correctly

INTRODUCTION

-

Flare is one of the key equipment in any refinery or LNG facility. Many relief valves relieve to flare, and if the flare is not permitted corrected – and functional - the plant is down. You cannot operate the plant without an active flare or flare field. The permitting of a flare is one of the prime activities in the permitting process. But the permit is dependent on the type of flare. This presentation covers various Federal Regulations involved in the permitting process, whether general or specific to each type of flare.

FLARE TYPES AND THEIR DESIGN FUCTURES

-

There are mainly three types of flares. The most common and most economical one is an elevated flare. There are elevated flares as high as 660 ft. The can be classified many ways; self-supported, guy wire supported, derrick supported, demountable type etc. One of the most important is the type of burner, whether a pipe tip or "furnace" burners. The second type is a ground flare with rows of furnace burners arranged in arrays at grade and surrounded with a wind fence. The last type is an enclosed flare with furnace burner enclosed with a tall stack. This type can be large as 70 ft in diameter and 150 ft tall. The flare type on a given project is selected based on the plot availability, location of the plant, sparing philosophy etc.

PERMITTING REGULATIONS

This presentation will discuss cover various Federal regulations that are applicable to all types of flares and the various parameter that the regulators are needing when apply for the permit. The presentation will cover the required destruction efficiency of the flaring gases, exit velocity for an elevated flare, alternate means of emission limitation for ground flares. In particular, the tip velocity, burner type, backpressure, emission testing and smokeless capacity are specific to each type of flare.

WHAT YOU WILL LEARN

-

· This paper will provide sufficient information to ensure the correct type of flare is selected and permitted

· This paper will cover various flares types employed in refinery and in LNG plants

· The limitations of these flare

· Importance of flare in any facility

· Application Federal Regulations

· Parameter required in the permitting applications

· Ground flare testing to establish minimum destruction efficiency

· Alternate means of emission limitation requirements in lieu of testing

548232 A Case Study on Application of HAZOP for a Brewery

Abstract

Beer is the most consumed alcoholic beverage in the world, and most popular beverage after water and tea. It's an alcoholic beverage produced by the saccharification of starch, often derived from the malted grains such as barley and wheat, and the fermentation of the resulting sugar. The brewing process can be broken into four main constituents before the bottling line, Mashing, Lautering, Wort boiling and Fermenting. Breweries commonly rely on ammonia refrigeration to keep their supplies cold. However, some brewers fail to realize that ammonia is one of the dangerous chemicals covered under 29 CFR 1910.119, OSHA's regulation for preventing or minimizing the release of toxic, reactive, flammable or explosive chemicals. "Safety matters at all levels", even for companies that may not be PSM level facilities.

A HAZOP study identifies hazards and operability problems. The study involves investigating how the plant might deviate from the design intent. It is based on the principle that several experts with different backgrounds can interact and identify more problems and failure scenarios working together and combining their results. This paper presents a review of consequences identified in brewery HAZOP. It also outlines some of the recommendations and action items identified from the case study of a brewery.

Keywords: Process Safety Management (PSM), PHA, HAZOP, Brewery, Ammonia, OSHA.

548410 A Basic Review of Fundamental of Thermodynamics and Its Application in the Analysis of Consequences and Accident Investigation

Once the stage of hazard identification has been concluded in a risk analysis, the next stage corresponds to select and specify the risk scenarios that will be modeled their effects or damages associated. This stage correspond to the consequences analysis and usually takes place in three basic steps: a) source term definition, b) dispersion model and d) effect or damage model.

Depending on the risk scenario to be modeled, several thermodynamic fundamentals or principles are applied in each of these steps supporting or sustaining the behavior and expected results of the consequences analysis model. In this work we propose to review the main thermodynamic fundamentals that support each of the models applied in each of the three steps that define a consequence analysis. This review of basic thermodynamic principles defines a set of different thermodynamic basic principles to apply to representative and different cases of consequence analysis. These basic thermodynamic principles are applied in the consequence modeling of a pair of cases of accident inventions to corroborate their validity.

548331 Evaluating Safe Location Criteria for Vents and Relief Devices Discharging to Atmosphere

In some process technologies, it may not be feasible to dispose the effluents from safety valves and vents to a closed system and discharging directly to the atmosphere can be the only alternative. Some examples include High Pressure Polyethylene Processes, Olefins and Storage units. The discharge can be directed to the atmosphere if:

- Local regulations permits it;
- Does not harm the environment;
- Does not harm the neighborhood;
- And is directed to a Safe Location.

The criteria for defining what a safe location is can vary, but it must include at least a consequence assessment of the flammable, toxic and thermal radiation effects. Some guidance can be found on industry standards and engineering references such as API 521 and the CCPS Guidelines for Pressure Relief and Effluent Handling Systems

One aspect to discharge hydrocarbons to a Safe Location is to evaluate the dispersion of the discharged hydrocarbons and ensure that it is diluted bellow the lower flammable before reaching the ground level or elevated structures. These areas are typically with people circulation and ignition sources. Another aspect that must also be evaluated is the thermal radiation generated due to an ignition of the discharge affecting personnel or damaging equipment or other assets. Lastly, if there is a possibility of the flammable mixture reaching confined spaces, the explosion possibility must also be assessed.

API 521 gives some guidance on chapter 5.8 for estimating the dispersion and concentration profile from a typical relief device discharge. It also describes that if the vapor relief velocity is high (> 30 m/s), the air entrance rate is also high and the vapors typically would be diluted below lower flammable limit in a very short distance.

This phenomenon explains the standard recommendation of locating the discharge of the relief device at least 3 m above any adjacent equipment, building or structure located within 15 m horizontally from the discharge point.

Experience has shown, however, that for some special cases these typical recommendations does not apply, such as:

- Heavy gas releases;
- Cryogenic releases;
- Control valve releases;
- Discharge piping with unconventional design (rain cap, horizontal, etc.);
- Discharge location within a building wake.

Adding to the complexity of the analysis, the relief devices can also be fitted with steam or other purge gas injection that could positively affect the dispersion.

After some case studies, which included the analysis of High Pressure Polyethylene plants Vent Stacks and Olefins unit relief and venting devices, Braskem developed an internal engineering standard with guidance and criteria for evaluating existing projects and new designs in order to ensure they comply with safe location. The guidance includes:

- Relief cases selection;
- Simplified and rigorous dispersion analysis;
- Consequence assessment endpoints;
- Thermal radiation effects to assets;
- Liquid, vapor and two-phase releases;
- Process Hazard Analysis;
- Diluting/snuffing steam;
- New systems and retrofitting criteria;

Applying the criteria in the developed standard allows the evaluation if the relief is in a safe location or not. If the discharge is not in a safe location, a risk ranking and recommendations to change the system must be made. One kind of a typical recommendation is to install the relief in a higher location. However, for some cases, achieving Safe Location is not possible and changes in the system are needed, such as relocating to a closed system or design changes to eliminate the relief scenarios associated with that relief device.

The development of an internal engineering guideline ensures a standard approach throughout the company for evaluating the Safe Location of hydrocarbon discharges and improves the quality of the risk evaluation during the PHA cycles.

548407 Experimental Study of Electrostatic Hazard inside Scrubber Column Using Response Surface Methodology

Experimental Study of Electrostatic Hazard inside Scrubber Column using Response Surface Methodology

Jingyao Wang[a,b], Yue Sun[a,b], Xiaodan Gao[a,b], Benjamin Wilhite[a,b], M. Sam Mannan [a,b]

[a]Mary Kay O'Connor Process Safety Center, Artie McFerrin Department of Chemical Engineering, Texas A&M University, College Station, Texas 77843-3122, USA

[b] Artie McFerrin Department of Chemical Engineering, Texas A&M University, 3122 TAMU, College Station, Texas 77843, United States

Abstract

Electrostatic ignition has triggered significant fire/explosion incidents in liquid processing industries including chemical, oil/gas, food, pharmaceutical and transportation. In this work, the impact of operating flowrate and temperature on the electrostatic energy accumulation in scrubber columns was studied. An *in-situ* charge measurement apparatus combined with a particle scattering laser function was installed inside a pilot-scale scrubber column, which can measure the aerosol particle size distribution, concentration and charge simultaneously. Four electrostatic ignition scenarios were considered including discharge from a single aerosol droplet, aerosol clouds, ungrounded conductor charged by the aerosol and ungrounded conductor charged by the liquid. Response surface methodology was applied to construct a quadratic empirical model correlating the electrostatic energy response with the flowrate and temperature. The obtained electrostatic response shows that corona from a charged single droplet was too small for safety concern. However, the ungrounded conductor polarized by charged aerosol or liquid can accumulate significant energy that is higher than the minimum ignition energy of some flammable chemicals. Temperature has two-sided effect where high temperature generates more concentrated charged aerosol but also facilitates the charge relaxation in the liquid phase. In all scenarios, increasing gas/liquid flowrate can cause significant electrostatic energy accumulation and can increase the hazards of many incidents. To reduce the electrostatic hazards, it is recommended to reduce the gas/liquid interaction flowrate and eliminate the presence of ungrounded conductors in liquid process operations.

Keywords: Scrubber column design, Electrostatic hazards, Response surface methodology, Process safety, Hazard identification, Aerosol hazards.

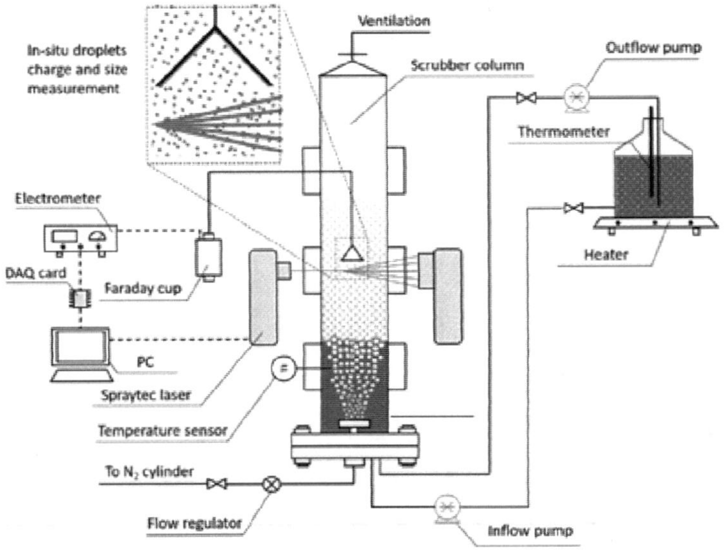

550852 Process Safety in Oil Refineries: Vacuum Distillation Column Hazop Study

Errors that can occur in the identification studies for hazards and risks which is one of the elements of a process safety management cause resulting major accident having undesirable long - term effects on human and environment. For preventing major accidents, identification and analysis of process hazards due to equipment failures, design or montage defects, improper maintenance, misuse, material defects, metal fatigue, corrosion, welding defects, human failure etc. are the main principles of process safety management. The aim of present study is to investigate possible accidental risks occurring in Vacuum Distillation Column of oil refining within contributing to the process safety frame.

In this cross - sectional study, semi quantitative HAZOP analysis was used for identifying accidental risks. By means of HAZOP studies 66 deviations, 238 deviation causes, 287 consequences, 651 barriers and 113 recommendation were identified. In the determination of accidental risk in the vacuum distillation column by applying semi quantitative HAZOP analysis method, effective results were obtained. The most common causes of deviations in the column have been found as control loop failure, leakage and mechanical failure. It is stated that the most common accidental risks are as a portion of 66.9 % operational deterioration, fire, explosion, release known as major accident risk. It is identified that the accidental risks mostly cause existence damage and the highest accidental risk occurs in the "vacuum column overhead system" and "bottom of the vacuum column system" 87.7 % of 113 recommendations are modification in the design, 5.3% is about modifications in the process safety management system.

551179 API Process Safety Site Assessment Program - Trends and Good Practices in Process Safety

The Process Safety Site Assessment Program is a collaborative industry effort managed by API and is focused on improving process safety in petroleum refining and petrochemical facilities. The program has conducted over 65 process safety assessments and 38 HF RP-751 assessments since its inception in 2012. The assessments are conducted by independent and credible third-party teams of industry-qualified process safety expert assessors using industry-developed protocols.

The presentation will review trends, both positive and low scoring, seen in the extensive benchmarking data collected. Good practices observed in the assessments will be identified and discussed. The emphasis will be to have the attendees leave with an understanding that typically some site(s) has developed a way to handle challenging areas of process safety.

545445 Emerging Trends in Process Safety Management Relating to Contractor Workforce Compliance

Following a Six Sigma Black Belt process improvement session conducted 2016 to 2018 for petrochemical complexes along the Gulf Coast the presentation identifies several emerging trends for Best Practices in industrial workforce compliance.

Safety Councils work for petrochemical plants and their contractors to facilitate innovative ways to reduce incident rates resulting in a 95% reduction in recordables. 2017 marked an evolution in Safety Council deliverables, slowly migrating away from site specific training to develop new workforce compliance and risk management processes. Working within the industry we developed several internationally recognized Best Practices for contractor compliance, risk management, labor cost management. ISTC collaborated with our partners bringing years of quantitative data to identify problems that are developing, have occurred in the workplace and formulating solutions, cost structures and most importantly compliance management to combat workforce issues. The presentation, best practices in contractor compliance highlights the journey from identifying the problems, and the associated issues with hiring, compliance, auditing, accident response and even Department of Homeland Security, CFATS inspections. Beyond the logistics of building solutions for a particular problem, the presentation reflects on how companies were able to share their issues among peers without violating confidentiality or company procedures to solve what was to be discovered as an industry trend. Using Six Sigma tools, the units were able to identify root problems, collaborate on a solution, model cost recovery and implement the solutions to the contractor workforce with a minimum of disruption to the business at hand.

Análisis Cuantitativo De Riesgo (QRA) Como Herramienta Para La Toma De Decisión Sobre La Infraestructura En Sistemas De Transporte De Fluidos Peligrosos

Daniel Mauricio Moncada Saavedra
CENIT -Transporte y Logística de Hidrocarburos
Cll 113 #7-80 Piso 13 Bogotá – Colombia
daniel.moncada@cenit-transporte.com

Maria Isabel Montoya Rodriguez
CENIT -Transporte y Logística de Hidrocarburos
Cll 113 #7-80 Piso 13 Bogotá – Colombia
maria.montoya@cenit-transporte.com

Prepared for Presentation at
American Institute of Chemical Engineers
2019 Spring Meeting and 15th Global Congress on Process Safety
New Orleans, LA
March 31 – April 3, 2019

Análisis Cuantitativo De Riesgo (QRA) Como Herramienta Para La Toma De Decisión Sobre La Infraestructura En Sistemas De Transporte De Fluidos Peligrosos"

Daniel Mauricio Moncada Saavedra
CENIT -Transporte y Logística de Hidrocarburos
Cll 113 #7-80 Piso 13 Bogotá – Colombia
daniel.moncada@cenit-transporte.com

Maria Isabel Montoya Rodriguez
CENIT -Transporte y Logística de Hidrocarburos
Cll 113 #7-80 Piso 13 Bogotá – Colombia
maria.montoya@cenit-transporte.com

Palabras clave:

Abstract

La pérdida de contención en activos utilizados para el transporte de hidrocarburos puede ocasionar impactos considerables a personas, medio ambiente, económicos y de reputación. Con el fin de mitigar la ocurrencia este tipo de eventos, se implementan estrategias sobre los activos, siendo la identificación y análisis de riesgo un elemento clave durante el proceso.

El presente análisis utilizó un enfoque cuantitativo, tanto para probabilidad como para consecuencia, con el fin de poder estimar el riesgo que genera la operación de una línea de transporte en un área de alta consecuencia clasificado como sitio identificado, planteando posibles alternativas de control y mitigación del riesgo.

Como resultado del estudio realizado se pudo establecer el nivel de riesgo actual asociado a la operación de la línea en el sitio de interés. Como actividades de mitigación para reducir el riesgo asociado se identificó: la reparación de los defectos, así como el realineamiento y reubicación de la tubería.

1 Introducción

Para el presente análisis se implementó un cálculo de la probabilidad de falla a partir del análisis de integridad estructural del sector de interés. Adicionalmente, se consideró el análisis de la susceptibilidad que se presenta a partir de las amenazas presentes, la cuales son identificadas a partir de las actividades de monitoreo e inspección desarrolladas sobre la infraestructura. El cálculo de la interacción entre las amenazas a la integridad; así como el resultado final de probabilidad de falla, se puede realizar mediante diferentes enfoques, el presente estudio utilizo cálculo de probabilidad de excedencia y aplicación de redes bayesianas para la estimación.

Para el cálculo de la consecuencia se buscó verificar la posible afectación a personas, producto de la ignición del producto transportado. Dentro de los sucesos finales se consideró: incendio de piscina, incendio por chorro de fuego, llamarada y la sobrepresión debida a la explosión. El cálculo de estos efectos se realizó en el sitio donde se presenta la perdida de contención, así como también para el caso en el que el hidrocarburo es transportado por un cuerpo de agua o se desplaza en tierra.

Una vez determinada tanto la probabilidad de que se desarrollen los sucesos finales identificados como la consecuencia que cada uno de ellos puede generar, se procedió a realizar el cálculo del riesgo expresando el resultado en términos de riesgo social e individual, comparando el resultado contra criterios de tolerabilidad referentes en la industria. Para la condición actual establecida se identificaron alternativas las cuales buscaban disminuir el nivel de riesgo, implementando actividades de mitigación del nivel de riesgo.

2 Análisis de Riesgo Cuantitativo

3 Frecuencia Final de Ocurrencia de los Sucesos Finales

En el sector bajo interés se tienen que la línea es susceptible a las siguientes amenazas:

3.1 Análisis de Condición.

En este aparte se relacionan los resultados de análisis de probabilidad de falla, obtenidos en el sector de interés, PK 43+000 al PK 45+000 del poliducto, para las amenazas de corrosión externa, corrosión interna, geotecnia y daños por terceros involuntarios:

3.1.1 Corrosión Externa (CE).

A continuación, se describe el estado de salud de las barreras de CE del poliducto en el sector solicitado (ver Ilustración 1), así como el análisis de probabilidad de excedencia de las anomalías ILI identificadas con la corrida del 2014. En la Ilustración 2 se observa la disponibilidad de la URPC y se reporta que la tubería cuenta con recubrimiento en buen estado. Bajo las condiciones y premisas consideradas para el cálculo y proyección de la POE (Probabilidad de excedencia; entre ellas una rata de corrosión de 2-3 mpy aproximadamente; no hay riesgo de excedencia de los criterios de profundidad y ERF (Factor Estimado de Reparación) de las anomalías ILI hasta el año 2021 en el sector 43+590 y 42+746 (Anomalías ILI con DR(Distancia de registro): 2646.76, 3490.08, 3492.38, 3492.71) (Ilustración 2).

Ilustración 1. Perfil de riesgo y salud de barreras CE.

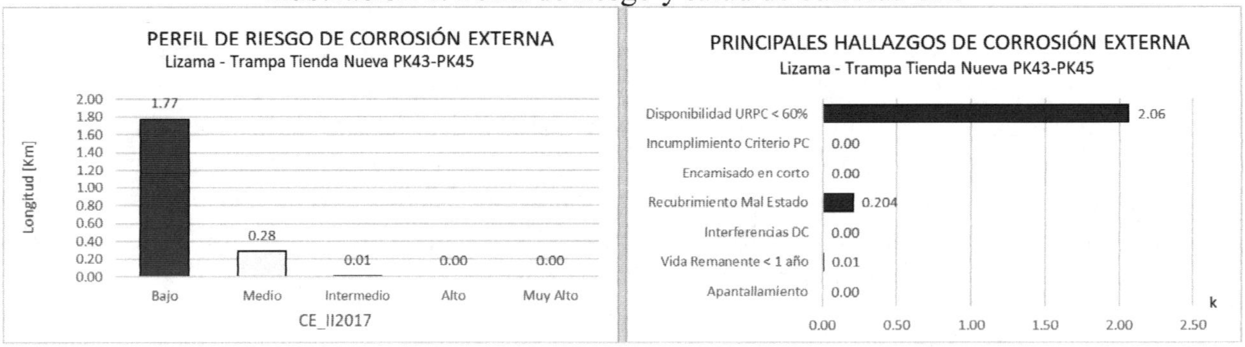

Ilustración 2. Análisis de POE de anomalías de CE en el tiempo.

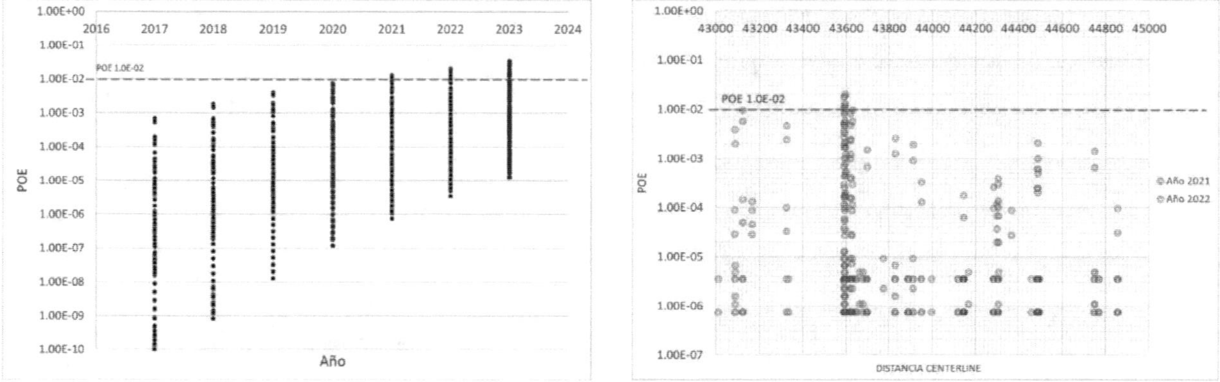

3.1.2 *Corrosión interna (CI).*

A continuación, se describe el estado de salud de las barreras de CI de la tubería en el sector solicitado (ver Ilustración 3), así como el análisis de probabilidad de excedencia de las anomalías ILI identificadas en la corrida del 2014. De acuerdo con los resultados de valoración de riesgo, se presenta un correcto estado de salud de las barreras de corrosión interna. Adicionalmente, de acuerdo con las condiciones y premisas consideradas para el cálculo y proyección de la POE, no hay riesgo de excedencia de los criterios de profundidad y ERF de las anomalías ILI en el mediano plazo (Ver Ilustración 4).

Ilustración 3. Perfil de riesgo y salud de barreras CI.

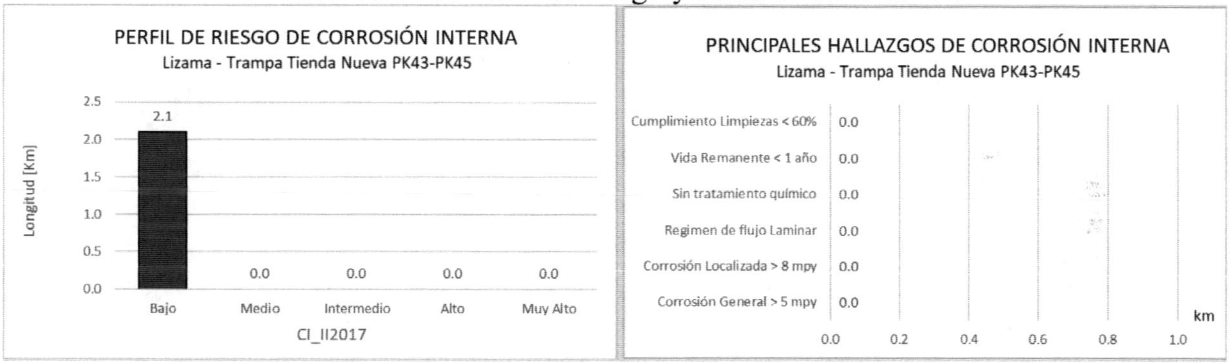

Ilustración 4. Análisis de POE de anomalías de CI en el tiempo.

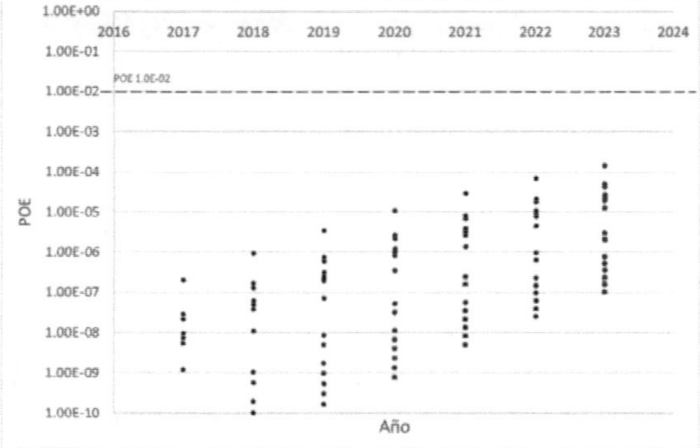

3.1.3 Geotecnia.

En el sector de interés se reporta hallazgo de geotecnia en la distancia centerline 44643 – 44703 (PK 44+400), donde se presenta proceso erosivo incipiente. Actualmente, no se identifica avance de la condición, razón por la cual, se continuará seguimiento a través de la inspección visual del derecho de vía (IVDDV).

Ilustración 5. Perfil de riesgo y salud de barreras de Geotecnia.

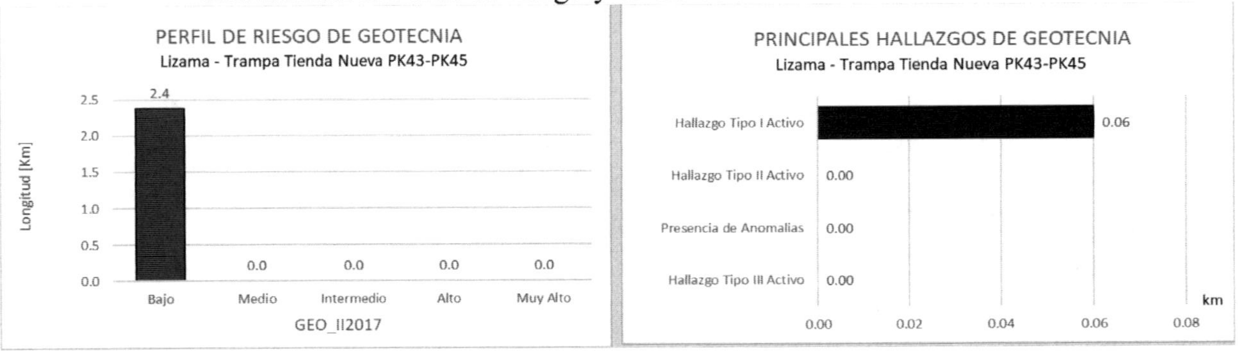

3.1.4 *Daños por Terceros Involuntarios (DPT).*

Debido a la condición de la tubería y a la gestión realizada por el operador para el control de la amenaza, ésta presenta una susceptibilidad baja de ser golpeada por un tercero sobre el derecho de vía. A continuación, se relaciona el perfil de riesgo de la amenaza de DPT en el sector analizado:

Ilustración 6. Perfil de riesgo y salud de barreras de Daños por Terceros Involuntarios.

De acuerdo con el análisis ejecutado, el evento más probable obedece a una pérdida de contención generada por la amenaza de corrosión externa; donde el análisis estructural identifica que el modo de falla se daría por fuga. Sin embargo, las anomalías encontradas no presentan riesgo de excedencia de los criterios de profundidad y ERF hasta el año 2021.

Con el fin de poder estimar la probabilidad de falla de la línea en el sector de interés, se realizó el análisis estructural considerando principalmente defectos por perdida de espesor, bajo estas premisas se obtuvo un valor de probabilidad de excedencia de los criterios de profundidad y ERF de 7.09E-4. Del análisis de condición se concluye que el iniciador más probable en este sector es la falla de la tubería por fuga asociado a corrosión externa, para el cual a continuación se presenta el análisis de riesgo.

4 Análisis de consecuencia

A continuación, se presenta el análisis de consecuencia para los sitios en los cuales se ha identificado que se puede llegar a presentar una pérdida de contención.

4.1 Identificación de escenarios

4.1.1 Selección de sustancias.

En el poliducto se transportan las siguientes sustancias: Gasolina, GLP, Jet A1 y B2

4.1.2 Identificación de iniciadores.

Los iniciadores representan las situaciones iniciales de pérdidas de contención o exposición que se pueden presentar en los das diferentes equipos. Los iniciadores pueden ser de dos tipos, genéricos o específicos. Los iniciadores genéricos, involucran una pérdida de contención debida a causas no específicas del proceso u operación, como pueden ser corrosión, errores de construcción, fallas de la soldadura, etc. Los iniciadores específicos se caracterizan por tener causas concretas debido a las condiciones de proceso, el diseño, los materiales utilizados, etc.

A continuación, se describen los iniciadores tenidos en cuenta para cada uno de los sitios en los cuales se identificó la inestabilidad de taludes.

- **Sector PK 43+590.** Este sitio tiene la menor distancia desde la infraestructura a la escuela bajo análisis, 30 metros aproximadamente, sumado a esto en las cercanías de este se ha identificado la presencia de anomalías las cuales, bajo las condiciones actuales de operación y deterioro hacia el año 2021, pueden exceder los límites de profundidad critica a la falla.

- **Sector PK 43+805 – PK 44+110.** En este sitio se tiene que ante la posible pérdida de contención del producto se puede llegar a presentar la formación de una ruta de derrame en tierra, por la cual se desplazaría el producto hacia el interior de la Escuela.

4.1.3 Identificación de sucesos finales.

La identificación de los sucesos finales que pueden tener lugar después de que se da el iniciador, se debe realizar utilizando los árboles de eventos o sucesos. Dichos árboles permiten ver el comportamiento de la sustancia tras la descarga y estimar la probabilidad de ocurrencia de cada suceso, teniendo en cuenta los sucesos intermedios que condicionan la evolución del suceso final.

Esta identificación de escenarios se lista en Tabla 11 para los escenarios que involucran sustancias inflamables tales como Gasolina o Nafta. Los posibles sucesos finales identificados en el poliducto son: derrame continental, incendio de piscina, incendio de chorro y llamarada.

Tabla 1. Identificación de escenarios.

Escenarios	Descripción	Iniciadores	Sustancia	Tipo de pérdida
Afectación directa				
E1	Poliducto	L3	Gasolina	Incendio de piscina
				Incendio de chorro
				Llamarada
E2	Poliducto	L3	GLP	Incendio de piscina
				Incendio de chorro
				Llamarada

Escenarios	Descripción	Iniciadores	Sustancia	Tipo de pérdida
Afectación directa				
Afectación indirecta				
E3	Poliducto	L3	Gasolina	Incendio de piscina
				Llamarada

4.2 Estimación del área de afectación por los efectos que involucran producto

4.2.1 Condiciones para la simulación.

Para determinar las consecuencias, es necesario definir las condiciones propias del entorno de la infraestructura que servirán como base para la simulación de los efectos. Estos parámetros son presentados en la Tabla 2. Las estabilidades atmosféricas fueron determinadas a partir de la rosa de vientos basada en datos estadísticos y modelos meteorológicos, la cual es presentada en la **¡Error! No se encuentra el origen de la referencia.**.

Tabla 2. Condiciones atmosféricas

Parámetro	Valor	Estación de Referencia
Temperatura promedio (°C)	31	Colosal
Humedad relativa promedio (%)	83	
Rugosidad el terreno (m)	0,08	
Estabilidad atmosférica más probable y velocidad del viento (m/s)	2F	
Estabilidad atmosférica F y velocidad del viento (m/s)	2F, 5D	

4.2.2 Distancias de afectación directa

Las distancias de afectación representan las longitudes dentro de las cuales se pueden manifestar los efectos correspondientes a los niveles de afectación, para cada suceso final (incendios de piscina o llamaradas). Se denominan distancias de afectación directas, ya que los sucesos finales de que dependen se originan y desarrollan su efecto directamente en el sitio en el que se produce la pérdida de contención del producto

4.2.3 Estimación de Descargas.

Para la simulación de los posibles efectos producto de una pérdida de contención, es necesario conocer cuánto producto podría derramarse en cada punto de estudio sobre la infraestructura. El volumen derramado corresponde al estimado teniendo en cuenta el llenado de la tubería, las condiciones de operación y el tiempo de respuesta operativo, es decir, el tiempo de detección más el tiempo de actuación (generalmente el seccionamiento del sistema o parada de bombeo).

Tabla 3. Tiempos de actuación

Equipo	Tiempo de Detección (min)	Tiempo de Actuación (min)
	Fuga	Fuga
Poliducto	30	180

Tabla 4. Volumen y Tasa de descarga

Volumen (US bls)	Tasa de descarga (bls/hr)
59	22

4.2.4 *Simulación de Efectos de los Sucesos Finales.*

En este numeral se presentan los resultados de las distancias de afectación directa, obtenidas para cada uno de los niveles de afectación asociados a los sucesos finales. En la **¡Error! No se encuentra el origen de la referencia.** se presentan las distancias de afectación obtenidas para el evento incendio de piscina, es importante resaltar que este depende en gran medida del área de empozamiento que se forma una vez se produce la descarga del producto.

En la **¡Error! No se encuentra el origen de la referencia.** se presentan las distancias de afectación para incendio de chorro obtenidas durante las simulaciones, es importante resaltar que la distancia obtenida depende de las propiedades químicas de la sustancia y de las condiciones en las cuales se produce la descarga.

Tabla 5. Distancia de afectación incendio de piscina.

Equipo / Abscisado	Iniciador	Sustancia	Volumen Derrame	Niveles de Afectación por Incendio						
				Incendio de Piscina						
				1.6 Kw/m^2	5.0 Kw/m^2	7.3 Kw/m^2	9.5 Kw/m^2	14.5 Kw/m^2	20.9 Kw/m^2	37.5 Kw/m^2
			(Bbl)	Distancia (m)						
K43+590	Rotura mínima	GLP	59	46,94	28,63	24,58	21,98	18,01	14,55	8,38
K43+590	Rotura mínima	Gasolina	59	94,63	49,69	35,17	27,20	20,98	20,98	-

Tabla 6. Distancia de afectación incendio JET FIRE.

Equipo / Abscisado	Iniciador	Sustancia	Volumen Derrame	Niveles de Afectación por Incendio						
				Chorro de fuego						
				1.6 Kw/m^2	5.0 Kw/m^2	7.3 Kw/m^2	9.5 Kw/m^2	14.5 Kw/m^2	20.9 Kw/m^2	37.5 Kw/m^2
			(Bbl)	Distancia (m)						
K43+590	Rotura mínima	GLP	59	45,9	33,61	30,93	29,33	27,13	25,23	23,42
K43+590	Rotura mínima	Gasolina	59	44,43	31,20	28,46	26,82	24,58	22,96	20,84

En la **¡Error! No se encuentra el origen de la referencia.** se presentan las distancias de afectación por llamarada; las cuales están relacionadas con la volatilidad del producto junto con las condiciones climáticas de las zonas del evento.

Tabla 7. Distancia de afectación Llamarada.

Equipo / Abscisado	Iniciador	Sustancia	Volumen de Derrame	Niveles de Afectación por Llamarada más probable		Niveles de Afectación por Llamarada más estable	
				LII	LII/2	LII	LII/2
			(Bbl)	Distancia (m)		Distancia (m)	
K43+590	Rotura mínima	GLP	59	24,12	58,53	26,48	66,7
K43+590	Rotura mínima	Gasolina	59	-	-	-	-

4.2.5 *Distancias de Afectación Indirecta*

Las distancias de afectación indirecta representan las longitudes de potencial afectación por sucesos finales que se originan en una trayectoria o ruta de derrame (en tierra o en agua) o, a partir de estancamientos o acumulaciones de producto. En este numeral se presentan los resultados de las distancias de afectación obtenidas para cada uno de los niveles de afectación asociados a los sucesos finales que se pueden presentar en la ruta de derrame. Estos resultados corresponden a la radiación generada por piscina de fuego, evento que en los cuerpos de agua depende de la sección del cuerpo de agua, así como por el frente de llama generado por el volumen derramado.

En la Tabla **88 y ¡Error! No se encuentra el origen de la referencia.**9 se referencian los corredores de afectación generados por incendio de piscina sobre las rutas de derrame en tierra identificadas en el poliducto, en el sector comprendido entre los PK 43+000 - PK 45+000.

Tabla 8. Distancias de afectación indirectas POOL FIRE.

Abscisado	Medio de Transporte	Sustancia	Incendio						
			Incendio en piscina						
			1.6 Kw/m^2	5.0 Kw/m^2	7.3 Kw/m^2	9 Kw/m^2	14.5 Kw/m^2	20.9 Kw/m^2	37.5 Kw/m^2
			Distancia (m)						
Km 43 - 45	Ruta en tierra	Gasolina	82,02	54,68	27,34	24,61	19,14	13,67	8,20

Tabla 9. Distancias de afectación indirecta FLASH FIRE

Abscisado	Medio de Transporte	Sustancia	Niveles de afectación por llamarada más probable		Niveles de afectación por llamarada más estable	
			LII	LII/2	LII	LII/2
			Distancia (m)		Distancia (m)	
Km 43- 45	Ruta en tierra	Gasolina	2,14	4,74	8,64	32,24

4.2.6 Áreas de Afectación

En sistemas de ductos, el área de afectación directa se define por un corredor medido desde el eje central del ducto. Este corredor está dado por la mayor distancia probable de afectación obtenida en la simulación de efectos para todas las sustancias analizadas. Por otro lado, el área de afectación indirecta está dada por varios corredores a lo largo de las posibles rutas de derrame tanto en suelo como en agua. En la Ilustración 77 se presenta el área de afectación identificada.

Ilustración 7. Área de afectación identificada.

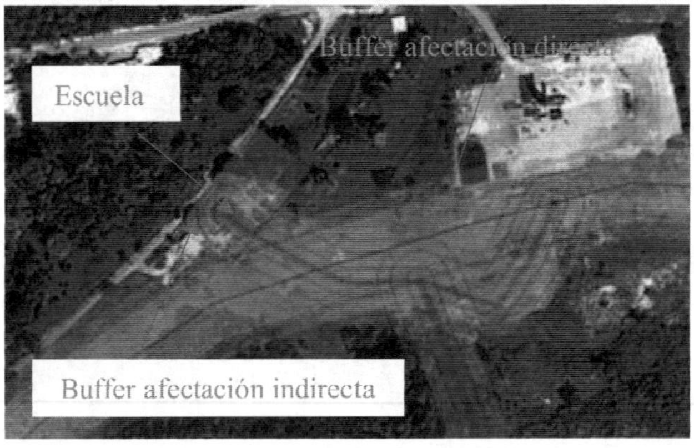

4.2.7 Identificación de Elementos Vulnerables

Para calcular el riesgo de la población respecto de la posible afectación producida por la pérdida de contención descrita, se requiere el número y localización de las personas, en la Tabla 100 se presenta la población identificada al interior de la escuela.

Tabla 10. Presencia de población Escuela

Presencia de Población							
Escuela - Dia				Escuela - Noche			
Niños	63	Profesores	6	Niños	0	Profesores	2

5 Evaluación del Riesgo.

A continuación, se presentan los resultados del análisis de riesgo el cual tuvo en cuenta la información presentada en los numerales anteriores asociada a las posibles causas de falla, los modos en que puede darse cada falla identificada, la frecuencia de ocurrencia de la falla, la probabilidad de ocurrencia de los sucesos finales, y finalmente, la estimación de la magnitud de las consecuencias de tales sucesos finales en el entornó en términos cuantitativos.

5.1 Riesgo Individual.

Los resultados de riesgo individual expresados en número de fatalidades por año (o año-1), sí permiten establecer zonas de aceptabilidad del riesgo según los diferentes criterios que se utilizan en la industria. En la Ilustración 8 se presentan los resultados de riesgo individual en el sector comprendido entre los PK 43+000 y 45+000.

Ilustración 8. Riesgo Individual.

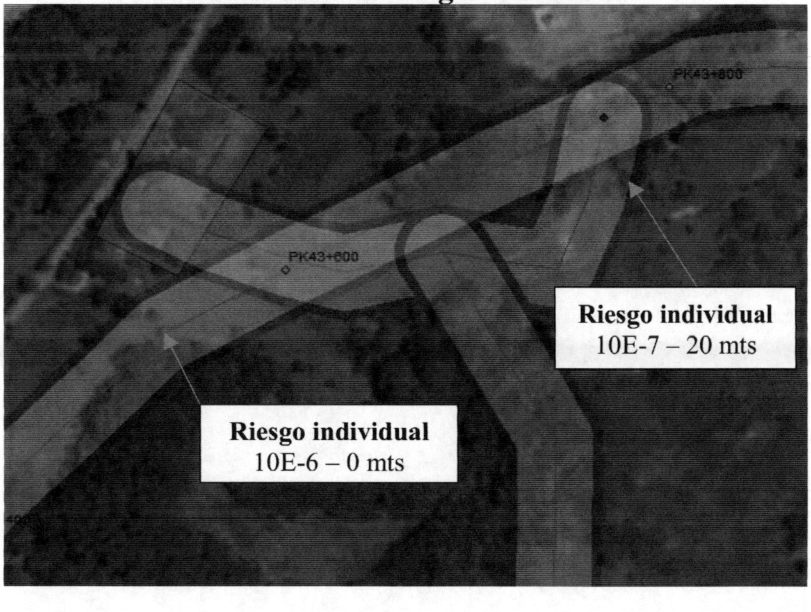

5.2 Riesgo Social

El riesgo social representa el número de fatalidades que pueden llegar a ocurrir en los sitios evaluados en el poliducto en el PK 43+000 – PK 45+000, al revisar la información contenida en la Ilustración 9 se puede apreciar como el riesgo se mantiene en la región tolerable, muy cercana al límite inferior ALARP. El evento dimensionante corresponde a una llamarada en la ruta de derrame, con el ducto transportando gasolina, la cual puede llegar a afectar a ocho (8) personas.

Ilustración 9. Riesgo social Escuela PK 43+000 – PK 45+000.

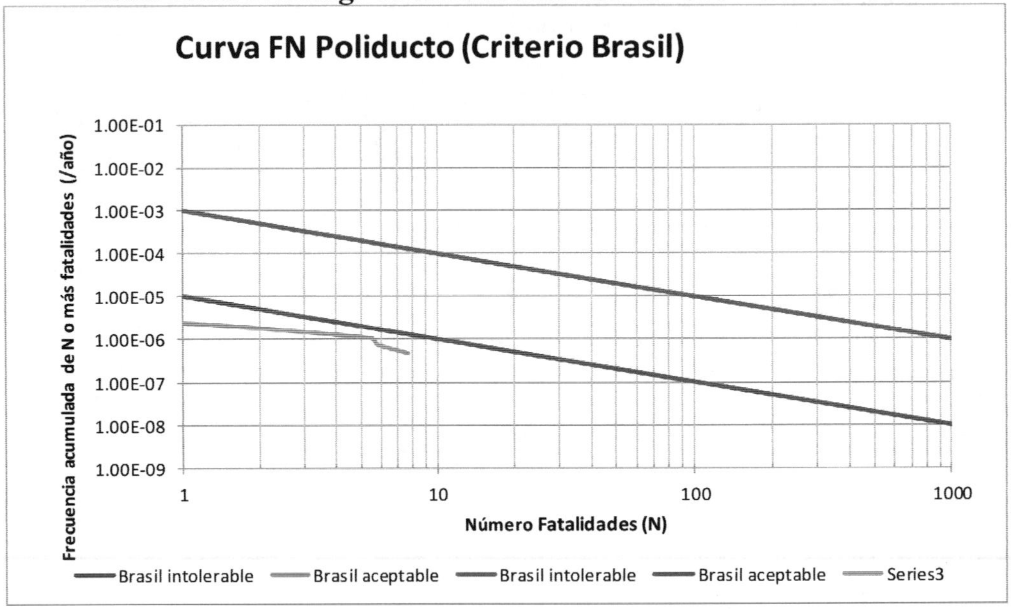

6 Alternativas de mitigación de Riesgo

Considerando los resultados del análisis de riesgo cuantitativo sobre el Poliducto para el sector PK 43+000 - PK 45+000, se tiene que ante una rotura total el evento dimensionaste corresponde a una llamarada con una distancia de afectación de 135 metros, para el caso de una fuga por corrosión el evento dimensionaste corresponde a una llamarada con una distancia de afectación de 32,24 metros, finalmente como resultado de una perdida de contención se puede llegar presentar la formación de una ruta de derrame en tierra la cual ingresaría a la escuela, con una distancia de afectación de 8.6 metros generada por una llamarada, tal como se presenta en la Ilustración 10.

Ilustración 70. Distancias de afectación

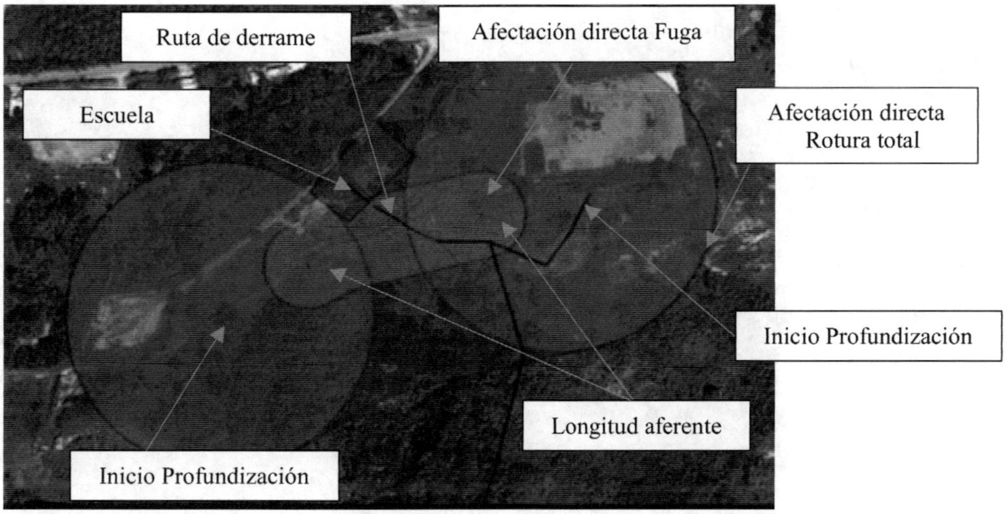

Teniendo en cuenta que el evento corresponde al presentado en la ruta de derrame, se hace necesario considerar la evaluación de alternativas de mitigación de riesgo. Con base en lo anterior, se consideró que la actividad de realineación y profundización cubra la longitud aferente de esta ruta, la cual es de 155 metros e inicia en la coordenada 74°33'35.466"O - 6°9'8.213"N y finaliza la coordenada 74°38'15.025"O - 6°9'8.737"N. Lo anterior, tiene como objeto de eliminar la ruta de derrame actual que conduce a la escuela en caso de una falla y perdida de contención, dado que el vertimiento se genere en sentido contrario al actual.

Sumado a lo anterior es necesario considerar una distancia de protección adicional para el dado caso en el que se llegue a presentar la rotura total de la línea, esta distancia se establece a partir del radio de impacto del evento llamarada, localizando los puntos sobre la línea que no presentan afectación a la escuela. A partir del análisis se estableció que esta distancia es de 316 metros e inicia en la coordenada 74°31'33,57146"O - 6°9'8,22442"N y finaliza en la coordenada 74°32'28,98241"O - 6°9'9,70756"N, esta distancia cubre la longitud aferente descrita anteriormente.

Dado que la alternativa de mitigación tiene como fin eliminar la afectación indirecta y el impacto directo sobre la escuela a continuación se presentan dos alternativas de trazados propuestos, que aseguran que las rutas de derrame en caso de una perdida de contención tendrán dirección opuesta a la ubicación de la escuela.

Ilustración 80. Opción 1 realineamiento entre PK 43 + 695 a PK 44 + 205.

Ilustración 91. Opción 2 realineamiento entre PK 43 + 625 a PK 44 + 460 de 752 metros.

En la ilustración 12 se presenta las dos alternativas propuestas y la distancia de separación del trazado actual de tubería, así como de la escuela.

Ilustración 102. Alternativas propuestas para mitigación de riesgo

7 Conclusiones

Del análisis realizado se tienen las siguientes conclusiones.

- El máximo número de fatalidades a generar es de ocho (8) y es generado por una llamarada en la ruta de derrame en tierra la cual, en el dado caso de que se presente el evento, ingresaría a las instalaciones de la escuela.
- La mayor susceptibilidad de la línea es por la amenaza de corrosión externa, especialmente en algunos sectores en los cuales el recubrimiento se encuentra en mal estado, esta condición junto con las anomalías encontradas en la línea, impactaron la frecuencia de falla establecida para el sector.
- En términos generales el nivel de riesgo se encuentra en la zona tolerable, muy cercana al límite inferior ALARP.

8 Recomendaciones.

- Es recomendable realizar actividades tendientes a disminuir el volumen disponible a desplazar en la ruta de derrame, para ello se puede realizar la profundización de la línea lo cual además de disminuir el efecto de los eventos por JET FIRE y FLASH FIRE disminuye la cantidad de producto disponible a desplazar al favorecer la evaporación de este.

- A partir del análisis realizado se tiene que la distancia a realinear y profundizar es:

  Opción 1 realineamiento entre PK 43 + 695 a PK 44 + 205 con una longitud de 516 metros
  Opción 2 realineamiento entre PK 43 + 625 a PK 44 + 460 con una longitud de 752 metros.

- Con ambas alternativas se elimina la afectación indirecta y la afectación directa, sobre la escuela, estas distancias cubren todos los posibles eventos que se pueden presentar y tienen afectación sobre la escuela en mención.

- Se pueden construir cortacorrientes los cuales eviten que la ruta de derrame ingrese o haga su tránsito en las cercanías de la escuela, junto con la construcción de estructuras de retención las cuales detengan el avance del producto mientras se realizan actividades de recuperación de este.

- Es recomendable rehabilitar el recubrimiento en aquellos sectores en los cuales se encuentra en mal estado, junto con la atención de aquellas anomalías que sean críticas.

9 Referencias

[1] Fingas, M. (2011). Oil Spill Science and Technology. Kidlington, UK: Gulf Professional Publishing Elsevier.

[2] D.O.T. Cornell Law School. Obtenido de 49 CFR 192.905 - How does an operator identify a high consequence area?: https://www.law.cornell.edu/cfr/text/49/192.905 Accesado 15 Enero, 2019

[3] D.O.T. Cornell Law School. Obtenido de 49 CFR 192.5 - Class locations: https://www.law.cornell.edu/cfr/text/49/192.5 Accesado 15 Enero, 2019

[4] Yellow Book. TNO. Methods for the calculation of Physical Effects Due to releases of hazardous materials (liquids and gases) – Third edition Second revised print 2005

[5] Purple Book. TNO. Guidelines for quantitative risk assessment – First edition 1999/2005

Referencias no citadas:

[6] ASME.ASME B31.8S. Managing System Integrity of Gas.New York, USA.2005.

[7] ASME.PIPELINE TRANSPORTATION SYSTEMS FOR LIQUID HYDROCARBONS AND OTHER LIQUIDS ASME B31.4-2002. New York, USA.2002.

[8] API.API Std 1160. Managing System Integrity for Hazardous Liquid Pipelines. Washington, USA.2008.132 Pags.

548036 Best Practice in MOC and Pssr Automation for Opex

Management of Change (MOC) and Pre Start-up Safety Review (PSSR) are still the most challenging elements of OSHA's Process Safety Management (PSM) standard and the EPA's Risk Management Program (RMP) rule. Now MOC and PSSR are requirements of the US Department of Interior's Bureau of Safety and Environmental Enforcement's Safety and Environmental Management Systems (SEMS). Even though the PSM standard has been around since 1992 and the industry has been managing change for several decades, we still can get it wrong, sometimes with disastrous results. The mindful or diligent efforts of many working in concert are necessary to ensure that change is identified, analyzed and executed in a quality way. But just when we get the process on paper and get the workforce up-to-speed, we have employee turnover, complacency sets in and sometimes there is regulatory change.

Many high-profile companies have addressed this challenge by creating and automating standard processes across their manufacturing or production facilities. MOC and PSSR are two of these major processes that have been successfully automated or **digitized**. Technology has been used to **automate** each step of the MOC workflow, enabling the collection of change ideas, formalizing the analysis of the ideas, and documenting the approval and execution of the change. Reports and dashboards can provide visibility of the health or current state of your MOC and PSSR processes. The use of **software** can enable the standardization of best practices.

On a well-meaning quest to standardize on a MOC/PSSR software solution and process, many companies are stifling creativity and innovation. Operational Excellence cannot be achieved in such ways. This paper discusses how the chemical process industry has defined the MOC and PSSR best management practices and how they should be automated for operational excellence.

Chemical Process Safety Requirements in the Fire Code

Ronald Melucci
Engineering Planning and Management, Inc.
959 Concord Street
Framingham, MA 01701
rbm@epm-inc.com

Prepared for Presentation at
American Institute of Chemical Engineers
2019 Spring Meeting and 15th Global Congress on Process Safety
New Orleans, LA
March 31-April 3, 2019

Ronald Melucci
Engineering Planning and Management, Inc.
959 Concord Street
Framingham, MA 01701
rbm@epm-inc.com
Keywords: NFPA Fire Code, International Fire Code, Process Safety

Chemical Process Safety Requirements of the Fire Code

Abstract

When most Chemical Process Safety professionals think about regulation and enforcement, they think about the applicable Occupational Safety and Health Administration (OSHA) and Environmental Protection Agency (EPA) regulations. However, the majority of US states adopt NFPA 1 or the International Fire Code (IFC) as their state fire code which contain requirements for chemical process safety. These requirements are applicable for processes of all sizes and levels of complexity below PSM and RMP thresholds and they address both fire and non-fire related chemical hazards. These requirements are enforced by municipal or state fire officials/marshals, who are becoming more aware of these requirements and chemical process safety in general. This paper will review some of the code requirements with a focus on requirements that may be commonly overlooked in the industry.

Introduction

The OSHA Process Safety Management (PSM) and EPA Risk Management Program (RMP) regulations, enforced by OSHA and EPA, respectively, are familiar to most safety professionals who work in industries that include processing of large quantities of chemicals. However, the applicable fire code requirements related to chemical process safety are often more restrictive than the federal regulations and are generally lesser known to such safety professionals.

The two model fire codes adopted by states in the US, are: *NFPA 1, Fire Code,* published by The National Fire Protection Association (NFPA), and the *International Fire Code* (IFC) published by the International Code Council (ICC). The states that do not adopt the IFC will typically use NFPA 1 and each state may also institute their own NFPA 1 amendments. Local amendments are almost always more restrictive than the model code verbiage, and many states have legislation that prohibits fire code amendments that reduce the level of safety/standard of care intended by the model fire code.

Both of these model fire codes contain many requirements that are similar to PSM and RMP regulations and contain references to them for additional guidance. Generally speaking, the model fire codes do not regulate all hazardous materials, but focus on hazards that are acutely dangerous to people or property. The PSM and RMP regulations address these hazards as well but additionally include materials that present a risk of chronic health or environmental hazards.

However, many states and municipalities have local fire code amendments that often include lower chemical quantity thresholds, such as those of the Commonwealth of Massachusetts as described in the following sections. The enforcement of these requirements is therefore not limited to OSHA and the EPA, as is commonly believed. The requirements are also enforceable by the local (municipal) fire authorities through the fire code references to these regulations.

Fire Code Requirements

The Commonwealth of Massachusetts currently adopts NFPA 1 – 2015 edition, with amendments, as 527 CMR 1.00: *Massachusetts Comprehensive Fire Safety Code*. Since 2006, 527 CMR has included chemical process safety requirements that are more restrictive than the federal regulations. The impetus for this was several accidents that occurred involving chemical process quantities below the OSHA and EPA thresholds. These Massachusetts (MA) amendments have been added to Chapter 60 of NFPA 1 – 2015 edition and include specific reference to OSHA PSM and EPA RMP regulations. Chemical processes with quantities above the PSM/RMP thresholds are defined in MA as "Category 5" processes. However, the 527 CMR amendments have also defined processes with lower quantities, i.e. Category 1, 2, 3 and 4, each with specific and unique requirements as described herein.

The NFPA 1 committee is currently reviewing proposals to adopt requirements with a "tiered" quantity threshold approach similar to MA. This is in recognition of the fact that severe hazards often exist, and severe accidents have occurred, in chemical processes with quantities below the PSM/RMP thresholds. It is likely that such a "tiered" approach will be incorporated into NFPA 1 and subsequently adopted by many US states. A preview of the applicable NFPA 1 and 527 CMR requirements are summarized herein.

Massachusetts Amendments to NFPA 1, Section 1.12.8.34.2 introduces the requirements as: *Process or Processing of Any Hazardous Material at any Facility* and it states the following:

(1) "A permit shall be required for the process or processing of any hazardous material at any facility identified in this *Code* as Category 2 through Category 5, as described in Chapter 60.
(2) The Authority Having Jurisdiction (AHJ) may require technical assistance in accordance with *1.15* to evaluate the adequacy of a Category 3 or Category 4 facility process safety conditions, programs, procedures, and practices undertaken at the facility, but only after a notice of denial has been properly served upon the person making application."

As part of the MA amendments to NFPA 1 – 2015 edition, Section 60.8, ***Hazardous Material Process or Processing,*** was added to 527 CMR. Section 60.8 applies to both new and existing facilities that process hazardous materials, with a list of exceptions such as:

- Motor vehicle service stations
- Refrigeration systems using a refrigerant other than ammonia or LPG,
- Certain wastewater treatment operations
- Storage of hazardous materials in atmospheric vessels.

Definitions:

NFPA 1 defines "**Hazardous Materials**" as "A chemical or substance that is classified as a physical hazard material or a health hazard material, whether the chemical or substance is in usable or waste condition."

OSHA and EPA define "**Process**" as "Any use, storage, manufacturing, handling, or the on-site movement of hazardous chemicals, or combination of these activities." The MA state fire code defines "Process" as "A sequence of operations in which the sequence can be inclusive of physical operations such as heating, cooling, mixing, distilling, compressing, and pressurizing, and chemical operations, such as polymerization, oxidation, reduction, and other chemical reaction processes. The sequence can involve but is not limited to: preparation, separation, combination, purification, or any actions that cause a change in state, energy content, or chemical composition."

MA Amendments to NFPA 1 Section 60.8 categorizes hazardous material processes into five (5) categories, as follows:

Table 1: Hazardous Material Threshold Quantities in the MA Amendments to NFPA 1

Category	**Quantity (gallons)**
Category 1	< 2.5
Category 2	2.5 – 60
Category 3	60 – 300 (or Group H)
Category 4	300 and not Category 5
Category 5	≥ the threshold quantities stated in 29 CFR 1910.119

The MA Amendments also add Section 60.8.5.1 to NFPA 1, which requires that facilities operating hazardous material processes must maintain the required documents and procedures for periodic inspection and review by the head of the Fire Department to remain in compliance with the state fire code. Requirements for each process category are as follows:

Table 2: Outline of Regulatory Requirements in the MA Amendments to NFPA 1

Category	Requirements
Category 1 (Vessel Capacity ≤ 2.5 gal)	• OSHA 1910.1450 (Occupational exposure to hazardous chemicals) • OSHA 1910.1200 (Hazard communication to employees) • Policy per NFPA 1, Ch. 66 (Flammable liquids) and Ch. 67 (Flammable solids) • Requirements of NFPA 1, Section 60.1 – 60.5
Category 2 (Vessel Capacity ≤ 60 gal)	• All requirements of Category 1 • Permit required per NFPA 1, Sections 1.12.8.34.2 and 60.8.1.1
Category 3 (Vessel Capacity ≤ 300 gal OR Group H Occupancy)	• All requirements of Category 1 & 2 • Category 3 Hazard Evaluation • Requirements of NFPA 1, Section 60.6
Category 4 (Vessel Capacity ≥ 300 gals AND not Category 5)	• All requirements of Category 1-3 • Limited Safety Program (PSM "lite")
Category 5 (Vessel Capacity ≥ the threshold quantities stated in 29 CFR 1910.119)	• All requirements of Category 1-4 • Process Safety Management (PSM) Program 29 CFR 1910.119 (OSHA) • Risk Management Program (RMP) 40 CFR Part 68 (EPA)

Category 1 – Requirement Details

Category 1 processes, as stated in Table 2 above, must possess documentation of compliance with OSHA 1910.1450, *Occupational Exposure to Hazardous Chemicals in Laboratories*, OSHA 1910.1200, *Hazard Communication,* maintain and implement a policy in compliance with NFPA 1, Ch. 66, *Flammable and Combustible Liquids*, and Ch. 67, *Flammable Solids*, demonstrate compliance with NFPA 1, Section 60.1.5.1, *Emergency Action Plan*, and Sections 60.5.1.4.3.2(5) through 60.5.1.4.3.2(9) and comply with permitting requirements in Sections 1.12 and 60.8.4 of the MA Amendments to NFPA 1.

OSHA 1910.1450 and 1910.1200

Compliance with OSHA 1910.1450 requires facilities to set and adhere to exposure limits, monitoring plans, hygiene plans, precautions, training, medical plans, and other plans, as found to be acceptable by the AHJ, to react to and minimize potential exposure to hazardous chemicals. OSHA 1910.1200 requires hazard communication methods for employees such as inventory documentation, hazard programs, labels, safety data sheets, and training.

NFPA Chapter 66 and 67

In addition to demonstrating compliance with OSHA standards, facilities meeting Category 1 requirements must implement and maintain a written policy in compliance NFPA 1, Ch. 66 and Ch. 67.Chapter 66 is divided into 29 sections, most of which contain multiple requirements. Section 66.6 contains the prescriptive requirements for fire and explosion prevention and risk control, and Section 66.6.4 goes on to list requirements related to a facility's hazards analysis, which is required to be reviewed to ensure that fire and explosion hazards are addressed by fire prevention, fire control, and emergency action plans. The AHJ should be consulted with for the development of the hazards analysis to ensure that the extent of the fire prevention and control is deemed acceptable or an engineering evaluation of the operation and application of fire protection and process engineering principles is soundly applied. The hazards analysis must include, but not be limited to, analysis of the following:

(1) Fire and explosion hazards
(2) Emergency relief from process vessels
(3) Applicable facility design requirements in Chapters 17, 18, 19, 28, and 29 of NFPA 30
(4) Applicable requirements for liquid handlings, transfer, and use, as covered in Chapters 17, 18, 19, 28, and 29 of NFPA 30
(5) Local conditions/exposures
(6) Emergency response capabilities of the local emergency services

Additional guidance for conducting a fire risk assessment can be found in NFPA 551, *Guide for the Evaluation of Fire Risk Assessments.* Facility management must repeat the hazards analysis when any of the following occur:

(1) Changes occur in the materials in process
(2) Changes occur in process equipment
(3) Changes occur in process control
(4) Changes occur in operating procedures or assignments

This is consistent with the requirements of OSHA 1910.119 Sections (e) and (l). Not all sections of Chapter 66 may be related to processes; however, Section 66.17 is for processing facilities. Section 66.17 includes prescriptive requirements for processing facilities such as location of process vessels and equipment (Section 66.17.4), containment, drainage, and spill control (Section 66.17.10), and management of operations hazards (Section 66.17.15). Later sections of Chapter 66 include additional requirements for dispensing, handling, transfer, and use of liquids (processes) (Section 66.18) and specific operations such as recirculating heat transfer system, vapor recovery and vapor processing system, solvent distillation units, and cooking oil storage tank systems in commercial kitchens (Section 66.19).

NFPA 1 Chapter 60

In addition to Chapters 66 and 67 of NFPA 1, Category 1 processes must also adhere to the requirements of 60.1 – 60.5 of NFPA 1 and the MA Amendments to NFPA 1. Depending on whether the quantities of materials at the facility exceed the MAQ or not, determines which requirements of Section 60.1 are applicable (Section 60.1.3).

Section 60.1.5 of NFPA 1 requires that an emergency action plan, consistent with the available equipment and personnel, shall be established to respond to fire and other emergencies in accordance with requirements set forth in NFPA 1. Section 10.8 of NFPA 1 includes additional requirements and guidance on generating emergency action plans, such as what must be included in the plans, purpose, responsibilities, etc. Section 60.4 of NFPA 1 contains requirements related to permissible storage and use locations and sets prescriptive requirements on the maximum allowable quantities for each individual control areas based on the area's occupancy type.

Section 60.5 contains the fundamental requirements for hazardous materials, including requirements for personnel training, ignition sources, systems, equipment, and processes (Section 60.5.1.4, 60.5.1.5, and 60.5.1.6 respectively). In Massachusetts, Section 60.5 includes multiple requirements that come from both NFPA 1 and the MA Amendments to NFPA 1. Despite Category 1 processes having a vessel capacity much less than Category 2-5 processes (Category 1 processes have a vessel capacity ≤ 2.5 gallons), there are numerous NFPA and state fire code requirements that must be followed. Therefore, the importance of fire protection engineers and their understanding of the NFPA and state fire codes cannot be understated for processing facilities. Each subsequent category process contains additional fire code fire protection/chemical process safety requirements of increasing complexity.

Category 2: Requirement Details

Category 2 processes must comply with all Category 1 requirements and additionally submit for a permit to the AHJ, as required per 1.12.8.34.2 and 60.8.1.1 of MA Amendments to NFPA 1. Permits submitted to the AHJ must comply with the requirements of Section 1.12 of NFPA 1 – 2015 edition. The nature of the operations and materials at the facility determine the permit requirements. Depending on the nature of the process, the AHJ may require technical assistance to evaluate the adequacy of facility process safety conditions, programs, procedures, and practices undertaken at the facility, but only after a notice of denial has been properly served upon the person making application.

Category 3: Requirement Details

While meeting all the requirements of Category 1 and 2 processes, Category 3 processes must also provide the following documentation and evaluations:

- Compliance with the requirements of Category 1 and 2 processes;
- Category 3 hazard evaluation; - per Section 60.8.2
- Implementation of process safety controls to mitigate hazards of normal and abnormal conditions identified in the hazard evaluation;
- Maintain documentation for review for a minimum of two (2) years;
- Third-party review (if required by the fire department/AHJ);
- Post-Incident Analysis per 60.8.6 (to be initiated within 48 hours of incident);

MA Amendment to NFPA 1, Section 60.8.2.2 provides requirements and details for the hazard evaluation that is required for Category 3 processes. **Hazard evaluations** are meant to identify hazards, including adjacent vessels that contain hazardous materials, and determine the required preventative, protective, and safety control measures in conformance with **r**ecognized **a**nd **g**enerally **a**ccepted **g**ood **e**ngineering (and safe work) **p**ractices (RAGAGEP) associated with a particular process or condition. Section 60.8.2.5 requires that a person with *specialized knowledge* beyond that of an average person, about risk assessment, process hazard analysis, and/or process safety management principles, for the process or processes being evaluated *conduct the necessary evaluations*. Section 60.8.2.5 further emphasizes the importance of both chemical process safety specialists and fire protection engineers for analysis of hazardous processing facilities.

In addition to generating a hazard evaluation, Category 3 processing facilities (vessel capacity ≤ 300 gal OR Group H Occupancy) must also meet the requirements of NFPA 1, Section 60.6, *Emergency Action Planning, Fire Risk Control and Chemical Hazard Requirements for Industrial Processes.* Where quantities of materials in use require compliance with Protection Level 1, Protection Level 2, Protection Level 3, or Protection Level 4 based on materials exceeding the Maximum Allowable Quantity (MAQ) in the following categories shall comply with the requirements of Chapter 7 of NFPA 400, *Hazardous Materials Code*:

(1) Unpackaged organic peroxide formulations that are capable of explosive decomposition in their unpackaged state
(2) Oxidizer Class 3 and Class 4: solids and liquids
(3) Pyrophoric solids, liquids, and gases
(4) Unstable reactive Class 3 and Class 4: solids, liquids, and gases
(5) Highly toxic solids, liquids, and gases
(6) Water-reactive liquids, Class 3

NFPA 400, Chapter 7, *Emergency Planning, Fire Risk Control, and Chemical Hazard Requirements for Industrial Processes,* contains requirements and guidance on topics such as:

- Process review and plan preparation
- Operating and maintenance procedures
- Safety reviews
- Incident investigation plans

Section 7.2.3 contains specific requirements for Emergency Action Planning that facilities must comply with. Some requirements of Section 7.2.3 include:

- Procedures for notifying and coordinating with the fire department and other applicable agencies
- Procedures and schedules for conducting drills
- Alternative measures for occupant safety
- Maintenance of fire protection equipment

Lastly, in lieu of complying with Chapter 60 in its entirety, occupancies containing high hazard Level 1 to high hazard Level 5 contents shall be permitted to comply with Chapter 10 of NFPA 400, as stated in NFPA 1, Section 60.7. MA Amendments to NFPA 1 further states that complying with NFPA 1, Section 60.7, is subject to an independent review in accordance with Section 1.15 and a copy, including its recommendations, shall be submitted to the Building Official.

Category 4: Requirement Details

Category 4 processing facilities (vessel capacity ≥ 300 gal) must meet:

All the requirements of Category 1, 2, and 3 processes

And also create and maintain a **Limited Safety Program (PSM "lite")**. The Limited Safety Program documents evaluation, policy and/or required procedures to ensure compliance with the following requirements of the MA Amendment to NFPA 1:

(1) Process information including, but not limited to, MSDS for the chemicals and products being processed, process chemistry, piping and instrumentation diagram, safety relief design, process control safety alarms and interlocks;

(2) Facility suitability including, but not limited to, the *Building Code* compliance, electrical hazard classification, ventilation design, fire alarm and fire protection, spill containment and control;

(3) A process hazard safety analysis including, but not limited to, effects in the event of failure, suitable administrative and engineering controls to minimize failure and to control unanticipated releases, and emergency responses to safeguard life and property;

(4) Written procedures, including routine operating and maintenance, as well as precautionary, shut-down and emergency response measures;

(5) A written training program for operating and maintenance personnel and outside contractors whose work or activity may affect process safety;

(6) A written records management protocol, which tracks any changes, including but not limited to changes to chemicals, equipment, operating procedures training program. Such records shall include the date of such change and the name of the manager responsible for such change; and an internal review at a maximum every three years.

These requirements are similar to those found in OSHA 29 CFR 1910.119, *Process Safety Management of Highly Hazardous Chemicals*.

Category 5: Requirement Details

Category 5 processing facilities must provide evaluation and documentation per 29 CFR 1910.119 OSHA, *Process Safety Management of Highly Hazardous Chemicals*, and 40 CFR Part 68 EPA, *Risk Management Program Rule*, all while meeting the requirements for Categories 1-4. OSHA's *Process Safety Management* (PSM) *of Highly Hazardous Chemicals* was issued in February 1992 and contains 14 elements that focus mainly on worker safety and catastrophic loss of containment. The Process Hazard Analysis (PHA) is the "heart" of the standard and is the most time consuming. The PHA must be completed every five (5) years and *is required to be completed by someone knowledgeable with the PHA methodology*. Additionally, compliance audits for the standard must be completed every three (3) years and may be completed in-house or by a third party (preferred). EPA's *Risk Management Program* (RMP) *Rule* consists of 12 elements and is focused on public safety and the environment. The program contains Chemical Accident Prevention Provisions (CAPP) and Accidental Release Prevention/Risk Management Plan Rule. The RMP must be submitted to the EPA every fire (5) years or if there are any changes.

Conclusion

Some fire code requirements for chemical process safety apply to processes with quantities much lower than the PSM/RPM thresholds. Failure to comply with these requirements can result in fire department citations, fines, remedial work and in extreme cases, shutdown of operations. Between 2013 and 2018, 214 OSHA PSM violations were recorded in Massachusetts alone, which resulted in over $485,000 in penalties. There are far more non-PSM facilities than PSM facilities in MA, so the magnitude of the problem can be seen by simple extrapolation. Failing to comply with the fire code and the federal requirements can also result in significant property loss, injuries and fatalities with much higher costs than the violation penalties. Compliance with the local fire code and federal regulations should be determined by experienced chemical process safety specialists and fire protection engineers to minimize the risk of significant loss.

Fire protection engineering and science is its own discipline and chemical process safety is its own discipline usually practiced by the chemical engineering field. Finding ways to create awareness and bridge the gap between the two disciplines, and sharing chemical process safety management concepts with those in the fire protection arena, will likely improve the oversight, understanding and enforcement of process safety resulting in better overall safety. There may be a way of joining these disciplines into one overarching safety category and therefore giving a better sense of how interrelated they really are.

546769 Poly-Butadiene Formation in Olefins Debutanizer Condenser

Abstract Title : Poly-butadiene formation in Olefins Debutanizer condenser

Discipline : Process Safety

Topic : Poly-butadiene formation in Debutanizer condenser

Author Name : Ali Bazroun

Participating team: Fahad AlShammari, and Eldridge Monette, Bertrand, Da, Cruz, Fahad AlSaad, Abdullah Al-Ismail

ABSTRACT

In one of Sabic olefins plant Plant, the condenser of the debutanizer tower was fouling. This was evident through the pressure increase in the tower and the cooling water control valve reaching the maximum opening (69%) limited by the mechanical stopper. The condenser process service is butadiene, which has the tendency of polymerizing and formation of popcorn, which could plug relief piping or accumulate in the condenser until the vessel shell is over stressed causing loss of containment and leading to fire or explosion.

The risk assessment team evaluated that the risk as significant, meaning that it could result into fatality and multiple lost time injury with a likelihood of 10E-3. With the propagating shell deformation, the likelihood could increase to L3 (10E-2) over a longer period. The risk could become C2-L3 and remain Significant. To ensure effective risk mitigation, a major plant modification was executed by constructing a new process line that bypasses the debutanizer tower and sends the material to the flare. The project execution was remarkable as it was completed in about a week. 15 tons of C4 is being flared per. This, however, allowed the olefins unit to mitigate the risk effectively without any interruption of the production. The change was not done before in the local industry, and it reflected how effective measures can be taken to mitigate risk through strong leadership and excellent engineering. Interim mitigations, moreover, were in place until the modification was completed: for example, inhibitor injection in the exchanger to inhibit the popcorn formation as soon as possible, and an emergency procedure in case loss of containment. Without the effective mitigation, Sabic operation team would have had to shutdown the olefins unit for over 2 months. This olefins unit feeds two poly ethylene that would shutdown consequently, causing significant losses over the mentioned period.

After isolating the exchanger, borescope test proved the formation of polymer. A root cause analysis was initiated, along with cleaning the tower and fabricating a new exchanger. The details of the RCA will be shared in the full paper.

19 AIChE Spring Meeting
15TH GLOBAL CONGRESS ON PROCESS SAFETY
NEW ORLEANS • 2019

Industrial process plant safety by collaboration of a Safety Instrumented System and a Dynamic Simulator

Tadaaki Ando
Yokogawa Corporation of America
12530 West Airport Blvd, Sugar Land, TX 77478, USA
Tadaaki.Ando@us.yokogawa.com
(Tadaaki.Andou@jp.yokogawa.com)

Yasuhiko Yamashiro
Yokogawa Electric Corporation
2-9-32 Nakacho, Musashino-shi, Tokyo, 180-8750 Japan
Yasuhiko.Yamashiro@jp.yokogawa.com

Prepared for Presentation at
American Institute of Chemical Engineers
2019 Spring Meeting and 15th Global Congress on Process Safety
New Orleans, LA
March 31 – April 3, 2019

Industrial process plant safety by collaboration of a Safety Instrumented System and a Dynamic Simulator

Tadaaki Ando
Yokogawa Corporation of America
12530 West Airport Blvd, Sugar Land, TX 77478, USA
Tadaaki.Ando@us.yokogawa.com
(Tadaaki.Andou@jp.yokogawa.com)

Yasuhiko Yamashiro
Yokogawa Electric Corporation

Keywords: Industrial process plant, SIS (Safety Instrumented System), Dynamic Simulator

Abstract

Safer HMI design for operators' quick and appropriate response is very important. Allowed maximum delay for operator's action from the alarm is not same all the cases because it is depends on many factors, i.e. ambient temperature, density of material, O2, and etc. So design of SIS includes HMI should be based "worst case". However, the 'worst case' is difficult to be found.

A "Dynamic Plant Simulator" can precisely simulate behaviors of the real process by visualizing how a hazard is diffused at various stages of a chemical reaction. And combination of DCS/SIS/HMI simulator with "Dynamic Plant Simulator", safety logic design is optimized about trip point and alarm threshold, and HMI design is modified safer manner.

The effective result of application of "cyber model by Dynamic Process Simulator" to a vinyl acetate monomer (VAM) plant and its value for operator training will be discussed.

1 Introduction

Industrial process plants in Japan have been seeking for the highest productivity and optimization of their processes, same with any process plant all over the world. These plants were renovated and expanded as time passed for the purpose, the plants themselves became complicated and troublesome as a result. Few years ago, severe process plant incidents occurred in Japan. The most probable causes of these severe incidents are out of "incorrect judgement and operation" and "shortage of prediction for technical aspects" during exceptional manual work that has been done by incidents. Understanding the hazardous situation by risk analysis and designing safety measures utilizing a SIS (Safety Instrumented System) are important and effective for resolving "incorrect judgement and operation" and "shortage of prediction of technical aspect". Safety measure includes human intervention via SIS man-machine interface which showing alarm, event, and process data.

The safety measure and its function are developed on the desk by brainstorming or so-called "hazard and operability study (HAZOP)." The verification of the safety function hardware could be established by reliability based analysis, SIL calculation in other word. However, the verification of safety logic is the issue because it has only been verified "on the desk". Especially the determination of the criteria for the emergency shutdown is the most important, but it is not possible to be confirmed with "try and error" on a running plant.

2 Safe HMI Design

Safer HMI design with considering different conditions and situations in the real process for operators' quick and appropriate response is very important. To design it, knowing the behavior of the process itself and related factors which affect the process is necessary.

For example, a reaction speed varies on the ambient temperature, density of material, O_2, catalyst, and etc., so allowed maximum delay for operator's action from the alarm is not same all the cases. So design of SIS includes HMI should consider "worst case". However, finding the worst-case scenario with how much is the allowed operator reaction is difficult because of many factors.

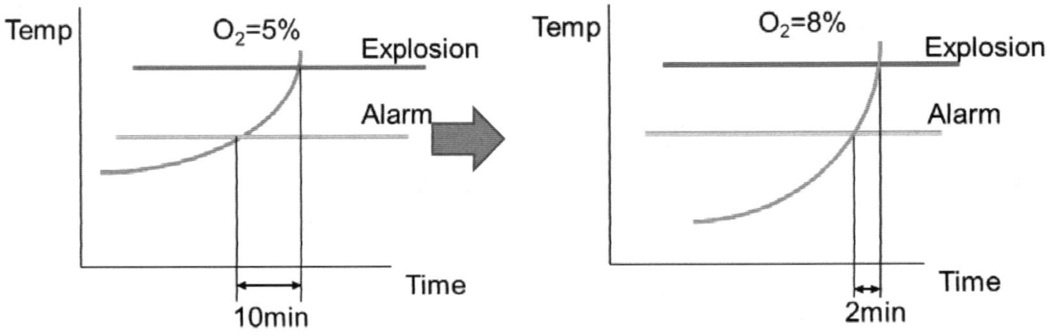

Fig.1 Example: Difference of O_2 density vs. allowed reaction time

Another example is the 'thermal runaway'. Even if feeding of material, heat, and catalyst were stopped by ESD because of a high temperature of a reactor, reaction continues due to a feed-forward loop of heating, means reaction generates heat to continue the reaction itself.

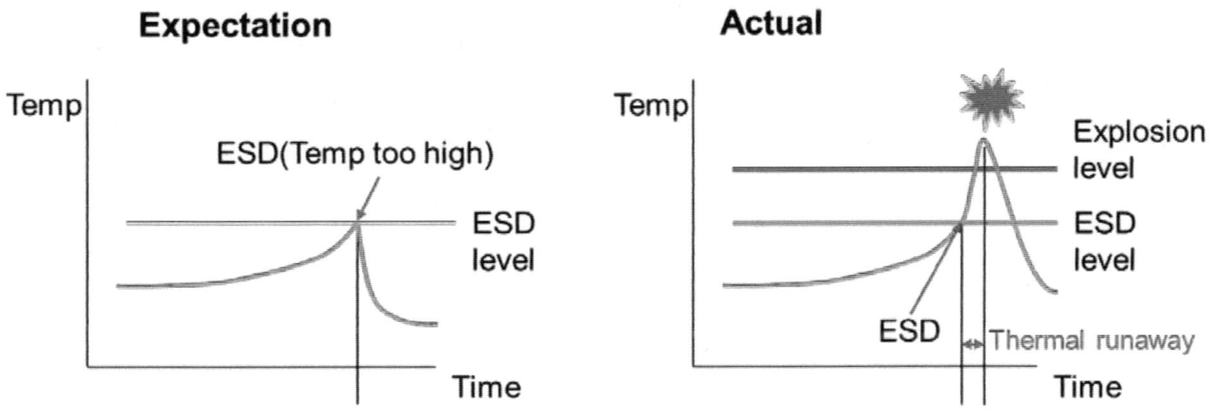

Fig.2 Even after ESD, thermal runaway leads the process to explosion

Thus, considered 'worst case' could not be a 'worst', or could be 'too much pessimistic'.

A "Dynamic Plant Simulator" can precisely simulate behaviors of the real process by visualizing how a hazard is diffused at various stages of a chemical reaction. And combination of DCS/SIS/HMI simulator with "Dynamic Plant Simulator", safety logic design is optimized about trip point and alarm threshold, and HMI design is modified safer manner.

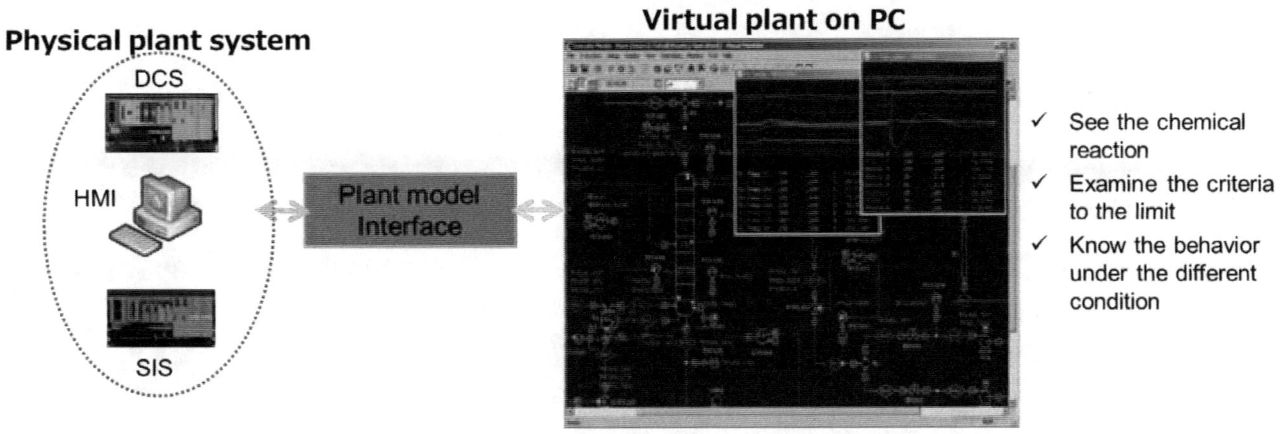

Fig.3 Dynamic Plant Simulator

3 "Dynamic Plant Simulator" at VAM (Vinyl Acetate Monomer) process

"Dynamic Plant Simulator" with "DCS/SIS/HMI simulator" was applied for several shutdown conditions of VAM plant. "Dynamic Plant Simulator" can simulate a variation of factors like ambient temperature and thermal runaway, because is based on a theory of chemical reaction. And "DCS/SIS/HMI simulator" is copying a real function and operation of DCS/SIS/HMI.

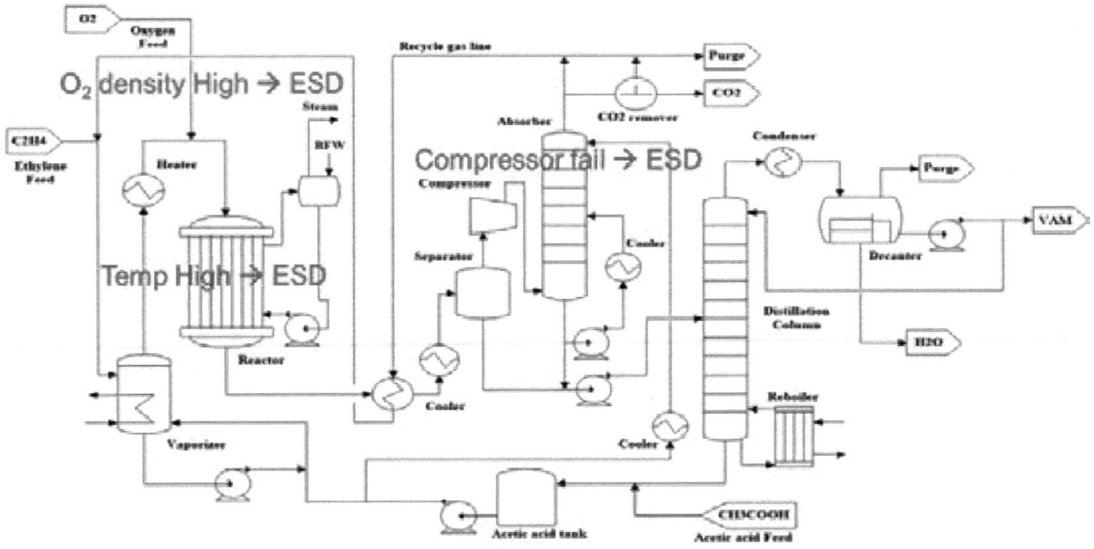

(Main Reaction) $C_2H_4 + CH_3COOH + 1/2O_2 \rightarrow CH_2{=}CHOCOCH_3 + H_2O$

(Side Reaction) $C_2H_4 + 3O_2 \rightarrow 2CO_2 + 2H_2O$

Fig.4 VAM process model

3.1 Shutdown by high temperature to avoid a damage to reactor

Reactor temperature must be monitored and controlled lower than 200 Degrees C (392 Degrees Fahrenheit) to avoid a damage of the reactor. If the temperature reached to a set point for emergency shutdown, emergency shutdown action stops feeding of Ethylene and O_2.

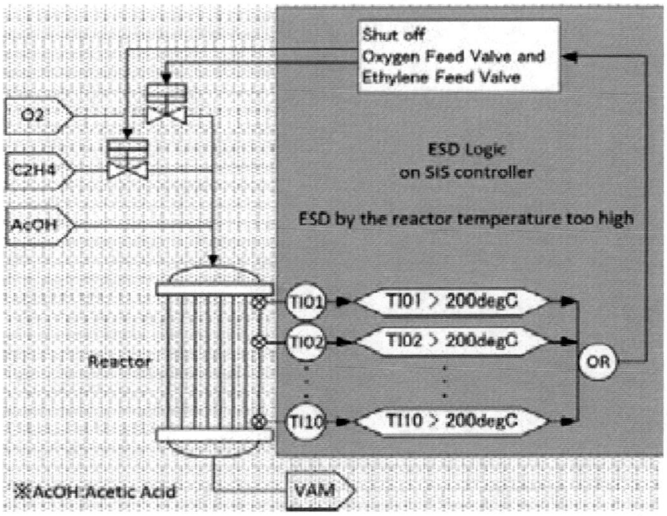

Fig.5 ESD logic for high temperature shutdown

When 190 Degrees C (374 Degrees Fahrenheit) is selected as the set point to shutdown, the result of simulation is, the reactor temperature must become over 200 Degrees C due to a runaway reaction in the reactor. It means reactor must be damaged. When 180 Degrees C (356 Degrees Fahrenheit) is selected, the simulation shows the reactor temperature can be kept lower than 200 Degrees C, so no damage to the reactor.

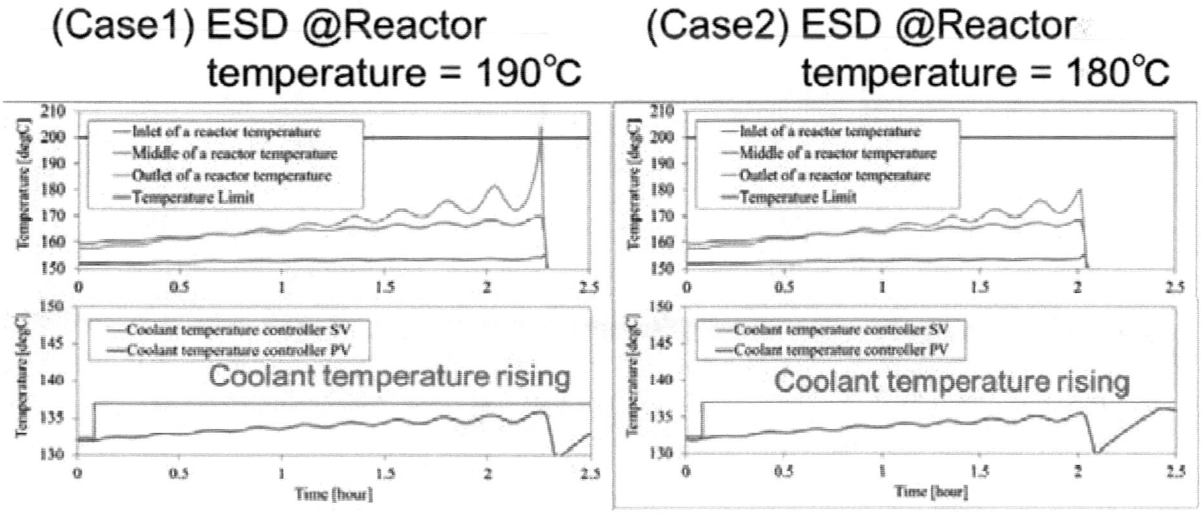

Fig.6 Simulation results

And simulation suggested the reactor temperature must be monitored in multipoint, middle of the reactor and top of the reactor. When the reaction temperature is lower than 160 Degrees C (320 Degrees Fahrenheit), temperature of a middle of the reactor is higher than other points, but when the reaction temperature is higher than 160 Degrees C, the temperature of top of the reactor is higher than other points.

3.2 Shutdown by high O_2 density to avoid explosion

If the density of O_2 in the reactor exceeds 8%, there is a risk of explosion. However, the O_2 density varies depends on other factors, i.e. feed pressure of ethylene to the reactor. Simulation has been done about O_2 density when the feed pressure of ethylene is suddenly changed. Simulation of some cases of dumping the feed pressure to lower side which increase the density of O_2 in the reactor were done.

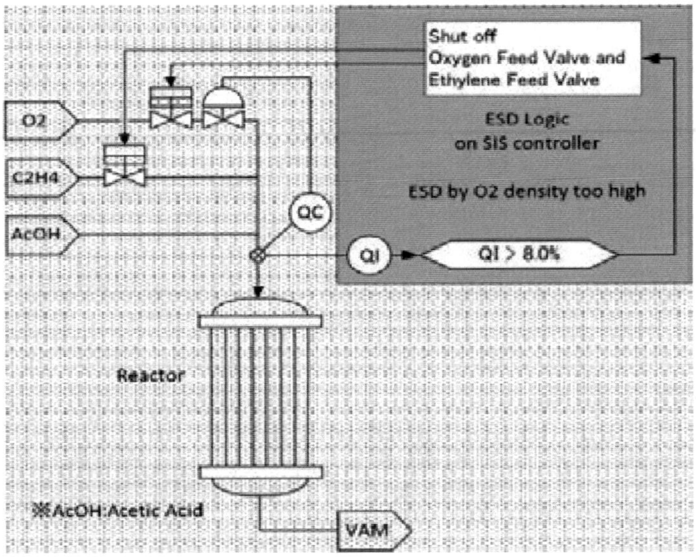

Fig.7 ESD logic for high O_2 density shutdown

When the feed pressure of ethylene dumped 10%, density of O_2 in the reactor doesn't reach to 8%. So there is no risk of explosion. However, the dumping of ethylene feed pressure is 20%, O_2 density will be higher than 8%, then an emergency shutdown must be activated. Shutdown does affect to business.

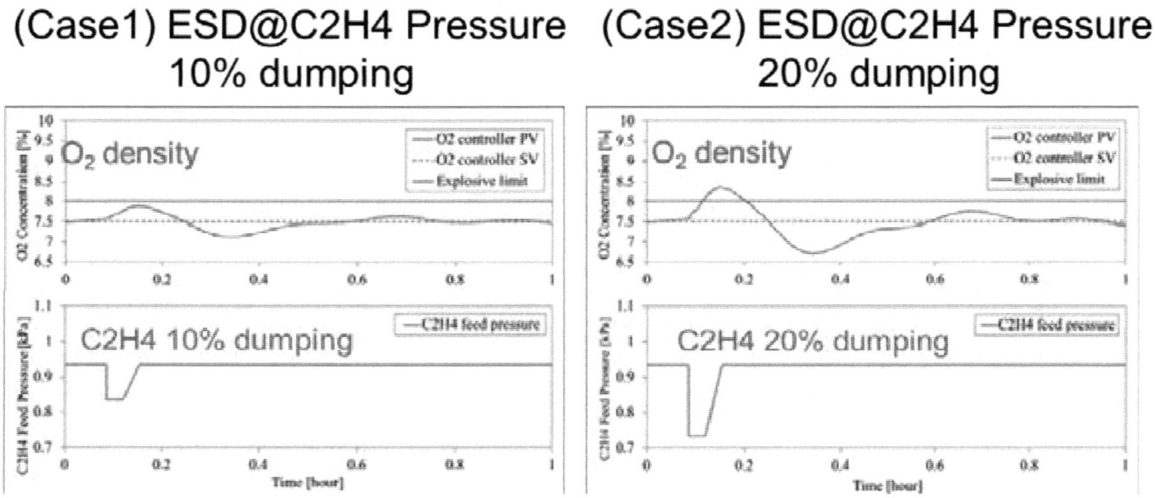

Fig.8 Simulation results

If there is a possibility of dumping ethylene feeding pressure, feeding O_2 in normal operation should be lower to avoid the risk of reactor explosion. On the other hand, lower O_2 density leads lower reaction speed, less production. Finding a best balance with safety and productivity is the key.

Using dynamic process simulator, many factors of process can be manipulated not only within normal operation range, but also close to abnormal, emergency situation or more un-realistic range, and results can be confirmed. This type of "try-and-error" approach is not possible on the real process, but possible in the "cyber" world to find a best balance of setting for maximize production with keeping or improving safety.

4 "Dynamic Plant Simulator" for Operator training

The training of operators is important in many reasons. Especially training for response to critical / emergency situation is indispensable, because operator rarely or never face such a situation and consequent secondary action to DCS/SIS or response to alarm flooding in a real plant. Therefore operator are impossible to familiarize with their required response in right procedure to critical / emergency situation. If operators are not capable or not knowing how to response such a situation, risk must propagate real hazard or accident. Furthermore, operators need to re-start the process after confirming the cause of emergency doesn't exist anymore and checking all the conditions of materials are steady, status of equipment are ready to re-start. This procedures are also not normal for operators.

"Dynamic Plant Simulator" realizes operator training with "equal to real" environment. Operators will be trained for correct and timely manner response until proficient. And operators will have more understanding for process itself by playing on the simulator.

5 Conclusion

The effectiveness of "Dynamic Process Simulator" for safer design of DCS/SIS/HMI is confirmed on VAM plant. The best balance of safety and productivity can be found and optimization of safety action is created by simulation. The created new design of DCS/SIS/HMI is possible to use an actual system directly. And operator training with "Dynamic Process Simulator" will increase a total safety of the plant.

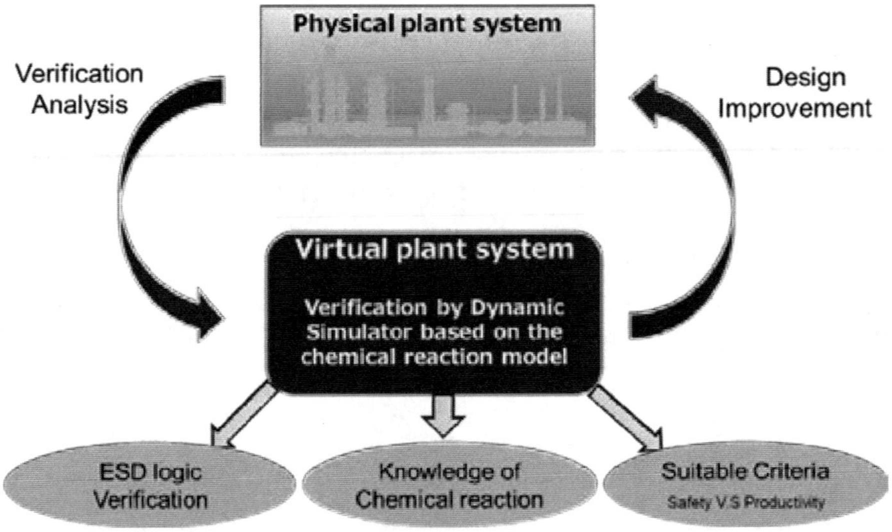

Fig.9:Virtual plant system realizes well balanced plant operation

Analysis of Chemical Effluent Discharge through the Emergency Relief System using Air-Water Properties

Basar Ozar
Jensen Hughes
600 W Fulton St. Suite 500
Chicago, IL 60661
bozar@jensenhughes.com

Jens Conzen
Jensen Hughes
600 W Fulton St. Suite 500
Chicago, IL 60661
jconzen@jensenhughes.com

Damian D. Stefanczyk
Jensen Hughes
3610 Commerce Dr. Suite 817
Baltimore, MD 21227
dstefanczyk@jensenhughes.com

Prepared for Presentation at
American Institute of Chemical Engineers
2019 Spring Meeting and 15th Global Congress on Process Safety
New Orleans, LA
March 31 – April 3, 2019

Analysis of Chemical Effluent Discharge through the Emergency Relief System using Air-Water Properties

Basar Ozar
Jensen Hughes
600 W Fulton St. Suite 500
Chicago, IL 60661
bozar@jensenhughes.com

Jens Conzen
Jensen Hughes
600 W Fulton St. Suite 500
Chicago, IL 60661
jconzen@jensenhughes.com

Damian D. Stefanczyk
Jensen Hughes
3610 Commerce Dr. Suite 817
Baltimore, MD 21227
dstefanczyk@jensenhughes.com

Keywords: two-phase flow, hydrodynamic load, emergency relief system, chemical effluent

Abstract

A sudden opening of a pressure relief device (typically a pressure relief valve or a rupture disk) during an overpressure event will cause the chemical effluent to accelerate through the emergency relief system (ERS) piping. The rapid change of momentum of the effluent creates transient hydrodynamic loads on the piping. ASME B31 in the United States and European Directive 2014/68/EU requires these transient loads to be considered along with other applicable loads (i.e., dead weight, thermal) in the system design. Additionally, the pressure distribution along the piping may be of importance, particularly when other vessels are connected to the same ERS. The back pressures generated on these vessels should be evaluated and compared with the vessels' MAWP (maximum allowable working pressure) as well as the set pressure of any isolation devices. As such, a detailed time history of the effluent pressure and hydrodynamic loads are required to perform a proper evaluation. The effluent may be composed of proprietary chemical mixtures containing various materials. Therefore, the exact material properties of the effluent during the transient might not always be known. In this study, a previously published methodology is used for calculating the dynamics of the effluent and the resultant hydrodynamic loads for a typical ERS by using the RELAP5 (Reactor Excursion and Leak Analysis Program) software. Air-water properties are assumed for the effluent. Key output parameters, such as pressure drop and hydrodynamic loads, are then compared with established hand calculation methods using the actual effluent properties. Given the mechanical nature of the hydrodynamic load event, it was found that the use of air-water properties provides a reasonable representation of the important characteristics of the actual effluent discharge in mechanical design space.

1 Introduction

Pressure build up in a system may occur as a result of a process upset, instrument failure or fire. These upset scenarios can result in either, or both, vaporization due to exothermic runaway reactions or generation of gases due to the decomposition process, which are of particular interest in terms of system and public safety. The extent of the potential hazards caused by such incidents are generally determined by the rate and amount of the energy being generated, the chemical substance makeup, and the quantity of the effluent.

The vessels or systems, which contain such chemicals (e.g. chemical reactors or storage tanks), are typically protected by an ERS that prevents from over-pressurization of the equipment as a consequence of the vaporization of chemicals and/or generation of gases. The process for designing of an ERS was previously discussed by Conzen and Wood (Ref. [1]) and Ozar (Ref. [2]). The design process consists of a process hazard analysis (PHA), defining credible and limiting upset scenarios, adiabatic calorimetry testing, vent sizing calculations, fluid transient analysis, pipe stress analysis, and a dispersion analysis. The central task in the design process is the fluid transient analysis (Refs. [1] and [2]), which is the main focus of this study.

The ERS is typically equipped with either a rupture disk or a pressure relief valve (PRV). Some systems may have both pieces of equipment. The sudden opening of the relief device during an overpressure event will typically result in a two-phase discharge and cause the chemical effluent to accelerate through the ERS piping. The rapid change of momentum of the effluent creates transient hydrodynamic loads on the piping. ASME B31 (Ref. [3]) in the United States and European Directive 2014/68/EU (Ref. [4]) requires these transient loads to be considered along with other applicable loads (i.e., dead weight, thermal) in the system design. Additionally, the pressure distribution along the piping may be of importance, particularly when other vessels are connected to the same ERS. The back pressures generated on these vessels should be evaluated and compared with the vessels' MAWP as well as the set pressure of any isolation devices. As such, a detailed time history of the effluent pressure and hydrodynamic loads are required to perform a proper evaluation.

The two-phase behavior of the chemical effluent has been studied extensively. Following earlier developments by Epstein et al. (Ref. [5]) and Grolmes and Leung (Ref. [6]), a simplified correlation was suggested by Leung (Ref. [7]) to predict flashing choked flow in a frictionless duct or nozzle. The correlating parameter, ω, was based on an isenthalpic (constant enthalpy) assumption.

Leung and Grolmes (Ref. [8]) integrated the momentum equation for homogenous two-phase flow in a horizontal duct using "ω" as a closure relationship for two-phase mixture specific density. Leung and Epstein (Ref. [9]) extended the derivation to an inclined duct configuration. Leung (Ref. [10]) provided the derivation of the reaction forces during a two-phase discharge from ERS piping.

Computational methods have also been used by various industries to investigate two-phase flow behavior in pipes and the hydrodynamic loads due to a sudden acceleration or deceleration of the fluid and the propagation of pressure waves. RELAP5 is a tool for analyzing small-break loss of coolant accidents and system transients in nuclear power plants. The program has the capability to model one dimensional single and two-phase fluid flow and heat transfer problems. The software has undergone extensive review and validation by the US Nuclear Regulatory Commission (Ref. [11]). It has also been benchmarked (Refs. [12], [13] and [14]) for predicting pressure wave propagation and hydrodynamic loads in piping systems. Refs. [15] and [16] provide a discussion about some of the nuclear power plant applications. However, RELAP5 is limited to non-condensable gas-water and steam-water two-phase flow applications.

This study provides a discussion about using the analytical methods developed by Leung and Grolmes (Ref. [8]) and Leung (Ref. [10]) to demonstrate the similarity between the two-phase mixture of common industrial chemical effluents and air/steam-water two-phase flow. Computational and analytical calculations for three different discharge scenarios are also provided. The comparisons between the computational and analytical results are used to demonstrate the applicability of using air/steam-water properties to simulate chemical effluent discharge to reasonably predict the hydrodynamic loads and pressure distribution in the ERS following the actuation of a pressure relief device.

2 Theoretical Background

The governing steady-state conservation equations for one-dimensional homogenous two-phase flow in a horizontal duct with a constant flow area are (Ref. [8]):

Mass

$$G = \text{constant} \tag{1}$$

Momentum

$$vdP + G^2\left(vdv + \frac{4fv^2dL}{2D}\right) = 0 \tag{2}$$

Energy

$$h_0 = h + \frac{G^2v^2}{2} \tag{3}$$

G, v, P, f, L, D and h are the mass flux, specific volume, pressure, two-phase friction factor, length of the duct, hydraulic diameter of the duct, and specific enthalpy, respectively. The specific enthalpy and specific volume are expressed in terms of quality (x) and phasic properties, i.e., $h_0 = h_{f0} + x_o h_{fg0}$, $h = h_f + xh_{fg}$, and $v = v_f + xv_{fg}$. Subscript 0 denotes stagnation conditions and fg the difference between vapor and liquid properties.

Epstein et al. (Ref. [5]) and Grolmes and Leung (Ref. [6]) suggested a simple relation between the two-phase specific volume and pressure as given by:

$$\frac{v}{v_0} - 1 = \omega\left(\frac{P_0}{P} - 1\right), \tag{4}$$

where ω is a non-dimensional parameter. Leung (Ref. [7]) correlated ω for flashing choked flow in a frictionless duct or nozzle. The correlating parameter was shown to be in the following simplified form:

$$\omega = \frac{x_0 v_{fg0}}{v_0} + \frac{C_{f0} T_0 P_0}{v_0} \left(\frac{v_{fg0}}{h_{fg0}}\right)^2 \tag{5}$$

C_{f0} is the heat capacity of the liquid. The last term on the right-hand side is zero for non-flashing gas-liquid two-phase flow and the non-dimensional parameter, ω, is provided by:

$$\omega = \frac{x_0 v_{fg0}}{v_0} \tag{6}$$

Leung and Grolmes (Ref. [8]) considered a two-phase horizontal discharge from a large reservoir (Figure 1). They integrated Equation 2 over the length, L, by substituting Equations 4 and 5 and using the isenthalpic (adiabatic and reversible) flow approximation. The momentum equations after integration yields to:

$$N \equiv 4f\frac{L}{D} = \frac{2}{G^{*2}}\left[\frac{\eta_1 - \eta_2}{1-\omega} + \frac{\omega}{(1-\omega)^2} ln\left[\frac{(1-\omega)\eta_2 + \omega}{(1-\omega)\eta_1 + \omega}\right]\right] - 2ln\left[\frac{(1-\omega)\eta_2 + \omega}{(1-\omega)\eta_1 + \omega}\left(\frac{\eta_1}{\eta_2}\right)\right] \tag{7}$$

The non-dimensional parameters are defined by:

$$\eta_1 \equiv \frac{P_1}{P_0}, \eta_2 \equiv \frac{P_2}{P_0}, \eta_3 \equiv \frac{P_3}{P_0}, \; G^* \equiv G/\sqrt{\frac{P_0}{v_0}}, N \equiv 4f\frac{L}{D} \tag{8}$$

η, $G^*$and f are the dimensionless pressure, dimensionless mass flux and two-phase friction factor, respectively. Leung and Grolmes (Ref. [8]) recommended a range between 0.003 and 0.005 for the friction factor.

G^* and η_1 are related via Bernoulli's equation for compressible fluid (Ref. [8]):

$$G^* = \frac{\sqrt{-2[\omega ln\eta_1 + (1-\eta_1)\omega - (1-\eta_1)]}}{\omega\left(\frac{1-\eta_1}{\eta_1}\right) + 1} \tag{9}$$

The exit condition can be either subcritical, in which case $\eta_2 = \eta_3$, or critical where the local choking statement would require (Leung and Grolmes, Ref. [8]):

$$G^* = \frac{\eta_2}{\sqrt{\omega}} \tag{10}$$

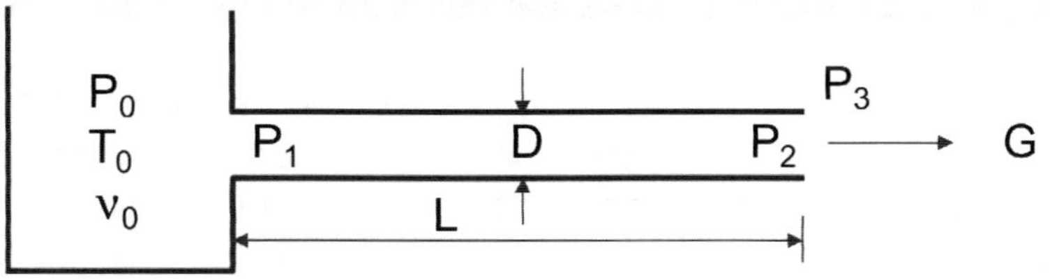

Figure 1. Two-phase horizontal discharge from a large reservoir.

With $\eta_1 = \eta_3 = \eta$, the above two equations satisfy the following transcendental equation (Leung and Grolmes, Ref. [8]):

$$\eta^2 + (\omega^2 - 2\omega)(1-\eta)^2 + 2\omega^2 ln\eta + 2\omega^2(1-\eta) = 0 \tag{11}$$

Similarly, the generalized thrust expression (T_R) is provided by Leung (Ref. [10]):

$$\frac{T_R}{P_0 A_e} = \frac{G^2}{P_0 \rho_0}\left(\frac{v_2}{v_0}\right) + \left(\frac{P_2}{P_0} - \frac{P_3}{P_0}\right) \tag{12}$$

The first term represents the fluid momentum expulsion rate and the second term represents the exit plane pressure (imbalance) force. For the case of unchoked exit, the exit pressure imbalance term vanishes, and the trust is only dependent on the exit momentum rate (Ref. [10]):

$$\frac{T_R}{P_0 A_e} = \frac{2[-\omega ln\eta_2 - (\omega-1)(1-\eta_2)]}{\omega\left(\frac{1}{\eta_2}-1\right)+1} \tag{13}$$

This approximate analysis shows that for two-phase flow in ducts (e.g. piping systems) ω is the sole independent variable and hence is the scaling parameter.

3 Transient Fluid Analysis using RELAP5

3.1 Description of the benchmark scenarios

A chemical reactor with a volume of 2.7 m^3 is considered for the benchmark scenarios. The reactor vessel is rated for 3 bars-g. The reactor is equipped with an emergency relief nozzle and a rupture disk of 41.1 mm inner diameter (DN40, 3.6 mm thickness). The nozzle is located near the top of the vessel. The piping downstream of the rupture disk has an inner diameter of 80.9 mm (4 mm thickness) and is 20 m long. It is a straight and horizontal section of piping.

Three different relief scenarios are used for the benchmarks. A summary of the postulated scenarios is provided below:

- Scenario 1: Polyurethane (PUR) and toluene diisocyanate mixture at 60 °C.
- Scenario 2: N-methyl pyrrolidone, formic acid, trimellitic anhydride and methylene diphenyl diisocyanate (MDI) at 90 °C.
- Scenario 3: Acetone under-fire scenario with vessel initially filled to 20% of its volume.

The chemical mixtures for Scenarios 1 and 2 are gas (CO_2, carbon dioxide) generating reactions. Alternatively, the scenarios are also called gassy reactions. A foamy mixture is expected inside the reactor vessel for these scenarios. Therefore, the chemical effluent discharging through the ERS is a two-phase gas-liquid mixture.

Scenario 3 is the boiling of a solvent due to a fire load applied on the vessel. The level of the boiling mixture is low, and liquid acetone is not entrained into the ERS. Only acetone vapor discharges through the ERS for this scenario.

3.2 Description of the computer model

The calculation was performed using the RELAP5/MOD3.3 (Patch 05) computer code, which is a light water reactor transient analysis code developed for the United States Nuclear Regulatory Commission (US NRC). It is suitable for the simulation of a wide variety of hydraulic and thermal transients in both nuclear and nonnuclear systems involving mixtures of steam, water, non-condensable gases and solutes under single phase and two-phase conditions (Ref. [17]).

A RELAP5 model was developed for the simulation of the venting scenarios from the reactor vessel into the ambient. The model was used to determine the hydrodynamic loads resulting from the sudden opening of the rupture disk. The mass flow rate of the two-phase effluent and the pressure distribution along the ERS piping were also calculated. Figure 2 provides an overview of the nodalization diagram of the model.

Time dependent volumes (TDVs) were used for the pressure boundary conditions and the thermo-physical condition of the fluid at those boundaries. Pipe components (P) were used to model the piping. Single junctions (SJs) connected the components such as pipes with pipes or pipes with TDVs. The valve junctions (V) was similar to the SJs but with the additional ability of the junction to open and close.

The length-to-diameter ratios (L/D) were set close to 1.0 for each pipe node and the time step was selected to be below the acoustic Courant limit. This is recommended for simulations involving propagation of acoustic waves and more generally for models where rapid fluid momentum changes are anticipated. The rupture disc was assumed to open linearly over a period of 10 msec, which is consistent with typical data (Ref. [18]). A steady-state calculation was performed for the first 5 seconds of each simulation to initialize the model. The transient portion of the simulation was initiated at 5 seconds, when each rupture disc started to open and was fully open at 5.01 sec. No heat transfer through the pipe walls was considered for the hydrodynamic fluid analysis. This is often a reasonable assumption for industrial processes due to the short duration of the relief event.

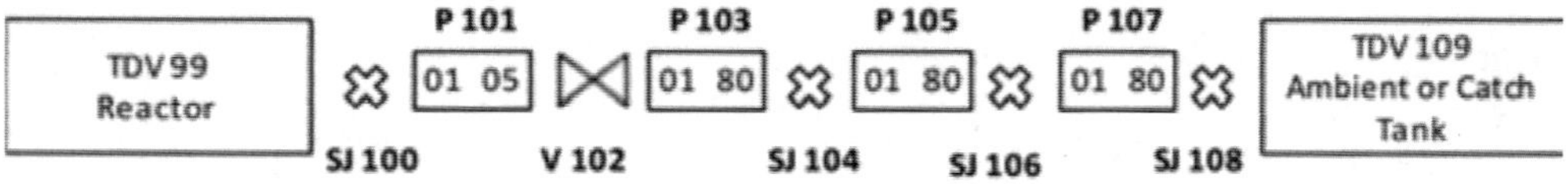

Figure 2. Nodalization diagram for the RELAP5 model.

The hydrodynamic loads (also commonly referred to as fluid force) on the modeled piping were computed through command files by using the RELAP5 output data. The generalized force equation in one-dimensional form can be resolved by a pipe segment bounded by two elbows as shown in Equation (14) (Ref. [19]). Note that the sign convention for the force is considered positive in the nominal direction of the flow as defined within the physical system.

$$F_W = -\frac{\mathrm{d}}{\mathrm{dt}} \int \rho A V dx \tag{14}$$

F_W is the hydrodynamic load, ρ is the fluid density, A if the inner flow area of the pipe, V is the fluid velocity, and x is the distance along the pipe. For pipe segments with an open end an additional force component needs to be considered, which is regarded as the thrust load. Examples of open ended pipe segments are pipes that connect to pressure vessels, large manifolds, or a catch tank. The total thrust load is calculated as shown in Equation 15, which is made up of the momentum flux (due to fluid momentum expulsion) and pressure terms:

$$F_T = -(P_E - P_A)A - \dot{m}_E V_E, \qquad (15)$$

where P_E, P_A, $\dot{m}_E$ and V_E are the pressure at the pipe exit, pressure of the reservoir the open piping is connected to, and the mass flow rate and fluid velocity through the pipe exit, respectively.

3.3 Boundary conditions for the RELAP5 model

A summary of the upstream boundary (reactor vessel) conditions for the RELAP5 models of the three scenarios are provided in Table 1. 10% over-pressure (10% greater than the set pressure) was used for Scenarios 1 and 2, which is typical for relief system sizing of reactive chemicals. 21% over-pressure was used for Scenario 3 per API 520's (Ref. [20]) recommendation for under-fire scenarios. Two-phase gas liquid discharge was assumed for Scenario 1 and 2, where foamy mixtures inside the reactor vessel were established. Single-phase gas discharge was assumed for Scenario 3. The gas-liquid mixture of chemical effluent was simulated by using the air-water properties. The quality values of the air-water mixtures were adjusted so that the ω value of the gas-liquid mixture of the chemical and that of the air-water mixture would match. The quality and the ω value of chemical mixtures were determined based on the initial fill ratio and the homogenous two-phase mixture assumptions. A single-phase acetone vapor release scenario (Scenario 3) was simulated by using air. The determination of the choking condition at the rupture disk was made based on the analytical model, which was described in Section 2. The downstream boundary conditions were set to be at typical atmospheric conditions (air at 20 °C and 1 bar-a).

Table 1. Upstream Boundary Conditions for the RELAP5 Model

Upstream Boundary Conditions	**Scenario 1**	**Scenario 2**	**Scenario 3**
Pressure	4.4 bar-a	4.4 bar-a	4.8 bar-a
Flow regime	Two-phase gas-liquid	Two-phase gas-liquid	Single phase vapor
Simulant	Air-water	Air-water	Air
ω	0.085	0.336	Not calculated
Choking at the burst disk	No	Yes	Yes

4 Results and discussion

The time history of the mass flow rate at the rupture disk calculated using RELAP5 for Scenario 1 is provided in Figure 3. The flow quickly accelerates following the opening of the rupture disk. The effluent experiences the frictional and form losses only for the sections of the piping, which it has already occupied at a given time. Therefore, the transient mass flow rate is slightly greater than the steady-state mass flow rate while the effluent is filling up the ERS piping. Once the entire ERS piping is filled, the mass flow rate achieves a steady-state value.

Figure 4 shows the pressure-time histories downstream of the rupture disk and at the end of the ERS piping for Scenario 1. The pressure downstream of the rupture disk gradually increases to its steady-state value as the effluent fills the piping. The pressure at the exit of the piping remains near atmospheric pressure (downstream boundary condition).

Figure 5 shows the dynamic load calculated on the ERS piping. The total load and its constituents are provided. There is a spike for the wave load as the rupture disk opens. This spike is caused by the depressurization wave traveling from the rupture disk to the reactor vessel. Typically, this section of the piping is filled with gas or vapor for which the speed of sound is well known. Therefore, the initial spike is captured accurately and the speed of sound of the two-phase mixture (which is typically unknown) does not have a significant role on the results. The wave load continues to be the dominant mechanism of the load as the effluent continues to fill up the ERS piping. Once the piping is filled up, the flow momentum term replaces the wave load. The flow momentum is due to the discharge of the effluent from the ERS piping into the ambient or a catch tank and it is the first term on the right-hand side of Equation 15. The pressure force is zero since the exit of the ERS piping is at the same pressure as the ambient; there is not choking at the exit of the piping.

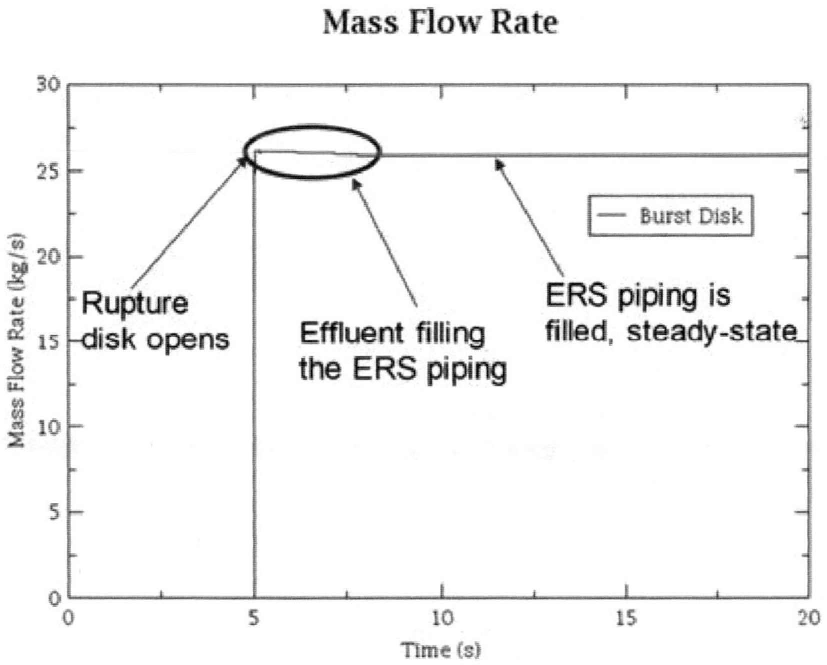

Figure 3. Mass flow rate for Scenario 1.

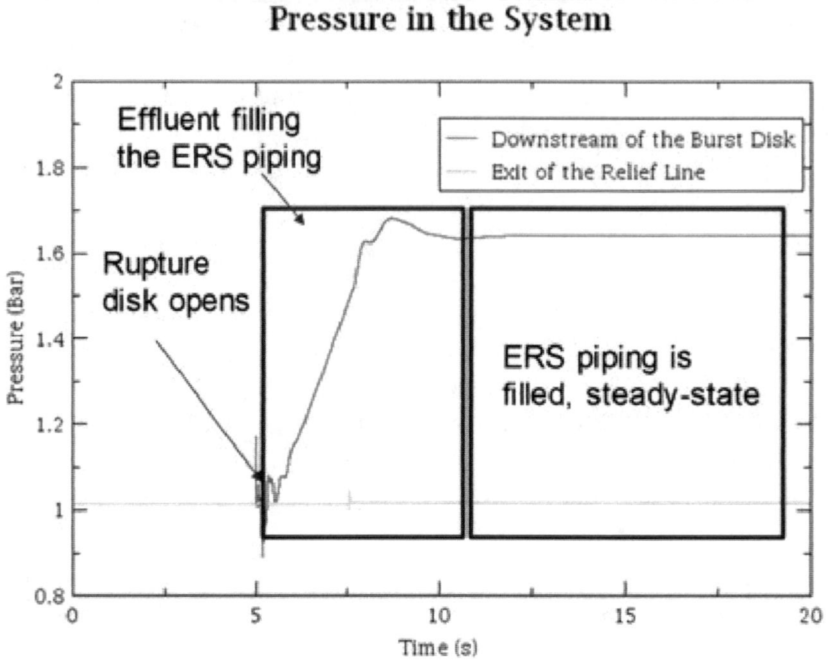

Figure 4. Pressure in the ERS piping for Scenario 1.

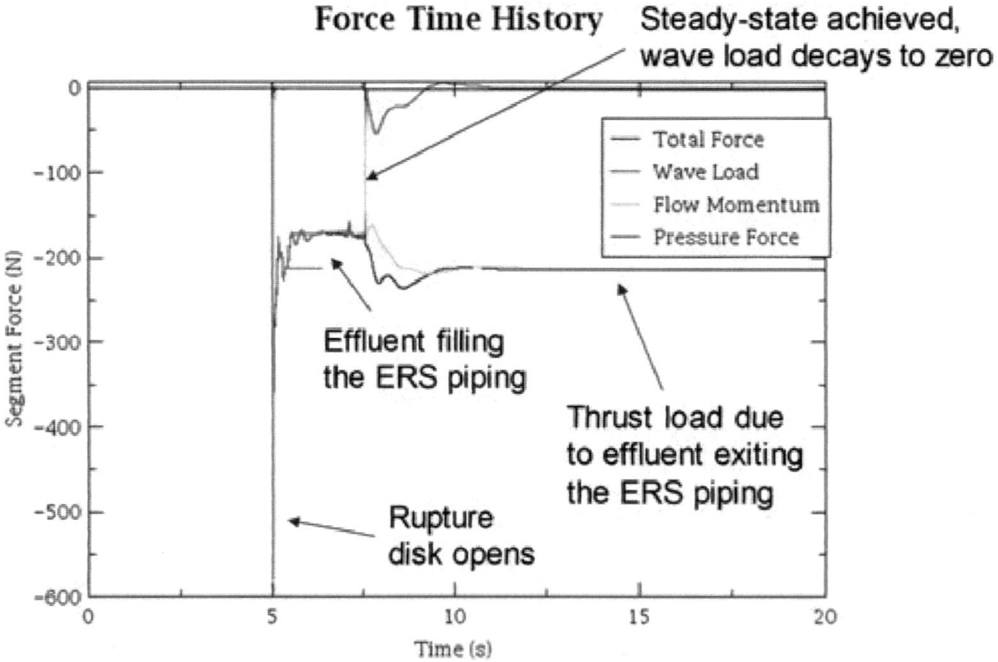

Figure 5. Dynamic loads for Scenario 1.

It is interesting to note that the wave load (while the ERS piping is filling up) and the load due to flow momentum (at steady-state) are almost similar. This is explained using the mathematical relations for the wave and thrust loads. The wave load for an effluent while filling up an empty piping (using Equation 14) is mathematically same as the flow momentum term in Equation 15.

The steady-state pressure, mass flow and thrust load values were compared between RELAP5 calculations and the analytical model, which was provided in Section 2. Actual chemical properties were used for the theoretical model. The comparisons between the two methods are provided in Table 2. The results demonstrate that the two-phase chemical effluent can be simulated with air/steam-water properties as long as the ω values are matched for these mixtures.

Table 2. Comparison between calculations using air-water (RELAP5) and actual two-phase chemical properties

Parameter	Scenario 1	Scenario 2	Scenario 3
Pressure downstream of the burst disk (bar-a)	1.6\1.7	1.6\1.7	1.9\1.6
Mass flow rate (kg/s)	25.9\26.1	16.3\16.4	1.2\1.2
Steady-state thrust load (N)	212.5\191.4	176.3\199.9	219.3\235.0

Black: RELAP5 results   Red: Theoretical model

In the chemical manufacturing industries, the effluent can be composed of proprietary chemical mixtures containing various materials. Therefore, it can be challenging to know the exact material properties at any given time during an emergency relief event. The method presented in this study requires only the stagnation properties of the chemicals at the venting conditions for predicting key parameters, such as the pressure distribution and wave loads during the transient. These properties are typically established by adiabatic calorimetric testing or similar characterization of the chemicals. The method presented in this study may also be used with other commercially available software, which can predict transient fluid dynamics and wave loads.

5 Conclusions

In this study, a previously published methodology is used for calculating the dynamics of the effluent and the resultant hydrodynamic loads for a typical ERS by using the RELAP5 software and assuming air-water properties for the effluent. Key output parameters, such as pressure drop and hydrodynamic loads, are then compared with established analytical methods using the actual effluent properties. The following conclusions were observed:

- The fluid transient analysis is a key aspect of the design of the emergency relief systems. It provides the wave loads required for the dynamic pipe stress analysis. It also provides key parameters such as the pressure distribution, mass and volumetric flow rates through the system, and wave loads. These key parameters are often used to determine back pressure effects on other vessels, for chemical dispersion analysis and dynamic pipe stress analysis.

- The analytical approach shows that there is a similarity between the air/steam-water and the gas/vapor-liquid chemical mixtures if the "ω" values of the two mixtures are matched. This allows for an emergency relief event to be simulated with air/steam-water properties.

- Similar pressure distribution in the piping system and loads can be achieved through the similitude of "ω". Therefore, "ω" of the air/steam-water mixture should be matched with the "ω" of the chemical effluent at the stagnation conditions. The quality of the air/steam-water mixture can be adjusted until a similar "ω" is achieved.

- The sonic velocity of the effluent is typically not known. Also, it is practically difficult to match the sonic velocity of the air/steam-water mixture with the sonic velocity of the chemical effluent, which is a major requirement for tracking pressure waves and wave loads. This study demonstrates that the emergency relief scenarios are pipe filling events, which only require the momentum flux to be calculated for a reasonable prediction of the initial wave load. Therefore, the sonic velocity of the effluent is not required to capture the initial wave load.

- The effluent can be composed of proprietary chemical mixtures containing various materials. Therefore, it can be challenging to know the exact material properties at any given time during the transient. The method presented in this study may be used with most commercially available software and air/steam-water properties for a reasonable representation of the important characteristics of the actual effluent discharge in mechanical design space.

- The study needs to be extended further to identify applicability and limiting conditions for this method.

6 References

[1] Conzen, J., Wood. R.G., (2015), "Modeling and Computation of Reaction Forces on Relief Piping During Depressurization", AIChE 11th Global Congress of Process Safety.

[2] Ozar, B., (2018), "Analysis of Chemical Effluent Discharge through the Emergency Relief System using Stream-Water Properties," DIERS Fall Meeting, 15-17 October, Houston, USA.

[3] ASME. *ASME Code for Pressure Piping, B31.* The American Society of Mechanical Engineers, New York, NY, 2016.

[4] "Directive 2014/68/EU of the European Parliament and of the Council," Official Journal of the European Communities, May 2014.

[5] Epstein, M., Henry, R.E., Midvidy, W., Pauls, R., (1983), "One-dimensional Modeling of Two-phase Jet Expansion and Impingement," Thermal-hydraulics of Nuclear Reactors, 2nd International Topical Meeting Nuclear Reactor Thermal-hydraulics, Santa Barbara, CA.

[6] Grolmes, M.A., Leung, J.C., (1984), "Scaling Considerations for Two-phase Critical Flow," Multi-phase Flow Heat Transfer III, Part A: Fundamentals, T.N. Veziroglu and A.E. Bergles, eds., 549, Elsevier, Amsterdam.

[7] Leung, J.C., (1986), "A Generalized Correlation for One-Component Homogenous Equilibrium Flashing Choked Flow," American Institute of Chemical Engineers, Vol. 23, pp. 1743-1746, October.

[8] Leung, J.C., Grolmes, M.A., (1987), "The Discharge of Two-phase Flashing Flow in a Horizontal Duct," American Institute of Chemical Engineers Journal, Vol. (33), p. 524.

[9] Leung J.C., Epstein, M., (1990) "The Discharge of Two-phase Flashing Flow from an Inclined Duct," Transactions of the ASME, Vol. 112, pp. 524-527, May.

[10] Leung, J.C., (1992), "Reaction Forces During Two-phase Discharges," American Society of Mechanical Engineers Journal of Fluids Engineering, Vol. (114), p. 689-692.

[11] Information Systems Laboratories, (2016), "RELAP5/MOD3.3 (Patch03) code manual volume I to VIII, NUREG/CR-5535/Rev P5-Vol I to VIII", prepared for USNRC.

[12] International Agreement Report, (2007), "Simulation of the Propagation of Pressure Waves in Piping Systems with RELAP5/MOD 3.2.2: Comparison of Computed and Measured Results", TÜV SÜD - NUREG/IA-0206, June.

[13] Sokolowski, L., Koszela, Z., (2012), "RELAP5 Capability to Predict Pressure Wave Propagation Phenomena in Single- and Two-Phase Flow Conditions", Journal of Power Technologies 92 (3), page 150 – 165.

[14] Miller, J.S., Ramsden, K.B., (2005), "Validation of a Thermal Hydraulic Computer Code to Perform Two-Phase Multi-Component Force Calculations for Structural Evaluations", Proceedings of 13th International Conference on Nuclear Engineering (ICONE13), 16-20 May 2005, Beijing, China.

[15] Ozar, B., Henry, C.E., Harvill, R., Norton, D.A., (2012), "Analysis for Low Pressure Cooling Injection System Suction Hydrodynamics for a Boiling Water Reactor," Proceedings of 20th International Conference on Nuclear Engineering (ICONE20), July 30-August 3, 2012, Anaheim, CA.

[16] Henry, C.E., Lim, J., Ozar, B., (2012), "Investigation of Hydrodynamic Loads Associated with Pyrotechnic Valve Actuation," Proceedings of 20th International Conference on Nuclear Engineering (ICONE20), July 30-August 3, 2012, Anaheim, CA.

[17] Riemke, R.A., Davis, C.B., Schultz, R.R., (2009), "Recent Hydrodynamics Improvements to the RELAP5-3D Code", Proceedings of the 17th International Conference on Nuclear Engineering, ICONE17-75380, July 12-16.

[18] Health and Safety Executive, "Offshore Technology Report 2002/023: Testing and Analysis of Relief Device Opening Times," Pipeline Simulation and Integrity Ltd., Yorkshire/UK.

[19] Moody, F.J., (1990), "Introduction to Unsteady Thermofluid Mechanics", John Wiley and Sons, New York, Section 7.4, p. 365-366.

[20] API Standard 520, (2008), "Sizing, Selection, and Installation of Pressure-relieving Devices in Refineries," 8th edition.

547437 Evaluation of the Thermal Stability of Styrene in Contact with Impurities

Safe handling and storage of reactive monomers have long been an issue for polymer manufactures. The unsaturated double bond may self-react to initiate unintended polymerization reactions under many scenarios, for example, completely depletion of inhibitors, exposure to fire or external heat, and in contact with impurities. Usually, a trace amount of incompatible species is able to trigger unintended reactions and lead to catastrophic consequences. Thus, understanding how the impurities promote self-reaction is expected to provide valuable guidance to safe monomer-handling practices. Our goal in this study is to quantify the contamination effects of some most common contamination chemicals on styrene self-polymerization. Thermal analysis tools including DSC and ARSST are applied to examine the exothermic characteristics of styrene mixed with contaminating substances at different concentration levels and mixing ratios. From the preliminary results, we observed the onset of self-polymerization shifted to a lower temperature when in contacted with water and also alkaline. A trace amount of concentrated strong acid initiated unwanted exothermic activity even at room temperature. Self-heating rate was also elevated when diluted acid was present. Additional research is needed to understand what is the impact in of other types and concentrations of contaminants such as rust, metal ions. Our study demonstrates significant implications in the prevention of runaway incidents during transportation and storage of styrene.

549038 A Mixed-Integer Programming Framework for Placement of Fire Detectors in Chemical Processing Facilities

In chemical processing plants, gas and fire detectors provide an important layer of protection for personnel. However, given the large variety of chemical and physical sensors and number of factors influencing their capabilities to successfully detect hazards, the decision of where to place sensors is challenging. Gas detectors (e.g. ultrasonic, catalytic bead) must be in contact with gas leaks to be effective, while fire detectors (e.g. point infrared, open path infrared, optical) require a visual path to the fire, free of obstructions, for successful detection. Numerical optimization techniques provide an opportunity to overcome these challenges and determine effective sensor placement.

The optimal gas detector placement problem was formulated by [1] as a stochastic program based on the p-median facility location problem and relied on scenarios generated from CFD simulations within a petrochemical facility. Their results showed that coupling coverage with expected detection time using a scenario-based method outperformed the existing coverage-only approaches and has inspired formulations based on MINLP approaches for non-uniform failure probabilities [2].

But while gas detectors must be placed in the path of gas leaks for effective detection, fire detectors must be placed within unobstructed visual distance of flames. The optimal placement of visual fire detectors in processing plants can be particularly difficult due to the large number of physical obstructions (e.g. piping, valves, tanks).

This work focuses on an optimization formulation for placement of fire (visual) detectors. The scenario-based, p-median approach demonstrated by [1] for gas detectors is translated to a maximum coverage model for fire detectors in a 3D space with consideration to failure probabilities. We show that this approach allows for the placement of fire detectors for a fixed sensor budget while under uncertainty. We demonstrate the effectiveness of this formulation on several case studies. We further discuss how the framework is readily scalable for larger problem sizes and amenable to various objectives.

We would like to thank the Purdue Process Safety and Assurance Center for their support in this work.

References:

[1] Legg, S. W., Benavides-Serrano, A. J., Siirola, J. D., Watson, J. P., Davis, S. G., Bratteteig, A., and Laird, C. D. (2012). A stochastic programming approach for gas detector placement using CFD-based dispersion simulations. Computers and Chemical Engineering, 47:194–201.

[2] Liu, J., and Laird, C.D., A Global Stochastic Programming Approach for the Optimal Placement of Gas Detectors with Nonuniform Unavailabilities, to appear in Journal of Loss Prevention in the Process Industries, 2017.

546855 Advance Alarm Management Applied to DCS Improvement in a Petrochemical Plant: A Case of Study

Alarms use in process industry is fundamental for avoiding process accidents in which people, the environment or process equipment could be damage. Currently, Carbon Black production plants are equipped with distributed control systems (DCS) for setting alarms values and to provide indication of abnormal events to the control rooms operators. However, the basic control system could not meet the needs against the different production conditions and the presence of improper alarms could send a bad signal to control rooms and it could be the start of a process safety event.

In the present study, an alarm system assessment was performed in a petrochemical plant, located at Mamonal industrial park, in Cartagena – Colombia; based on data analysis, it was observed that when process conditions changes occured, the DCS was not be able to adjust alarms settings automatically, according to the load fluctuation and/or process conditions. In order to solve this problem, in first place the DCS performance was studied and it was reported an alarms list, which provided information regarding to the alarm flooding phenomena, alarm chattering and others alarm system issues.

On the other hand, the information earned and alarm rationalization methodology were used for establishing an action plan in order to improve DCS performance. Results showed the utility of these tools and it was identified the need to perform betters in the alarm system logic to avoid flooding alarm phenomena and to make operations safer.

548361 Overcoming Motivational Biases in Hazops

Hazard and Operability Studies (HAZOPs) are one method of assessing the risk in a facility. Based on the risk ranking of a scenario, recommendations may be required to reduce the risk. Often, these recommendations require additional capital expenditure and/or operational resources. In aging plants where resources are not readily available or in greenfield projects where there is a tight timeline to start up, there is sometimes a bias to risk rank scenarios such that there is no need to add additional safeguards. Motivational biases can outweigh common sense such that scenarios are not ranked according to their real risk. We have developed a method to overcome some of these motivational biases with the use of a polling system to engage all participants in the HAZOP. Participants are asked to vote anonymously on the severity and likelihood for a given scenario, then the results are discussed. During this discussion time, arguments for all the different risk rankings are discussed, including those voted on by the minority in the room. This facilitates an open discourse about the real risks for a given scenario rather than shifting the attention to the most vocal person in the room, who may be strongly influenced by a desire to reduce spending or minimize the effort required for maintenance. Following this discussion a second poll is performed to reach a consensus. We have found this to be an effective method for reducing motivational biases and ensuring that the right decisions are made based on the most logical argument instead of based on budget or time constraints. Tracking polling results increases the accountability of participants and makes them less likely to skew the risk ranking. The requirement for consensus ensures that all opinions are heard and considered, even if they are not popular. Together these make the polling system an effective method for overcoming motivational biases in HAZOPs and ensuring that safe decisions are made.

546515 Integrating Culture into Process Safety Assurance

Purpose: PETRONAS Chemicals Group Berhad (PCG) has embarked on a group-wide culture transformation, introducing the concept of the Results Pyramid® and a Culture of Accountability® i.e. a workplace culture where people take accountability to think and act in the manner necessary to achieve results.

Methodology: This paper will discuss two such activities anchoring on the Results Pyramid® that have been deployed across several plants and units within PCG.

The first activity is an enhancement of existing assurance programmes, where culture subject matter experts are assigned to observe plant leadership teams and front-liners during interviews and site verifications conducted by the technical assessors. The observations are synthesised into belief statements, which are presented together with the technical findings.

The second activity is Power Up, a session designed to incorporate powerful and specific feedback exchange between various teams, culminating in a specific list of actions to improve delivery of each team's activities in order to achieve the shared key results. This workshop can be done either as a follow-up to an assurance activity, in preparation for a project/turnaround activity or as a part of a periodic performance review.

Results: Examples of observations, feedback exchanges and specific actions from these activities will be shared, together with the impact on process safety leading and lagging indicators. Recommended practices and lessons learnt for continuous improvement will also be included.

Conclusion: Applying the Results Pyramid® concept to the design and implementation of Process Safety-related programmes has helped to identify the underlying leadership, mindset and behaviour issues impacting their success. By doing so, more specific interventions can be put in place to ensure full compliance to safety requirements and maximise effectiveness of these programmes.

548442 Emergency Plans Integration By Diferent Companies

In Brazil, different companies with pollution potential must have an emergency plan. However, each company analyzes its risks and contingency measures separately, based on the Conama 398 law. With the establishment of the Area Plan, companies from the same territory with pollution potential caused by oil in waters began to analyze how to combat a major accident quickly and effective together. This work shows how the integration of the individual contingency plans of each company and the cooperative result of the area plan in territory of significant scope and considerable environmental sensitivity.

No Brasil, diferentes empresas com potencial de poluição devem ter um plano de emergência. No entanto cada empresa analisa seus riscos e medidas de contingência separadamente, com base na lei Conama 398. Com a instituição do Plano de Área, empresas do mesmo território com potencial de poluição causadas por óleo em águas passaram a analisar como atender conjuntamente de forma rápida e eficaz a um grande acidente em seu território de abrangência. Este trabalho mostra como foi estruturada e organizada a integração dos planos de emergência individuais de cada empresa e o resultado cooperativo do plano de área em território de abrangência significativa e considerável sensibilidade ambiental.

548111 Moving Back to Flammable Refrigerants - Facility Siting and Personnel Safety Considerations

"Everything old is new again – Paul Allen"

Not just a catchy tune from the 1970's, this jingle summarizes a current technological shift in the world of refrigeration. Historically, refrigeration cycles used toxic and/or flammable materials as the working fluid. As the technology matured and made its way into transportation and consumer applications, the working fluid also evolved. Halogenated hydrocarbons, such as chlorofluorocarbons (CFCs), were developed as replacements to conventional refrigerants such as ammonia and propane. CFCs, which may have toxic effects, are desired for their tunable working range and for being non-flammable. However, the environmental drawbacks to the use of CFCs has been well researched and documented. As we embark into the future, where climate change and environmental quality are at the forefront of concern to consumers and industry alike, we are seeing a move back to flammable refrigerants. Everything old is, indeed, new again.

New construction and expansion projects can certainly consider and plan for the use of flammable refrigerants in commercial and industry facility design and siting, but what about retrofit applications? What are the risks and hazards associated with converting a CFC refrigeration system, designed and sited for the use of non-flammable refrigerants, to a system now utilizing flammable refrigerants? Beyond the need for additional fire protection systems, what risks are posed to the facility and personnel? In this paper, we seek to answer these questions by comparing the hazards and risks of mid-scale refrigeration systems using either non-flammable or flammable refrigerants. We explore how facility siting constraints may change with respect to the use of flammable refrigerants, and perform a quantitative risk assessment (QRA) for an example retrofit project in order to assess the potential for increased risk that should be considered in such projects.

548210 Making Sense of Risk Results

Risk based studies are very useful, but at the same time, the results are extremely complicated. While we have a basic set of results that can be used to illustrate the risks to people, these are relatively simple for risk practitioners and sometimes too complicated for site managers.

FN curves show the risk to all people from all sources, and individual risk contours illustrate the risk from all scenarios to people across specific locations, but how do we translate these results into actions. Generally, the action is to relocate people to lower risk areas, or upgrade buildings to mitigate the impacts identified. However, this is not always feasible. What if the site footprint is limited; there are public receptors who cannot be relocated; or building upgrades are cost-prohibitive or infeasible?

We must better quantify and evaluate the **sources** of risk rather than focusing on mitigating the impacts from the events. We need to evaluate the type of risks, the potential initiating events, and then we can assess the potential methods to effectively mitigate the risks. Existing risk result presentations are often very poor at illustrating the sources of risk. To effectively identify the sources of risk requires that we carefully dissect the risk results, while maintaining their validity and intent.

This paper will focus on how we find these risk drivers and make enough sense of them to create actionable recommendations that protect people.

551205 Ammonia Dispersion Modeling for Ammonia Storage with Multi-Scenario Failure System

PT Pupuk Kalimantan Timur is one of the largest fertilizer companies in Southeast Asia where located in Bontang City, East Kalimantan Province, Indonesia. The total production capacity of PT Pupuk Kalimantan Timur are 2.7 million Tons / year Ammonia, 3.43 million Tons / year Urea and 350 thousand tons / year NPK.

To accommodate ammonia production, PT Pupuk Kalimantan Timur currently operates 3 ammonia tanks with a capacity of 2 x 25,000 tons and 1 x 50,000 tons. In order to manage the hazard risk contained in the ammonia tank, it is necessary to do a modeling to map the potential and ammonia leakage potential areas and the impact of their exposure to the workforce and the community around the factory area.

The results of this modeling are expected to be used by Pupuk Kalimantan Timur's management in carrying out corrective actions for occupational health and safety performance, especially in controlling the potential hazard of ammonia, especially in ammonia tanks.

In this paper modeling is done using the AERMOD® Modeling System software. AERMOD Modeling System was chosen because it can be used on:

1. U.S. EPA regulatory Gaussian plume air dispersion model.
2. For near-field impacts (50 km max).
3. Prefered model for complex terrain.
4. Improved treatment of urban effects.

For modeling, we choose the AS-T-101 ammonia tank which has the largest capacity, which is 50,000 tons. Ammonia distribution models will be simulated based on weather conditions and ammonia leak source scenarios that can occur in the AS-T-101 ammonia storage tank system. Ammonia leakage sources will be displayed in 5 scenarios, i.e:

1. 1 Vent opens
2. 1 PSV poping
3. 2 PSV poping
4. 3 PSV poping
5. 3 PSV and 1 vent poping

The AERMOD Modeling System program is used to make ammonia distribution models for each event scenario assuming an exposure radius of up to 10 km. Based on the results of the calculation and modeling of the potential distribution of ammonia from AS-T-101 using the AERMOD Modeling System, the highest exposure of ammonia is in the 5th scenario, which is up to 36611 μg / M3.

546560 De-Mystifying the Concepts of Hazardous Area Classification

Purpose

The purpose of this topic is to address and clarify major items of confusion associated with hazardous area classification assessments and compliance auditing in classified locations. The ultimate objective is developing documentation that provides a framework for the control of electrical energy ignition sources in environments made hazardous by the potential presence of flammable vapors.

Background

In February of 1992 Occupational Safety and Health Administration (OSHA) 29 CFR 1910.119 Process Safety Management of Highly Hazardous Chemicals became law in the United States and confusion reigned supreme. Over the next several years industry professionals began the process of sorting through the responsibilities, regulations and impending deadlines that had been placed squarely in their laps. Acronyms such as PSM (Process Safety Management), PSI (Process Safety Information), RAGAGEP (Recognized and Generally Accepted Good Engineering Practices), MOC (Management of Change), and PHA (Process Hazards Analysis) became commonplace in the vocabulary of process safety professionals nationwide.

Employers now were required to meet stringent deadlines regarding the creation and management of a plethora of process information associated with equipment in the process covered by this law, i.e. any covered process. This information would be used to support the exhaustive efforts needed to create a baseline PHA for each covered process by May of 1997. In addition OSHA required all PHA's be placed on a five (5) year life cycle requiring their review and revalidation. Along with the revalidation of PHA's came the task of also revalidating all process safety information on the same five (5) year cycle.

As a result, every willing and able bodied employee was thrust into the quagmire of supporting, participating in, facilitating, or conducting PHA's. The required information used to support PHA's included the materials of construction, Piping and Instrument Diagrams (P&ID's), Relief System design and design basis, ventilation system design, material and energy balances, safety shutdown systems, and last but not least **electrical classification**. Yes, "electrical classification" was and still is legally considered an integral part of required process safety information.

This abstract addresses the following and provides an interactive discussion regarding the three (3) major myths surrounding electrical classification:

- assigning a classification to a location
- proper use of the classification information
- keeping the documentation evergreen

  The following topics will be addressed and clarified:

- What Is Hazardous / Electrical Classification?
- The Process of a Hazardous Area Classification Assessment
- The Concept of Electrical vs Hazardous Area Classification
- The Compliance Assessment (where does it fit in the process of controlling electrical energy ignition sources)
- Common Compliance Assessment Issues

546108 Maintaining Safety Excellence By the Implementation of Process Safety Inspection in Badak LNG

Badak LNG as one of the most experienced LNG Producers in the world, has undergone safety culture change throughout its operation. The goal of this change is to achieve excellent safety performances, indicated by the lower rate of Lost Time Injury Frequency (LTIF), lower number of Loss of Primary Containment (LOPC) Tier-1, High Potential Incident and multiple Train trip. This change was triggered by the occurrence of a fatal incident in 2005 as well as lost time injury incident in 2006. Since then, Badak LNG has integrated Process Safety Management (PSM) into its Company Management System, strengthened the roles and responsibilities of Managements, and pursued for interdependent safety behaviour of each personnel. One of PSM implementations is through process safety inspection.

Process safety inspection is an activity to assess the current equipment integrity based on visual observations compared with equipment's standard condition. The standard condition checklist for process safety inspection in Badak LNG is based on Physical Barrier Assessment tools, owned by Det Norske Veritas (DNV). The assessed items are divided into 5 categories: Inherent Safety, Hazard Prevention Measure, Hazard Control, Mitigations, and Support System. The overall score is determined by total number of standard items in all categories compared to the total number of verified items in all categories. The minimum acceptable score is at 80%.

During the year of 2012-2017, the average score of process safety inspections in Badak LNG is 90%. As a result of these inspections, Badak LNG had also identified the need of improvement in some items such as blinds, pipework and flanges, passive fire protection, fire & gas detection system and lock-open/lock-close block valves. In overall, by the implementation of PSM, Badak LNG has aligned on the right path to maintain its safety excellence achievement in LNG business, such as zero LTIF since 2007, zero multiple Train trip since 2012, zero LOPC Tier-1 since 2016, and zero High Potential Incident in 2017.

548411 Using HAZOP/LOPA to Create an Effective Mechanical Integrity Program

Many people view the conduct of a HAZOP/LOPA to address regulatory requirements as a chore, and stop there. However, the implementation of a quality HAZOP/LOPA has the potential to provide a framework for addressing numerous safety and operational optimization issues at plants, including the formulation/refinement of the Mechanical Integrity Program. The purpose of this paper is to focus on the mechanical integrity program, illustrate how a quality HAZOP/LOPA can support the effective implementation of some of the new requirements for California Refineries, and optimize key elements of an effective Mechanical Integrity Program, e.g.:

- Inspection/testing methods

- Testing intervals

- Maintenance outage periods

- Repair prioritization and allowable outage

- Identification of low priority equipment

544678 *"Identifying Critical Process Parameter to Ensure Asset Integrity" or "Pulse of Mechanical Integrity Management System"*

The management of equipment begins with an awareness that ageing is not about how old the equipment is, but is about what is known about its condition and the factors that influence the onset, evolution and mitigation of its degradation. The presentation provides an overview of Integrity Operating Window (IOW) parameters of any process that are key to understand the asset degradation and prevent Process Safety Incidents.

IOW are the situations in the process industries where a process parameters (temp., pressure, dosing rate, conc. etc...) when gets out of operating windows could result in significant loss of primary containment leading to potential severe consequences.

Arrazi - SABIC developed an IOW system that covers IOW identification, associated risk, mitigation (interim – long term), stewardship/management of IOWs and elimination.

The presentation shall cover Insight concept of the Topic, major Process Safety accidents related to breeching IOWs, actual example, IOW Management Process, its influence on PM plans etc.

three key bullet points that we can include in the agenda:

- Knowing operating limits with associated mechanical integrity or process safety risk to ensuring proper actions to be in operating envelope
- Developing the engineering leadership to systematically guide the operators, superintendents and engineers to be in safe zones
- Dynamic inspection frequency based on plant conditions and sharing some industrial major incidents for this critical item

The Practice of Process Safety Analysis and Management System Based on Safety Reliability
(THE NEW TITLE: The PHAMS system in SINOPEC)

Jiang Xue
SINOPEC Research Institute of Safety Engineering,
QingDao,China
Jiangxue.qday@sinopec.com

The PHAMS system in SINOPEC

Jiang Xue
SINOPEC Research Institute of Safety Engineering,
QingDao,China
Jiangxue.qday@sinopec.com

Keywords: Safety Barriers Reliability, PHAMS, HALOPA, SIL, Safety Management Online.

Abstract

The big data technique has been in trend for the process safety assessment. In China, it's the very need to transfer the qualitative safety management to quantitative safety management, which should rely on the professional risk analysis technique, as well as the standard safety reliability database. Based on the above, it's very important to integrate several professional risk analysis technique, as HAZOP, LOPA, SIL etc. to realize the integral risk identification and evaluation, grading and management plan making, supporting by the proprietary SINOPEC risk metrics and grading standards, as well as safety equipment reliability database,which is developed as the Process Safety Analysis and Management System(PHAMS) for SINOPEC, to support all SINOPEC branches to make dynamic risk analysis and management online

1 Introduction

With the smart factory's development, the static, periodic risk assessment and management can not satisfy the identification and real-time safety control of plant's risk events. The PHAMS which establishes the plant-wide dynamic risk analysis, can effectively solve the risk rating and monitoring of the unit and realize the plant-wide or corporate-wide management optimization.

The key data for HALOPA, SIL and Management module in PHAMS is the dependent protection barriers, which we rely on RBI correction, Personnel exposure probability and ignition probability to optimize the six mainly safety barriers' failure probability, simultaneously, do use of the Bayesian Correction Modelling to verify the instant failure probability on site, as well as the SIL identification, which can validly save the grading cost on time and personnel, and all of these are towards to the deeper research on dynamic risk analysis for PHAMS.

2 HALOPA Analysis System

2.1 SINOPEC RISK MARTIX

According to international common matrix criteria, as well as statistics from the SINOPEC accident data and the maintenance data, the frequency of risk occurrence and severity classes are coupled to the 8×7 matrix by a linear exponential model, and be assigned risk index to each risk level.

SINOPEC RISK MARTIX		Occurrence Frequency							
		1	2	3	4	5	6	7	8
		$\leq 10^{-6}$/y	10^{-6}-10^{-5}/y	10^{-5}-10^{-4}/y	10^{-4}-10^{-3}/y	10^{-3}-10^{-2}/y	10^{-2}-10^{-1}/y	10^{-1}-1/y	>1/y
Severity Classes	A	1	1	2	3	5	7	10	15
	B	2	2	3	5	7	10	15	23
	C	2	3	5	7	11	16	23	35
	D	5	8	12	17	25	37	55	81
	E	7	10	15	22	32	46	68	100
	F	10	15	20	30	43	64	94	138
	G	15	20	29	43	63	93	136	200

Figure 1 SINOPEC Risk Matrixp[1]

Based on this matrix, SINOPEC has established four-level risk list for establishing targeted management of branch network. For example, the executive board of the group is in charge of the red-level risk list, the branch directors is in charge of the orange-level risk list, etc.

2.2 HALOPA Analysis Modelling

2.2.1 System risk analysis based on equipment operation cycle

The HALOPA analysis model can be used to identify the hazard of the device or process unit, clarifying the initial event, process deviation, independent protective layer and consequence events, combined with the event tree analysis method to analysis the reduction influence of the occurrence frequency of initial events. The protection layer performs a failure frequency evaluation role to determine the final frequency of the on-site device risk events during each operating cycle under the coupling reduction of the existing protective layers at each level [2].

An important assumption of the model is that each protection layer is independent, that is, the failure of a single protection layer will not affect the risk reduction impact of other existing protection layers [3]. Each time a protective layer takes effect, the frequency of accidents is reduced by the risk reduction of the protective layer.

When the frequency of the initial event does not exceed 1 time/year or is less than twice the test frequency of the first protective layer, the evaluation of the consequence frequency can be referred to the following formula [4]:

$$f_I^C = f_I \times P_{enable} \times \prod_{j=1}^{J} PFD_{Ij} \qquad (1)$$

where

f_I=Occurrence frequency of initial event I annually;

Penable=Moderation factors by enabling conditions and other conditional probabilities.

PFDIj=Failure probability of dependent protection layer j of initial event I.

When the frequency of the initial event exceeds 1 time/year or exceeds the test frequency of the first protective layer twice, the first protection layer failure shall be used as the initial event, and the calculation formula of the consequence frequency is as follows:

$$f_I^C = f_{\mathrm{IPL\_first}} \times P_{enable} \times \prod_{j=2}^{J} PFD_{Ij} \qquad (2)$$

where

$f_{\mathbf{IPL\_first}}$= The failure probability of 1st dependent protection layer

The HALOPA analysis model in PHAMS designed a modified calculation model for equipment failure frequency and personnel exposure probability, and realized BPCS, SIS, physical protection, leakage storage and collection, fire and explosion suppression for equipment operation cycle and device in service. The system's six major types of protection layer failure frequency have the correction factors of risk reduction, and realize system risk assessment based on the device production cycle. For example, according to the system risk assessment of a diesel hydrogenation unit based on the PHAMS system, on-site data collection of various protective layers and key correction conditions under one cycle of the driving operation to date, online preliminary analysis of the risk overall value and grading of the feeding unit The results are as follows:

Figure 2 HALOPA Analytical Function Online

2.2.2 Dynamic risk assessment based on real-time data of safety control systems
According to the protection layer frequency analysis model, the failure probability of the protection layer lowering the frequency of a certain initial event can be regarded as θ , and the a priori risk probability distribution function is:

$$f_0(\theta) \propto \theta^a(1-\theta)^b \quad (3)$$

Its likelihood function $L(E|\theta)$ is

$$L(E|\theta) \propto \theta^m(1-\theta)^{n-m} \quad (4)$$

where

a, b=the prior distribution parameter of the failure probability of the protective layer;
m=The number of the posterior cumulative event occurred when the protective layer is valid;
n=The number of the posterior cumulative event occurred when the protective layer is failed.

With the posterior probability distribution function corrected by dynamic risk analysis as$f_1(\theta|E)$, the posterior probability distribution can be obtained by:

$$f_1(\theta|E) = \frac{f_0(\theta)L(E|\theta)}{\int f_0(\theta)L(E|\theta)d\theta} \propto \theta^a(1-\theta)^b\theta^m(1-\theta)^{n-m} \propto \theta^{a+m}(1-\theta)^{b+n-m} \quad (5)$$

Since m and n are parameters of the age change and the posterior probability distribution function seems to be continuous, the dynamic failure frequency of the protective layer at a certain moment can be obtained by the following formula[4-6].

$$P_i(\theta) = \int_{i-1}^{i} f(\theta)d\theta = (1-F_i(\theta|E)) - (1-F_{i-1}(\theta|E)) = F_{i-1}(\theta|E) - F_i(\theta|E) \quad (6)$$

In where $F_i(\theta|E)$ is $f_i(\theta|E)$ the cumulative failure frequency function as

$$F_i(\theta|E) = \int_0^i f(\theta|E)d\theta \quad (7)$$

And then bring in the following formula to get the dynamic risk frequency of the device or process unit:

$$f_I^C = f_I \times P_{enable} \times \prod_{j=1}^{J} P(\theta)_{Ij} \quad (8)$$

and

$$f_I^C = f_{IPL\_first} \times P_{enable} \times \prod_{j=2}^{J} P(\theta)_{Ij} \quad (9)$$

Where

$P(\theta)_{Ij}$=Posterior failure probability value of the jth independent protection layer

Based on this, the PHAMS system realizes the real-time update of the protection layer risk reduction frequency, and intelligently analysis the overall risk value and grading of the unit to plant.

3 SIL Intelligence Identification Online System

3.1 SIL Identification Modelling

The analysis data of the SIS protection layer of the HALOPA evaluation system in PHAM is updated and interconnected by means of the safety reliability database. After the HALOPA dangerous event using the same safety instrument loop is automatically brought into the SIL identification, the PHAMS system takes the same kind of consequences. C, to calculate without the cumulative frequency F^C before the SIF

$$F^C = \sum_{i=1}^{n} f_i^C \qquad (10)$$

where

$$f_I^C = f_I \times P_{enable} \times \beta \times \prod_{j=1}^{J} PFD_{IJ} \times \prod P_{condition} \qquad (11)$$

$$\beta = \frac{2^N - 1}{N} \qquad (12)$$

where

f_I^C= The occurrence frequency of the dangerous event I with consequence C

β= the common factors caused by several SIFs

N=the number of SIFs for the dangerous event I

Based on the modelling, the PHAMS system can intelligently group common risk events and assess the risk reduction factor RRF of SIF in each group [7] :

$$\text{RRF} \geq \frac{F^C}{TMEL} \qquad (13)$$

and

$$1 \times R_{\text{Factor}} < RRF_{SIL1} \leq 10 \times R_{\text{Factor}} \qquad (14)$$

$$10 \times R_{\text{Factor}} < RRF_{SIL2} \leq 100 \times R_{\text{Factor}} \qquad (15)$$

$$100 \times R_{\text{Factor}} < RRF_{SIL3} \leq 1000 \times R_{\text{Factor}} \qquad (16)$$

where

R_{Factor}= risk reduction factor between1 to 10.

Based on the SIL identification results of each group, the optimal solution is intelligently selected as the RRF of this safety instrument loop.

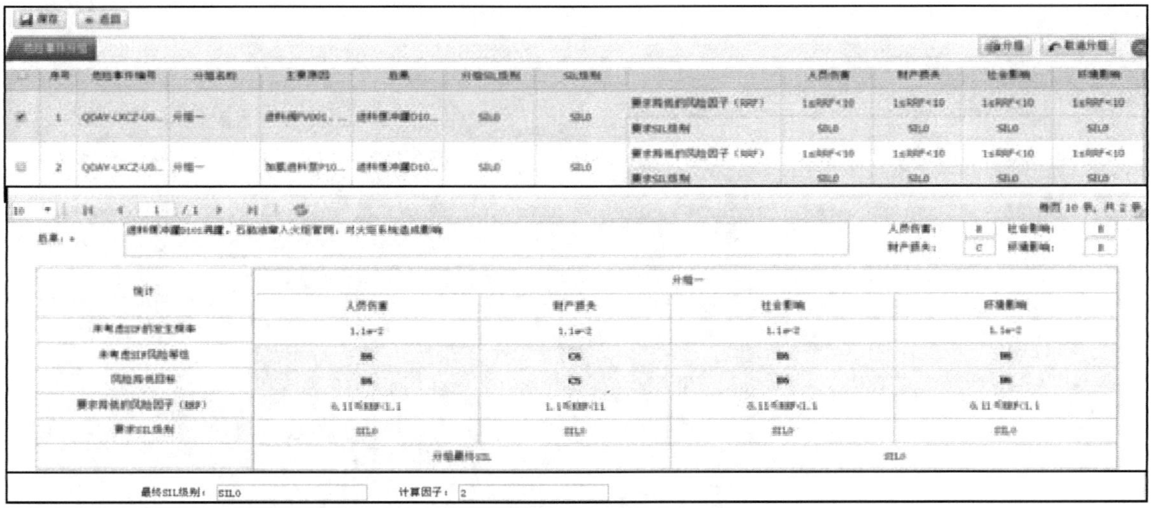

Figure 3 SIL Intelligence Identification Online

3.2 The SIL Identification Based on SINOPEC Risk Matrix in PHAMS

With the aid of data warehouse technology, PHAMS system can compare HALOPA assessment data which involved with SIS protection layers of the initial situation, the consequences classes and non SIS protective layer of data, and automatically take the residual risk of the events into the risk matrix. Then based on SINOPEC risk matrix, the SIL identification module can intelligently couple and assess the residual risk and RRF.

according to Sinopec's safety management focusing on high severity impact, PHAMS sets an independent TMEL requirements (as shown in the following table), and can automatically selects the most serious impact event as the final SIS identification data of this event. .

Table 1 TMEL of SIL Intelligence Identification[8-10]

Severity Classes	*Health & Safety (r/y)*	*Property Damage (r/y)*	*Social Impact (r/y)*
A	$\leq 1\times 10^{-1}$	≤ 1	≤ 1
B	$\leq 1\times 10^{-2}$	$\leq 1\times 10^{-1}$	$\leq 1\times 10^{-1}$
C	$\leq 1\times 10^{-3}$	$\leq 1\times 10^{-2}$	$\leq 1\times 10^{-2}$
D	$\leq 1\times 10^{-5}$	$\leq 1\times 10^{-4}$	$\leq 1\times 10^{-4}$
E	$\leq 1\times 10^{-6}$	$\leq 1\times 10^{-5}$	$\leq 1\times 10^{-5}$
F	$\leq 1\times 10^{-7}$	$\leq 1\times 10^{-6}$	$\leq 1\times 10^{-6}$
G	$\leq 1\times 10^{-7}$	$\leq 1\times 10^{-6}$	$\leq 1\times 10^{-6}$

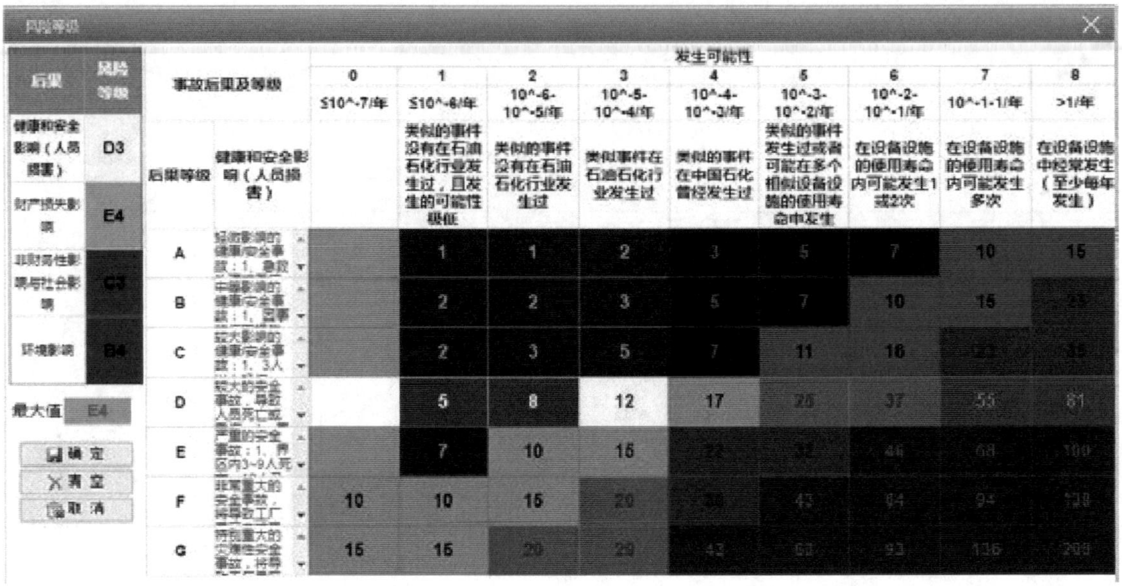

Figure 4 SIL Intelligence Identification on SINOPEC Risk Matrix

On the basis of SIL identification, the PHAMS system can also implement SIL verification based on multi-factor model, and evaluate the structural constraints, PFDavg and MTTF respectively. The verification result data is consistent with the accuracy requirements of IEC61508 and IEC61511.

4 Risk Management and Decision Support System

With the powerful safety and reliability database and expert experience database of PHAMS system, as well as the data mining of the result data of HALOPA and SIL system, the Risk Management and Decision Support System can implement the cumulative risk frequency calculation of independent devices and plants in construction or in service, to assess the overall risk based on different impact severities through the linear comparison [11-12].

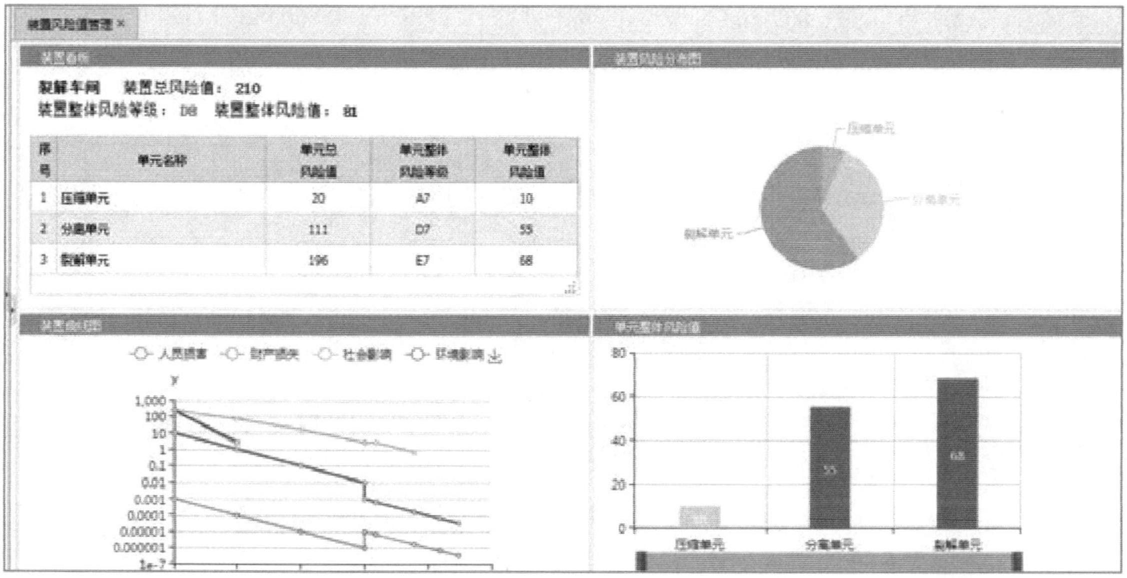

Figure 5 Overall Risk Rating and Risk Event Distribution Analysis of Ethylene Cracking Facility

5 Conclusion

1. PHAMS system sets up 9 categories of safety reliability database, expert experience data and Sinopec risk matrix, etc, which can support unified SINOPEC risk assessment and management standards, risk analysis template, process data storage and data mining for PHAMS system , as well as the interconnection and real-time update of data of each system.
2. HALOPA system implements plant overall risk analysis based on protection layer data, and performs data coupling evaluation from static analysis based on equipment cycle and dynamic risk correction to realize real-time evaluation of risk assessment technology and device operating conditions. The dynamic risk analysis research will be further strengthened to realize the system risk prediction function.
3. The SIL identificaiton and verification system can realize the evaluation of safety instrument function of in service and in construction. In the long run, it helps to improve the safety management level of the plant's overall safety instrument function
4. The risk management and decision support system can realize the overall risk assessment and risk distribution based on various evaluation results, through in-depth mining and statistical analysis, the PHAMS can realize the big data analysis management of Sinopec branches' risk management.

6 References

[1] The Bureau of Safe Production Supervision and Administration of SINOPEC. Guide for Safety Risk Identification and Assessment of SINOPEC[S].2018

[2] AQ/T 3054-2015. Guide for Application of Protective Layer Analysis Methods. Chapter 8： Scenery Frequency Calculation [S].

[3] Khan,Seyed Javad Hashemi,Nicola Paltrinieri,Paul Amyotte,Valerio Cozzani,Genserik Reniers.

Dynamic risk management: a contemporary approach to process safety management[J]. Faisal Current Opinion in Chemical Engineering.

[4] CCPS. Risk-based Process Safety [M]. 2013.

[5] Maryam Kalantarnia,Faisal Khan,Kelly Hawboldt.Dynamic risk assessment using failure assessment and Bayesian theory[J].Journal of Loss Prevention in the Process Industries . 2009 (5)

[6] Rakesh Ranjan,Sonam Singh,Satyanshu K. Upadhyay. A Bayes analysis of a competing risk model based on gamma and exponential failures[J] . Reliability Engineering and System Safety . 2015

[7]Zugravu-Soilita N. How does Foreign Direct Investment Affect Pollution? Toward a Better Understanding of the Direct and Conditional Effects[J]. Environmental & Resource Economics, 2017:1-46.

[8] Khakzad N, Yu H, Paltrinieri N, et al. Chapter 5 – Reactive Approaches of Probability Update Based on Bayesian Methods[J]. Dynamic Risk Analysis in the Chemical & Petroleum Industry, 2016:51-61.

[9] Villa V, Paltrinieri N, Khan F, et al. Towards dynamic risk analysis: A review of the risk assessment approach and its limitations in the chemical process industry[J]. Safety Science, 2016, 89:77-93.

[10] Mark Adrian van Staalduinen,Faisal Khan,Veeresh Gadag. SVAPP methodology: A predictive security vulnerability assessment modeling method[J]. Journal of Loss Prevention in the Process Industries.

[11] Podofillini L, Dang V N. Conventional and dynamic safety analysis: Comparison on a chemical batch reactor[J]. Reliability Engineering & System Safety, 2012, 106(4):146-159.

[12] R A F. Problems with risk analysis in the chemical industry. A detailed examination of the theoretical and practical problems faced by the risk analyst in the study of a chemical plant[J]. Plant/operations Progress, 1983, 2(3):185–190.

549550 Process Hazard Analysis

Process Safety Management System starts with managing the process related risks and from the first step of establishing a new process unit that is needed to evaluate the risks and define the necessary measurements. Process Hazard Analysis methodologies are the systematic tools which help to evaluate the risks in a process. For different needs there are several methodologies like DOW FETI, HAZOP, What if, HAZID, Fault Tree and Event Tree Analysis, Scenario Modelling, Facility Siting etc. Some of these methodologies are semi quantitative some are quantitative methodologies. In order to define and prevent major industrial accidents in our refinery we use DOW FETI, HAZOP, FTA and ETA and Scenario modelling. Moreover for some kind of projects we may also apply What if and HAZID analysis instead of HAZOP depending on the complexity. Facility Siting is another tool for risk related studies which is used to evaluate the risks for buildings in order to protect people inside the buildings. The workflow for a full unit PHA study starts with DOW FETI calculation which is used for risk ranking of al static equipment inside the unit. DOW FETI calculation is done in accordance with the equipment temperature, pressure, hold-up amounts, components and also reaction types if there is any. Then risk assessment process continues with HAZOP studies which is for scenario identification. In HAZOP, first step is chopping the unit into small parts to be evaluated which is called nodes. After that all nodes are evaluated using P&IDs and PFDs by keywords like temperature, pressure, flow, level etc. HAZOP helps to understand what kind of scenarios are possible inside the unit and these critical scenarios are taken through FTA and ETA studies. HAZOP is a semi quantitative risk assessment method where FTA and ETA are quantitative methods. FTA is applied for critical scenarios in order to find out the probability of top event then ETA is then applied in order to find out the probability of scenarios like fire, explosion, toxic dispersion etc. caused by this top event. FTA and ETA analysis all use the statistical PFD and failure rate values of all initiating events and barriers defined in HAZOP for the related scenario. Scenario Modelling is the following step which is used for understanding the effect of the probable scenario. The most probable scenarios coming from ETA is modelled to see the effect zones. These effect zones are used to plan Emergency Response Studies for all these potential major accidents.

548517 Site-Wide Asset Integrity Analytics & Benchmarking

All too often, reliability engineers focus on the equipment that is deemed as most critical or covered by some regulatory reference, while sometimes forgetting about those areas where something as simple as water can cause major issues. Non-destructive testing technicians and API inspectors alike can help facilities gather and compile data to perform predictive analytics and even benchmark sites with similar assets. This data can be used to identify assets where others may be seeing issues. This also allows facilities to compare reliability engineering effectiveness. This presentation will share some sample results of non-destructive testing and inspections of both regulatory required and non-required assets in the fuel ethanol production sector as well as an example of how that data can be assembled and shared across any sector to educate and learn from each other. Afterall, there are no secrets in safety!

548330 Reliability, Availability, and Maintainability (RAM). What Does It Provide?What Should You be Aware of before You Conduct a RAM Study?

Reliability, Availability and Maintainability (RAM) studies can significantly add value to your Process Safety Management (PSM) activities, especially with regards to Mechanical Integrity (MI). RAM studies are an excellent way to determine which equipment limits you from achieving optimum availability and throughput. Using industry failure rate data or even site-specific data can help you identify areas for improvement and determine how best to maintain your equipment. But RAM studies can go beyond purely theoretical studies by using the experience and knowledge gained from operating and maintaining similar facilities to develop a pragmatic approach to solving mechanical integrity-related issues. This paper will cover the pros and cons of site versus industry data and will present a case example of utilizing experience to better define the proposed problems and solutions. If done correctly, RAM studies can be more than statistics; they can be roadmaps to process reliability improvements.

546707 Driving Asset Integrity and Process Safety Consistency across a Multinational Organization

Mexichem is in the process of implementing world-class process safety management (PSM) and Asset Integrity Management (AIM) programs globally for their facilities. This is a major undertaking as Mexichem, to a large extent, has grown through acquisitions. Therefore, each plant has a safety culture and management system unique to that asset. As we all know, it's difficult to drive global consistency when dealing with different types of operations in various regions throughout the world under differing regulatory regimes. This is especially difficult when there is no common understanding of process safety. One of the first steps Mexichem has taken is to develop a "fit-for-purpose" management system with an emphasis on developing a common understanding of process safety and ensuring the culture is robust enough to support the system when it is implemented.

To aid Mexichem in this overwhelming task, Risktec Solutions were engaged to develop and deliver a bespoke set of interactive training workshops on a variety of PSM / AIM topics. The first of which was related to the identification and management of safety critical equipment (SCE). This presentation will outline the plan and methodology Mexichem and Risktec have used to transfer such knowledge and discuss some of the successes and challenges along with way.

548216 Alarm Floods and Their Connection to Plant Incidents and Near Misses

Investigations of many catastrophic plant events have shown alarm flooding to have been significant distractions for the operators just prior to the incident. Alarm floods have also been determined as a cause or contributing factor to smaller incidents and near misses.

In fact, many of the more recent findings by U.S. Chemical Safety and Hazard Investigative Board (CSB) include alarm management as one of the key issues in the investigative reports.

Even after several years of trying, many plants still struggle with controlling alarm floods. Many also have problems generating consistent and accurate alarm metric reports for the entire plant. Now that the ISA 18.2 Alarm Performance Metrics are considered "generally accepted good engineering practices," we should make haste to demonstrate where each of our plants fall (particularly the PSM plants) versus these metrics before the regulating authority moves in this area. If we are not able to prove we meet the metrics (particularly for alarm flooding) under all operating conditions, we should remediate.

Static rationalization can reduce your average number of alarms, but without controlling the alarm floods, there is no help for the operator when he needs it the most. For most, we must begin the process by recognizing that alarm floods occur most of all during a change of process state.

The means of controlling alarm floods has been available to us for many years. Unfortunately, only a select few have implemented the solution and are achieving the desired results today. These plants have achieved equal to or better than ISA 18.2 since before it was published.

This session will cover the justification for true alarm management from the safety and economic perspective with several practical steps that everyone should take if they are not currently meeting the ISA 18.2 specifications during all operating conditions.

548318 Do Incident Data & Lessons Learned Really Facilitate Process Improvement?

Regulations are a set of guidelines for the industry for protecting the people and environment, when it comes to process safety systems. While some companies tend to find that complying with regulatory requirements are always a challenge, on the other hand some tend to go overboard in their bid to stay compliant. In spite of the route taken, there are incidents – major, minor and near misses.

Other than the accumulation of this data, the investigations also result in lessons learned. Over the past four decades, lot of incident data and lessons learned has accumulated, been assimilated, and also analyzed. Complementing that are the audits, system reviews, hazard identification exercises, changes to regulations, and knowledge sharing forums.

It is a very efficient and effective mechanism to leverage these vast sources of current information developed, and implement its learnings, in the operations of existing processes. While the core concepts of safety are universal across the various industrial sectors, its application may differ. But once again, the lessons learned databases are effective sources of making this leveraging process more efficient.

This paper will discuss the some case studies of how the concept of lessons learned and the knowledge of the incident data has been leveraged by the authors in various manners ensuring success from one industry to another – example (1) Learnings from an incident due to gaps in the Emergency Response Plan, bringing about success at another chemical manufacturing site; (2) Incident data from a steel plant, when analyzed and applied to a chemical plant resulted in improved of effectiveness.

546673 Structured & Scrutinized Approach Towards HSE Excellence; A Success Story

Author: Usman Ali, Process Safety Engineer

Co-Author: M Ahsan Sarfraz, Unit Manager HSE

This paper studies the practices & programs that support reporting of Near misses, minor unsafe conditions, incidents and quantitative risk assessment strategies in our organization for analyzing and learning from past events of operational failures, injuries, safety and environmental incidents.

A near miss is a leading indicator to an incident that, if scrutinized appropriately, can prevent injuries and accidents. Some persons may be tempted to write off near misses as "no harm, no damage" situations, but organizations who track near misses, determine how and why they occurred, and take corrective action can prevent similar or more serious incidents from occurring in the future. Fatima Fertilizers Limited believes that, "We don't look at what happened; we look at what could have happened".

Keeping in mind the importance of near miss reporting, FFL initiates drive and implement systems for

1. Capturing and increasing near miss reporting at whole site.
2. Logging data for analysis and using this valuable data for identifying gaps in existing process safety system.
3. Implementing sustainable improvement plans and safety campaigns (Safe hands, Safe Routines, Job Cycle Checks, Safe Drive etc) in our process safety systems for making our site a safer workplace.

These actions has improved the site performance which is reflected as Site has now over 45 Safe Million Man Hours, Total Record-able Injury Rate (TRIR) is below 0.2 since last 03 years and No Loss time injury since last 06 years.

548162 Events Report Database Analysis for SHE/PSM Programs Improvements

Incident databases contain records of events that have occurred in a petrochemical facility during recent years. The most of chemical industries report databases are filled with accident valuable data (frequency and severity). Harnessing data mining to explore incident databases, it was used data mining to analyze the incidents database of the studied plant. Inferent approaches are ineffective for low frequency, high consequence events because of their rarity. Given this information limitation, this work uses database analyses to forecast incident frequencies, their relevant causes, equipment involved, and their possible consequences, in specific petrochemical plants.

Databases information analysis, also help to avoid future accidents, thereby reducing the risk. More specifically, this study presents dynamic analyses of incidents in a petrochemical plant ubicated in Mamonal industrial park, in Cartagena- Colombia. Differences in the chemical-specific predictions with the overall predictions are demonstrated. Furthermore, value theory is used for consequence of rare events by formulating distributions for events over a threshold value. Finally, the accidents number is used to estimate the capital at risk within an industry utilizing the frequency and loss-severity distributions.

In conclusion, it was recommended to update de maintenance program (predictive, preventive and corrective) in cases of abnormal changes in production rates, comparing the current production levels, with the analyzed plant production forecast. Also, it was suggested to improve the report culture, due to gaps founded regarding to events classification.

549483 The Process Safety Analysis and Management System in Sinopec

The big data technique has been in trend for the process safety assessment. In china, it's the very need to transfer the qualitative safety management to quantitative safety management, which should rely on the professional risk analysis technique, as well as the standard safety reliability database. Based on the above, it's very important to integrate several professional risk analysis technique, as HAZOP, LOPA, SIL,BOWTIE etc. to realize the integral risk identification and evaluation, grading and management plan making, supporting by the proprietary SINOPEC risk metrics and grading standards, as well as safety equipment reliability database,which is developed as the Process Safety Analysis and Management System for SINOPEC, to support all SINOPEC branches to make dynamic risk analysis and management online

The key data for HAZOP, SIL and Management module in PHAMS is the dependent protection barriers, which we do focusing on the failure probability, at the same time, we do use of the Bayesian Correction Modelling to verify the failure probability of safety equipent and protection barriers on site, which can validly save the grading cost on time and personnel, all of this is towards to the deeper research on dynamic risk analysis for PHAMS.

Thanks to the PHAMS, we has set up and implemented several standards on HAZOP, LOPA and SIL for SINOPEC.

552170 Analysis and Improvement of Safety Integrity Level Considering Human Factors

Safety Instrumented System (SIS) is an irreplaceable active protection layer that prevents the propagation of process accidents and plays an important role in reducing the risk of accidents. In addition to the sensor response triggering the SIS, a human emergency button is often provided in practice to ensure that the SIS is triggered by personnel in the event of the sensor part failure. Due to human factors added, the emergency button will improve the reliability of the SIS. However, there is no research to study the human reliability of the SIS upon sensor part failure in a continuous operation chemical plant, and how human factors can improve and affect the overall reliability of the SIS still remain undiscovered. In this paper, a verification methodology was proposed which integrated with Cognitive Reliability and Error Analysis Method (CREAM) and Reliability Block Diagram (RBD) to emphasize human reliability analysis of the SIS verification. The operator behavior is subdivided into several steps and the Human Error Possibility (HEP) is calculated by a modified CREAM. Then the RBD method is used to verify the SIS reliability considering human factors. An SIS system of one reacting furnace was taken as a case study using this method to show how human factors can improve the reliability of the SIS. The proposed methodology allows SIS verification considering human factors to improve safety integrity level in real practice thus help stakeholders saving costs and making more decisions on SIS implementation.

Siri, How to Design a Procedure? Integration of Natural Language Processing in Procedure Design

Nilesh Ade
Mary Kay O'Connor Process Safety Center (MKOPSC), Texas A&M University
College Station, Texas 77843-3122, USA
nilesh14@tamu.edu

Noor Quddus, Pranav Kannan, S. Camille Peres, Ahmed Harhara, Trent Parker and M. Sam Mannan
Mary Kay O'Connor Process Safety Center (MKOPSC), Texas A&M University

Prepared for Presentation at
American Institute of Chemical Engineers
2019 Spring Meeting and 15th Global Congress on Process Safety
New Orleans, LA
March 31 – April 3, 2019

Siri, How to Design a Procedure? Integration of Natural Language Processing in Procedure Design

Nilesh Ade
Mary Kay O'Connor Process Safety Center (MKOPSC), Texas A&M University
College Station, Texas 77843-3122, USA
nilesh14@tamu.edu

Noor Quddus, Pranav Kannan, S.Camille Peres, Ahmed Harhara, Trent Parker and M. Sam Mannan
Mary Kay O'Connor Process Safety Center (MKOPSC), Texas A&M University

Keywords: Human Factors, Machine Learning, Procedures, Natural Language Processing

Abstract

Procedures play a vital role in high-risk systems in ensuring that the required level of system performance is achieved while maintaining the associated risks below an acceptable level. An effective procedure has a low level of task complexities *i.e.,* elements contributing to a multiplicity of paths/outcomes, uncertainty in information, interdependency, a multiplicity of instructions/object of instructions and an excess amount of information. As a part of this study, we have developed a Natural Language Processing algorithm that can identify these elements in a procedure. The algorithm was tested for a dataset consisting of 20 procedures and the results were found to be promising.

1 Introduction

1.1 Procedures in the high-risk domain:

In the oil and gas and chemical industry, multiple incidents have been attributed to procedural deficiencies such as a lack of procedural adherence, the presence of outdated procedures and the absence of procedures. A study focusing on heater operations in process industry found that around 14% of operation failures resulted from procedure deficiencies [1]. Similarly, Since the inception of U.S. Chemical Safety and Hazard Investigation Board (CSB) in January 1998, amongst the 815 recommendations made by CSB for approximately 100 investigated incidents, 6% of the suggested recommendations involve procedural deficiency [2]. These incidents include the Formosa Plastics Vinyl chloride explosion, 2004, the Texas City refinery explosion, 2005 and the Deepwater Horizon explosion, 2010 cumulatively leading to a death toll of 31.

The United States Environmental Protection Agency (US EPA) defines Standard Operating Procedures (SOPs) as "a set of written and detailed instructions that document a routine or repetitive activity followed by an organization to achieve uniformity of the performance of a specific function" [3]. In simpler terms, a procedure provides instructions to an operator to perform a specific task/function, generally in a step-by-step manner. An ideal procedure facilitates consistency in the quality of task outcome, addresses task hazards and safety concerns, minimizes opportunities for miscommunication and provides the right level of information to the operator

[3]. Thus, procedures are important for achieving the required levels of performance and safety in high-risk domains. Acknowledging the importance of procedures, the Occupational Safety and Health Administration (OSHA) has included operating procedure as one of the elements of the Process Safety Management (PSM) [4]. The United Kingdom Health and Safety Executive (UK HSE) generally recommends a procedure for all complex tasks [5].

1.2 Task level complexities in procedures:

The relationship between complexity and operator performance has been addressed in multiple studies in the past [6]–[8]. Task complexity has been found to have a profound effect on task performance [9]. This negative impact of task complexity has also been attributed to the knowledge of the operator performing the task [10]. Interestingly, by knowing the complexity associated with a particular task, the performance for that task can be predicted to a certain degree [11]. Task complexity is classified into 3 independent categories; worker level complexity, task level complexity and interaction complexity [12]. The worker level complexity relates to the attributes of the worker/operator (*e.g.,* experience) whereas, the interaction complexity relates to the interaction of the operator with the task (*e.g.,* the familiarity of the operator with the task) [12]. The most relevant type of task complexity in terms of procedure design and analysis is the task level complexity, which relates to the attributes of the task alone while disregarding the external factors of the system [12]. This allows the quantification of complexity through the analysis of the written procedure for a given task [12]. Task level complexity arises due to the presence of a multiplicity of paths/outcomes, uncertainty in information, interdependency, a multiplicity of instructions/objects of instruction and excess amount of information in procedures [12]. These elements can be detected in a written procedure with the help of linguistic identifiers.

1.3 Natural language processing in process safety:

Natural Language Processing (NLP) involves the application of computer science, information engineering and artificial intelligence to process and analyze large amounts of natural language data. NLP has been successfully implemented in various domains for multiple applications. Interactive virtual assistants (*e.g.,* Siri, Alexa, Google assistant), spam filtering in emails and recommendation of products/services for consumers based on their online behavior are some of the well-known applications of NLP. However, the application of NLP in the safety domain is still in a relatively nascent stage. NLP has been applied with unsupervised machine learning for automated analysis of Arizona OSHA injury reports for categorization of each reported incident amongst the 4 categories; falls, struck by objects, electrocutions, and trench collapse [13]. Similarly, NLP has been applied to determine root causes and contributing factors in aviation incidents by processing of aviation safety narratives [14].

In this study, we have implemented NLP for identification of elements contributing to task level complexity. The developed algorithm identifies such elements based on linguistic identifiers. The application of NLP for the described objective ensures consistency, reproducibility and minimizes the associated subjectivity in task complexity identification.

2 Methodology

2.1 Base vocabulary:

The first step in this study was to define the linguistic identifiers for detecting elements that contribute to task complexity. These linguistic identifiers were based on the 5 types of task complexity *i.e.,* judgment, decision, interdependency, step-size, and step-information [12]. Judgment complexity arises in a step of a procedure when there is uncertainty in the information provided and the operator must counter this uncertainty based on personal judgment [12]. Decision complexity is present when there is a multiplicity of paths/decisions for a given step in a procedure and each path/decision can lead to a different outcome [12]. Interdependency complexity arises in a step when the step itself is dependent on other step/steps [12]. Step-size complexity is present when multiple instructions are provided in a single step [12]. Lastly, step-information complexity arises when an excess amount of information is provided in a step [12]. Since each of these types of task complexity was defined for a single step, the algorithm in this study was developed to detect step level complexity as opposed to procedure level complexity. The detection of step level complexity provided a higher resolution in the obtained results. More importantly, the linguistic identifiers for task complexity identification are defined in the form of individual words/tokens [12]. Hence, the algorithm analyzes individual words/tokens in every step. The linguistic identifiers for the described task complexities are represented in Table 1. The listed linguistic identifiers were considered as base-vocabulary for the NLP algorithm for detecting task level complexity in a step for a procedure.

Table 1. Base vocabulary for NLP algorithm [12]

Task complexity	**Linguistic identifiers**	**No of linguistic identifiers**
Judgment	'raise', 'reduce', 'lower', 'minimize', 'hold', 'establish', 'maintain', 'ensure', 'allow', 'verify', 'inspect', 'isolate', 'lockout', 'all', 'slowly', 'until', 'higher', 'lower', 'above', 'sufficient'	20
Decision	'if', 'then', 'while', 'as', 'either', 'or'	6
Interdependency	'go', 'proceed', 'repeat'	3
Step-size	'and', ',', ':'	3
Step-information	'warning', 'note', 'caution', '(', ')'	5

2.2 Algorithm development:

The NLP algorithm was primarily developed using the open source library NLTK [15] in the programming language Python 3.6 [16]. After defining the base vocabulary, the next step in the algorithm was to process the text data in procedures. The algorithm analyzed individual steps in a procedure and broke every step into tokens. A token is a sequence of characters that generally include words, numbers, punctuation marks *etc.* [15]. This can be understood from the following example:

Example A:

Reduce the temperature of the reactor R101.

Tokens for Example A:

['Reduce', 'the', 'temperature', 'of', 'the', 'reactor', 'R101', '.']

Thus, every token for the step was compared to the base vocabulary and if a linguistic identifier from the base vocabulary was detected as a token, then the step was considered to contain the type of complexity for which the linguistic identifier was detected. For Example A, the step will contain *judgment* complexity due to the presence of the linguistic identifier 'Reduce' from the base vocabulary.

However, this simple text matching of tokens and base vocabulary may not always detect task complexity in a step as it is limited by the number of linguistic identifiers defined in the base vocabulary (Table 1). Thus, it was essential to incorporate a more advanced approach that could compensate for the inexhaustive nature of the base vocabulary. This was achieved using the unsupervised machine learning algorithm GloVe [17]. GloVe is a machine learning model that allows vector representation of words. Specifically, we implemented the pre-trained 'GloVe-6B-200d model' [17] in the NLP algorithm. This model has been trained using a corpus of 6 billion tokens and has a vocabulary of 400,000 words that are used more frequently [17]. Each word in the model vocabulary is represented using a 200-dimensional vector [17]. The advantage of representing a word as a vector is that it allows the quantification of linguistic similarity between 2 words by calculation of the cosine similarity (Euclidian distance) between the vectors that represent those 2 words. An example of linguistic similarity quantification between 2 words is represented in Figure 1:

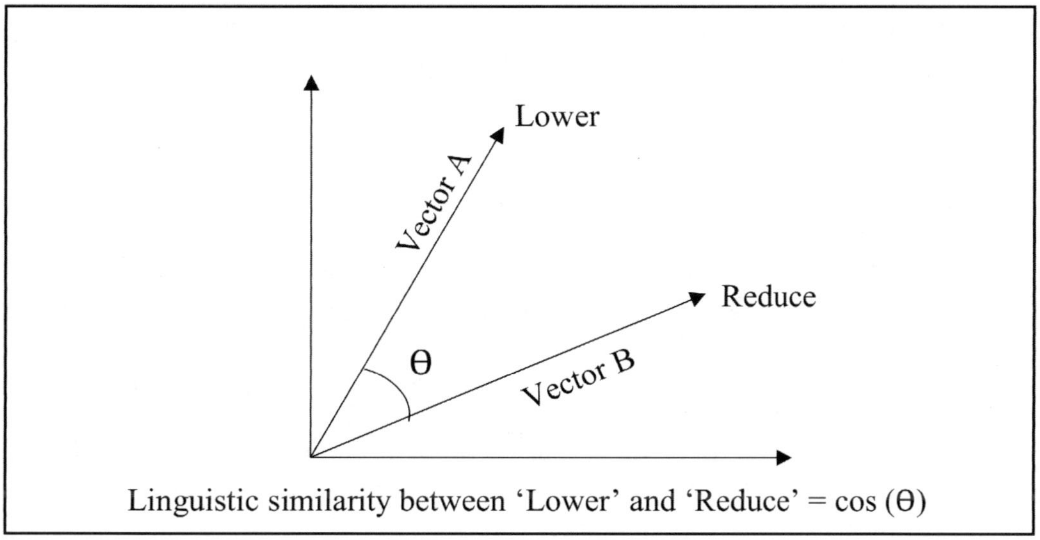

Figure 1. The linguistic similarity between 'Lower' and 'Reduce'

The value of linguistic similarity between 2 words varies from 0 to 1 (being a cosine of an angle). Thus, 2 words can be deemed to have a similar linguistic sense if their linguistic similarity exceeds a pre-defined critical value. The critical values of the linguistic similarity were determined such that they capture the essence of that type of task complexity and do not consider words as linguistically similar if they are not relevant to that type of task complexity. Specifically, the GloVe model was tested on an initial set of procedures and the critical values were determined for each type of complexity such that false positives amongst linguistically similar words for identifiers in Table 1 were minimized. This was achieved using a trial and error approach. The critical values for each of the type of task complexity are defined in Table 2:

Table 2. Critical values for linguistic similarity

Task complexity	Critical values for linguistic similarity
Judgment	$d_j = 0.78$
Decision	$d_d = 0.85$
Interdependency	$d_i = 0.78$
Step-size	$d_s = 0.72$
Step-information	$d_w = 0.6$

In the developed NLP algorithm, every token of a step is compared with every linguistic identifier in the base vocabulary and their linguistic similarity is evaluated. If the linguistic similarity exceeds the critical value for a specific task complexity (Table 2), then the step is determined to contain that type of task complexity. The application of GloVe model for identifying task complexity can be understood from the following example:

Example B:

Increase test pressure until PSH is tripped.

Although the token 'Increase' is not present in the base vocabulary (Table 1), it is linguistically similar to the identifier 'Reduce' defined in the base vocabulary (Linguistic similarity between 'increase' and 'reduce' is 0.79 which is greater than d_j defined in Table 2). Therefore, the step is determined to contain *judgment* complexity (Table 1).

The overall structure of the NLP algorithm is represented in Figure 2:

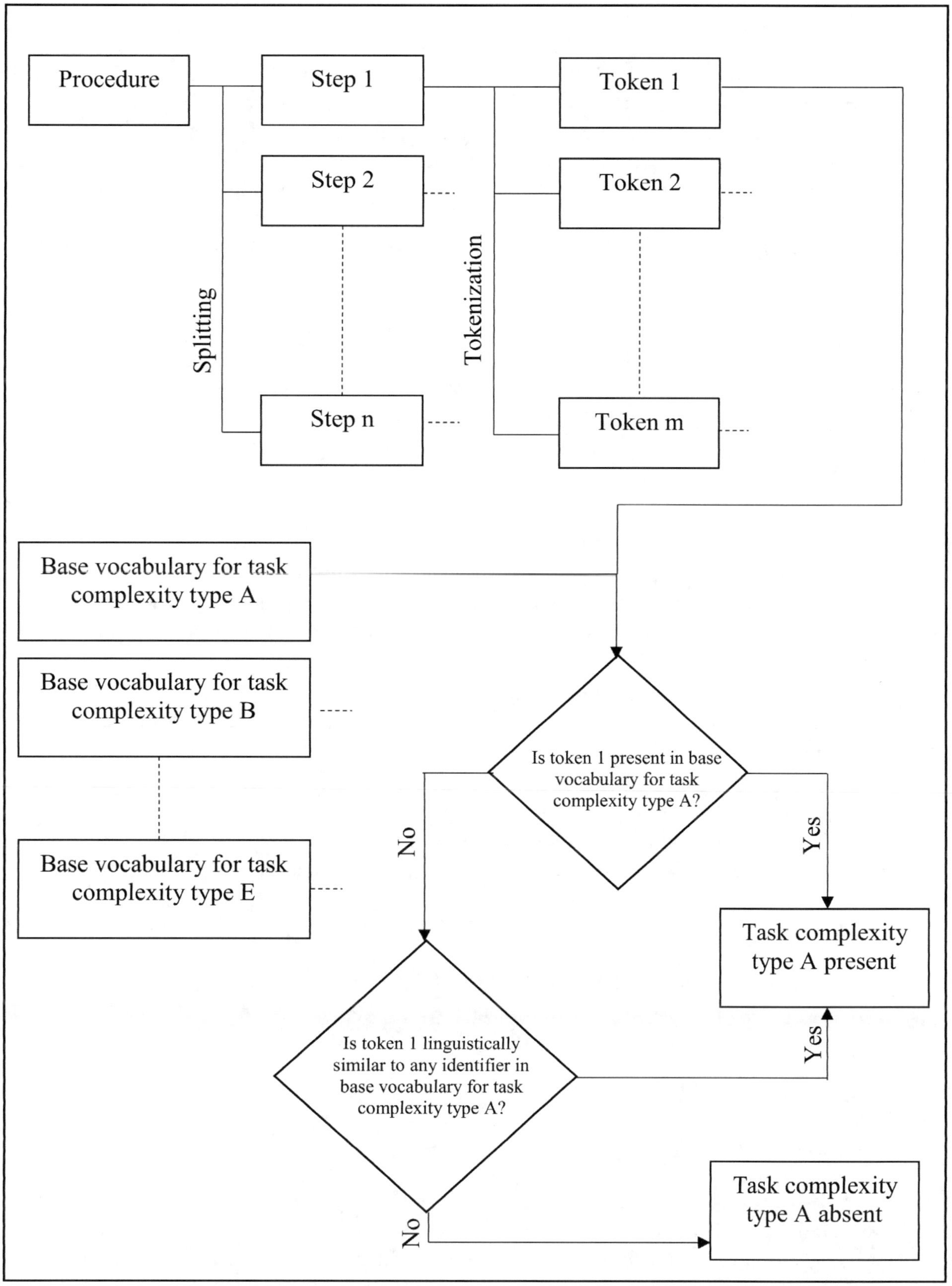

Figure 2. Structure of the NLP algorithm

2.3 *Application of the algorithm:*

The algorithm was tested on a set of 20 procedures reflecting various tasks in the oil and gas and chemical industry. These procedures were obtained from multiple stakeholders in the Advanced Procedures consortium [http://advancedprocedures.tamu.edu]. Overall the dataset comprised of 550 steps from these 20 procedures. The algorithm effectively identified all 5 types of task complexities while pin-pointing the tokens that were contributing to task complexity and the type of task complexity associated with it. The results derived from the algorithm for one such procedure of centrifuge separation task used at Shell's Robert training facility is shown in Figure 3:

1. Take a sample from the oil outlet of the bulk oil treater. The sample tap is located just upstream of LCV – 301A/302A/303A. (Last number correlates to train number).
2. Samples should be taken in two 200 ml certified centrifuge tubes. Fill tubes to the 100 ml mark.
   Wear proper PPE.
   Note: Vapors may ignite, causing flashing fire. Use Gas Monitor to detect LEL.
3. Place tubes into heater and let the samples heat to 115-120 degrees.
   Ensure applied heat does not exceed the vapor temperature of the oil and water.
4. Place tubes onto opposite sides of the centrifuge to maintain balance.
5. Spin the centrifuge for five minutes tests at 70 percent power.
   Ensure centrifuge is fully stopped before opening.
6. Obtain readings of combined basic sediment and water in percentages from the tube to the nearest 0.1 ml.
   Note: Read the bottom of meniscus.
7. Average the results from the two different tubes.
8. If your results are questionable, take two more samples.
   Note: Samples that are further than .3 ml apart are questionable.

Legend for type of task complexity:

Judgment Decision Interdependency Step-size Step-information

Figure 3. Structure of the NLP algorithm

3 Results & Discussion

3.1 *Algorithm results for centrifuge separation procedure:*

Based on the results obtained for the centrifuge separation task, it can be observed that the NLP algorithm is not perfectly accurate and does exhibit false positives (for *e.g.,* 'Take' in step 1 for interdependency complexity). However, the number of false detections (*i.e.,* 5) was relatively low as compared to the true detections (*i.e.,*17). More importantly, two of the correctly identified tokens ('should' and 'before') were not present in the base vocabulary and thus, were identified through the GloVe model.

3.2 Overall algorithm results:

The NLP algorithm results for the entire dataset of 552 steps were fascinating. The algorithm was allowed to store the detected tokens with an objective to create a dictionary of words that contribute to task complexity. As the algorithm processed steps in the database, this dictionary of words increased in size with respect to each of the type of task complexity. This learning nature of the NLP algorithm is represented in Figure 4.

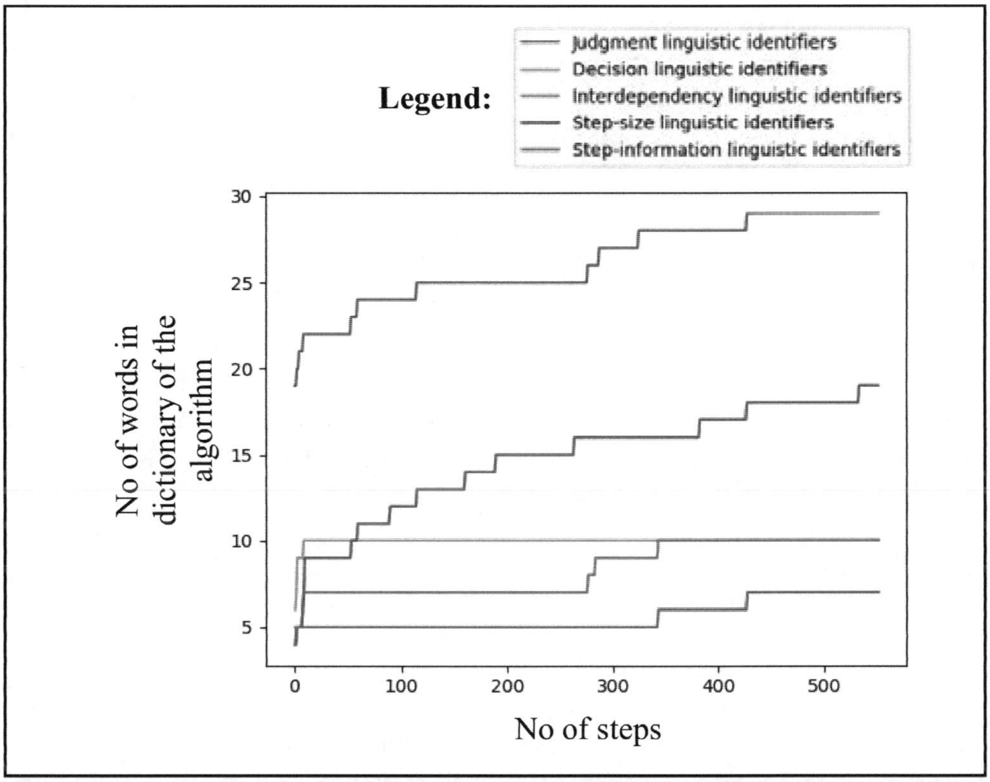

Figure 4. Learning nature of the algorithm

As observed from Figure 4, the NLP algorithm learns new words/tokens contributing to different types of task complexity as it processes more steps. This denotes that the GloVe model is effective in capturing those word/tokens leading to task complexity and not reflected in the base vocabulary of the algorithm. The dictionary of the NLP algorithm obtained after preprocessing the entire dataset of 552 steps is shown in Table 3. The NLP algorithm learns significantly for step-size complexity (850% increase in the number of linguistic identifiers) while its learning process for decision complexity ceases after a certain number of steps. It is important to note that the learning behavior of the NLP algorithm is controlled by the critical values defined in Table 2. Lower critical values result in faster and improper/inaccurate learning whereas, higher values inhibit the process of learning completely.

Table 3. Dictionary of the NLP algorithm

Task complexity	Linguistic identifiers	No of linguistic identifiers
Judgment	'lockout', 'hold', 'lower', 'all', 'minimize', 'not', 'only', 'there', 'isolate', 'verify', 'both', 'increase', 'maintain', 'slowly', 'raise', 'ensure', 'come', 'other', 'establish', 'inspect', 'well', 'above', 'before', 'until', 'reduce', 'below', 'higher', 'sufficient', 'allow'	29
Decision	'either', 'or', 'not', 'if', 'when', 'then', 'does', 'while', 'should', 'as'	10
Interdependency	'proceed', 'you', 'if', 'just', 'take', 'repeat', 'go', 'come', 'do', 'let'	10
Step-size	'now', ':', 'as', 'including', 'with', '.', 'when', 'where', 'and', ',', 'both', 'all', 'only', 'that', 'while', 'other', 'but', 'well', 'in'	19
Step-information	'issued', '(', 'caution', 'note', 'warning', 'alarm', ')'	7

As seen in Table 3, the dictionary of the NLP algorithm is substantially larger than its base vocabulary (Table 1). However, this dictionary does contain false positives that are not necessarily associated with that type of task complexity (*e.g.,* 'you' in interdependency). More importantly, certain words/tokens are found to be common in multiple types of task complexities (*e.g.,* if appears in both interdependency and decision complexity). This indicates that certain elements/words can contribute to multiple types of task complexities.

4 Application

An important point to address is how the developed algorithm can help procedure writers to develop procedures that improve the performance and safety of processes. In a simpler sense, the developed algorithm in this study can identify words (tokens) that might negatively affect the performance of the operator. These words essentially result in multiplicity of paths/outcomes, uncertainty in information, interdependency of a step on other step(s), multiplicity of instructions/object of instructions and excess amount of information. By identifying the words through the developed algorithm, these deficiencies can be identified and subsequently eliminated. Elimination of such deficiencies can help the industry to lower risk and improve productivity of their processes. The application of the algorithm for better procedure design is shown through examples in Table 4. The examples are part of the tested dataset and thus, are indicative of current procedural practices.

Table 4. Example applications of the developed algorithm

Sr. No	Example step in a procedure	Type of complexity identified by the algorithm	Identified token resulting in complexity and associated drawback	An improved alternative for the mentioned step by addressing the identified complexity
1	Increase test pressure until PSH is tripped	Decision complexity	The token 'until' introduces a decision (multiple outcomes) that needs to be made by the operator and doesn't provide enough knowledge about the tripping point itself.	Increase test pressure above 120 psi to achieve tripping of PSH
2	Open block valve downstream of the control valve CV101 and close gas inlet valve upstream of the control valve CV101	Step-size complexity	The token 'and' indicates that there are multiple instructions in a single step. This multiplicity of instructions might lead to errors while performing this step.	Open block valve downstream of the control valve CV101 Close gas inlet valve upstream of the control valve CV101 (separate steps)
3	Adjust operating parameters to maintain mainline rate and pressure	Judgment complexity	The token 'maintain' indicates that there is a judgment involved due to the uncertainty in information provided with respect to the values of rate and pressure required	Adjust the operating parameters to maintain rate and pressure at 20 bbl/min and 120 bar

4	a) If fan belt is acceptable, proceed to step c. b) If fan belt is unacceptable, proceed to step d. c) Open safety guards d) Perform LOTO procedure and readjust the belt.	Interdependency complexity	The token 'proceed' introduces interdependency amongst steps and might lead to errors resulting from misinterpretation	If the belt is acceptable, open safety guards. If the belt is unacceptable, perform LOTO procedure and readjust the belt.
5	Look over job Note: check worksite hazards. equipment needed for disassembly and containment	Step-information complexity	The toke 'note' indicates that there might be excess information provided in the step that might not truly be an action that needs to be carried out.	Look over job (The information pertaining to worksite hazards and equipment should be preferably provided at the start of the procedure and not within a step of the procedure)

5 Conclusion & Future Work

It is concluded from this study that NLP can help in identifying elements in a procedure that contribute to task complexity and eliminating these elements can be helpful towards the better design of procedures. The algorithm can be tested against human coders for the considered step data and more insight can be obtained into its accuracy. The algorithm can also be improved further by feeding the dictionary of words (Table 3) as base vocabulary. This iteration might help in detection of additional words/tokens contributing to task complexity.

6 Acknowledgments

This research was co-funded by the Next Generation Advanced Procedures Initiative (http://advancedprocedures.tamu.edu/). We would further like to acknowledge the stakeholders of the initiative for access to the procedures that helped in the testing of the algorithm.

7 References

[1] P. Bullemer, "Understanding Process Safety Challenges associated with Heater Operations in the Process Industry," in *Mary K O'Conner Process Safety Center 17th Annual International Symposium*, 2014.

[2] U.S. Chemical Safety and Hazard Investigation Board (CSB), "Completed Investigations," 2018. [Online]. Available: https://www.csb.gov/investigations/completed-

investigations/?F_FromDate=1997&F_ToDate=2000.

[3] U.S. EPA, "Guidance for Preparing Standard Operating Procedures (SOPs)," *United States Environ. Prot. Agency*, no. April, pp. 1–55, 2012.

[4] Occupational Safety & Health Administration, "Process Safety Management Guidelines for Compliance," *Osha 3133*, pp. 1–16, 1994.

[5] UK Health and Safety Executive, "HSE Human Factors Briefing Note no 4: Procedures," 2015. [Online]. Available: http://www.hse.gov.uk/humanfactors/topics/04procedures.pdf. [Accessed: 09-May-2018].

[6] D. J. Campbell, "Goal Levels, Complex Tasks, and Strategy Development: A Review and Analysis," *Hum. Perform.*, vol. 4, no. 1, pp. 1–31, 1991.

[7] S. H. Chan, Q. Song, and L. J. Yao, "The moderating roles of subjective (perceived) and objective task complexity in system use and performance," *Comput. Human Behav.*, vol. 51, no. PA, pp. 393–402, 2015.

[8] J. Park and W. Jung, "Identifying objective criterion to determine a complicated task - A comparative study," *Ann. Nucl. Energy*, vol. 85, pp. 205–212, 2015.

[9] H. Topi, J. S. Valacich, and J. A. Hoffer, "The effects of task complexity and time availability limitations on human performance in database query tasks," *Int. J. Hum. Comput. Stud.*, vol. 62, no. 3, pp. 349–379, 2005.

[10] H. T. Tan, T. B. P. Ng, and B. W. Y. Mak, "The effects of task complexity on auditors' performance: The impact of accountability and knowledge," *Auditing*, vol. 21, no. 2, pp. 81–95, 2002.

[11] D. C. Maynard and M. D. Hakel, "Effects of objective and subjective task complexity on performance," *Hum. Perform.*, vol. 10, no. 4, pp. 303–330, 1997.

[12] P. Kannan, N. Quddus, S. C. Peres, and M. S. Mannan, "Can We Simplify Complexity Measurement ? a Primer Towards Usable Framework for Industry Implementation," *Proc. Hum. Factors Ergon. Soc. Annu. Meet.*, pp. 853–857, 2018.

[13] A. Chokor, H. Naganathan, W. K. Chong, and M. El Asmar, "Analyzing Arizona OSHA Injury Reports Using Unsupervised Machine Learning," *Procedia Eng.*, vol. 145, pp. 1588–1593, 2016.

[14] S. Robinson, "Multi-Label Classification of Contributing Causal Factors in Self-Reported Safety Narratives," *Safety*, vol. 4, no. 3, p. 30, 2018.

[15] S. Bird, E. Klein, and E. Loper, "Language Processing and Python," *Computing*, p. 479, 2009.

[16] Python Software Foundation, "Python Software Foundation. Python Language Reference,

version 3.6.," 2016.

[17] J. Pennington, R. Socher, and C. Manning, "Glove: Global Vectors for Word Representation," *Proc. 2014 Conf. Empir. Methods Nat. Lang. Process.*, pp. 1532–1543, 2014.

THE FRONTLINE REALITY OF PROCESS SAFETY RISK – A CASE STUDY ON IMPLEMENTING OPERATIONAL RISK MANAGEMENT SYSTEMS

Simon Jones
Sphera – Director, Solutions Consulting
Simon.Jones@sphera.com

Andrew Bartlett
Sphera – Solutions Consultant
ABartlett@sphera.com

Prepared for Presentation at
American Institute of Chemical Engineers
2019 Spring Meeting and 15th Global Congress on Process Safety
New Orleans, LA
March 31 – April 3, 2019

THE FRONTLINE REALITY OF PROCESS SAFETY RISK – A CASE STUDY ON IMPLEMENTING OPERATIONAL RISK MANAGEMENT SYSTEMS

Simon Jones
Sphera – Director, Solutions Consulting
Simon.Jones@sphera.com

Andrew Bartlett
Sphera – Solutions Consultant
ABartlett@sphera.com

Keywords: Operational risk management. Frontline operational risk. Process Safety Management. Process safety barrier health.

Abstract

Today's energy and chemical industry facilities face a challenge to truly understand the frontline operational reality and impact of all site activity and equipment conditions on process safety risk. While process safety methods and techniques developed over the years have improved the design of our facilities, many industry operators do not provide systems and solutions for maintenance and operations - where decisions continuously take place that impact well-designed process safety barriers and increase exposure to Major Accident Hazard (MAH) risk.

The dynamic nature of risk at the frontline demands new ways to make MAH risk exposure visible and prominent, so every operational decision is informed by real-time events and managed with care.

There is a shift in the hazardous industries to look at how digitalization and so-called Industry 4.0 techniques can connect business systems and change how risk and process safety are measured, monitored and managed.

An emerging category of Operational Risk Management (ORM) business systems offers the potential to make companies and their work teams safer by delivering enterprise-wide insight into how planned work activities and plant equipment health impact process safety barrier health. This early warning system makes risk exposure visible, prominent and available to all in real-time, reducing vulnerability to high-potential near-misses and major hazard events.

This paper sets out the principles and models behind this new approach to operational risk management and includes a case study sharing how a major international refinery operator is using digitalization to connect a real-time view of process safety barrier health to the frontline to support more effective hazard management.

1 Introduction

The process safety practices and guidelines for designing, managing and operating the facility are well-known and documented by regulatory and engineering bodies. The rules of how to run facilities, maximize production, and manage risks are typically encoded in organizations' Operational Management systems.

But, how do we know good practice is being applied? How do we ensure it is effective, and how do we connect this practice to frontline operations? Traditionally, KPIs and audits are our principal tools, yet despite investments in performance monitoring and continuous improvement, the industry continues to experience catastrophic accidents[1].

2 Survey of process safety and operational risk engineers

According to Petrotechnics' 2018 industry survey[2], 86% of industry leaders report a gap between how process safety is intended and the reality of operations. Only 38% believe industry operators proactively manage process safety risk. Sixty-nine percent say challenges to delivering effective process safety are due to safety culture and overall risk awareness.

56% of respondents said that they found an increase in risk on their plant when undertaking periodic process safety reviews (for example, 3-5 year periodic reviews of PHAs, HAZOPs, HAZIDs, LOPA studies, etc.). Indeed, it is good news that periodic reviews are finding these issues, but this statistic does beg the question – what was happening to the risk level on the plant in-between times? Is this a blind spot for operators in terms of continuous monitoring? It appears that when our good engineering design and process enters into service at the frontline, it is immediately subject to attrition; the operational environment causes processes, systems, and equipment to degrade over time, thereby increasing risk. It seems this change in risk is only highlighted when periodic process safety reviews are undertaken

3 What Process Safety-related information is important for frontline operations? – or "the trouble with KPIs"

In any typical day on a facility maintenance work is performed; it is usually scheduled as either repairs or preventative maintenance routines. Technicians obtain a permit to work (PTW), and good practice says the permit will have information on it related to work hazards and mitigating measures that must be put in place to safely carry out the job.

But this is not the whole story. The work crew must understand more about the facility, equipment and the status of process safety barriers in the area they will work. Today, managers rely on KPIs collected for the facility to provide some insight, coupled with knowledgeable and experienced frontline supervisors.

For example, on a typical production facility, the following KPIs are likely collected:

- Lost time injury statistics and reports
- Loss of Primary Containment (LOPC) statistics and reports - Tier 1 and 2 process safety incidents per ANSI/API RP 754

- Safety-critical maintenance plan attainment against target
- MOCs open for longer than 90 days
- Key policy compliance issues (typically found through audit)
- Emergency response drill completion and lessons learned

Do these kinds of KPIs and measures help inform the frontline reality of operational risk for those going to work on a specific unit or area of plant?

Imagine a work crew heading out to the plant to carry out a maintenance task that involves breaking containment. The work crew will probably be aware of the Tier-2 process safety events that occurred on the facility and may also know about specific LOPC events on the Unit in the last quarter.

But the facility-wide KPIs may have little relation to the reality on the ground within the Unit itself. For example – would maintenance know that changes were made to the equipment that are to work on, and that the drawings are not updated? Would the work crew know of work to be performed on a faulty gas detector array on a different craft work schedule, and how that simultaneous activity may affect their ability to execute their own work safely?

From a process safety barrier perspective, would the work crew have a view of deferred maintenance on other safety critical elements (SCE) of the facility as well as an understanding of how any associated risk impacts the unit? If something does go wrong, will the work crew know how to respond? Is all the necessary emergency response and recovery equipment in place and in service? What if emergency response drills were not completed as scheduled on the Unit due to the repairs required from damage caused by the Tier 2 event?

The point being; there is a lot of specific local detail behind KPIs that are relevant for those at the frontline, but it may not be readily available. The challenge is; how can we present this information in a simple, meaningful way to the frontline worker? How can we provide a real-time view of risk vulnerability?

4 Digitalization – Operational Excellence and Industry 4.0

In many ways, the dynamic nature of the frontline is ripe for technology to better support decision-makers. In fact, according to a 2018 Verdantix study, Operational Excellence and Industry 4.0 strategies are among the top factors triggering operational risk management (ORM) system implementations today[3]. Over 73% consider digital technology valuable, if not essential, for effective operational risk management[4].

An emerging category of ORM enterprise software seeks to support safe and effective operational decisions by providing critical innovations[3], including:

- An enterprise-wide approach to process safety barrier management, permit to work, management of change (MOC), incident management, risk assessments, and process safety management - all accessible via desktop web applications and mobile devices for use in the field
- Wearable technology as a source of data for use in the field

- Industrial Internet of Things (IIoT) as a source of data from critical devices in the field
- Advanced modelling capabilities to create dynamic "digital twin" asset models to deliver what/if simulation
- Advanced analytics from big data to provide actionable insights on operational risk status and trends

Operational risks arise from a complicated set of interrelated parameters and are viewed and managed in differing ways depending on the role and level in the organization. The challenge lies in simplifying this complexity and enabling all levels of the organization to collectively focus on the elements of risk that are most important.

5 Mapping information sources to a process safety barrier model

The sources of operational risk – including activities, deviations, and non-conformances – will be particularly relevant to work crews in the field. Expanded, these include the items referenced in Table 1.

Table 1: Activities, equipment, and management system status as a source of barrier impairment

CRITICAL EQUIPMENT STATUS			RISK ARISING FROM ACTIVITIES			MANAGEMENT SYSTEM STATUS	OTHER CONDITIONS
Performance Standard Failure	*Overrides*	*Maintenance Schedule*	*Maintenance*	*Turnaround*	*Simultaneous Activities*	*Procedural Deviation*	*Plant Status*
Degraded or impaired major accident hazard barriers Degraded or impaired Safety Critical Elements	Long-term override of instrumented safeguarding systems	Deferment of scheduled safety-critical maintenance or inspection Preventative maintenance backlog	Hot work - either planned or underway Breaking containment tasks – either planned or underway Confined space entry tasks – either planned or underway Temporary inhibits or overrides to safeguarding systems for maintenance task execution	Scheduled shutdown or turnaround at Unit or facility	Simultaneous Operations (SIMOPS) Conflicting work tasks	Operation of plant or equipment outside normal parameters Operating without critical team competencies on shift MOC overdue Other temporary, risk-justified changes to policies and procedures	Temporary equipment in place of failed permanent equipment Degraded or impaired well barriers Temporary repairs Process leaks

This diverse set information resides in multiple business systems spread right across the organization. The following characteristics of this information are also important from a frontline perspective:

- Time or duration – when the activity is planned to happen, or for how long the impaired condition will last;
- Location – what area is impacted - does the activity or condition have, for example, a very localized Unit impact, or does it impact multiple Units or even the whole facility
- Risk – to what extent does the activity or condition impact a risk model

Operational Risk Management systems seek to gather this information from whatever business system it resides in and present it based on time, location and risk impact (Figure 1).

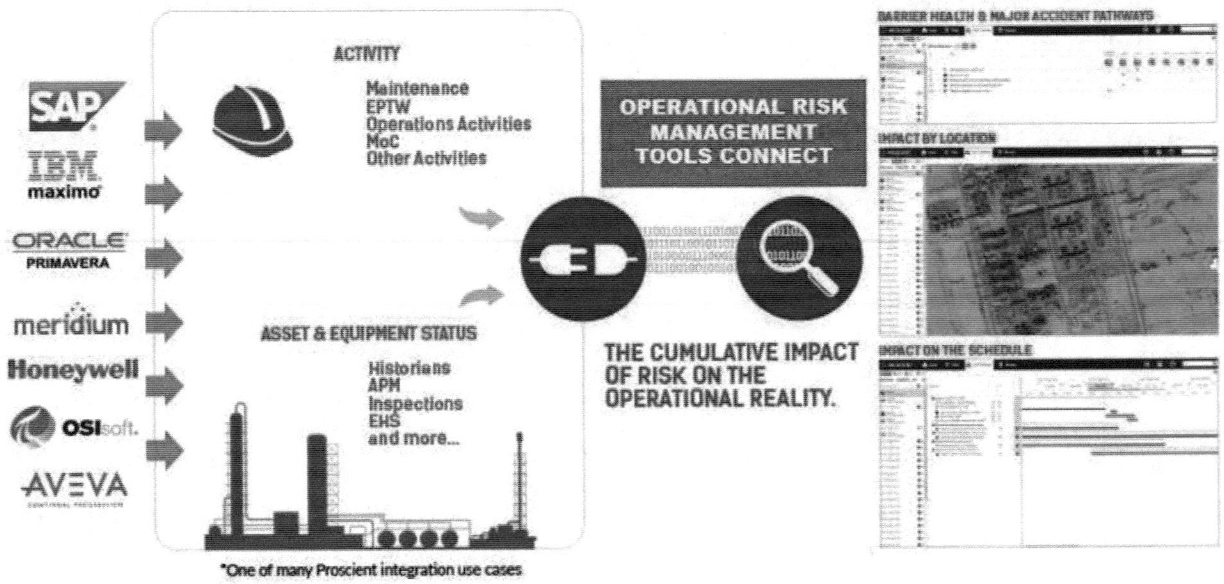

Figure 1 Connecting diverse sources of data to the frontline using ORM systems

Focusing on dynamic risk modelling for this data, Operational Risk Management systems show the cumulative impact of all these dynamic operational risks by mapping their impact on a set of fundamental barrier groupings and by associating them with the major accident hazards under management at the site.

This fundamental barrier model is based on multiple layers of protection or mitigation. Note that these barrier groups are not singular barriers. Each is a collection of equipment, instrumentation or people-driven processes which collectively fulfill the function of the barrier for a location. The idea is simple; if there are impairments in several barrier groups, risk increases because a major accident is more likely to occur – see Figure 2.

The fundamental barrier model has its origins in James Reason's Swiss Cheese metaphor; the holes in the barriers represent impairments, and when the holes line up, an event can occur and escalate into a major accident. The degree to which risk pathways develop represent the potential level of risk.

For example, if there is impairment on the 'containment' barrier happening at the same time and within the same location as impairment in the 'ignition control' barrier, the risk of having an uncontrolled release of hydrocarbons to the atmosphere is higher, as is the risk of ignition. Combined, they can result in a fire or explosion as the result of the subsequent failure of more than one barrier group. If the 'detect' barrier (e.g. gas and fire detectors) is also compromised, and the ability to protect is also compromised (e.g. say the water deluge system is inoperable), these additional breaches can result in the potential occurrence of a major event. The degree of escalation and the scale of the consequences will depend on the mitigation barriers or the ability to respond to the incident.

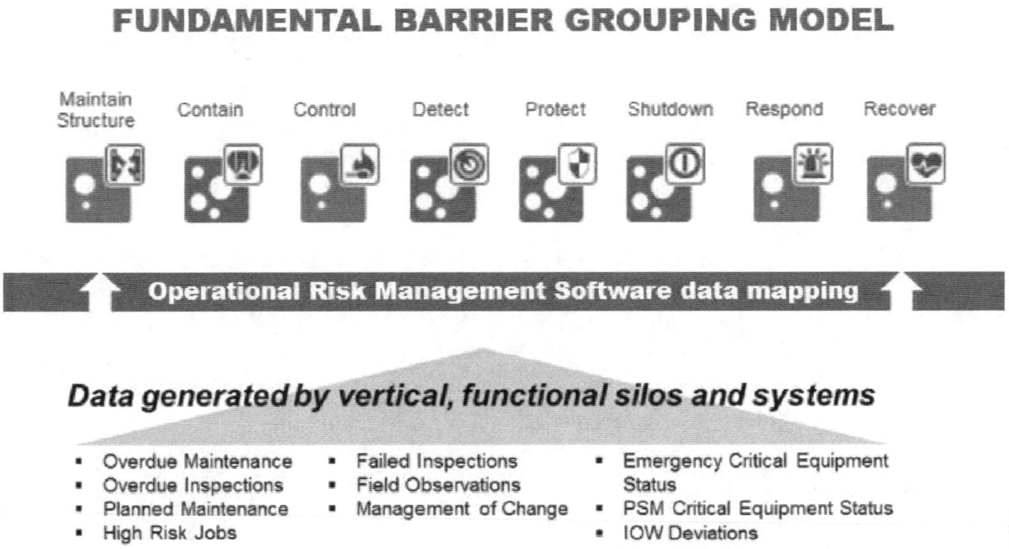

Figure 2 Mapping risk-related data to a process safety barrier model

Similarly, the potential barrier impact of any maintenance or operational activity can be modelled – for instance, if a planned positive isolation is needed to prepare for confined space entry, it is reasonable to assume there is a potential impact on the process containment barrier for the period in which first line break is undertaken. Open flame hot work in a unit represents a degradation of the ignition control barrier for the period the permit is issued.

This simple, elegant barrier-based approach enables operators to predict and better manage the outcome – whether that means to postpone a particular planned activity or accelerate maintenance to address deviations or non-conformances on the facility that are increasing risk levels overall.

6 Enabling the connected industrial worker

Enabling the connected industrial worker using mobile devices is the next opportunity in the digital transformation journey. From a process safety standpoint, we must take care to ensure the data presented to the frontline - through intrinsically safe mobile devices - offers the dynamic information work teams need to support process safety hazard management.

If we think about the daily activities of the frontline worker, there are many situations where providing a common view of the operational reality of the facility - that is, an understanding of where equipment conditions or planned activities may impact operational risk – will help support effective decision-making. Example uses for mobile devices to support frontline risk management of the are summarized in Table 2.

The principle design basis for any connected worker app needs to be to easily understand the cumulative impact of all activity and all risks on the facility, in order to support effective decision-making.

Table 2: Uses for mobile devices in support of frontline risk management

User group	Uses for mobile devices in support of frontline risk management
Work crew	• See where work is to be carried out by accessing a plot plan view of work location – a digital twin-view of the asset • Understand risk status on facility by looking at risk heat maps layered on plot plan views • Record work progress in the field and update status on activity records • Update the progress of isolations in the field – recording isolation status, lock, and lockbox information in real-time • Scan equipment tags to confirm correct equipment is being worked on • See where simultaneous work is happening in neighbouring units • Understand real-time critical equipment status • Create task risk assessment (JSAs) and other assessments for emerging work • Perform and record gas tests & audits
Supervisor	In addition to work crew use cases: • See when and where work and activity is happening on the facility • Check for risk status and conflicts before approving or issuing work • Review GPS history for when and where work and inspections were carried out • Monitor real-time work execution and performance • Access reports on the real-time health status of the asset • See specifically what is contributing to elevated levels of operational risk
Asset leader	In addition to Supervisor use cases: • Understand real-time risk status on asset(s) • Understand where and when critical work is happening • Understand risk trend on asset(s)

7 Case study: delivering a real-time view of critical equipment status and its impact on risk

A major national oil company is building and will operate a world-scale refinery in the Middle East. Currently in greenfield phase, the operator has commissioned a significant Industry 4.0 initiative to develop and deliver a technology-driven approach to integrate a suite of business systems to better support asset and operations processes and management. The operator has a sophisticated safety management system and clear corporate standards and practices. The operator wishes to make a real-time view of operational and process safety risk a central element of decision-making from plant start-up onwards.

The operator is implementing an Operational Risk Management system to manage the following process safety related work processes (Figure 3)

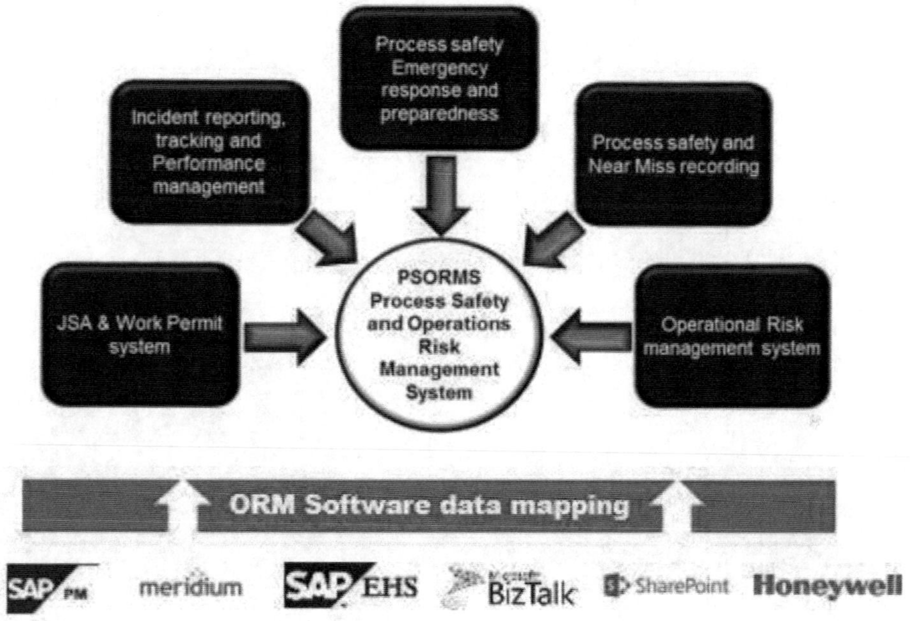

Figure 3 Process Safety-related work processes supported by ORM System

In addition to managing the barrier risk impact of these processes, the ORM system also connects out to other key business systems to deliver a near real-time view of the risk status associated with critical equipment:

- Data historian – for near real-time IOW (integrity operating window) health data associated with critical equipment (this data historian is itself tracking the operational distributed control system (DCS) and critical alarms systems)
- Asset Performance Management (APM) system – for status of critical equipment
- Maintenance Management System (MMS) - for inspection and maintenance records and associated plans and schedules

The relationship between the various business systems and data sources is summarized in Figure 4.

Since the project is still in the design phase, the project team was able to access design package materials from the EPC contractors responsible for each refinery unit to identify the critical equipment. This includes bringing together a variety of information in useful formats:

- From design phase Hazop and asset integrity studies
- Health parameters associated with specific items of identified safety-critical equipment
- Critical equipment types/categories mapped to the fundamental process safety barrier model
- Records representing all critical equipment are set up in the ORM software
- Through integration, the ORM software "listens" to the health of critical equipment – as illustrated in figure 4
- Non-conformances are mapped to the fundamental process safety barrier model

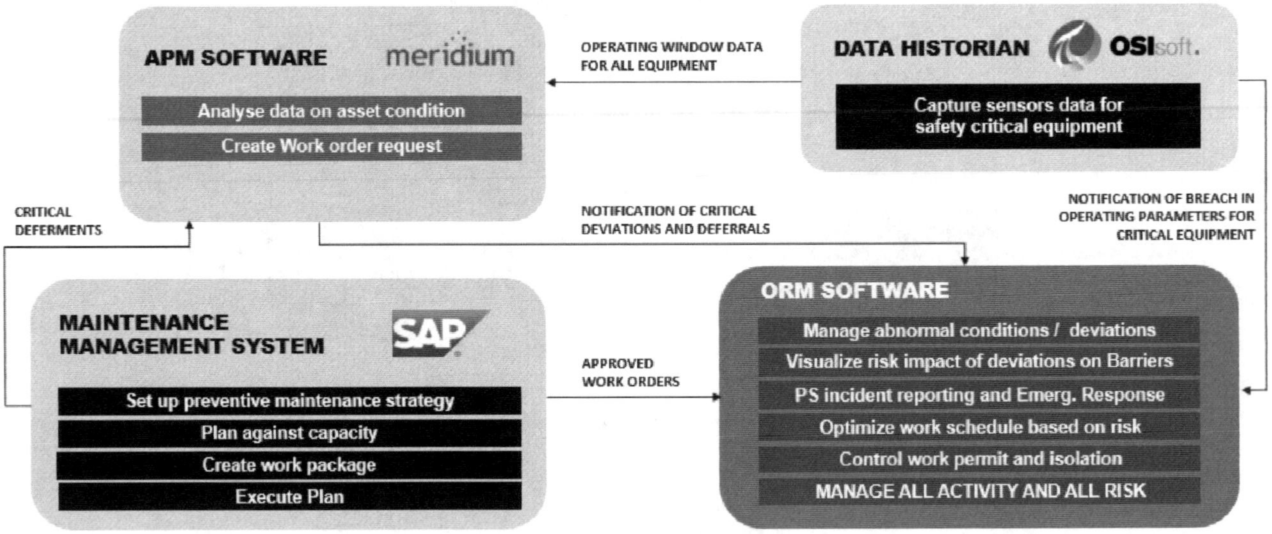

Figure 4 Business system integrations to deliver near real-time risk view of critical equipment

In addition, the ORM system is used to manage all permitted activity on the facilities. The integration described provides a combined real-time view of all equipment risks and all activity risks on the barrier model to support effective decision-making. Figure 5 provides screenshots illustrating how information from these business systems appears in the ORM application in time, location and barrier health screens.

See the risk on the asset – location view:

Understand the impact on the schedule – time view:

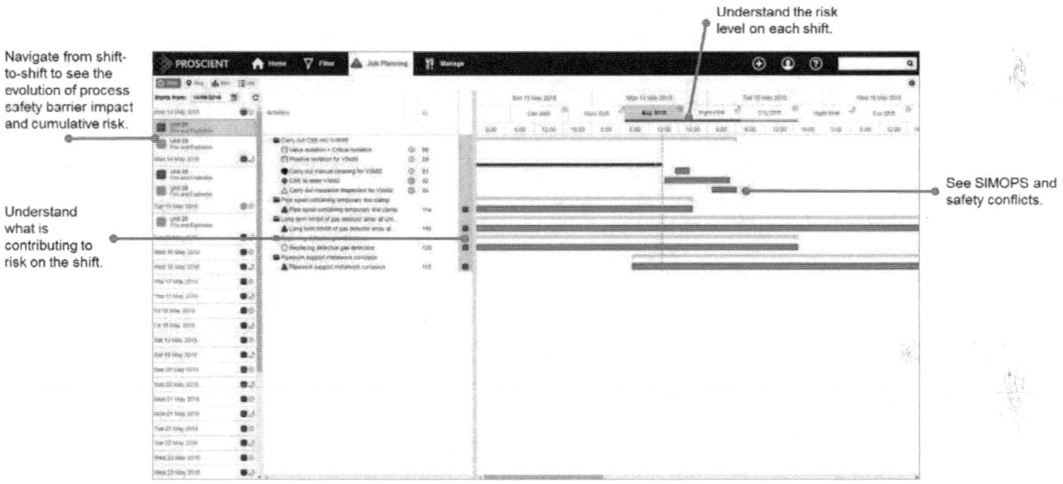

Understand what's driving the risk – barrier health view:

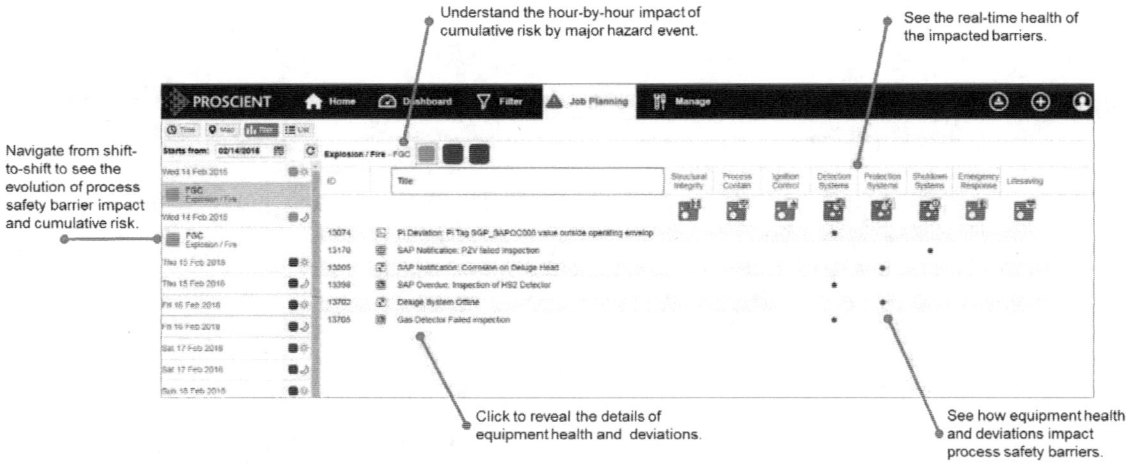

Figure 5 ORM software - location, time and barrier health views of frontline risk

8 Conclusion

Industry-leading operators are adopting strategic approaches to operational risk management that drive safer, more sustainable, more productive operations overall. Industry 4.0 technologies are connecting business systems and changing how risk and process safety are measured, monitored, and managed. These easy-to-use systems and solutions for maintenance and operations deliver timely, relevant information to protect process safety barrier health and reduce exposure to Major Accident Hazard (MAH) risk at the frontline.

9 References

[1] The 100 Largest Loss events 1978 - 2017 - Large property damage losses in the Hydrocarbon Industry, 25th edition, Marsh & McLennan, 2018

[2] 2018 Survey on Process Safety and Operational Risk Management, Petrotechnics 2018 – https://www.petrotechnics.com/insights/psm-orm-part-2/

[3] Verdantix, 2018a, Smart Innovators: Operational Risk Management Software, August 2018, London, UK

[4] Operational Excellence Index 2017 - https://www.petrotechnics.com/insights/operational-excellence-index-2017-survey-report/

Just the Facts

A Bad Way to Communicate Process Safety

Dr TJ Larkin
Partner – Larkin Communication
230 Park Avenue, 3rd Floor West
New York, New York 10169

Larkin@Larkin.Biz
www.Larkin.Biz

Prepared for Presentation at
American Institute of Chemical Engineers
2019 Spring Meeting and 15th Global Congress on Process Safety
New Orleans, LA
March 31 – April 3, 2019

Just the Facts

A Bad Way to Communicate Process Safety

Dr TJ Larkin
Partner – Larkin Communication
230 Park Avenue, 3rd Floor West
New York, New York 10169

Larkin@Larkin.Biz
www.Larkin.Biz

Keywords: communication, process safety, supervisors, frontline, incidents, illustrations, graphics, fear, behavior change, employee communication

Abstract

You should scare your employees. Communicate major incidents in a way that demands employees' attention and makes them afraid. Why? A properly constructed fear appeal can double the amount of behavior change caused by your communication. You do not need to terrify your employees. Only a small amount of fear (discomfort) is enough to maximize the behavior change caused by your communication.

1. Fear Appeals

"Fear appeals" are a type of communication designed to scare the people who see them. Communicators use fear appeals because they work. Research into fear appeals show they create significantly more behavior change than similar messages without the fear; sometimes double the rate of behavior change (1).

Fear appeals have two parts: disturbing photos or drawings; and a recommended action giving the receiver a way to avoid the bad consequences shown in the photo or drawing.

It is essential that the disturbing photo or drawing causes an unpleasant physiological reaction in the receiver. This unpleasant feeling is what fuels the behavior change. Without this discomfort—you get no change. It is equally important to give the receiver a "way out" (2). A "way out" is some recommended action they can do to avoid becoming the next victim.

Figure 1 is a model showing how fear appeals work.

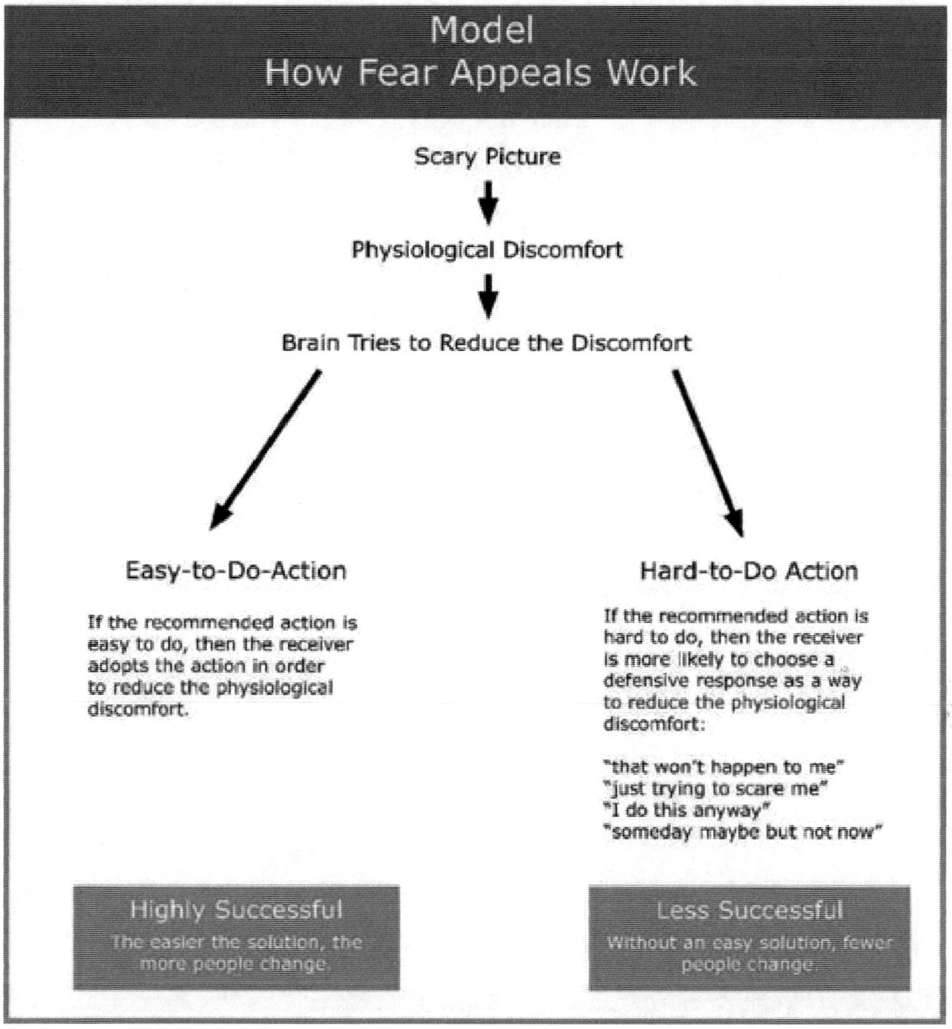

Typical topics where fear appeals are used:

- drunk driving
- bicycle helmets
- child seats for cars
- sunscreen
- home protection systems
- flossing teeth
- anti-smoking
- car seat belts
- texting while driving
- smoke alarms

Figure 2 (on the following page) shows how a fear appeal is used while communicating a major incident. In this case the major incident was a fatality caused by an exploding storage tank at onshore oil and gas site.

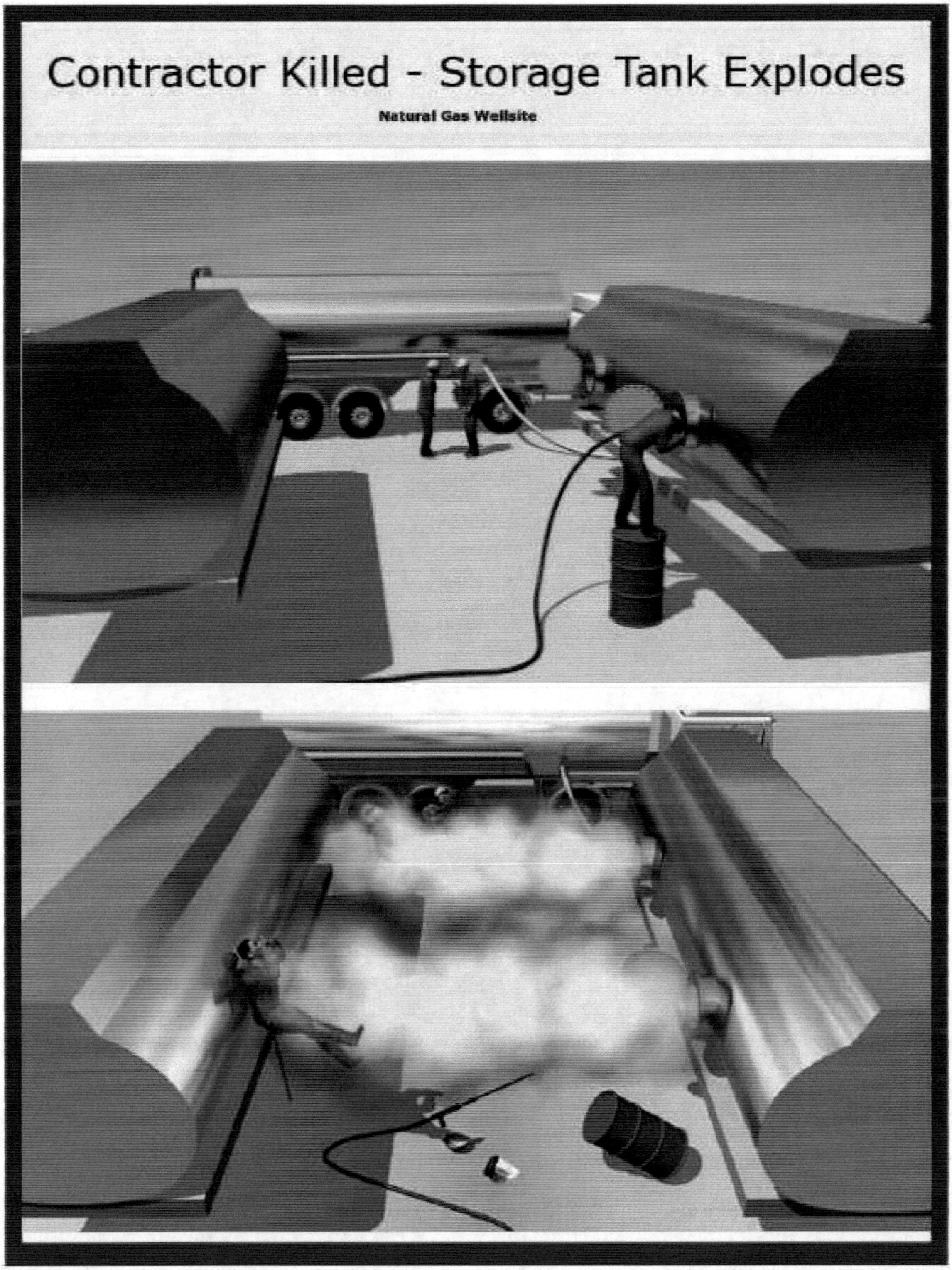
Contractor Killed - Storage Tank Explodes
Natural Gas Wellsite

Background Information Figure 2: Contractor Killed – Storage Tank Explodes	
Background	Onshore oil & gas site. Contractor cleaning a storage tank filled with sand and fluids. Before cleaning, an employee tested the atmosphere inside the tank and his detector alarmed for an explosive environment. No one told the cleaning contractor about the explosive atmosphere inside the tank. During cleaning the tank exploded. Cleaning contractor was killed. Supervisor was seriously burned. Ignition source was never found.
Easy-to-do Solution	Post signs warning about the explosive atmosphere discovered in the tank.
Document Details	Complete document is two pages. Only the cover is shown here. The complete document can be downloaded *Lessons Learned – Communicating Major Incidents* at: www.Larkin.Biz

2. Fear appeals work

In general, fear appeals work. Across a broad range of topics, putting a fear appeal (scary picture) into your communication increases the amount of desired behavior change (3). However, there is one situation when fear appeals work especially well: fear appeals produce above-average amounts of behavior change when the fear appeal supplies an "easy-to-do" action showing the receiver how to avoid becoming the next victim (4).

For example, a scary image of tooth decay combined with a recommendation to floss will get more behavior change (flossing) than a scary image of lung cancer with the recommendation to stop smoking. Why? Because "flossing" is much easier to do than "stop smoking." For example, fear appeals do not work well with drug addiction. Ending a drug addiction is not an "easy-to-do" behavior.

If the action recommended in the fear appeal is "hard to do," people will often use defensive methods to reduce the physiological discomfort created by the scary image. And they will do this defensive reaction instead of adopting the recommended action. Defensive reactions to a fear appeal are responses like:

- "this will never happen to me"
- "just trying to scare me"
- "that's exaggerated"
- "I'll stop someday"

This, for example, is why fear appeals are rarely used to reduce obesity rates. Changing what you eat and beginning a long-term exercise program are "hard-to-do actions," and consequently they are usually meet with defensive reactions. With hard-to-do recommended actions, the response becomes, "Someday, I'll do that," instead of the adopting the recommended action (changing diet and beginning exercising).

On the other hand, fear appeals with easy-to-do recommendations work especially well. Fear appeals showing frightening images of AIDS patients resulted in 48% more students using condoms (a relatively easy-to-do recommendation) (5). In another study, scary images of people suffering from tetanus combined with a phone number, hours, and directions to the tetanus injection location caused 30% of everyone who received the message to go get the injection (6). With the fear appeal removed (pictures of scary tetanus victims taken out of the communication) only 3% got the injection. The fear appeal delivered a ten-fold increase in behavior change.

Fear appeals are most effective when recommending easy-to-do actions, such as: wearing seat belts, no texting while driving, taking a multivitamin, getting a mammogram, slowing car speed in a school zone, etc.

Figure 3 (on the following page) shows another example how a fear appeal is used while communicating a major incident. In this case, a painter is killed when his scissor truck rises unexpectedly and presses his head against a water sprinkler pipe.

Painter Dies When Scissor Truck Goes Up Unexpectedly

Background Information Figure 3: Painter Dies When Scissor Truck Goes Up Unexpectedly	
Background	Contractors painting a sprinkler pipe. Scissor truck could not get immediately under the sprinkler pipe. Scissor truck had a safety feature called: "two-hands on" operation. To move the truck, or raise or lower the platform, required two hands: one hand depressing an "enable" button with the other hand operating the joystick. Contractor overrode this safety feature by permanently depressing the enable button with a nail. After overriding the enable button, the platform could rise by moving the joystick alone. Painter's safety harness became accidently wrapped around the joystick. As the painter leaned over the platform to reach the sprinkler pipe, the joystick was inadvertently pressed forward. The rising platform squeezed his head between the top rail of the guard rail and the sprinkler pipe running along the ceiling. The painter's head was crushed, and he died.
Easy-to-do Solution	Don't override safety features.
Document Details	Complete document was four pages. Only the cover is shown here. The complete document can be downloaded *Lessons Learned – Communicating Major Incidents* at: www.Larkin.Biz

3. Fear appeals are perfect for the chemical industry

Fear appeals are a perfect fit for communicating major incidents in the chemical industry. In chemicals, we have disastrous events many of which can be avoided with relatively "easy-to-do" actions.

For example, in 70% of lockout/tagout fatalities, the worker did not even attempt any sort of lockout (7). While locking out may be a hassle or inconvenient, it is much easier to do than changing the way you eat, beginning an exercise program, or kicking an addition to cigarettes or drugs.

Or, consider that approximately half the people who die in working-at-height incidents die wearing a safety harness (8). They are wearing the harness, but they have not attached the lanyard to an anchor point. While attaching the lanyard may be inconvenient, it is still a relatively easy-to-do solution compared with kicking a drug addiction or completely changing your diet (hard-to-do actions).

We want to prevent major incidents, and many of the actions needed to do so are relatively "easy to do," for example:

- using a fire watch when doing hot work
- atmospheric testing before entering a confined space
- inspection checklist before using a fork lift
- choosing the right chemical-resistant gloves for a specific job
- using a lock box
- doing a risk analysis just before starting a job

It is the combination of disastrous events and easy-to-do recommendations that make fear appeals so perfect for our industry.

Figure 4 (on the following page) shows how a fear appeal is used while communicating a major incident. In this case a welder was killed when a pneumatic plug fired from an open pipe end and hit his head.

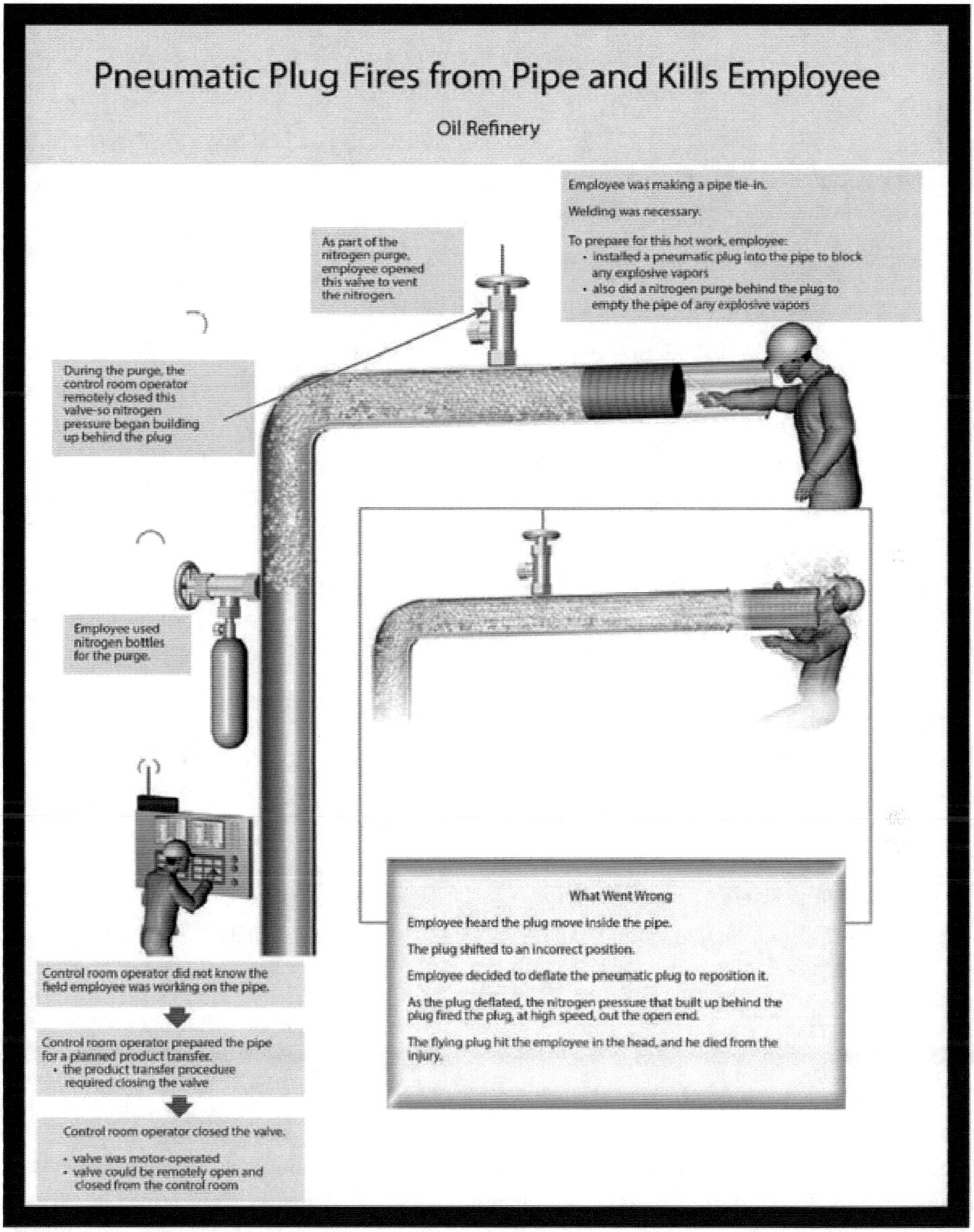
Pneumatic Plug Fires from Pipe and Kills Employee
Oil Refinery
Employee was making a pipe tie-in.
Welding was necessary.
To prepare for this hot work, employee:
• installed a pneumatic plug into the pipe to block any explosive vapors
• also did a nitrogen purge behind the plug to empty the pipe of any explosive vapors
As part of the nitrogen purge, employee opened this valve to vent the nitrogen.
During the purge, the control room operator remotely closed this valve-so nitrogen pressure began building up behind the plug
Employee used nitrogen bottles for the purge.
Control room operator did not know the field employee was working on the pipe.
Control room operator prepared the pipe for a planned product transfer.
• the product transfer procedure required closing the valve
Control room operator closed the valve.
• valve was motor-operated
• valve could be remotely open and closed from the control room
What Went Wrong
Employee heard the plug move inside the pipe.
The plug shifted to an incorrect position.
Employee decided to deflate the pneumatic plug to reposition it.
As the plug deflated, the nitrogen pressure that built up behind the plug fired the plug, at high speed, out the open end.
The flying plug hit the employee in the head, and he died from the injury.

Background Information Figure 4: Pneumatic Plug Fires from Pipe Killing Employee	
Background	Employee making a pipe tie-in. Employee decided to isolate the hot work area. Employee installed a pneumatic plug. Employee did a nitrogen purge behind the plug. Employee opened a valve to vent the nitrogen. Control room operators (not knowing any work was underway) remotely closed the value. With the venting valve closed, nitrogen pressure built up behind the plug. Employee heard the plug move. He decided to deflate and reposition the plug. As the plug deflated, it fired out the end of the pipe. Plug hit the employee in the head and killed him.
Easy-to-do Solution	Lockout the venting valve in the open position.
Document Size	Original document was two pages. Only the cover is shown here. The complete document can be downloaded *Lessons Learned – Communicating Major Incidents* at: www.Larkin.Biz

4. Fear appeals are not unprofessional

Engineers are sometimes uncomfortable with fear appeals. Their training as engineers attracts them to a "just the facts" approach for communicating major incidents. This is a mistake.

Is it "unprofessional" when Mothers Against Drunk Driving show a five-year-old girl saying "happy birthday" to her dead father while visiting his grave at the cemetery? https://www.youtube.com/watch?v=y-Tabvyrf0I

Is it "unprofessional when AT&T runs a commercial showing a dad dropping off his daughters and then, while reading a text, crashing into a young boy chasing a ball across the street? (https://www.youtube.com/watch?v=hVEBJxS2J_Y)

Is it "unprofessional" when home security companies show TV commercials where a hooded burglar kicks in the front door as a young wife and daughter try to escape up the stairs of their home? https://www.youtube.com/watch?v=cMpKDZDjBWM

Try telling these people their fear appeals are "unprofessional." They will respond by telling you their highest priority is protecting innocent people, and they are willing to use fear appeals to do so. Why is the chemical industry any different? Shouldn't we be using fear appeals to protect our employees from the dangers that exist in our industry? And especially, shouldn't we be using fear appeals to stop the same type incidents happening over and over? Do we value a "just-the-facts" approach more than we value the safety of our employees?

The "just the facts," approach—so essential to good engineering—is the wrong approach for communicating major incidents.

A study published in CEP found that this "stick to the facts" approach is not working. This approach is not stopping repeat incidents. This study, conducted in oil refineries, found most major incidents are only remembered for three years after the date of the incident. Unless you were directly involved in the incident, or directly involved in the corrective actions taken after the incident, after three years, it's gone. The collective memory of the incident is gone after three years and it is as if the incident never happened (9). So, it happens again.

5. Separate the analytic investigation of the incident from the employee communication

Major incident investigations use a variety of analytic tools such as "bow-tie analysis" and "root cause analysis."

These analytic tools may well form the basis of your *investigation*, but they should not form the foundation of your *communication*. They are too analytical.

Since Aristotle, we have known: communication that arouses emotion gets more behavior change than communication relying exclusively on reason. Emotion, not reason, drives behavior change (10). Your decision to avoid emotion makes your communication impotent.

This does not mean that analytic findings coming from your investigation should not be included in the communication, of course they should. But it does mean that a fear appeal is the first thing receivers see. It means the fear appeal is the centerpiece of your major incident communication. And then, in a supporting role, analytic information should be added.

6. Only moderate levels of fear are necessary

One particularly helpful finding in the fear appeal literature is that you do not need to fill the page with guts and gore to get the desired effect (3). This research shows that moderate levels of fear-inducing images produce the same amount of behavior change as high levels of fearful images. The relationship between the amount of fear in the image and the amount of behavior change begins with a steep positive curve, but then flattens quickly. After inducing a moderate amount of fear, adding more fear does not add more behavior change.

So, you do not need to do fear appeals that are sensational or shocking. All you need from your image is a small amount of physical discomfort elicited in the receiver. In the most practical terms, when you first show the image to someone, see if they make a calm "eww" sound. If so, that is all you need.

The sample communication shown in this paper are all examples of moderate fear appeals. There is no blood in these illustrations. There are no smashed skulls. There is no one screaming in pain. High-fear appeal illustrations would be unprofessional and more importantly unnecessary. You don't need high fear to get the maximum behavior change.

Since fear appeals work perfectly well with only moderate levels of fear, you can communicate the incident in a respectful and dignified way and still get the behavior change you need to avoid the incident happening again.

Figure 5 (on the following page) shows how a fear appeal is used while communicating a major incident. In this case, an employee cleaning a pipe flange died from nitrogen asphyxiation.

Nitrogen Asphyxiation During Turnaround

Employee killed. Supervisor seriously injured.

Nitrogen asphyxiation while working under a black plastic sheet.

Employee and supervisor cleaned the flange on a 48 inch pipe.

They did a black light inspection looking for any left over dirt or grease on the flange surface.

To see better in the bright sunlight, they covered the pipe opening with a black plastic sheet and climbed underneath.

The supervisor forgot he ordered a nitrogen purge the night before.

Background Information Figure 5: Employee Dies from Nitrogen Asphyxiation During a Turnaround	
Background	During a plant turnaround, an oxygen feeder was removed for cleaning. An employee and a supervisor were tasked with cleaning a large, 48-inch flange before reassembling the oxygen feeder. After cleaning the flange, they decided to do a blacklight inspection, which would reveal any leftover dirt or grease. In the bright noon sunlight, they could not see the blacklight results, so they decided to cover the end of the pipe with a large sheet of black plastic. The employee and supervisor climbed under the black sheet to continue the blacklight inspection. Unfortunately, the supervisor forgot he had ordered a nitrogen purge. The black sheet created a temporary enclosure and trapped the venting nitrogen. The employee died from nitrogen asphyxiation. The supervisor survived but was seriously injured.
Easy-to-do Solution	Post signs warning of nitrogen asphyxiation.
Document Size	Original document was three pages. Only the cover is shown here. The complete document can be downloaded *Lessons Learned – Communicating Major Incidents* at: www.Larkin.Biz

7. Answering the Objections to Using Fear Appeals in the Employee Communication

There are two common objections to using a fear appeal when communicating a major incident. First, that the communication is too dramatic and should be toned down. Second, that the communication will be used against the company in possible legal action.

The objection that the communication is too dramatic is understandable. In our view, this objection is rooted in a desire to avoid embarrassing managers at the site, and to avoid reopening the psychological pain felt by everyone involved. These tragedies involve people who are employee's coworkers, friends, neighbors, and sometimes family members.

This objection (Why cause any more pain?) is understandable but wrong. The ultimate goal here is not minimizing the pain of those at the site, but rather protecting still-living employees from a similar tragedy. The ultimate goal is making sure an incident similar to this one doesn't happen again.

The objectors will respond, the communication should get the behavior change (stopping a repeat of the incident) without reopening wounds at the site. This request is impossible. More than fifty years of research into fear appeals shows that communication without fear appeals produce much less behavior change. Removing the fear appeal often results in communication that results in no behavior change at all. The request to get fear appeal levels of behavior change without using a fear appeal is understandable but impossible.

The second objection is that the fear appeal will be used in possible legal action against the firm.

The problem with this objection is that the fear appeal communication is derived from the much larger, and more detailed, accident investigation. And, the accident investigation is discoverable.

Opposing attorneys will request the accident investigation, and once they have the accident investigation, they have much more information than is contained in the employee communication.

Where the legal objection has merit is making sure the employee communication is an accurate reflection of the details in the more formal accident investigation. These documents are both discoverable, and they should be consistent.

8. Conclusion

Safety professionals need expertise in many areas including engineering and human behavior. This paper hopes to extend their expertise into an additional area: employee communication.

What safety professionals most need to know is that communication with a fear appeal illustration is the best way to change employees. And "changing employees" is what you need if your goal is avoiding hurting or killing another employee in a similar incident.

9. References

1. **Albarracin, D.**, "Fear-Based Appeals Effective at Changing Attitudes, Behaviors After All," *American Psychological Association Press Release* (October 22, 2015).

2. **Ruiter, R.; L. Kessels; G Peters; and G Kok**: "Sixty Years of Fear Appeal Research: Current State of the Evidence," *International Journal of Psychology,* 49, (2) (2014), p. 63-70.

3. **Tannenbaum, M.B.; J. Hepler; R. Zimmerman; L. Saul; and S. Jacobs**: "Appealing to Fear: A Meta-Analysis of Fear Appeal Effectiveness and Theories," *Psychological Bulletin*, 141 (6) (2015), p. 1178-1204.

4. **Witte, K. and M. Allen**: "A Meta-Analysis of Fear Appeals: Implications for Effective Public Health Campaigns," *Health Education & Behavior*, 27 (5), (October 2000), p. 591-615.

5. **Tanner, J.; J. Hunt; and D. Eppright**: "The Protection Motivation Model: A Normative Model of Fear Appeals," *Journal of Marketing*, 55 (3) (July 1991), p. 36-45.

6. **Leventhal, J.; R. Singer; and S. Jones**: "Effects of Fear and Specificity of Recommendation Upon Attitudes and Behavior," *Journal of Personality and Social Psychology*, 2, (1965) p. 20-29.

7. **Bulzacchelli, M. T.; J. Vernick; G. S. Sorock; D.W. Webster; and P.S. Lees**: "Circumstances of Fatal Lockout/ Tagout-Related Injuries in Manufacturing," *Journal of Industrial Medicine*, (51) (2008), p. 728-734.

8. **McCurley, L.**: "Today's Goal is Tomorrow's Safety: Harness Based Work at Height," *Research and Practice for Fall Injury Control in the Workplace,"* Proceedings of International Conference on Fall Prevention and Protection, (November 2011) p. 237-241.

9. **Throness, B.**: "Keeping the Memory Alive, Preventing Memory Loss that Contributes to Process Safety Events," *Process Safety Progress*, AIChE, 33 (2) (June 2014) p. 115-123.

10. **Aristotle**: *Rhetoric*, Translator: W. R. Roberts, ReadaClassic, (2010)

19 AIChE Spring Meeting
15TH GLOBAL CONGRESS ON PROCESS SAFETY
NEW ORLEANS • 2019

The Butterfly Effect – a Case Study on One Company's Transformation of a Process Safety Program

Jeff DeWald
Mallinckrodt Pharmaceuticals
3600 North Second Street
Saint Louis, Missouri 63147

Jeff.DeWald@MNK.com

Prepared for Presentation at
American Institute of Chemical Engineers
2019 Spring Meeting and 15th Global Congress on Process Safety
New Orleans, LA
March 31 – April 3, 2019

The Butterfly Effect – a Case Study on One Company's Transformation of a Process Safety Program

Jeff DeWald
Mallinckrodt Pharmaceuticals
3600 North Second Street
Saint Louis, Missouri 63147

Jeff.DeWald@MNK.com

Keywords: Process Safety Culture, Stakeholder Outreach, Workforce Involvement

Abstract

As process safety professionals we spend a considerable amount of our time trying to sell a concept to company leadership which will help improve components of our PSM program. If valid justification is made and the concept is approved a more difficult challenge awaits us - implementing the idea and turning it into a sustainable program.

This paper tells the story of a process safety professional's dream come true - full approval and funding to transform their company's PSM program. The paper will describe the process for justifying the project as well as strategies (and lessons learned) for mobilizing and implementing the transformation amidst a culture that "has always done it this way and hasn't had any problems". This paper will also discuss challenges and strategies for ensuring a sustainable program once transformation is complete.

Components within the transformation include an entirely new Mechanical Integrity and Instrumented Safety Systems programs as well as revamped Process Safety Information, Hazardous Area Classification, Emergency Response and Pressure Relief Device Programs.

1 Introduction

The concept of the "butterfly effect" theory states that nearly imperceptible changes of a complex system can have large effects over time. Its name was derived from the idea that a butterfly flapping its wings can ultimately cause a cyclone half a world away. In other words, small acts can serve as a catalyst for much bigger outcomes.

Like the butterfly effect, small decisions made in a room years ago have cascaded and grown into tangible differences in the way our company approaches Process Safety. There wasn't a defining moment that tipped the scales for Project Monarch (a multi-year, multi-million dollar project to upgrade Process Safety within the company). Instead, the culmination of persistence, data gathering, and good timing influenced its direction.

This case study will review:

1) the pre-cursors that helped create the project;
2) the subsequent panic and planning that accompanied this endeavor;
3) the mobilization of the team involved in the project;
4) successes and lessons learned along the way;
5) a description of the various programs that have benefitted from the project; and
6) keys for ensuring sustainability once the project has finally ended.

Several important notes should be mentioned before beginning the paper:

1) Project Monarch is still in progress. The case study is meaningful because the project has reached a point where it is self-sustaining.
2) The session in which this paper and presentation resides is titled "Process Safety Culture: Working Through a Transformation." The exacting and observant reader may have noticed that culture was not detailed as a topic of conversation in the introduction, but only because culture is the key topic of the case study and will be highlighted throughout this paper as both a cause and effect of Project Monarch.
3) It appears that this paper has committed to extensive butterfly metaphors. The name of the project was specifically chosen to indicate a transformational change, so each phase of the project will be compared to the lifecycle of the butterfly.

2 Egg (The Creation of the Project)

Although it would be nice to say there was a secret formula to the successful creation of Project Monarch, it was merely a culmination of persistence, patience, data, good timing, and a feisty site director.

2.1 Process Safety Council and Awareness Committee

The formation of the Process Safety Council was a foundational component of the entire project. It was established well before Project Monarch as a result of an internal audit. The intent of the Council was to have functional groups develop programs from the ground up to address key areas from the audit, including Engineering, Operations, and Process Safety. The essential cross-functional basis of the group was positioned as follows:

> "No one functional group aligns with all of the elements. No one element aligns with any one functional group. Many functional groups can affect the success or failure of many elements."

The Process Safety Council reported through a Steering Committee (including Vice Presidents of Engineering; Manufacturing; Environmental, Health & Safety (EHS); Legal; and Research & Development), who were accountable for the success of the Process Safety Council. A designated leadership team consisting of the subject matter experts were responsible for completing the goals and milestones.

The budget for the Process Safety Council was enough to start a few projects, which were primarily driven by team members who were allowed to scale back their daily roles to participate. These team members had numerous opportunities to develop their skills through training and networking.

The Process Safety Council created a number of teams to address key areas of the audit. It was decided several months later that an additional team titled the Process Safety Awareness Committee was necessary. Aptly named, it was the goal of the Awareness Committee to broaden and deepen the knowledge base for process safety at all levels of the organization. Ultimately, this served as a main springboard for the formation of Project Monarch.

One of the Awareness Committee's projects was an intensive Process Safety Management (PSM) training session that was given to anyone in the organization who had the authority to create/change processes or equipment that could be affected by PSM, including Process/Project/Maintenance Engineers and Area Managers. The training provided many case studies (including on a local level) to show how simple decisions could have dramatic outcomes if not thoroughly reviewed.

A slightly less intensive version of the PSM training session was presented to plant management, operations/maintenance supervisors, and other plant functional groups that interconnect with process safety. This process safety basics course was not as detailed as the intensive version, but still demonstrated how process safety affects all functional groups through case studies.

A two-way dialogue was the key component to the intensive PSM training session and the process safety basics classes. During each session, trainers allotted time to address concerns from the audience. The most successful classes were those when the instructor sat back and allowed attendees to respond to their own concerns.

All employees throughout the organization were given monthly reminders from the Awareness Committee with the Center for Chemical Process Safety (CCPS) Process Safety Beacon, as well as reviewing key incidents or near misses within the company with an attempt to make it relatable to the area.

Finally, the Steering Committee and other executives within the organization were updated through periodic meetings where the Process Safety Council discussed progress metrics and small examples from its various teams.

On one occasion, the Awareness Committee conducted an all-day process safety training session for the Vice Presidents of all the functional groups within the organization. The executive training highlighted local case studies, but also justified the business need for process safety and demonstrated the types of decisions that would affect process safety, regardless of functional group. Engaging a large number of executives to participate in this mandatory training was a testament to how much the Awareness Committee had infused process safety into the organization, and it helped open up dialogue with this important group once Project Monarch was initiated.

2.2 The Catalyst

The Process Safety Council slowly built momentum and budget, but a limitation was becoming evident: the cross-functional teams had other day-to-day responsibilities as well. As Council members successfully increased awareness of process safety for the process owners, they also became more aware of the heavy responsibilities that lie with being a process safety advocate. People were getting restless with the slow progress. One such person was a Plant Manager who had endured an internal audit and wanted to prioritize projects. He wanted to ensure PSM concerns were addressed on a project level, while also helping the Process Safety Council meet their goals on a systemic level.

As luck would have it, this Plant Manager was a go-getter and eventually earned a promotion to Operations VP. Retaining the fundamentals of process safety, he asked the EHS group for a tool that would help him prioritize various initiatives, projects, and action items by risk. A concerted effort was made to gather information from all of the plants and put all necessary follow-up items from the various risk analysis methods into the newly developed centralized EHS Management System. By prioritizing risk with data, a story began developing on how to meet the goals of the Process Safety Council.

This is the point in the story where the idea for Project Monarch hatched, primarily because of diligence, data and informed people. An engineering study was ordered to analyze the time and money necessary to develop a full-time, concerted effort to improve the process safety program.

3 Larva (Planning the Project)

Awareness and support for process safety had reached a new peak. Although the Process Safety Council created a good foundation for the core of the process safety program, resources were limited and team members had to divide their responsibilities between the Council and their "real job."

When the new Operations VP took over, he wanted to completely upgrade the company's PSM program at any cost, but a detailed plan needed to be developed to convince the President and CEO of the company. For those in the Process Safety Council, the dream of holding a giant novelty check dedicated entirely to Process Safety quickly turned into a cold realization that a massive amount of planning and preparation had to be done. An engineering study would help determine the best way to coordinate and resource this project.

The EHS Management System helped to identify areas of improvement, but everything was highly interconnected, and it was difficult to determine the scope for each area. A process flow was developed to help define a boundary and starting point for each area. An example of this process flow can be seen in Figure 1.

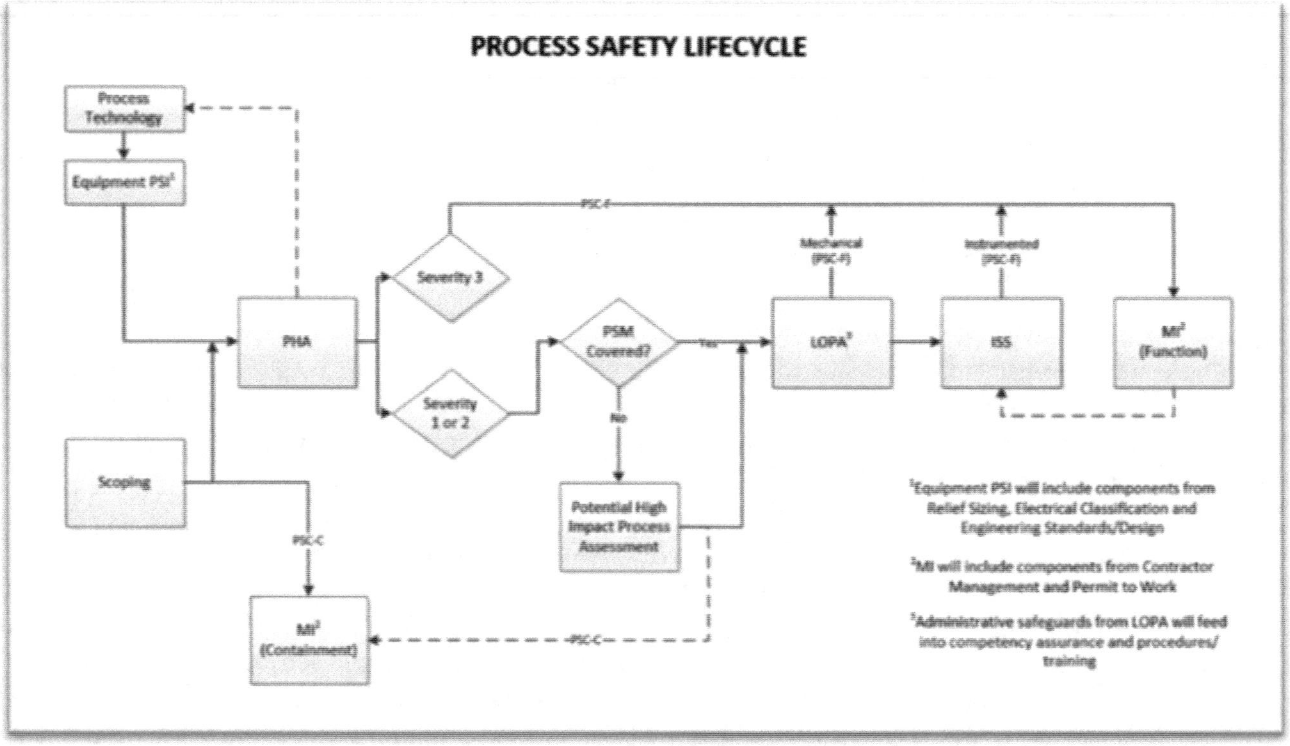

Figure 1. The Project's Workflow

Once the inputs and outputs from each technical area were defined, the dependencies and bottlenecks for each of the areas (called Workstreams) were determined and a crude project plan began to emerge.

The project plan started with laying out the Workstreams, breaking them down into milestones, and defining the dependencies between each Workstream. This was done for the Workstream team itself (i.e., developing the standards) and detailing how it would be implemented at each of the plants. Each milestone had to include an estimated length of time to complete, and flag whether the goal could be met with parallel teams. This was a challenging task because some sites run batch processes in which multiple products can be run on multiple lines.

Utilizing a singular team to complete a Workstream would develop that team to become subject matter experts on the topic. For example, having a single team upgrade the Process Safety Information (PSI) for the 80 processes at a plant, and then feed their output to a team that did the Process Hazardous Analysis (PHA) and Layers of Protection Analysis (LOPA).

Utilizing parallel teams to complete multiple Workstreams would develop the team to be a subject matter expert on the process. For example, simultaneously focusing multiple teams on their respective processes from PSI to PHA to LOPA.

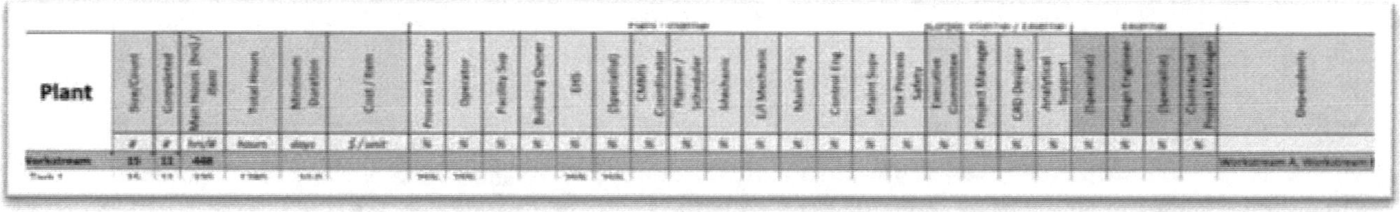

Figure 2. Resource Mapping the Milestones

Once the "easy" part of Workstream mapping at each plant was complete, the more difficult task of assigning a budget and timeline to the overall project loomed. The company's resident "Numbers and Spreadsheet Nerd" (honorary title only) helped the team determine outcomes on two different scenarios: how long would the project take if we had unlimited resources vs. unlimited money?

Unlimited resources assumed that a certain number of people would be hired internally to reduce costs. The challenge of this assumption was that the hired person would have to be occupied full-time and have a specialized focus on a limited number of Workstreams. It would not make sense to hire and train five full-time PHA facilitators, but only have enough work to keep them employed for one year.

Unlimited money assumed that the work would be contracted out to save time. The challenges of this assumption included a limited contractor pool that was qualified and available to complete specific tasks and a limited number of internal plant resources available to assist the contractors. It would not be practical to use 80 consultants to do 80 different PHAs in one week.

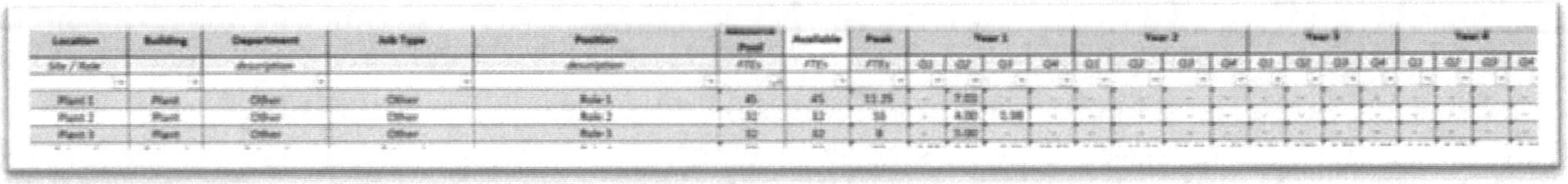

Figure 3. Resource Load by Role

Once the bounds were determined (money vs. people), many iterations were tested to find the most efficient use of time, resources, and money. This information was developed into an executive summary and pitched to the executive team.

It would be difficult to determine and divulge the details of how the project was presented and ultimately accepted in the executive conference rooms. The key point to highlight is that the pitch to the President, CEO, and Board of Directors was made by individuals who were well-informed from their involvement with the Process Safety Council and Process Safety Awareness Committee. Project Monarch gained approval, becoming a multi-year, multi-million dollar company-wide initiative, with its own department and own cost center.

4 Pupa (Transition)

After Project Monarch was approved, transitions began and things became chaotic for a period of time. If one tried to force this transitional phase into a butterfly metaphor, it could almost be said the Project Monarch was cocooning. Many things occurred in a close timeframe during this phase, so this section will be told from the perspective of three different groups: the new Program Manager, the new Core Team and the somewhat established Workstream Leaders.

4.1 Program Manager

The EHS Sr. Director (Project Monarch's champion) had an important decision to make: selecting a Program Manager. The question was whether this person should be someone knowledgeable in the technical aspects of the program, or someone with established influence in the organization. The sage and wearisome reader may be shaking their head at this point, for they know this assessment was not entirely as black and white as stated. Someone with established influence could learn enough of the technical aspects to lead Project Monarch. Likewise, certain people with technical expertise could learn to navigate the political waters with mentoring. Ultimately, the appointed Project Manager was an individual with established influence based on the three primary goals of Project Monarch: (a) to move quickly/efficiently, (b) to challenge the status quo, and (c) to implement sustainable changes.

The newly established Program Manager had the unenviable task of trying to create a team for Project Monarch while learning the intricate technical details. Fortunately, he had a wealth of available technical resources, in addition to the detailed project engineering study, and the experience of the Process Safety Council to help form the strategy.

The existing pool of technical expertise led the Workstreams. Meanwhile, the Program Manager worked on creating a Core Team to help socialize this high-profile project and tie together Workstreams to deliver sustainable outcomes.

During this transition phase, the Program Manager spent significant time with the Project Monarch team and leadership teams at each of the various plant sites (Site Leadership) to stress the sudden importance of Project Monarch and determine how best to implement and demonstrate results at the site level.

4.2 Core Team

The magnitude of the Project Monarch made it critical to use additional resources to help socialize the project across the company, communicate the plan to the affected plants, and show successes to justify the project. This responsibility had previously fallen on the technical experts in the Process Safety Council, but most of these technical experts had to focus on the increased responsibility of leading the Workstreams. The Project Manager established a Core Team with a broader set of traits to ensure Project Monarch's goals would be appropriately socialized.

The Core Team involved two separate groups: Project Management and Risk Management. The five people on the Project Management side focused on project planning, change management, maintenance excellence, project administration and communication. The six people on the Risk Management side focused on tackling mitigation items that had been identified from the Process Safety Council, or to serve as extra resources for the Workstreams as they worked on their goals.

4.3 Workstream Leads

The Workstream Leads were primarily team leaders from the Process Safety Council; only now they had a larger scope, more budget, and dedicated resources at their disposal. Workstreams from this group included Process Hazard Evaluation (i.e., Scoping, PSI and PHAs); Functional Reliability (i.e., LOPA and Instrumented Safety Systems (ISS)) and Mechanical Integrity.

Other initiatives from within the EHS group were brought into Project Monarch as well. Workstreams from this group include Emergency Response and Occupational Safety.

The Workstreams were rounded out with expertise from the Engineering Department. Workstreams from this group include Electrical Area Classification and Pressure Relief Devices. Even though the Workstream Leads from Engineering were not dedicated full-time to Project Monarch, they became project managers of their Workstreams and used contractors to help complete their tasks. The Engineering group was given permission to focus on Project Monarch because the Governance Council agreed to limit their workload. The Engineers were able to charge their time to Project Monarch and were given performance goals associated with Project Monarch.

Embedding the Engineering group as Workstream Leads was a critical component of Project Monarch's success. These individuals eventually became technical experts, and naturally incorporated their expertise into their normal projects.

4.4 Challenges During the Transition Phase

A couple of challenges existed between the established technical experts and the newly developed Core Team. The first challenge was learning how to speak the same language. Most of the Core Team had to learn the process before they could start planning, which involved having technical conversations with the Workstream Leads. Many conversations between the Core Team and Workstream Leads in the early days of Project Monarch focused on explaining the history and projected path of the project before getting to the real issue.

Process Safety Acronyms

ANSI	American National Standards Institute
API	American Petroleum Institute
ASTM	American Society for Testing and Materials
BPCS	Basic Process Control System
CCPS	Center for Chemical Process Safety
IEC	International Electrotechnical Commission
IPL	Independent Layer of Protection
ISA	International Society of Automation
LOPA	Layer of Protection Analysis
OSHA	Occupational Safety and Health Administration
MI	Mechanical Integrity
MOC(A)	Management of Change (Authorization)
MOC	Materials of Construction
MSDS	Material Safety Data Sheet (under OSHA)
NFPA	National Fire Protection Act
NIOSH	National Institute of Occupational Safety and Health
PEL	Permissible Exposure Limit
PFD	Process Flow Diagram
PHA	Process Hazard Analysis
P&ID	Piping and Instrument Diagram
PPE	Personal Protective Equipment (such as gloves, respirators, etc.)
PSI	Process Safety Information
PSM	Process Safety Management
PSSR	Pre-Startup Safety Review
QRA	Quantitative Risk Analysis
RAGAGEP	Recognized and Generally Accepted Good Engineering Practices
RMP	Risk Management Planning
SIL	Safety Integrity Level
SIS	Safety Instrumented System
SIF	Safety Instrumented Function

Industry / Company Terms

CBI	Confidential Business Information
HV	Hand Valve
NESHAP*	Plant term - same as TOx
NRC	Nuclear Regulatory Commission
PSV	Process Safety Valve (pressure relief)
R&D	Research and Development
TOx	Thermal Oxidizer
UPS	Uninterruptable Power Supply
XV	Refers to an automatic valve

Quality Terms

CAPA	Corrective and Preventative Action
FDA	Food and Drug Administration(21 CFR)
IQ	Installation Qualification
OQ	Operational Qualification
QA/QC	Quality Assurance/Quality Control

Figure 4. Process Safety Translator

Another major challenge was establishing and transitioning responsibilities from the Workstream Leads to the Core Team. Many Core Team members were brought in to challenge the way Process Safety had "always been done" and to ensure outcomes could be sustainable. An appropriate balance had to be established between the two groups to ensure quality output, while being efficient and straightforward. It was critical that these individuals were action-oriented, innovative, flexible, and courageous.

5 Adult (Project Monarch)

Present day: the butterfly emerged from its cocoon. It has been three years since the inception of Project Monarch — the kinks were smoothed out and a rhythm was established. Various priorities ebb and flow, but Project Monarch has generally maintained its original scope and budget. The effort has established itself enough to extract key lessons to apply to future successes.

5.1 Project Management: Key Lessons

5.1.1 Turnover and Role Shifting

One discovery was that temporary projects tend to have a high turnover rate. The Core Team managing the project is no longer in place, and the third Program Manager is presiding over the Project Monarch. This is not a comment on the instability of the project, but rather the result of the project falling into a stride that no longer requires constant socialization and planning.

During the first year of Project Monarch, the Project Manager needed to be highly visible and constantly communicate with sites. Now, its justification is well accepted and communication processes are successfully integrated into the routine activities of each site. It is still the Program Manager's responsibility to demonstrate the overall progress of the project through metrics and goals, but most of the communication happens at the site level, where Site Leadership now holds Workstream Leads accountable to report their status.

5.1.2 Budget Cuts

High-profile projects are not immune from company-wide budget cuts. In lean times, it was important to keep core components of the Workstreams moving, while recognizing that other components may have a delayed start. During these periods, our Governance Council made decisions to allow Project Monarch to trade money for time (that is, to allow us to extend the project's timeline in exchange for an annual budget cut).

5.1.3 Communication

Communication cannot be overstated. The Project Manager regularly communicates program-level issues to Site Leadership, while Workstream Leads tend to work out the details with the technical experts at each site. It is critical not to forget to include middle management in the chain of communications. When area managers or functional group managers are not aware of Workstream plans, competing initiatives with their own goals quickly arise.

The best strategy is to assume that middle management is not aware of Workstream goals for the year, even if Site Leadership has approved plant resources for the goal. When specific communication is provided to middle management around Project Monarch goals, they can appropriately support the plans, and minimize arguments around priorities and competing initiatives. A wide number of operational excellence tools exist for this type of exercise that enable all departmental goals to roll up to the site for resource and budget planning.

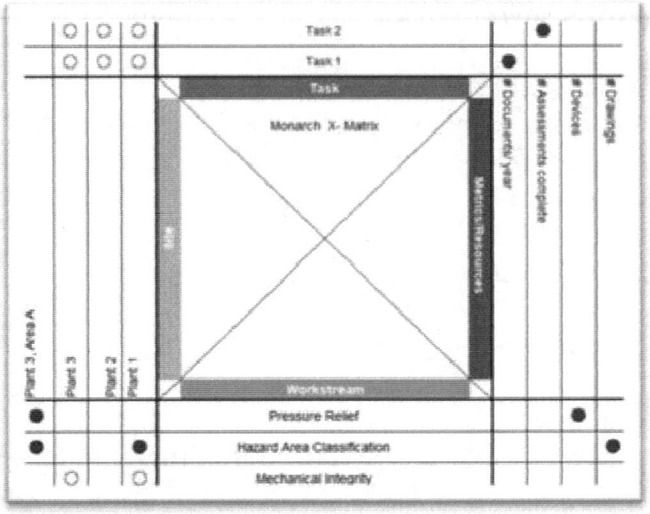

Figure 5. Site Level X-Matrix Project Planning Tool

5.1.4 Capital

Project Monarch was defined as a way to create resources and systems that help to mitigate risk. Individual sites have to provide the capital and related resources for any final installations, which ultimately gives the sites more ownership of Project Monarch (instead of having "them just do it all for us"). Arranging capital management this way also helps to ensure that Project Monarch-related funds are used appropriately, and not for useful but unrelated work that would otherwise expand the scope.

Money was set aside in the capital budget for Project Monarch-related activities at the plant sites, so it was clear that existing budgets would not have to cover the additional work.

5.2 Workstreams: Key Achievements and Lessons

This section highlights the various Workstream accomplishments to date, and calls out individual successes and lessons.

5.2.1 Process Hazard Evaluation

The Process Hazard Evaluation Workstream is dedicated to gathering information and making informed decisions to evaluate risk. The Workstream Lead resides in the corporate Process Safety group, and is assisted by the Risk Management group in the Core Team and process safety teams at the plant sites. There was a heavy focus on this Workstream during the first year, as the products from this group were highly visible and generally served as the starting point for most of the other Workstreams. Products from this Workstream included:

Process Technology Manuals – a standard format was developed for cataloging the process technology portion of Process Safety Information. Centralizing the information in a standard format on the company intranet site made it easier to determine what was appropriate.

Figure 6. PSI Document Center on Library Intranet

PHA Level Loading – as products were added to the sites over time, it created an uneven PHA schedule from year to year. An evaluation of the PHA schedule over the five-year period identified where there was an overabundance of PHAs in a certain area at one time. The Core Team helped facilitate extra PHAs in order to level load the 5-year schedule.

PHIP – Potentially High Impact Processes. This was defined internally as processes that are not technically covered by the PSM regulation, but still have high severity scenarios associated with them. PHIP are typically identified in a PHA or other risk analysis. A system was developed to identify and document these processes. Workstreams incorporated PHIP within their scope.

5.2.2 Mechanical Integrity

The Mechanical Integrity Workstream primarily involves the maintenance and engineering groups. Project Monarch-staffed positions include the Workstream Lead, a certified inspector, two site coordinators, a specialist, and a Computerized Maintenance Management System (CMMS) administrator. The group also uses a few contracted inspectors and a specialist. Products from this Workstream include:

Equipment Documentation – the central record of equipment resides in the CMMS. This record is enhanced by including asset information, the bill of materials, and links to supporting information. Drawings, such as isometric drawings and Piping & Instrumentation Diagrams (P&IDs), are stored in a drawing management system and linked to the CMMS record. Design files and repair reports are stored in a document repository and linked to the CMMS record. This information loads up the

CMMS with information that can then be used by anyone who is maintaining, operating or improving the equipment.

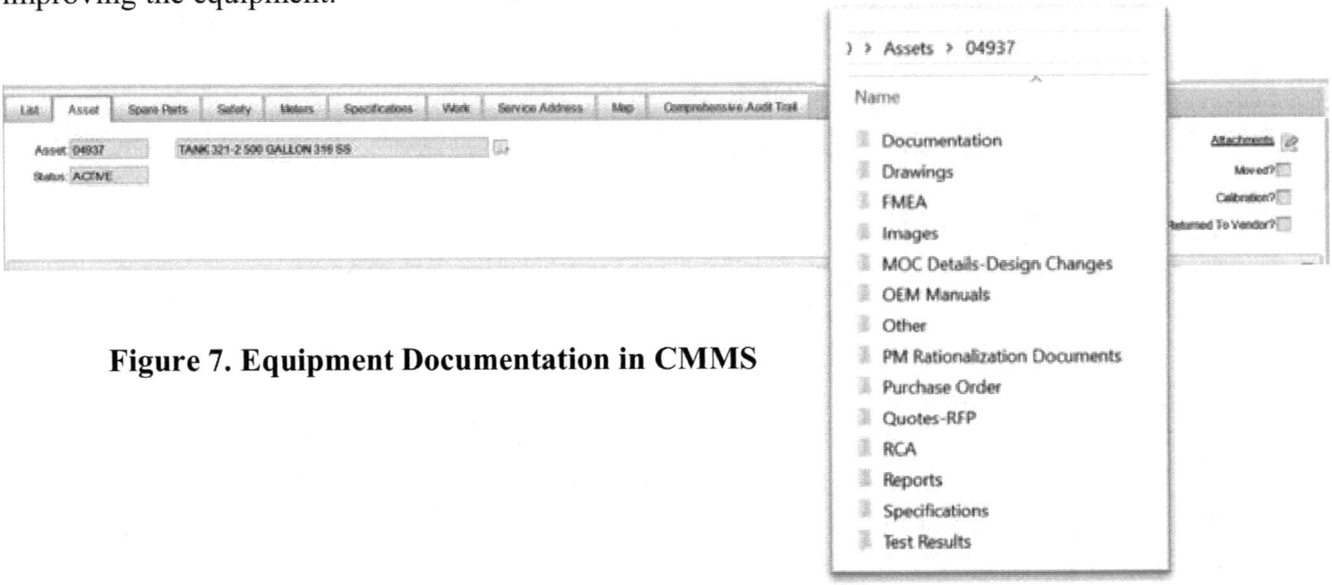

Figure 7. Equipment Documentation in CMMS

Definition of RAGAGEP – defined as the Recognized and Generally Accepted Good Engineering Practice for the maintenance of the equipment. In some cases, it was identified that the standard maintenance frequencies used by the plants were much more frequent than RAGAGEP, which helped improve efficiencies in the maintenance of the equipment.

Equipment Inspection and Testing – any inspection deficiencies are documented, and the criticality of the repair is communicated according to the risk, as determined by the inspector. This has been the most visible part of the Workstream, producing the most successes. The inspections helped prevent a number of loss of containment events that otherwise may not have been caught.

It is important to note that as the site maintenance groups become more familiar with the RAGAGEP, they also become more empowered to ask what the RAGAGEP requires and recommends. This knowledge enables the maintenance group to better define the boundaries of their work processes and optimize accordingly.

One key lesson was that familiarity with the RAGAGEP also opens up interpretation questions by individual sites. As a result, the strategy to follow a particular RAGAGEP becomes dependent on the individual site and is largely influenced by their local jurisdictions. Once the RAGAGEP is decided it is important to agree on and consistently adhere to the RAGAGEP strategy instead of changing course midway through implementation.

5.2.3 Pressure Relief

The Pressure Relief Workstream is led by a project engineer using a combination of consultants, internal engineers, and a full-time Project Monarch engineer. Products from this Workstream include:

Pressure Relief Devices – this team re-evaluates the pressure relief devices to ensure the design basis is documented and to provide a standard format. Working closely with the Process Hazard Evaluation Workstream, they ensure that appropriate scenarios have been defined and that reactivity data is available. Each device has a standard report and supporting documentation that is stored in one place and is easily accessible for PHAs. The Workstream has a defined scope and each area is treated like a project. A large amount of upfront work from other functional groups is required to ensure the scenarios are defined. After that occurs, the Workstream can do the rest of the work without assistance, unless recommendations are issued. A full-time person was hired by Project Monarch to help with continuity (i.e., to help in times of lean budgets, to review the contractors work and to become the company subject matter expert on relief devices).

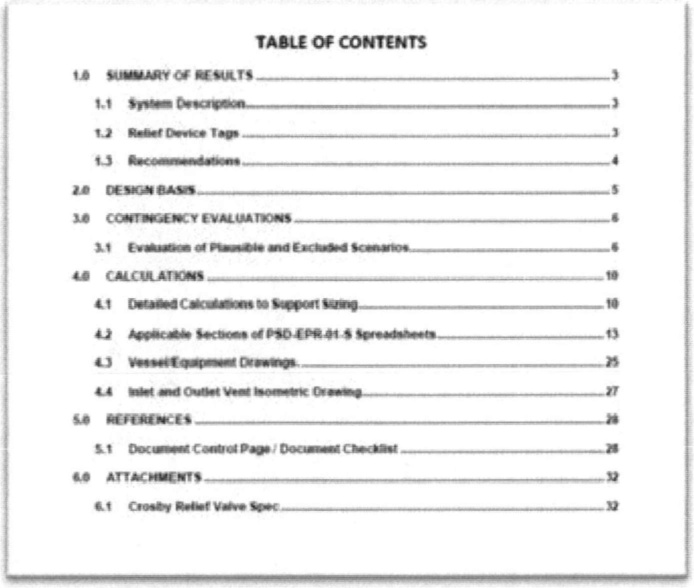

TABLE OF CONTENTS

1.0 SUMMARY OF RESULTS 3
1.1 System Description 3
1.2 Relief Device Tags 3
1.3 Recommendations 4
2.0 DESIGN BASIS 5
3.0 CONTINGENCY EVALUATIONS 6
3.1 Evaluation of Plausible and Excluded Scenarios 6
4.0 CALCULATIONS 10
4.1 Detailed Calculations to Support Sizing 10
4.2 Applicable Sections of PSD-EPR-01-5 Spreadsheets 13
4.3 Vessel/Equipment Drawings 25
4.4 Inlet and Outlet Vent Isometric Drawing 27
5.0 REFERENCES 28
5.1 Document Control Page / Document Checklist 28
6.0 ATTACHMENTS 32
6.1 Crosby Relief Valve Spec 32

Figure 8. Pressure Relief Device Report (Table of Contents)

Pressure Relief Discharge – A guideline was developed to define a term within the Pressure Relief Device standard stating that a device must vent to a "safe location." The Pressure Relief Discharge RAGAGEP is used with the Pressure Relief Device effort to define the design basis for the discharge of the relief device. The result is the establishment of an internal refined modeling network. The team developing this guideline reached out to an external expert to review the document and to help run a pilot program, which provided valuable information for the end product.

Combustible Dust – a project engineer who serves as the company's resident dust expert reviews and standardizes the design basis on combustible dust systems under a separate Workstream. This project engineer holds more of a technical services role than a project role. One of the deliverables from this project includes a drawing that shows the exclusion zone from a relief panel, so that future engineers know not to build in the relief discharge area. An unintended but pleasant secondary consequence of this arrangement allows the engineer to create projects from any recommendations that may arise.

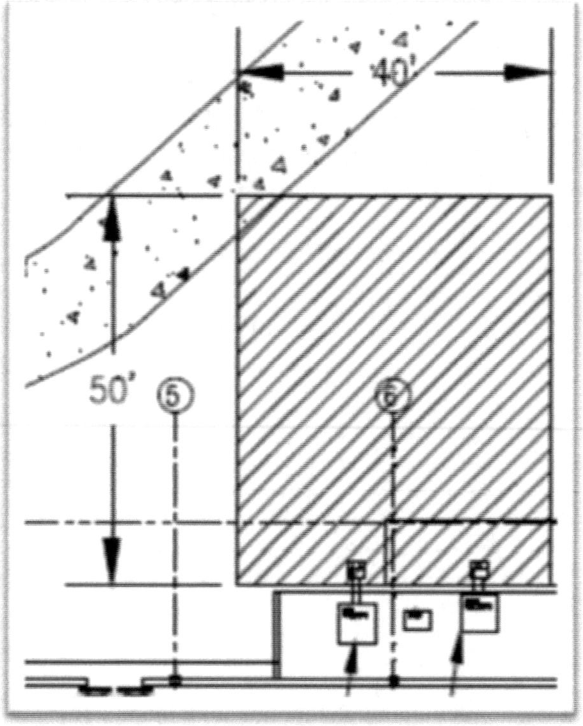

Figure 9. Pressure Relief Device Report (Table of Contents)

5.2.4 Occupational Safety

The Occupational Safety Workstream manages some projects that are initiated within the EHS Group. Products from this Workstream include:

Flame Resistant Clothing – the evaluation process for Flame Resistant Clothing was standardized and documented. This work led to a more streamlined procurement process for clothing at the sites, which was combined into one contract.

Work Permit Process – this is an ongoing project to document the work permit process and implement it into an electronic and mobile platform. The project affects a large number of people in the operations and maintenance groups. A lesson learned from this activity is that people accept technology in varying degrees of vocal intonation. To help get people more familiar with the new technology, a Community of Practice Forum was developed. This forum enables users to express and communicate issues and concerns, share improvement opportunities, and share best practices.

5.2.5 Electrical Classification

Similar to Pressure Relief group, the Electrical Classification Workstream is led by a project engineer using a combination of consultants and engineers at the sites. Products include:

Electrical Classification Report – this effort begins with a Determination phase to review and document the design basis into a standard format and update the electrical classification drawings according to RAGAGEP. Before beginning work in each area, the team conducts a kickoff meeting for the engineers and their managers to explain the process and the scope. An Audit occurs after the Determination and recommendations are compiled. Recommendation resolution philosophies differ from site to site, which necessitates the audit team and the site to engage with each other to determine an appropriate resolution. The Workstream Lead manages quite a few projects at any given time, so picking a reliable consultant has been key. This Workstream conducts awareness training after each project, which has been helpful for the sites as they inherit the program.

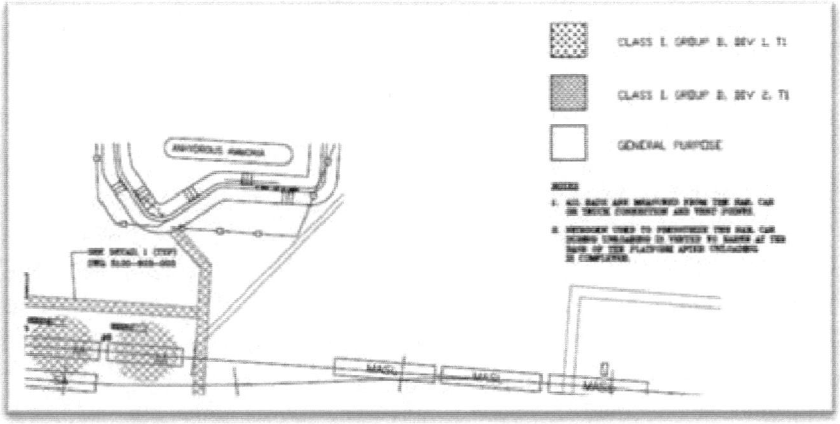

Figure 10. Electrical Classification Drawing

5.2.6 Functional Reliability

The Functional Reliability Workstream primarily works with the instrumentation and controls groups. Project Monarch staffed positions include the Workstream Lead and an ISS Engineer. The group also uses contracted assistance for logic review, Human Machine Interface (HMI) design, and CMMS integration. Products from this Workstream include:

Instrumented Safety Systems (ISS) – the definition expands beyond the traditional RAGAGEP of Safety Instrumented Systems (SIS) scope to include Instrumented Protective Functions (i.e., non-SIS instrumented safeguards credited in LOPA). The primary lesson from this Workstream was the importance of understanding the customer. The team decided to use the existing CMMS system for ISS asset management because a large, unrelated project was underway at the same time to add the Instrument Database to the CMMS. The ISS team learned that it was important to work with all levels of the instrumentation and controls groups to gain access to the control systems and modify within the allowed boundaries of existing plant standards.

Alarm Management – this includes a baseline assessment, bad actor analysis, and documentation/rationalization (D&R) according to the industry RAGAGEP. Similar to lessons learned from the ISS efforts, it is important for the alarm management group to work closely with the plant controls group since information is being pulled from the secure manufacturing network. Alarm philosophies vary for each plant and even within the plant, influenced by their control strategies and modes of operation. The D&R can be a time consuming process that is a similar structure to a PHA, but it should not be done with the PHA. Successful sites incorporated the D&R phase into existing projects, as internal resources were almost always an issue within the controls group.

5.2.7 Emergency Response

Similar to Occupational Safety, the Emergency Response Workstream is an initiative within the corporate EHS group and used project management tools to standardize the process. The main product from this Workstream was a detailed evaluation of the design basis of the fire protection systems, which were documented in a standardized format. Similar to most Workstreams, communication was key to ensuring sites were aware of the effort and to coordinate a collaborative effort to tackle any recommendations.

5.3 Collaboration: Key Lessons

After three years, each Workstream produced tangible contributions to the overall progress of process safety at the sites. Sites have shifted their mindsets from “we might as well listen to these Monarch people because they aren’t going away” to a genuine collaborative effort. Constant communication and awareness building was critical to the success of this collaboration. This is evident within Workstreams, where sites now reach out to Workstream leads for help with projects or planning. Having Workstream leads act more like consultants to the site gives facilities greater ownership and input on how to achieve common goals.

6 Migration

The Monarch butterfly perpetuates its lifecycle by migrating each year to certain areas in the South and Mexico. As Project Monarch begins to wind down, it must also transition to a phase where it can be a sustainable process for the sites. For example, if the original effort began as 80% Project Monarch and 20% sites, the distribution would look equal now; at the conclusion of the project, it should primarily be a site-driven effort. It isn’t to say that the Project Monarch effort is diminishing, but that the site’s involvement is increasing.

When Workstreams begin wrapping up their projects, the sites will start to absorb roles from the Project Monarch pool. It is important to maintain continuity during this transition and to ensure that sites still have appropriate resources. Even at this early stage, succession planning and training is underway to ensure that the entire fate of a Workstream’s stability does not rest on one person’s shoulders.

Smaller sites may not be able to justify absorbing a role because the workload isn't available. In these cases, the sites will have to rely on the corporate support system. Therefore, it is critical for EHS and Engineering to have enough expertise to keep the Workstream going.

6.1 Keys to Sustainability

6.1.1. Spot Audits

Many functional groups indicated that spot audits are necessary to keep the program sustainable. However, the spotlight from the obligatory tri-annual PSM audit can sometimes be too intense to provide an appropriate assessment on the status of a certain Workstream. Workstream Leads can become defensive about the state of their program during a tri-annual PSM audit, under the pressure of an auditor trying to find non-compliance items as Site Leadership get daily briefings.

Since Project Monarch is more of a continued improvement issue than a compliance issue, spot audits provide a more relaxed way of gauging the overall status, where the intent is to make course corrections on a honed program.

6.1.2 Managing Skepticism

Skeptics will always be a key challenge with any project that has sweeping changes. A single curmudgeonly voice of skepticism, especially if that voice is in a position of influence, can negatively impact the progress of a project or a Workstream. Establishing a network of people that understand and support the process on all levels of the organization can help minimize potential damage by skeptics. Communication is vital to maintaining the support of this network. There is no proven method to working with skeptics other than to have everyone within the Project Monarch, the Workstream and the sites deliver a consistent message and to stay vigilant.

6.1.3 Competency Assurance

Turnover is to be expected at the sites, and irreplaceable people will eventually need to be replaced. Because of the relentless efforts of the Awareness Committee to build a foundation of process safety knowledge within the company, Project Monarch can take that embedded knowledge and ensure it never lessens over time by developing specific curricula and training for the Workstreams.

6.1.4 Change Management and Data Management

As Project Monarch winds down, people with less time and expertise will be asked to keep a Workstream going. To help ensure sustainability, each Workstream within Project Monarch was asked to add a few questions to a Management of Change functional review checklist to ensure that documentation will be kept evergreen as changes are made. The questions are meant to prompt the change owner to think about the following aspects of Project Monarch:

- How the change will affect the Workstream?
- Who should the change owner contact as the subject matter expert?
- What documents should be updated as a result of the change, and where they are located?

On the topic of data management, it is difficult for a large group of people to decide how and where to store data. Short of having an extensive data management project, it was learned that the next best approach is to define and require specific storage areas for the data.

7 Conclusion

The intent of this case study was not to give a roadmap for overhauling a process safety program. As all organizations have different cultures and philosophies, it is acknowledged that there are different approaches to a successful journey. At the very least, this case study should inspire readers to think about how they would justify a process safety transformation, or develop a subsequent implementation plan if they were given freedom to make sweeping cultural changes. By actively thinking about the process, it can trigger better insights about where to start.

However, the major lessons learned from Project Monarch over the years are universal:

- Data drives decisions (especially for the people charged with approving the project). Gathering all recommendations from all sources of risk analyses in one place and determining a method for categorizing and prioritizing is the first step to creating a business case.
- Awareness and communication empower decision makers and give them more responsibility to speak up when process safety systems are lacking. The more people know, the more they can collaborate to solve issues, whether it is at the program-level or the project-level.
- Butterflies are a great metaphor for just about anything in life.

Where There is a Will There is a Way – A Story of Organizational Transformation

Richard Emerson
Bangkok Synthetics Company Ltd
Thailand
Richard_E@bst.co.th

Palaniappan Chidambaram
DuPont Company Singapore Pte Ltd
Singapore
Palaniappan.Chidambaram@dupont.com

Prepared for Presentation at
American Institute of Chemical Engineers
2019 Spring Meeting and 15th Global Congress on Process Safety
New Orleans, LA
March 31 – April 3, 2019

Where There is a Will There is a Way – A Story of Organizational Transformation

Richard Emerson
Bangkok Synthetics Company Ltd
Thailand
Richard_E@bst.co.th

Palaniappan Chidambaram
DuPont Company Singapore Pte Ltd

Keywords: Process Safety Implementation, Organizational Culture

Abstract

Successful organizational transformation requires clear vision, inspirational leadership, management commitment, effective communication, and engagement of employees at all levels. The Bangkok Synthetic Company Limited (BST), the largest producer of mixed C4 products (butanes, butenes, butadiene, etc.), including synthetic rubber, in Southeast Asia operates plants in two sites in Thailand and employs approximately 700 people. In 2013, BST embarked on a journey to turn their safety vision, "no harm to anyone anytime," into a reality with a focus on process safety management and cultural transformation.

This presentation will share some of the key leadership approaches to engage and transform the organization, and the outcomes and benefits realized in the process.

1 Case for change

In 2013 the Industrial Estate Authority of Thailand (IEAT), the governing body responsible for permitting and regulating the chemical industry in Map Ta Phut, Thailand, communicated that they were developing and issuing a new regulation requiring chemical companies to have a process safety management system in place as part of the renewal of operating permits. This communication, coupled with its safety performance history, provided BST with a case for change.

Fortunately, the management organization at BST recognized the need to take a stronger ownership role in both the development and implementation of the PSM program and provided the necessary resources and support. As part of the implementation, BST informally assessed where it had sufficient capabilities, and where external expertise was required. As a result of this assessment, BST decided to engage DuPont Sustainable Solutions (DSS) for improving the Process Hazard Analysis (PHA) capabilities.

2 PSM implementation initiative

With a desire to go beyond compliance in order to manage risks effectively, BST invested a significant amount of resources (people, time and effort) to strengthen their existing safety, health, and environmental (SHE) procedures, and develop internal competencies and capacity in Process Hazard Analysis, while focusing on implementation and improvement of other process safety management elements. A risk-based approach was adopted in the PHA implementation in three units with a focus on high hazard sections. Identified risks were prioritized, and action plans were drafted to implement the recommendations to mitigate risks. The BST Board of Directors approved significant capital investment to implement the recommendations proposed during PHA assessments over multiple years.

In addition to the PHA implementation program, BST also embarked on a two-year implementation journey for the rest of the PSM elements. Engineers and managers from the line manufacturing organization formed multiple teams to develop and implement the procedures for each element. A governance team, which included the PSM Element Team leaders and the senior leaders in manufacturing, met monthly to review progress. The manufacturing leadership team reviewed and approved each procedure and the implementation plans. At the end of two years, BST had completed its initial implementation of the PSM program.

With a focused effort on implementation, BST saw positive changes in attitudes, mindset, capability, awareness and confidence among the team. This prompted BST to consider the impact senior and middle management could have on the entire organization by better engagement with front-line supervisors and employees.

3 Culture transformation journey

3.1 Culture Assessment and Feedback

BST leadership agreed on the need to move the organization from a mindset focused on compliance to one focused on commitment. BST decided to engage DSS in late 2015 to conduct a formal safety culture diagnostic which included surveys, interviews, focus groups and site assessments. The results of this assessment were presented in the form of a gallery that enabled BST leaders to reflect on the current cultural state individually and discuss as a team. The diagnostics revealed a prevalence of a silo mentality, compliance driven mindset that was not fully compliant, one-way communication, and lack of alignment on the value of safety among the management team. In addition, safety was viewed as a "manufacturing issue", not a company issue. These revelations were somewhat of a shock to management, and they committed to change. A strong desire to shift from dependent state (compliance mindset) toward interdependent state (commitment mindset) was expressed by the leadership team.

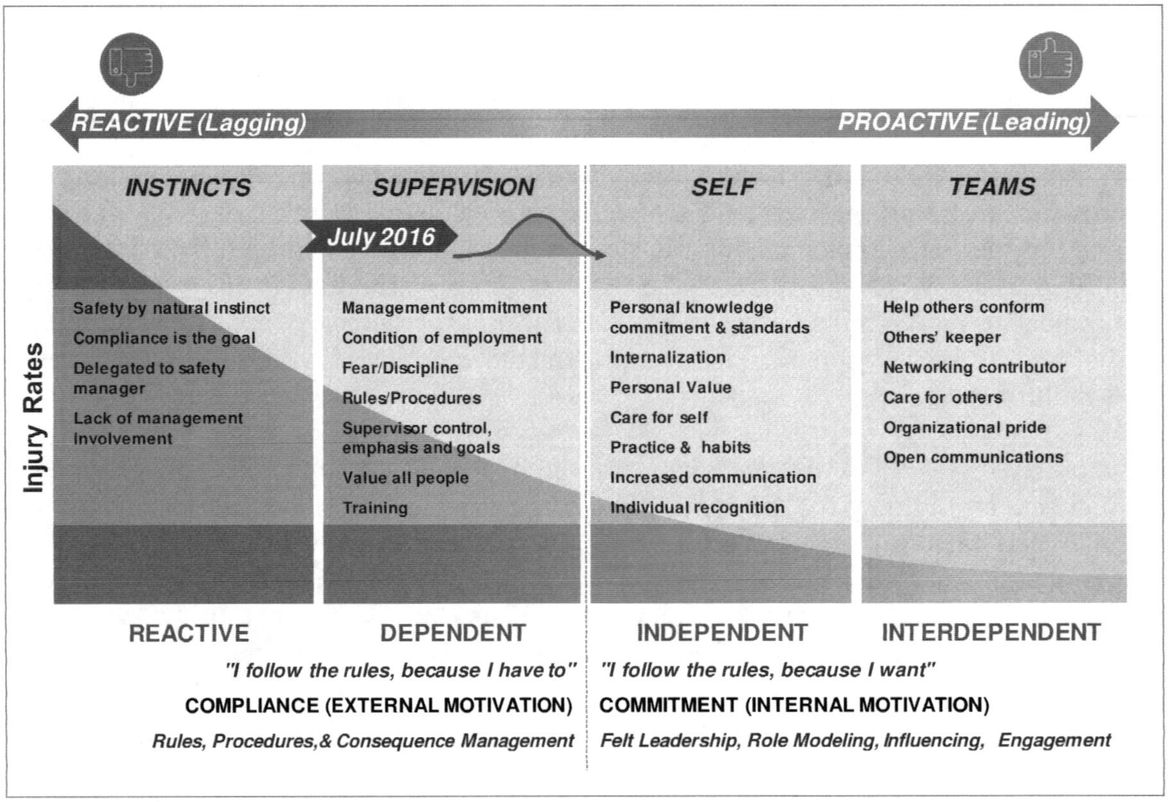

DUPONT BRADLEY CURVE – CULTURE MATURITY MODEL

Based on these insights, the key action taken by the leadership was to actively involve and engage the entire organization in the culture transformation process. This included understanding the current state, realizing the need for change and developing the vision for the future. Several workshops engaging the senior and middle management were conducted to envision the future with more clarity, and to lead and manage change effectively. During the workshops, BST's SHE Vision "No harm to anyone anytime" was reaffirmed. However, the feedback from the engagement indicated that the understanding, commitment, and practices were not as strong as they needed to be to achieve this vision. The workshops were designed and conducted to address this need for improvement.

3.2 "One Company Culture" and Leadership Alignment

Twenty-five senior and middle managers from different manufacturing and supply chain functions at the sites embarked on a coaching program with a focus on four core competencies: the value for safety, my role in safety, engaging others and felt leadership. In addition to improving the understanding of the function and importance of leadership in achieving safety performance improvement, leaders learned how to interact differently with subordinates and contractors. The approach to discussions and engagements shifted from a "tell" approach to an "ask" approach. This naturally reduced the tension felt during engagements and turned the tone from one of a "police interrogation" into a learning, engaging, and partnering environment. Leaders learned to listen first, and then to respond appropriately. Safety Observation Tour (SOT) interactions became much more engaging and became learning experiences for both leaders and others. It also resulted in new insights into existing observed behaviors. This engagement process enabled employees and contractors to see and mitigate risks themselves rather than having others point out the risk and tell employees what to do.

An operational discipline (OD) program was initiated with the front-line supervisors to improve the effectiveness and ownership of management system implementation. The operational discipline initiative focused on evaluating the individual and organizational characteristics, followed by front line employees identifying improvement areas and presenting the action plan to management for approval. Natural work teams were formed in each work area and focused on defining the standards and practices to be applied in their areas. The teams conducted OD performance self-assessments using the DSS framework and developed action plans to improve their performance. Progress was presented periodically to management by the front line. The increase in team and individual ownership and accountability for safety was a clear outcome. Progress and actions, including pictures of improvements, were posted in the teams' areas for the organization to see. The organization shifted from a top down approach to safety ownership at all levels and functions.

To address the perception that safety was only a manufacturing issue, an SHE cascade process was designed and executed for the first time in 2017. A concerted effort was made to align management throughout the company to understand the vision, values, principles and requirements of employees and managers associated with safety performance. The objective was to own them as a management responsibility, and then cascade this understanding all the way down through the organization to include annual contractors. A significant amount of time was invested by the senior management in the cascade process to ensure alignment in the organization. The cascade process encouraged feedback. Questions were collected and then analyzed to define improvements in both the clarity of messages and practices. These improvements were built into a second round of cascades, and feedback indicated a desire by the organization to make the cascades an annual practice.

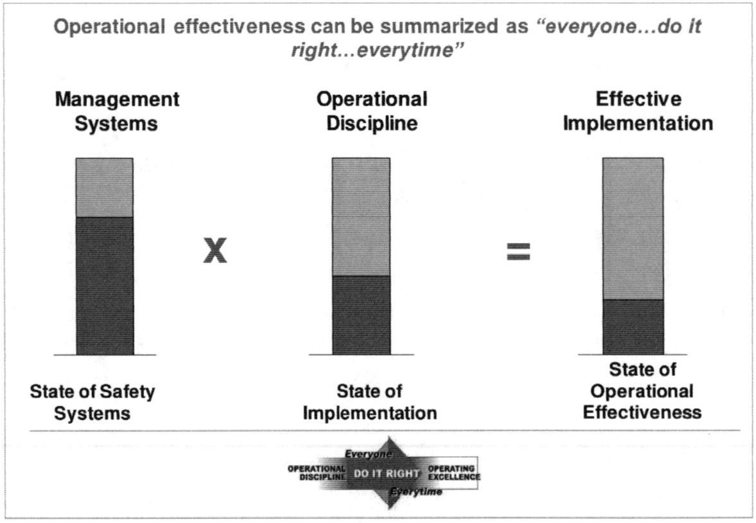

- ➔ Leadership by example
- ➔ Sufficient & capable resources
- ➔ Employee involvement
- ➔ Active lines of communication
- ➔ Strong teamwork
- ➔ Shared values
- ➔ Up-to-date documentation
- ➔ Practice consistent with procedures
- ➔ Absence of shortcuts
- ➔ Excellent housekeeping
- ➔ Pride in the organization
- ➔ Continuous Improvement
- ➔ Just and Fair
- ➔ Risk Sensitivity

3.3 Shift in Values, Attitudes and Beliefs

The overall change in mindset was felt in the organization at all levels and decisions made reflected the value given to safety over production, cost, or schedule. This was evident through an employee's courage to question the ability of the relief system to handle higher feed rates being proposed, and management's decision not to conduct tests outside the safe operating limits even though that meant sacrificing potential income with high market demand for the product. The previous culture discouraged such discourse.

Felt leadership and operational discipline were demonstrated during a full plant turnaround involving 4000 contractors without an incident. It was the best turnaround result BST has ever experienced with outstanding outcomes both from a timing and safety perspective. For many individuals and senior managers, the turnaround marked a turning point in their view of the new BST safety culture. It was proof of the theory that good safety is indeed good business. The organization started to view near misses and incidents as learning opportunities with a focus on improvements at the system level. Regular forums to engage employees and listen to their safety concerns were established, in addition to cascading the vision and involving all levels.

3.4 Culture Transformation Journey Continues...

At the end of 2017 a second survey of values, attitudes and beliefs was conducted. It showed leadership was more aligned, teamwork was better, senior managers had bonded together and were collaborating more effectively. Not only had technical competencies improved, but capabilities in general had developed and there was a very clear shift in mindsets and behaviors. Business leaders displayed greater self-confidence regarding safety and paid more attention to the social context of risk. By regularly discussing risks in an engaging way in all areas of the company, not only in manufacturing, the commitment to "no harm to anyone anytime" became one that people believed in and acted upon, both on and off the job. Specific initiatives were rolled out by BST in 2018 with a special emphasis on delineating individual and management behaviors, and risk awareness and recognition as part of their journey toward an interdependent culture.

3.5 Integration of safety in Capital Projects and Management Support

In 2017 BST began developing the scope of two significant expansion projects, one at each of its sites. These expansions will take place while the existing plants are in operation. Early in the scope development, BST management made a strategic decision to invest significant resources to proactively identify SHE aspects in all phases (engineering, procurement, construction, startup), and integrate them formally into the project plans and schedules to ensure safe execution and operation with the advisory support of DSS. This management decision further reinforced the value that management put on continuously improving the BST safety culture.

4 Conclusion

BST has undergone a tremendous change since 2013 – from viewing safety as a purely technical issue to realizing that people are at the crux of any safety effort. In the last four years, BST has changed its focus and discovered that a strong safety culture, modeled by senior leadership, and coupled with consistent process safety management systems not only improves safety performance, but can also have significant productivity benefits.

While a clear shift from a focus solely on production and profit to one that is safety-driven is evident, the organization realizes that the journey toward a culture committed to "no harm to anyone anytime" is a never-ending journey. Recent analysis of performance has indicated that the implementation of good systems is not enough, and that a focus on defining, measuring, and improving the underlying mindsets, behaviors and skills is required to continue the journey.

Frontline Supervisors are the Safety Culture

Renato Prestes
Global EHS Ops Leader/Ag Solutions
FMC
2929 Walnut St, Philadelphia, PA 19104
rpresstl@gmail.com

TJ Larkin
Partner – Larkin Communication
230 Park Ave. 3rd Floor West, New York, New York 10169
www.Larkin.Biz
Larkin@Larkin.Biz

Prepared for Presentation at
American Institute of Chemical Engineers
2019 Spring Meeting and 15th Global Congress on Process Safety
New Orleans, LA
March 31 – April 3, 2019

Frontline Supervisors are the Safety Culture

Renato Prestes
Global EHS Ops Leader/Ag Solutions
FMC
2929 Walnut St, Philadelphia, PA 19104
rpresstl@gmail.com

TJ Larkin
Partner – Larkin Communication
230 Park Ave. 3rd Floor West, New York, New York 10169
www.Larkin.Biz
Larkin@Larkin.Biz

Keywords: supervisors, safety, frontline, communication, incidents, behavior change, research

Abstract

This paper argues that no one and nothing is more important for safety performance than the frontline supervisor. Of course, other things matter—but nothing matters more. In most cases, when frontline supervisors insist on high-safety performance, it happens. When supervisors don't, safety performance declines and incidents increase.

If supervisors are the key to better safety, what is the role for corporate safety managers? We suggest that what corporate safety needs to do is redesign their safety communication making it easier for supervisors to use. Meeting supervisor needs should be the dominant force guiding our safety communication.

This paper will show three examples of improved safety communication. Safety communication designed with supervisor needs as the top priority:

1. NOV (redesign of safety policies and procedures)
2. Dow Chemical (new one-page summaries for Dow's Life Critical Standards)
3. ConocoPhillips Australia (redesign of their risk assessment tool: Step-Back 5X5)

1 Supervisors as the Safety Culture – Example (Storage Tank Explodes)

This section of the paper describes a tragic incident at an onshore natural gas wellsite. This incident demonstrates the critical role local supervisors play in determining safety practices.

A cleaning contractor was killed at a natural gas wellsite in Canada.[1] The contractor was cleaning a tank with a high-pressure water hose, which was draining to a nearby vacuum truck.

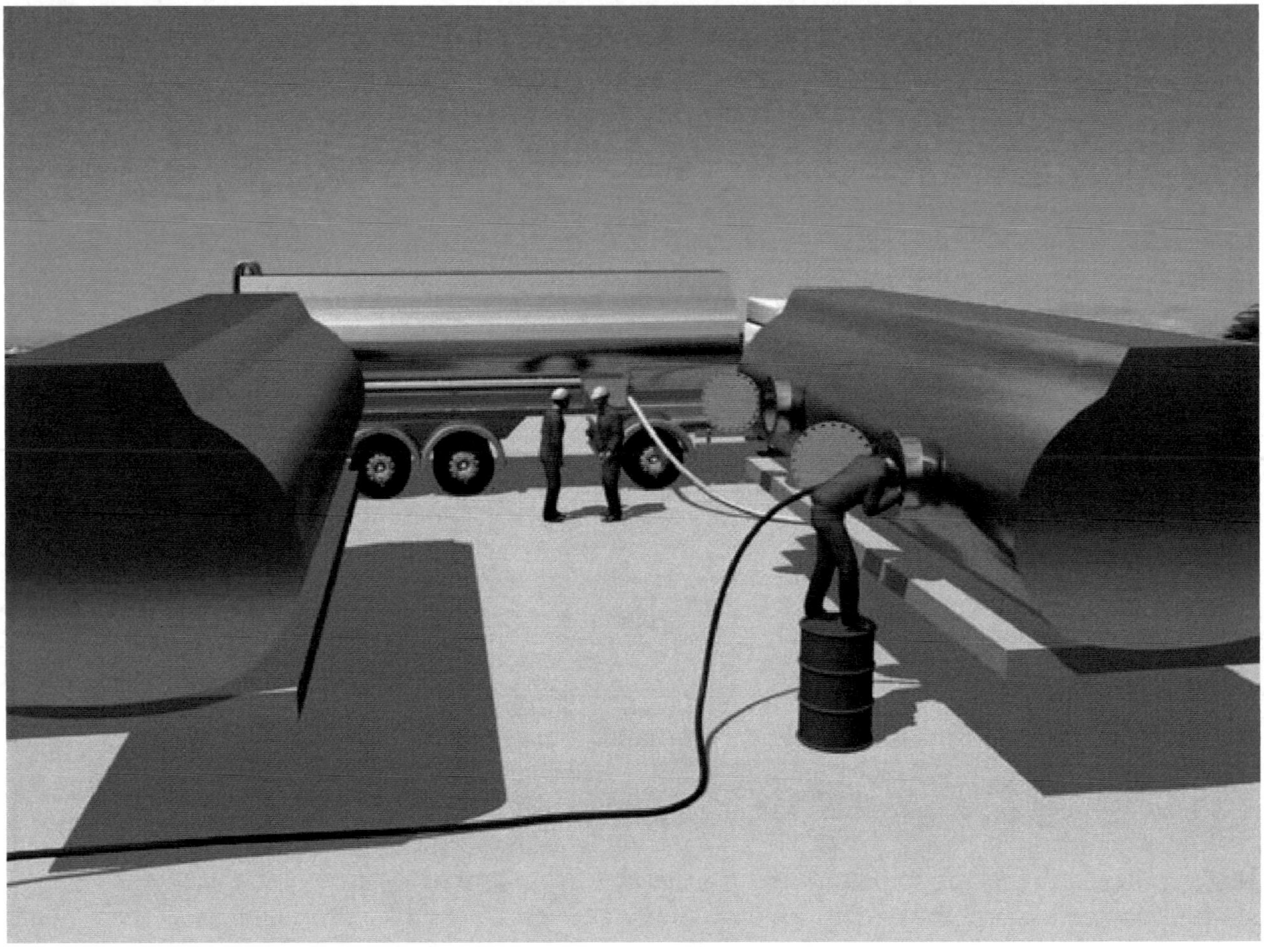

During the cleaning, the tank exploded. The contractor leaning into the tank died. Two others, onsite supervisor and another contract cleaner, were seriously burned.

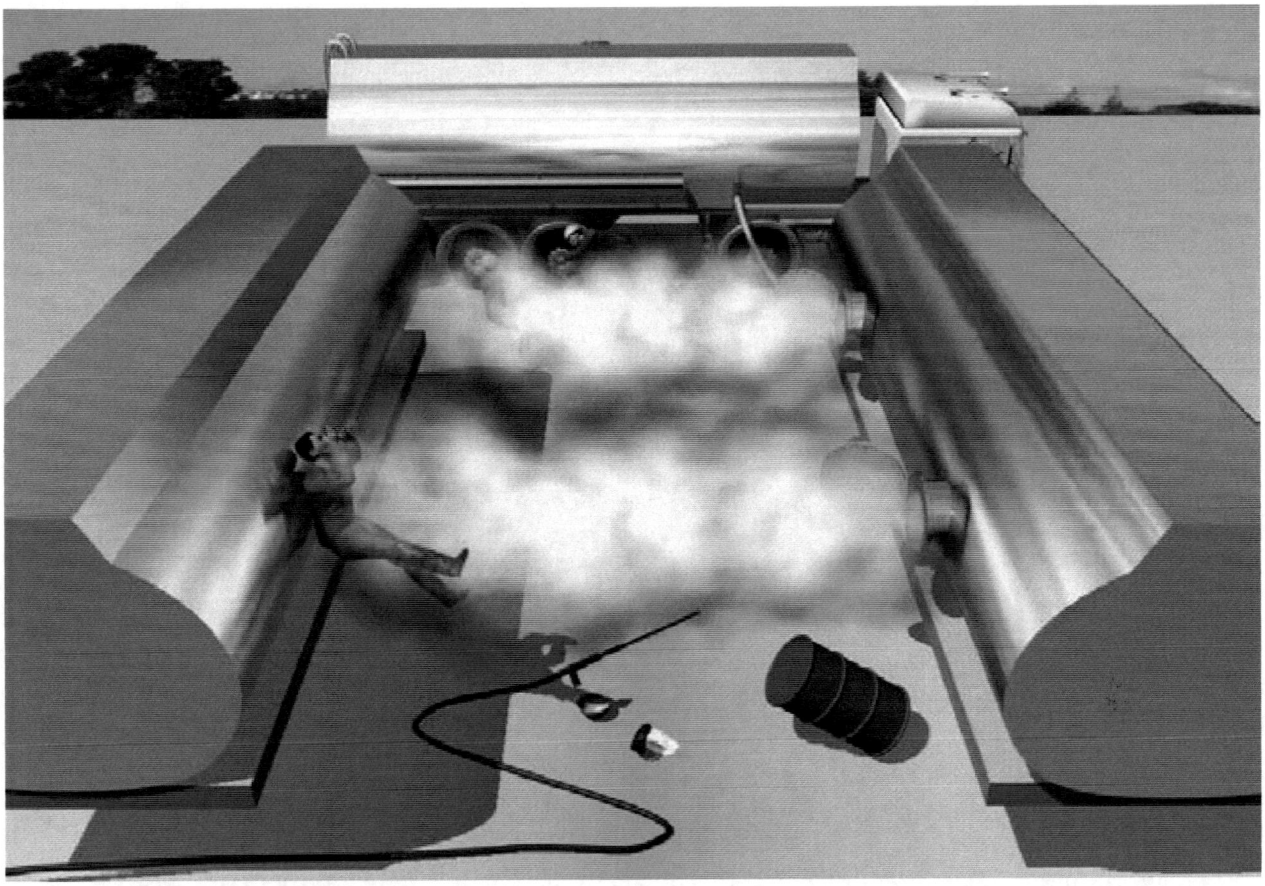

Many serious safety violations were involved in this tragedy. Perhaps, most significantly, an onsite employee tested the atmosphere inside the tank before the cleaning began. This testing revealed an explosive environment inside the tank. Tragically, no signs were posted warning of the explosive environment, and the explosive reading was not communicated to the cleaning contractors. The investigators could not determine the actual ignition source.

Additional safety deficiencies contributing to this tragedy include:

X	no isolation	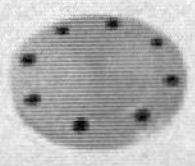	Values upstream and downstream of the tank were closed but no blinds or blanks were installed.
X	no lockout		Values were closed but no locks or tags attached.
X	no purging		Tank was not purged with nitrogen or water.
X	no grounding		The vacuum truck was not grounded or bonded by cable to the tank or the ground.

Why? Why didn't the cleaning contractors follow better safety procedures? The answer, revealed during the investigation, is that the supervisor at this site did not require better safety procedures.

During the investigations, the contractors said they varied their safety procedures at every site they visited—depending on what safety was required by the local supervisor. The contractors "knew" the correct safety procedures to follow before cleaning a tank, but only "followed" these correct procedures when demanded by the local supervisor.

This is what we mean when we say that supervisors **are** the safety culture.

2 When it Comes to Safety—Supervisors Matter More than the Company

An interesting study published in *Safety Science* looked at employees' safety perceptions on offshore oil & gas platforms operating in the Norwegian Continental Shelf.

This Norwegian study found that local platform managers had a greater effect on safety than the companies (corporations) that operated the platform.[2]

Specifically, this study looked at:

- 4,479 employees
- working on 52 different offshore platforms in the Norwegian Continental Shelf
- operated by 5 different oil & gas companies
- measuring safety based on a questionnaire asking employees to assess the quality of safety practiced on their platform

Researchers found employees' evaluation of safety was best predicted by the local platform managers and not by the companies they worked for. In other words, the variance in employees' perception of safety between platforms inside the same company was greater than the variance between the companies themselves.

Add to this Norwegian result, the work of Sicco van As. Doing research with eight Dutch chemical companies, Professor van As found a correlation of -0.65 between supervisors' informal face-to-face conversations about safety and the number of serious incidents (or near misses) at each site.[3] Squaring this correlation gives the explained variance, meaning that about 40% of the variance between these eight chemical companies can be explained by the amount of informal, face-to-face communication with supervisors about safety.

The more supervisors communicate safety—the fewer the incidents and near misses.

3 Role for Corporate Safety Managers

If supervisors matter more than the company, what, if anything, is the role for corporate safety managers?

Our answer is: corporate safety managers need to make it easier for supervisors to do the informal, face-to-face communication. Corporate safety can make new communication, or redesign existing communication, to better fit the needs of frontline supervisors.

Fitting the needs of frontline supervisors means changes such as:

- making procedures easier to read and follow
- creating risk assessment tools that are more relevant and easier to use
- summarizing safety documents, highlighting the most critical content

Here, we will give three specific examples:

1. National Oilwell Varco (redesign of their safety policies and procedures)
2. Dow Chemical (new one-page summaries for Dow's *Life Critical Standards*)
3. ConocoPhillips Australia (redesign of their risk assessment tool: *Step-Back 5X5*)

In each example, the company decided their existing communication was not easy enough for supervisors to use. And they decided to fix it.

4 National Oilwell Varco (NOV): Policies & Procedures

NOV has higher-level policies, for example, "Management of Change," "Risk Management," "Documents & Records." NOV also has lower-level procedures, "Crane Lifting & Rigging," "Confined Space," "Hydrogen Sulfide."

NOV decided none of these were easy enough for supervisors to read and understand.

NOV is not alone. The average safety procedure in oil & gas, mining, and chemicals is written at grade level 14. The problem is that only 17% of U.S. adults can read at grade level 14.[4]

NOV decided to rewrite all their policies and procedures. Without changing any of the content (all the content remained the same) but rewriting the content in a more comprehensible style: using graphic design (color and page layout), adding illustrations and photos, and reducing the reading complexity from grade level 14 to grade level 8 (50% of U.S. adults can rad at grade level 8).

Below is a typical procedure (*Crane Rigging & Lifting*):

1.0 General

1.1 This procedure was designed and developed to assist in the inspection and authorization of below-the-hook lifting devices. Below-the-hook lifting devices are defined as any device used to attach a load to a hoist. These devices include, but are not limited to: slings, chain, wire rope and synthetic; and structural and mechanical devices.

1.2 Safety to persons below the operation and to the material being moved is the concern of all persons who operate cranes or do rigging.

1.3 Safe and efficient rigging requires skill, exercising good judgment and constant application of safety rules and practices.

1.4 This policy shall be applicable to all operations that utilize rigging/lifting equipment.

1.5 This site procedure is in addition to Overhead Crane Operations.

2.0 Device Approval

2.1 Approved lifting devices are those which are: 1) engineering-approved, 2) made of lifting components (approved shackles, chains, D-rings, etc...), 3) are proof-tested, and 4) load rated (marked). Approved lifting devices are subject to further inspection as defined in these criteria.

2.2 A lifting device is considered unapproved if it fails any of the four criteria defined above. Unapproved lifting devices shall not be used.

2.3 Shop made devices that meet the criteria of 2.1 may be used.

3.0 General Controls

3.1 All purchased lifting devices shall have the rated capacity affixed. Any in house manufactured lifting devices shall have been designed by engineering with the load capacities determined by engineering calculations or by pull testing. All in house lifting devices shall be manufactured in accordance with latest ASME/ANSI/AWS or other governing standard(s). Capacities shall be affixed to all in house manufactured lifting devices.

3.2 Alloy steel chain slings shall have permanently affixed durable identification tags stating size, grade, rated capacity, and reach. Grade 80 chain shall be the minimum allowed.

Below is the same content after the rewrite:

Crane Rigging & Lifting Device Safety

Introduction (1.0)

Top Priorities	Your top two priorities when operating a crane or rigging: 1st protect people working in the area 2nd protect the load you are moving
When you use this procedure?	Use this procedure when you do any work that involves rigging or lifting equipment.
Other procedures you should look at.	Overhead Crane Operations
All below-the-hook lifting devices are covered in this procedure 	If you put *anything* (mechanical or structural) on a crane hook, you need to follow this procedure. For example:  sling chain wire rope synthetic rope

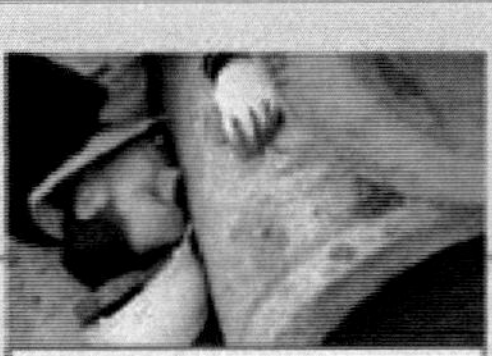

Needed for a safe lift:
- skill
- careful thinking
- follow procedures

Lifting Equipment Must Be Approved (2.0)

Not approved? Don't use it	If the lifting equipment you put on the hook is not approved, DO NOT USE IT.
"Approved" means four things 	**Approval: Four Requirements** #1 Engineering engineering inspected and approved the equipment #2 Parts Approved all parts approved (for example: shackles, chains, D-rings) #3 Load Rated equipment has label showing maximum weight it can lift #4 Proof Tested someone tested that it can lift what the label says it can lift  If the lifting equipment is missing any one of these four requirements... DO NOT USE IT
Shop-made equipment must be approved	It doesn't matter if you bought this lifting equipment or if you made it in a shop. If you are going to lift with it—the equipment must be approved.

"Company Name" HSE Management System			
Document Name:	Document Number:	Effective Date:	Page: 1 of 2
Version:	Author:	Web Location	

If you are using a printed (paper) copy of this document, you need to make sure you are using the most recent version. Check to see if you using the most recent version by going to the "Web Location" in the box above, and looking at the number in the "Version" box in the footer of the online version of this document. If the Version number shown on the online document is the same as the version number on your printed copy, you are good. If not, download and print the newer version before using this document.

The above page is the same content (before the rewrite and after the rewrite). The only difference is the communication. The difference is adding principles of graphic design and lowering the reading complexity to grade level 8. The page below summarizes some of the design changes.

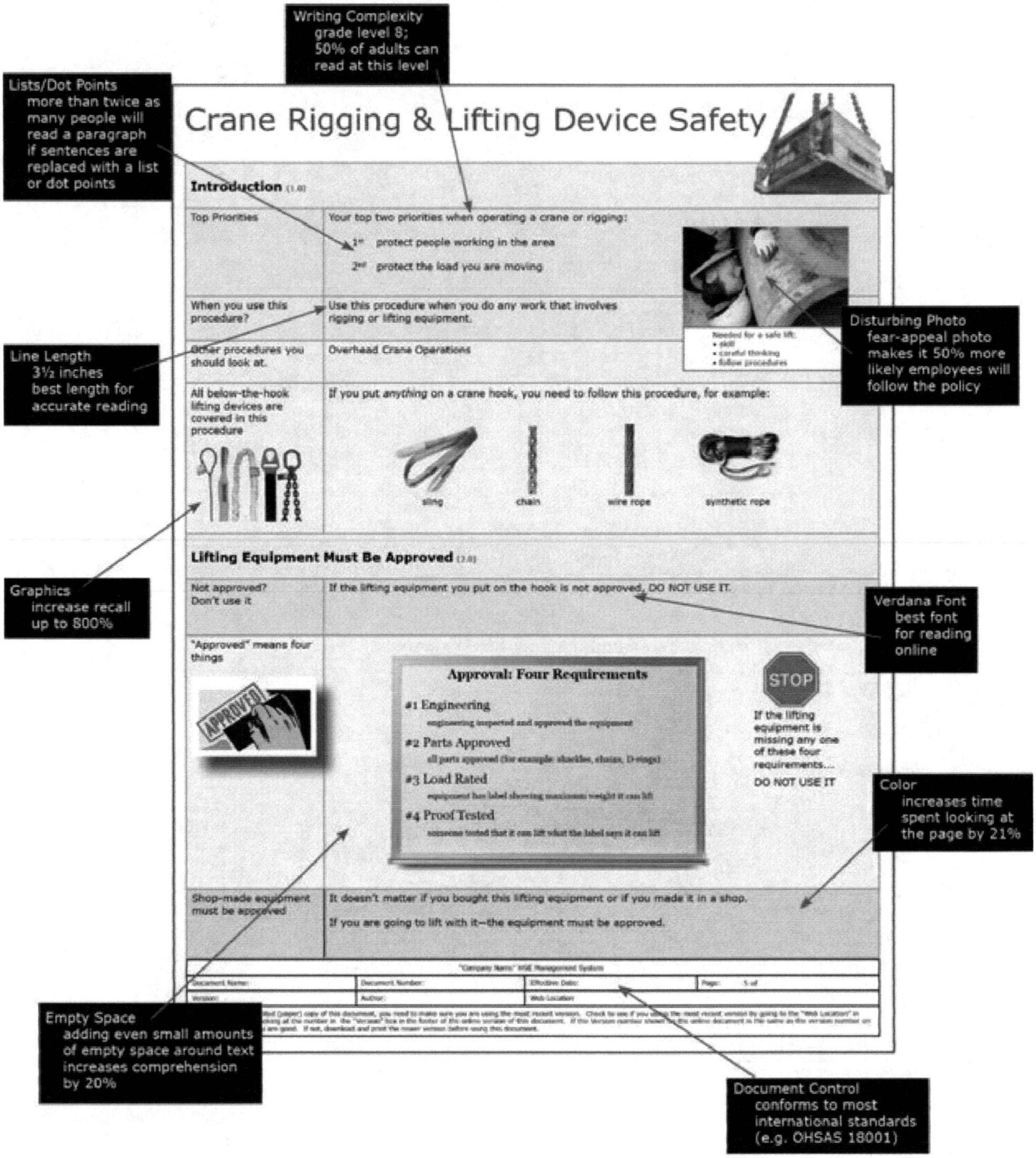

We know that supervisor communication correlates with improved safety performance, the next step is to rewrite our communication making it easier for supervisors to use, and more importantly to talk about.

NOV made it easier, much easier, for supervisors to understand, talk about, and follow their safety policies and procedures.

5 Dow Chemical: Life Critical Standards

Dow has identified eight safety practices as *Life Critical Standards*.

Dow's *Life Critical Standards* are:

1. Confined Space
2. Electrical
3. Hot Work
4. Elevated Work
5. Hydroblasting
6. Energy Isolation
7. Line & Equipment Opening
8. Safe Work Permit

Each of these practices has a complete document containing all the details associated with that standard.

But Dow recognized that these large documents are not easy for supervisors to use, so Dow decided to summarize each document in an easy-to-use communication style.[5]

For example, Dow took its "Line & Equipment Opening" standard and communicated the most critical content in a way that would be easier for supervisors to use.

This summary does not replace all the supporting documents concerning line & equipment opening. Instead it summarizes the most important points in an easy to understand style. A format that makes it easy for supervisors to talk about the standard.

Minutes before a line opening, when everyone may be in a hurry, the Dow supervisor can stop the crew and say, "Here, look at this page, I want you to review the most important parts of our standard before doing that line opening."

Dow has turned these graphical pages into posters (after translating them into local languages) and posted them in control rooms throughout the company.

Dow wants their supervisors to know and enforce these standards. Dow has also made the communication enabling supervisors to do so.

6 ConocoPhillips: In-the-Field Risk Assessment

ConocoPhillips Australia (Bayu Undan location) has several layers of risk assessment, including their policies, permit to work, and job hazard analysis. However, ConocoPhillips Australia was especially concerned with the final layer of risk assessment, called, *Step-Back 5x5 Risk Assessment*.

The *Step-Back 5x5* is not intended to be a long, detailed list of all possible hazards. Instead the intent of the *Step-Back 5x5* was to be the final risk assessment happening just before the work begins.[6]

The *Step-Back 5x5* was intended as a tool to help supervisors have a last critical look at the work site, searching for risks that were not, or perhaps could not be anticipated in the other layers of risk assessments (policies, permit to work, job hazard analysis).

ConocoPhillips Australia felt their existing *Step-Back 5x5* document was not easy for supervisors to use.

The original *Step-Back 5x5* form looked like this:

ConocoPhillips	STEP BACK 5 X 5 & 4 QUARTERS PROGRAM
Name:	Task:
Date:	

HAZARD MANAGEMENT QUESTIONS	Yes	OPERATING & PROCESS INTEGRITY QUESTIONS	Yes
Is the scope of work understood by the work group?		Could this activity result in a hydrocarbon release?	
Is the work scope the same as what is written on the permit?		Is an Operator required to be in attendance?	
Is a JHA required for this task? Has it been reviewed and understood?		Is continual gas monitoring/atmospheric testing required?	
Are all personnel trained and competent to complete the task?		Do you require a major equipment re-instatement check list?	
Are there other work activities around us that create hazards?		Have isolations been verified as free of energy?	

HAZARD IDENTIFICATION							
Work Environment	**Yes**	***Work Environment***	**Yes**	***How can we get hurt - Injury Causes***	**Yes**	***How can we get hurt - Injury Causes***	**Yes**
Housekeeping		Sea State		Line of Fire		Substance / Gas Exposure	
Waste Disposal / Recycle		Lighting in immediate area		Caught Between		Stored Energy / Struck By	
Hazardous Atmosphere		Loss of Containment		Strain / Sprain		Awkward / Unstable Body Position	
Spill Potential / Location of spill kit		Drains / Scuppers / Vents		Slips & Trips		Hand Injury	
Dust		Heat Exposure / Hot Surfaces		Falls from height		Pinch / Crush Points	
Noise		Dropped object risks/overhead work		Sharps & Sharp Edges		Other	

LIFE SAVING RULES, CONTROLS, & VERIFICATION							
Permit to Work	**Yes ✓**	***Isolation Verification***	**Yes ✓**	***Crane / Lifting Operations***	**Yes ✓**	***Working at Height***	**Yes ✓**
Is a PTW required for this task		Is an isolation required for this task		Equipment is certified / Test Tag in place *Jan. - June (**Blue**) July – Dec. (**Green**)* *Even Year (**Yellow**) Odd Year (**White**)*		Safety harness/Anchor point needed	
Have PTW conditions been met		Blue isolation lock & red tag attached		Dogman clearly identified		Scaffold approved / Scaff-Tag in date	
Work members signed the permit?		All personal locks attached		Communication plan in place & known		Ladders securely tied off	
New starters in the group?		Isolation tested to confirm no energy		Barriers in place where required		Edge protection / Tools secured	
Confined Space Entry	**Yes ✓**	***P.P.E.***	**Yes ✓**	***Tools & Equipment***	**Yes ✓**	***Emergency***	**Yes ✓**
Confined Space Entry Certificate		Correct gloves for the task		Suitable for the area classification		Location of Call Points & Phones	
Observers in place		Respiratory Protection required		Personal gas detector required		Nearest safety shower	
Access & Escape routes clear		Hot Work PPE		Tools inspected & certified		Escape Routes clear & known	
Rescue equipment in place		Sun Protection		Tools in good condition/working order		Location of emergency equipment	
Penetrations	**Yes ✓**	***Bypassing Safety Devices***	**Yes ✓**	***Fatigue & Dehydration***	**Yes ✓**	***Other***	**Yes ✓**
Is a penetration certificate required		Authorisation to bypass in place		Fatigue risks managed		Barriers & Signs in place	
Electrical discipline review		MOS's listed on permit		Drinking Water available		MSDS in use or required	
Mechanical piping discipline review		Recorded in facility register		Job rotation required		Manual handling controls	
P&ID attached		Mitigations identified		Shade protection		WAVES Observation requested	

ALL/HSE/PRO/040 Rev 4 – Attachment 3.2 – 4 QUARTERS   Page 1 of 2

ConocoPhillips Australia felt they could do better. They redesigned the form. And, they did so by starting with their supervisors. They asked their supervisors what they needed, they asked their supervisors: What do you need to do a good final risk assessment just before the work begins?

Redesigning the risk assessment form based on supervisors' advice produced a completely new document.

STEP BACK 5x5

Date:________________ Location:______________________________ Permit #_____________

Job Description:____

1 LIFE SAVING RULES

Check the Box if Rule Applies

☐ Permit to Work ☐ Confined Space ☐ Working at Height ☐ Lifting ☐ Isolation ☐ Excavation ☐ Bypassing Safety Devices

JSA/JHA Reviewed? ☐ Yes

Controls in Place? ☐ Yes

Everyone Knows Stop Work Authority? ☐ Yes

2 AROUND WORK AREA

Walk *Around* the Work Area Looking for Hazards

Work Area

SIMOPS

Above ↑	I checked above the area for hazards?	☐ Yes
Beside ← →	I checked all sides for hazards?	☐ Yes
Below ↓	I checked below the area for hazards?	☐ Yes

Hazards *Around* Work Area (not in Permit or JHA)

____

____

____

____

____

3 INSIDE WORK AREA

Walk *Inside* the Work Area Looking for Hazards

Work Area

People Hazards	Process and Electrical Hazards
Examples: Trips, Sharps, Falls	Examples: Hot Pipes, Fluids, Pressure, Cables
I checked for *People* hazards? ☐ Yes	I checked for *Process* hazards? ☐ Yes

Hazards *Inside* Work Area (not in Permit or JHA)

____

____

____

____

____

4 EMERGENCY

Emergency - Check Box After Showing the Team...

Muster Point Primary & Alternate ☐	Alarm Location ☐	Fire Extinguisher ☐
	Deluge Control ☐	Escape Route ☐

Write Details

☐ Emergency Radio Channel

Nearest Emergency Telephone

88

Nearest Emergency Shower

____

Nearest Emergency Life Ring

____

Rev 5

ConocoPhillips Australia measured the difference between the original and redesigned risk assessment forms.

- Reduced number of words from 680 words to 123 words (82% reduction)
- Reduced the number of questions from 102 to 48 (53% reduction)
- Reduced the average time to complete the form from 20 minutes to around 5 minutes (75% reduction)
- Finally, ConocoPhillips asked its supervisors if the new form was better at assessing workplace risks: 69% of supervisors said it was better.

The ConocoPhillips Australia redesign is an example of building communication with "supervisor use" as the dominant design principle.

7 Conclusion

The communication most likely to reduce serious incidents comes from supervisors talking informally, face-to-face, with their teams.

The job of corporate safety is to help supervisors do that communication.

This means designing communication with supervisors as the guiding principle:

- Reducing the reading complexity to grade level 8 to 10
- Adding graphics to communicate the content visually
- Making communication designed more to be "talked about" than to be "read"

Begin your communication with the question, "What do supervisors need?"

The best communication follows a simple principle: communication should be "receiver oriented;" meaning it is designed with the receivers in mind.

Most safety communication is "sender oriented." Most of our communication is designed to meet the needs of the senders. Sender-oriented communication asks, "What does the sender need?" And then designs the communication to meet those sender needs.

Safety communication should be about meeting the needs of the receivers, in this case the supervisor audience.

And, we will know if we succeed. When supervisors do more informal, face-to-face communication about safety, incidents go down.

References

1. WorksafeBC: "Accident Investigation Report: Explosion in Pressure Vessel during Well Completion," Report #: 2008124490126, 2008, p. 1-20.

2. Høivik, Dordi and Bente E. Moen: "What is More Important for Safety Climate: The Company Belonging or the Working Environment," *Safety Science*, December 2009, p. 1324-1331.

3. Sicco van As: "Communication and Industrial Accidents," SOM Research Report, University of Groningen, The Netherlands, available online: http://irs.ub.rug.nl/ppn/228120691.

4. Larkin, TJ and Sandar Larkin: *Silver Bullet – Using Supervisor Communication to Reduce Accidents*, Larkin Communication Consulting, available online: www.Larkin.Biz.

5. Dow Chemical Company: "Dow's Life Critical Standards: Making Them Easier to Follow," Presentation, 2019 Spring Meeting and 15th Global Congress on Process Safety, March 31 to April 4, 2019, New Orleans.

6. ConocoPhillips Australia (Bayu-Undan location): "Last Line of Defense – Improving Our Step Back 5x5," Presentation at the 2017 Spring Meeting & 13th Global Congress on Process Safety, San Antonio, Texas.

Enhancing Behavior-Based Safety Culture through Casual Safety Observations

Jennifer N. Boice
W. R. Grace and Company
1201 Gulf States Road
jennifer.boice@grace.com

Prepared for Presentation at
American Institute of Chemical Engineers
2019 Spring Meeting and 15th Global Congress on Process Safety
New Orleans, LA
March 31 – April 3, 2019

Enhancing Behavior-Based Safety Culture through Casual Safety Observations

Jennifer N. Boice
W. R. Grace and Company
1201 Gulf States Road
jennifer.boice@grace.com

Keywords: behavior, behavior based safety, BBS, culture, safety observations, observations

Abstract

Despite the influx of automated systems in the chemical manufacturing plant, human factors continue play a pivotal role in the day-to-day activities, which can consequently result in incidents and injuries. Even so, human error remains the cause of many safety incidents that continuously occur in the workplace. This study focuses on providing a simple tool that can be used to teach personnel in manufacturing facilities about identifying safety-critical items that impact routine work. Employees are trained to take a more personal and practical approach to decision-making on the job - by training the mind to focus on 'what could go wrong' and 'how could this be done differently'. This has resulted in a change in our work environment that is centered on a behavioral-based safety culture that is driven by the front-line users. Furthermore, this tool provides a method for implementing behavioral-based safety techniques and allows the users to take ownership for making sustainable improvements on their processes. We believe this effort has resulted in a paradigm shift and a gateway to deepen our understanding on how people can change their behavior to make safer choices.

1 Introduction

Historically, copious amounts of time have been invested to understand how to motivate employees to demonstrate safe behavior in the workplace [1-3]. There are numerous studies that focus on providing training, tools, and strategies for improving workplace safety [4-7]. Many of these reports are centered on using behavior-based safety approaches to modify safety climate [7]. In the specialty chemicals industry, new processes and chemistries are constantly being introduced; therefore, it can be challenging to incorporate behavior-based safety techniques effectively in this environment.

At our site, we rely heavily on our operations personnel for meeting production and safety targets. We have multiple semi-batch processes taking place simultaneously across the plant. The nature of our business requires that we have the flexibility to quickly modify our equipment setups to accommodate a frequently-changing production schedule. This creates a fast-paced and ever-evolving work environment where we must rely on our operators to make safe choices and quickly identify potential risks. Therefore, it is critical that we teach them how to recognize and anticipate risks.

As process safety professionals, it seems we are always searching for new ways to re-energize our sites' safety culture. We have focused on creating a Safety Observation tool from an existing practice to use at our site to improve the value of safety observations. Historically, the Safety Observations exercise had been treated as a form used to perform a quick scan of another employee's work. Since this was a familiar task for our wage-roll employees, we chose to focus on improving this exercise. We also took into consideration, the fact that the operators are the front-line users in the plant and are intimately involved with the day-to-day production activities. Additionally, they are responsible for routinely handling highly hazardous chemicals, and we want to make sure they are anticipating potential risks associated with their daily activities. Therefore, we must instill a strong and sustainable safety culture on the front line to provide a robust safety barrier as part of our operations. We believe that we can make small changes to the Safety Observation tool that will encourage them to focus on safe behavior and will teach them to identify high-risk situations before a task is started.

After evaluating the 'Old' Safety Observation Form, a method for modifying the form to create the new tool was conducted in four simple steps: 1) focus on identifying the area of improvement, 2) determine purpose for why the improvement is needed, 3) define metrics so progress can be tracked, and 4) set goals to meet targets. These steps mirror a Kaizen approach, as this will promote incremental improvements in day-to-day activities [8]. After this evaluation, we created a 'Revised' Safety Observation Tool so that the metrics could be compared.

2 Phase 1: Evaluating the Old Safety Observation Form

2.1 Focusing on Specific Areas of Improvement

We recognized the Safety Observation exercise as an Area of Improvement (AOI), but more specifically, there were multiple elements of the Old Safety Observation Form that we identified as AOIs that are described below. Before this initiative began, the Safety Observation forms were already being completed for over a year by various wage-roll employees across the site and were routinely reviewed by management. However, the safety observations that were reported were not necessarily being used as effective means of making safe changes or instilling safe behavior practices in the front-line users. We needed the operators to produce safety observations that would show an improvement in our safety culture.

We examined 141 Safety Observations reported over the course of one year. We had approximately 200 operations employees on average at our site at this time, which equates to less than one safety observation being performed per person over the course of the year.

An example of the Safety Observation Form is shown in Figure 1. As you can see on this form, there is a lot of text and not much space for an operator to record personal comments or observations. The feedback we received from some of the operators was that it resembled a permit and was overwhelming to fill out. The content of the text was not inspiring the participants to engage in the exercise, or consider any improvements resulting in reduced risk or safe behavior. With this feedback, we developed our first two AOIs:

AOI #1: The form is not user-friendly

AOI #2: The form does not engage the user to critically think about their own involvement in the observed task

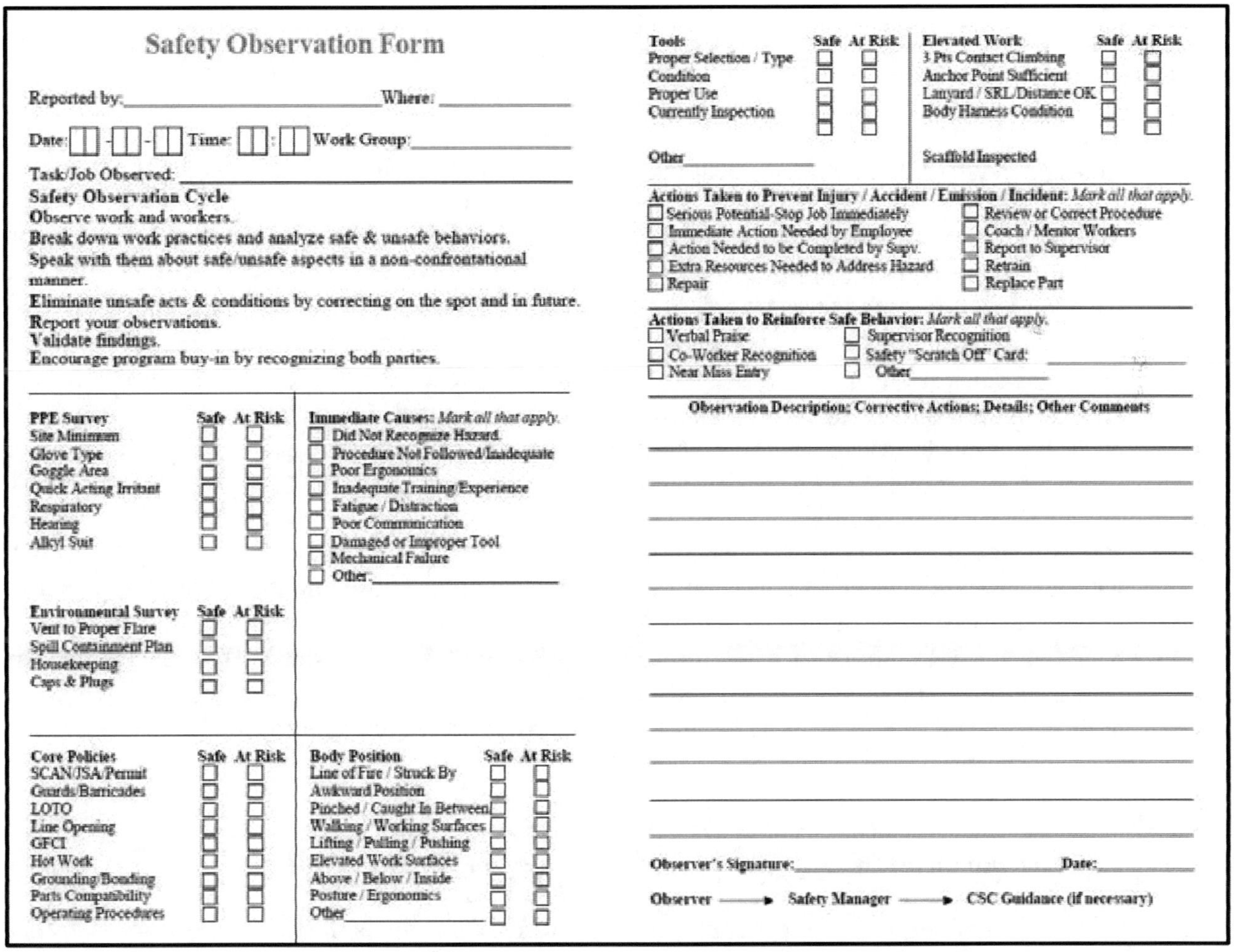
Safety Observation Form

Reported by: ____ Where: ____

Date: ☐☐-☐☐-☐☐ Time: ☐☐:☐☐ Work Group: ____

Task/Job Observed: ____

Safety Observation Cycle
Observe work and workers.
Break down work practices and analyze safe & unsafe behaviors.
Speak with them about safe/unsafe aspects in a non-confrontational manner.
Eliminate unsafe acts & conditions by correcting on the spot and in future.
Report your observations.
Validate findings.
Encourage program buy-in by recognizing both parties.

PPE Survey	Safe	At Risk
Site Minimum	☐	☐
Glove Type	☐	☐
Goggle Area	☐	☐
Quick Acting Irritant	☐	☐
Respiratory	☐	☐
Hearing	☐	☐
Alkyl Suit	☐	☐

Immediate Causes: *Mark all that apply.*
- ☐ Did Not Recognize Hazard.
- ☐ Procedure Not Followed/Inadequate
- ☐ Poor Ergonomics
- ☐ Inadequate Training/Experience
- ☐ Fatigue / Distraction
- ☐ Poor Communication
- ☐ Damaged or Improper Tool
- ☐ Mechanical Failure
- ☐ Other: ____

Environmental Survey	Safe	At Risk
Vent to Proper Flare	☐	☐
Spill Containment Plan	☐	☐
Housekeeping	☐	☐
Caps & Plugs	☐	☐

Core Policies	Safe	At Risk
SCAN/JSA/Permit	☐	☐
Guards/Barricades	☐	☐
LOTO	☐	☐
Line Opening	☐	☐
GFCI	☐	☐
Hot Work	☐	☐
Grounding/Bonding	☐	☐
Parts Compatibility	☐	☐
Operating Procedures	☐	☐

Body Position	Safe	At Risk
Line of Fire / Struck By	☐	☐
Awkward Position	☐	☐
Pinched / Caught In Between	☐	☐
Walking / Working Surfaces	☐	☐
Lifting / Pulling / Pushing	☐	☐
Elevated Work Surfaces	☐	☐
Above / Below / Inside	☐	☐
Posture / Ergonomics	☐	☐
Other ____	☐	☐

Tools	Safe	At Risk
Proper Selection / Type	☐	☐
Condition	☐	☐
Proper Use	☐	☐
Currently Inspection	☐	☐
Other ____		

Elevated Work	Safe	At Risk
3 Pts Contact Climbing	☐	☐
Anchor Point Sufficient	☐	☐
Lanyard / SRL/Distance OK	☐	☐
Body Harness Condition	☐	☐
Scaffold Inspected	☐	☐

Actions Taken to Prevent Injury / Accident / Emission / Incident: *Mark all that apply.*
- ☐ Serious Potential-Stop Job Immediately
- ☐ Immediate Action Needed by Employee
- ☐ Action Needed to be Completed by Supv.
- ☐ Extra Resources Needed to Address Hazard
- ☐ Repair
- ☐ Review or Correct Procedure
- ☐ Coach / Mentor Workers
- ☐ Report to Supervisor
- ☐ Retrain
- ☐ Replace Part

Actions Taken to Reinforce Safe Behavior: *Mark all that apply.*
- ☐ Verbal Praise
- ☐ Co-Worker Recognition
- ☐ Near Miss Entry
- ☐ Supervisor Recognition
- ☐ Safety "Scratch Off" Card: ____
- ☐ Other ____

Observation Description; Corrective Actions; Details; Other Comments

Observer's Signature: ____ **Date:** ____

Observer → **Safety Manager** → **CSC Guidance (if necessary)**

Figure 1. The 'Old' Safety Observation Form, before revisions

2.2 Determine the Purpose of Performing Safety Observations

The responses on the form did not provide specific solutions to reduce the risk of a task or improve the safety of a task, according to feedback received from the safety team. Therefore, the next AOI was identified:

AOI #3: Operators were not providing specific solutions to reduce or prevent risk

At our site, the employees are instructed to go out in the field, observe someone performing a routine task, and fill out the form accordingly to report their observations. We found they were simply commenting on what they saw (e.g., I watched the operator connect the portable tank to the reactor…). This presents a significant (perhaps arguably the most significant) AOI – using the safety observation exercise as an opportunity to encourage proactive and forward thinking. We want them to think about what could go wrong with the task they are observing and anticipate possible hazards. The structure of the old form did not provide any blank space for the operators to comment on what could go wrong because it was so cluttered. We would like the operators to take a more personal approach to observing the task, and they will do this if we directly ask them to write about what could go wrong while doing this job, as seen below in AOI #4.

AOI #4: Operators were not documenting what could go wrong with a task

Taking time to determine the purpose of the exercise is critical to get the most information from the results. For example, we thought about what we were trying to achieve with the Safety Observation exercise that would prevent this activity from being perceived as another checklist or more paperwork. The responses that are written by the operators on the form can reveal details about their behavior and decision-making strategies. If they are not responding to a question favorably, or ignoring it altogether, that could expose either an opportunity for training or a chance to revise the form. Improving the value of safety observations could be accomplished by incorporating each AOI into the revised tool.

The way the questions were presented on the form seemed to influence the responses that were generated. Presenting questions in a logical manner with simple language for the user to understand can add value to results [9]. In addition, special questioning techniques can be used to gather more information about behaviors and attitudes [10]. Therefore, the revised survey must contain carefully crafted questions that can influence how users are approaching tasks and provide us with the information we are seeking.

2.3 Define the Metrics to Assign 'Value' to the Safety Observations

After determining the significance and objectives of the safety observation exercise, metrics were defined to track progress. Metrics are an easy way to quickly and consistently communicate results. Monitoring real time progress drives motivation and maintains support to implement the change. There is no need to make the metrics complicated because the purpose of establishing metrics is to provide the ability to see a snapshot of the progress at any given time.

We developed metrics so we could determine the 'value' of using the Old Safety Observation form. Our metrics were largely influenced by purpose - AOI #3 and #4. The metrics were very simple: one point was given for identifying a problem and one point for providing a solution for a total of two points. Each of the safety observations were reviewed using these metrics.

A common theme throughout a bulk of the observations was the reporting of something they saw (e.g., "When driving the fork truck, I noticed the light in the alley was out."), reporting what task they had just completed (e.g., "I transferred the solvent through the overhead piping from reactor A to the portable tank"), or simply reporting on a task like the ones described here with a statement following the description that said, "The task was performed safely". None of these reports from the safety observations demonstrated much value. These reports contained very limited forward-thinking for evaluating possible risks.

Of the 141 Old Safety Observation forms reviewed, 57 percent of users identified a problem, while only 18 percent provided a solution. We examined results from the operators who performed three or more safety observations using the old form. These results are shown in Table 1. The results are listed in order from highest score at the top (best value) to lowest score. The highest score is 75 percent (out of 100 percent) while the lowest is 0 percent. An arbitrary descriptor (e.g., "Observer 1, Observer 2, etc.) was assigned to each operator to keep results anonymous. These results are interesting because the number of safety observations is high, but the value of each was determined to be considerably low. More than half of the reports scored 50 percent or below, as indicated on Table 1 under the bold line. Unfortunately, these results also reveal that our shift leaders (highlighted as the shaded regions in Table 1) are scoring in the lower 50 percent, as can be seen with Observer 13, 17, 18, and 20. Observer 20 is a shift leader, and while the quantity of reports he generated is very high, the 'value' of his reports was determined to be very low using the developed metrics.

Table 1. Summary of Results using the Old Safety Observation Form

Observer	Points	Possible Points	Score
1	9	12	75%
2	7	10	70%
3	4	6	67%
4	4	6	67%
5	4	6	67%
6	4	6	67%
7	5	10	50%
8	3	6	50%
9	3	6	50%
10	8	18	44%
11	2	6	33%
12	2	6	33%

13	2	6	33%
14	5	16	31%
15	3	10	30%
16	2	8	25%
17	2	8	25%
18	2	8	25%
19	6	26	23%
20	5	38	13%
21	0	8	0%

2.4 Set Goals for Change

After the feedback had been evaluated, we set a goal to target an increase of 20 percent in the number of problems identified and solutions provided to improve the overall value of safety observations. The AOIs which were focused on were the same as described in the previous sections. It was critical that the proposed change was achievable. Our proposed change was to use the AOIs listed above to create a Revised Safety Observation Tool. It was best to set small goals so that the group could target positive outcomes and remain motivated. If the change was too drastic or unrealistic, the group could become demotivated and uninterested A strategy that can be used to foster enthusiasm is to encourage group efforts so that the teams can determine how they want to approach the problem. It does not matter *how* the goal is reached, it just matters that the paradigm is shifting in the right direction. Sustainable change will occur if ideas are driven by the frontline users – in our case – the operators.

3 Phase 2: Implementing the Change and Surveying the Results

3.1 Incorporating the AOIs to Create a Revised Safety Observation Tool

AOIs identified and actions needed to address them:

AOI #1: The form is not user-friendly

- *Action - Modify the form to be more user-friendly*

AOI #2: The form does not engage the user to critically think about their own involvement in the observed task

- *Action - Revise the content of the form to engage the user to critically think about his own involvement the observed task*

AOI #3: Operators were not providing specific solutions to reduce or prevent risk

- *Action - Revise the form to engage the operators in identifying potential problems and providing solutions*

AOI #4: Operators were not documenting what could go wrong with a task

- *Action* - *Revise the form to include questions that promote forward-thinking*

These AOIs were the focal point for change as the Old Safety Observation form was revised into a new, functional tool. We found that we could easily incorporate all the AOIs into the new tool. AOI #1 was accomplished by removing unnecessary text and tables, which helped to cleanup and simplify the form. The Revised Safety Observation Tool is provided in Figure 2. When comparing it to the old form in Figure 1, the differences are obvious – it is much cleaner and more organized than the old form. To include AOI#3, we explicitly stated the words "Problem Identified" and "Proposed Solution" with blank spaces below each so that they could write in responses. Since we want them to identify the problem and think about solutions, adding a space on the form that tells them to do this should improve the feedback we get in these areas. By simply providing a space to write in responses, we are setting the expectation that we want this feedback. Setting expectations is a critical component for developing a strong behavior-based safety culture so that the users will have direction and know exactly what you are requesting of them.

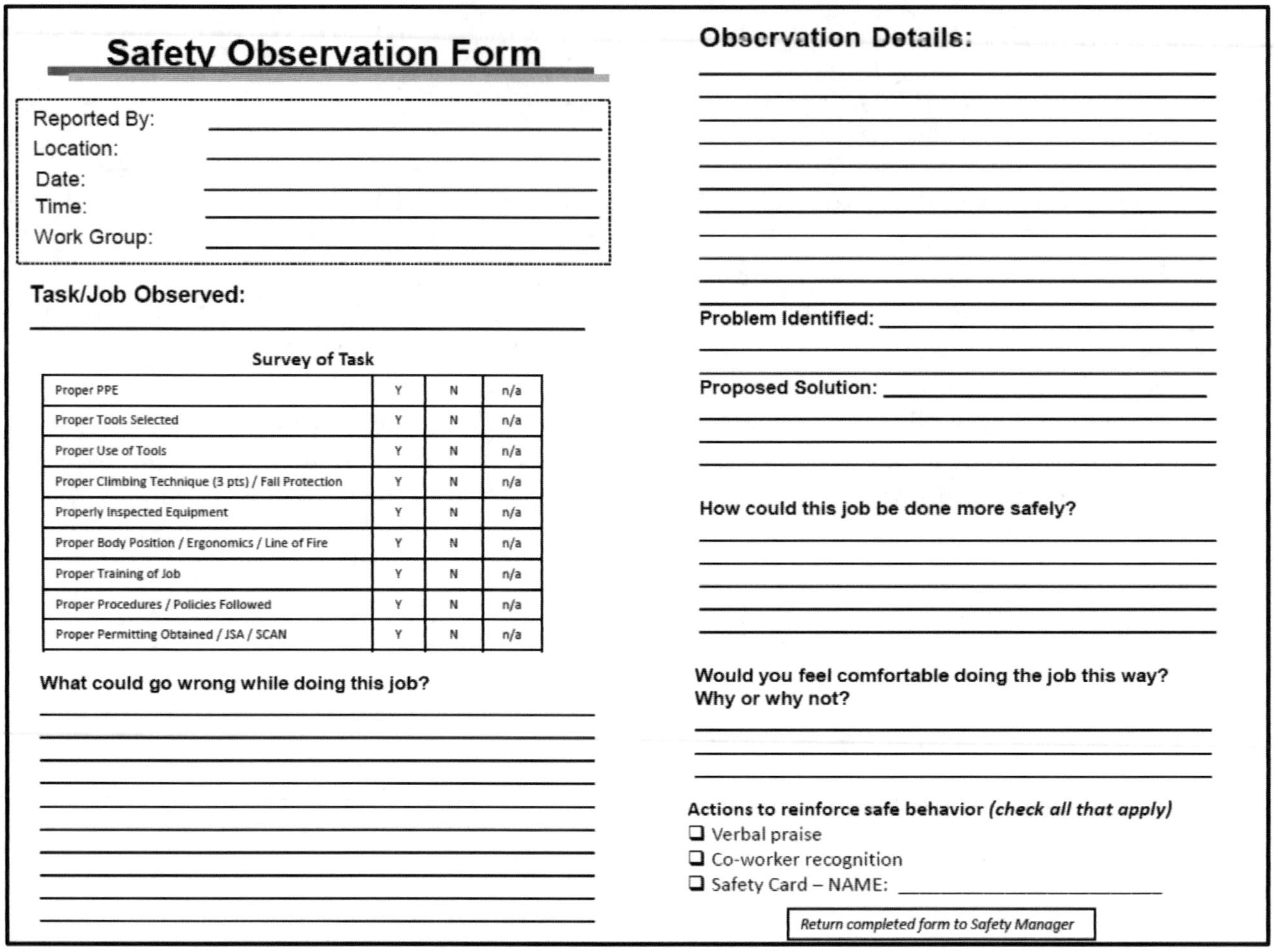

Safety Observation Form

Reported By: ______
Location: ______
Date: ______
Time: ______
Work Group: ______

Task/Job Observed:

Survey of Task

Proper PPE	Y	N	n/a
Proper Tools Selected	Y	N	n/a
Proper Use of Tools	Y	N	n/a
Proper Climbing Technique (3 pts) / Fall Protection	Y	N	n/a
Properly Inspected Equipment	Y	N	n/a
Proper Body Position / Ergonomics / Line of Fire	Y	N	n/a
Proper Training of Job	Y	N	n/a
Proper Procedures / Policies Followed	Y	N	n/a
Proper Permitting Obtained / JSA / SCAN	Y	N	n/a

What could go wrong while doing this job?

Observation Details:

Problem Identified: ______

Proposed Solution: ______

How could this job be done more safely?

Would you feel comfortable doing the job this way? Why or why not?

Actions to reinforce safe behavior *(check all that apply)*

- ❑ Verbal praise
- ❑ Co-worker recognition
- ❑ Safety Card – NAME: ______

Return completed form to Safety Manager

Figure 2. The 'Revised' Safety Observation Tool, after revisions

Incorporating AOI #2 and #4 required integrating more behavior-based safety thinking, which is at the core of what we are trying to achieve. For AOI #2, we wanted to get the operators thinking about how they felt, so we incorporated a question that asked them if they would feel comfortable doing the job this way. We believe that if they are taking a more personal approach to observing these tasks – putting themselves in the other person's shoes – they will carefully consider if the operator performing the task makes him, the observer, feel safe. If an observer feels his own safety is in jeopardy while watching someone else perform a task in a different way, this could reveal opportunities for discussion and improvement. For AOI #4, we believed the form needed to be revised to inspire them to think ahead and demonstrate forward thinking. To accomplish this, we created the question, "What could go wrong while doing this job?". This is a critical-thinking question that will train them to inherently consider and anticipate potentially dangerous situations before executing a task. Adding this question to the new tool is key because if we can change the way they are approaching safety observations by prompting them to stop and think when they *observe* a task, we can change the way they *perform* a task.

3.2 Analyzing the Results from using the Revised Safety Observation Tool

The new tool was evaluated using the same metrics that were used for evaluating the old form to determine if we could gain more 'value' from this revision: one point for Problem Identified and one point for Solution Provided for a total of two points. Additionally, the new tool was strategically structured to challenge them to think about what could go wrong with the task.

Table 2 shows the summary of results using the Revised Safety Observation Tool. The results revealed that 158 safety observations were performed, which shows an overall improvement in participation. Like Table 1, the results in Table 2 are shown in the same manner: the results are listed in order from highest to lowest score, arbitrary descriptors were assigned to each operator (e.g., Observer A, Observer B, etc.), and we focused on the results from the operators who performed three or more safety observations using this tool. These results revealed a remarkable improvement in overall scores, as they were all determined to score at 50 percent or higher, with over half of them achieving a score of 100 percent.

It is significant to include that this part of the study was conducted over 6.5 months as compared to one year for the previous evaluations using the old form. Our company also underwent an acquisition during this that impacted our plant site resulting in fewer employees participating in the safety observation exercise. Despite the shorter amount of time and the company transition, the operators' participation and enthusiasm significantly increased, which we believe demonstrates a shift in our culture that is centered around behavior based safety.

Using the new tool, 91 percent of the operators identified a problem and 91 percent proposed a solution, which is an immense improvement from the scores when the old form was used (57 percent and 18 percent, respectively). Therefore, we exceeded our goal of targeting a 20 percent increase in the number of reports for identifying a problem and providing a solution. Out of the 158 safety observations performed, 156 commented on "What could go wrong while doing this task?", which strongly indicates that they are exhibiting forward thinking and anticipating potential hazards of the task. Correspondingly, the value of the safety observations being performed by shift leaders also increased significantly, as shown by the shaded regions in Table 2. Three out of the four shift leaders achieved a score greater than 50 percent. The remaining shift leader took a different approach – he supported his team from behind the scenes, coaching them on how to effectively execute safety observations. The participants on his team consistently produced safety observations with high value, all scoring at 50 percent or greater. Changing supervisory practices has been shown to produce successful results for change in the workplace [4]. In this case, the practice by the supervisor catalyzed the change in his team to produce more safety observations with higher value.

Table 2. Summary of Results using the Revised Safety Observation Tool

Observer	Points	Possible Points	Score
A	12	12	100%
B	12	12	100%
C	26	26	100%
D	28	18	100%
E	6	6	100%
F	22	22	100%
G	10	10	100%
H	8	8	100%
I	6	6	100%
J	6	6	100%
K	6	6	100%
L	9	10	90%
M	25	28	89%
N	7	8	88%
O	12	14	86%
P	12	14	86%
Q	5	6	83%
R	8	10	80%
S	8	10	80%
T	6	8	75%
U	4	6	67%

3.3 Summary of Findings and Lessons Learned

We have learned that we can make simple modifications to an exercise we had been doing for years and measurable progress in our safety culture. The operators expressed a higher level of satisfaction when performing safety observations using the new tool. Simplifying the text on the Safety Observation tool and providing spaces to comment has improved overall participation and has improved the value of the safety observations. In addition, revising the tool to inspire them to think more critically by providing the leading questions like, "what could go wrong while doing this job," has made an impact on their behavior and attitude – they actively discuss hazard evaluation and incident response as part of their daily tasks. This has led to more open discussions taking place in the control rooms. We have been able to use the safety observation tool to cultivate an environment where discussing 'what could go wrong' is not considered to be pessimistic thinking, it is proactive and forward-thinking. It appears from this study that the safety behaviors of the front-line users may be improved when given the right tools and establishing appropriate expectations. This has been accomplished by taking a more personal and practical approach to safety when they are asked to consider what could go wrong on a job they routinely perform. In summary, we have learned two key insights from this study:

1) Provide the right tools for the job and the job will be done better

   - We provided a tool to influence our safety culture focusing on how routine operations are being conducted
2) Establish expectations to see positive results
   - We set the expectation of responding to the questions on the tool by reorganizing the form and providing spaces on the form so the operators could comment

4 Conclusion

We rely heavily on our operations personnel to efficiently and safely conduct the day-to-day activities in our plant. Many safety incidents are caused by human error, so there is a constant need to improve in this area. We developed a simple tool by modifying our existing safety observation exercise: the new tool targeted specific questions asking our front-line users to consider what could go wrong while doing a task and how could the task be improved. This method has resulted in a more personal and practical approach to conducting safety observations. The safety observation tool has empowered our front-line users to be more proactive about identifying risks associated with routine tasks and to take ownership of making sustainable changes to directly impact their safety. This study has suggested that we are on the forefront of a paradigm shift in our behavior based safety culture and has provided a new way to anticipate potential hazards.

5 Disclaimer

The views and opinions expressed in this publication are strictly those of the authors and do not necessarily reflect those of the authors' employer, organization, or any other group or individual. The information is presented without warranty or guarantee. Any action taken from the information provided herein (including recommendations) shall be at the user's sole risk.

6 References

1. J. Mullen, "Investigating factors that influence individual safety behavior at work." Journal of Safety Research 35 (2004) p. 275-285.
2. A. Ghahramani and H.R. Khalikhali, "Development and Validation of a Safety Climate Scale for Manufacturing Industry." Safety and Health at Work 6 (2015) p. 97 - 103.
3. M.A. Griffin and A. Neal, "Perceptions of Safety at Work: A Framework for Linking Safety Climate to Safety Performance, Knowledge, and Motivation." Journal of Occupational Health Psychology 5 (2000) p. 347 - 358.
4. K.J. Nielsen, "Improving safety culture through the health and safety organization: A case study." Journal of Safety Research 48 (2014) p. 7 - 17.
5. A. Neal and M.A. Griffin, "A Study of the Lagged Relationships Among Safety Climate, Safety Motivation, Safety Behavior, and Accidents at the Individual and Group Levels." Journal of Applied Psychology 91 (2006) p. 946 - 953.
6. M.D. Cooper and R.A. Phillips, "Exploratory analysis of the safety climate and safety behavior relationship." Journal of Safety Research 35 (2004) p. 497 - 512.
7. E.S. Geller, "Behavior-Based Safety and Occupational Risk Management." Behavior Modification 29 (2005) p. 539 - 561.
8. CCPS. *Conduct of Operations and Operational Discipline For Improving Process Safety and Industry*. Center for Chemcial Process Safety, American Institute of Chemical Engineers, New York, NY, 2011.
9. J.M. Converse and S. Presser. *Survey Questions: Handcrafting the Standardized Questionnaire*. Sage Publications, Thousand Oaks, California, 1986.
10. A. Fink. *How to Ask Survey Questions*. Sage Publications, Thousand Oaks, California, 2003.

Everything you need to Know about Human Reliability for Process Safety

William Bridges
Process Improvement Institute, Inc. (PII)
1321 Waterside Lane
Knoxville, TN 37922
wbridges@piii.com

William C. Rhodes II, P.E.
Process Improvement Institute, Inc. (PII)
82 Via Santa Maria
San Clemente, CA 92672
wrhodes@piii.com

Prepared for Presentation at
American Institute of Chemical Engineers
2019 Spring Meeting and 15th Global Congress on Process Safety
New Orleans, LA
March 31 – April 3, 2019

UNPUBLISHED

Everything you need to Know about Human Reliability for Process Safety

William Bridges
Process Improvement Institute, Inc. (PII)
1321 Waterside Lane
Knoxville, TN 37922
wbridges@piii.com

William C. Rhodes II, P.E.
Process Improvement Institute, Inc. (PII)

Keywords: human factors, process safety, PSM, human error, risk-based process safety, RBPS, conduct of operations, operational discipline

Abstract

Weak control of human factors leads directly to error. An error can result in an initiating event of an accident sequence (directly by the mistake made or indirectly, such as by a process control loop failing prematurely). Human error is also the cause of failure of each layer of protection. This paper discusses each of the 10 primary human factors and describes what we know about their relative importance in accident causation. The data presented is from basic research by the authors on the root causes of more than 15,000 accidents and near misses; and also based on the review of thousands of accidents analyzed by others and on summary data from many companies. This paper lists where focus should be placed (i.e., which human factors tend to be key) and provides proven ways to optimize these human factors so that the base human error rate at a site is as low as possible. Tie in to other applications, such as determining the probability of human error for risk analysis is also presented. Case studies and examples are used to illustrate key points. This paper is a follow-on to Human Factors Elements Missing from Process Safety Management (PSM), presented in 2010 at the 6th Global Congress on Process Safety, AICHE.

I. Introduction

All accidents (or nearly all, if you consider that there are some natural phenomena that we either cannot guard against or choose not to guard against) result from human error. This is because humans govern and accomplish all of the activities necessary to control the risk of accidents. Humans influence other humans in the process – not only do humans cause accidents (unintentionally) by making errors directly related to the process itself, but they also cause errors by creating deficiencies in the design and the implementation of management systems (i.e., we make errors in authorities, accountabilities, procedures, feedback, proof documents, and continual improvement provisions). Ultimately these management systems govern the human error rate directly contacting or directly influencing the process. The process-related activities where errors have the most influence include:

- Designing of a process
- Engineering of a process
- Specifying the process components
- Receiving and installing equipment
- Commissioning
- Operating the process
- Predicting safeguards necessary to control the risk at an acceptable level and sustaining these safeguards for the life of the process
- Maintaining, inspecting and repairing the process
- Troubleshooting and shutting down the process
- Managing Process Changes

99% of accidental losses (except for natural disasters) begin with a human error ***(supported by data from more than 1500 investigations)***

Root causes of accidents are management system weaknesses
(Center for Chemical Process Safety, American Institute for Chemical Engineers, "Guidelines for Investigating Chemical Process Incidents", 2003) – OSHA agrees

Weak Management Systems → Human Error → Accidents

Figure 1: Relationship between Accidents and Management Systems

Implementing human factor engineering and policies to prevent accidents is not a new concept. Nearly all (or all, from a more complete perspective) of the causes and root causes of major accidents in the past 30 years have been the result of poor control of human factors. This has been cited in many root cause analysis reports and papers concerning these major accidents. Although governmental regulations and industry-specific guidelines have been developed for Process Safety Management (PSM) systems, these documents lack the needed guidance in the critical area of effectively managing human error. This paper fills this gap by summarizing the key factors that directly lead to human error. Further, it presents this information in a practical manner that can be incorporated into any company's PSM program to effectively improve human reliability.

Definitions for Use in This Paper

- This paper will use the term **Human Error** to mean the errors that are made during direct interface or direct influence of the process.
- **Human Factors** are those aspects of the process and related systems that make it more likely for the human to make a mistake that in turn causes or could cause a deviation in the process or could in some indirect way lead to the increased probability of an accidental loss.

- **Management systems** are the administrative controls an organization puts in place to manage the people and workflow related to the process under consideration, and so these inherently attempt to control human factors

II. Human Error Fundamentals

Types of Human Error

In simplest terms, there are only two types of human error: *Errors of Omission* (someone skips a required or necessary step) and *Errors of Commission* (someone performs the step wrong). But in addition, these errors occur either inadvertently (unintentional error) or they occur because the worker believes his or her way is a better way (intentional error, but not intentional harm). Intentional errors can usually be thought of as errors in judgment. Some believe a "lack of awareness of the risk" causes these errors, but in actual practice, the worker who commits an intentional error may well be aware of the risk. They instead believe they know a better way to accomplish a task or they believe there are already too many layers of protection (so bypassing one layer will not cause any harm).

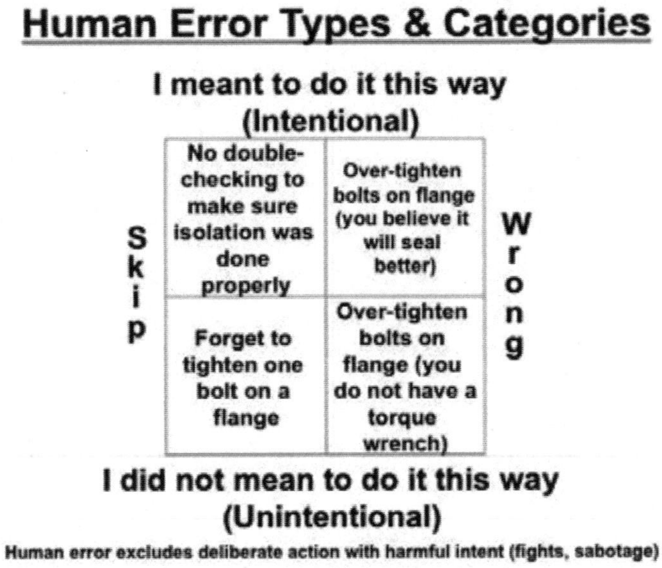

Figure 2: Types and Categories of Human Error

Regardless of type or category of human error, the organization can and should exert considerable control of the errors.

Controlling Human Error through Management Systems

Generally speaking, if an organization does not directly control risk factors, the organization cannot directly control quality, safety, environmental impact, or production to acceptable levels. A common example is mechanical reliability, where an organization seeks to understand the factors that lead to mechanical failures and implements systems to reduce the impact of these failure mechanisms. Similarly, an organization must understand the factors leading to human error and build management systems to sustainably control human error, thereby managing the risk of accidental losses that impact quality, safety, the environment, production, or assets.

In order to sustainably control the human error risk of a complex process (such as an oil/gas operation, refinery, chemical plant, steel plant, automobile manufacturing, aircraft manufacturing, etc.), the organization must design and implement management systems to:

- **Understand the Risk** – This involves predicting problems; which in turn includes predicting the risk of possible accident/loss scenarios, establishing the appropriate design and the right layers of protection to control risk to a tolerable level
- **Control Risk Factors Day-to-Day** – This involves controlling the original design by maintaining the established layers of protection and managing changes to the design using integrated management systems
- **Analyze Actual Problems and Determine Weaknesses in the System** – This involves identifying weaknesses in designs and management systems and weaknesses in risk understanding through root cause analysis of actual problems (losses and near-losses)

These three elements are illustrated in the following figure:

© Process Improvement Institute, Inc. (2004-2010)
Figure 3: Controlling Risk through Human Reliability

Management systems control the interaction of people with each other and with processes. They are the high level procedures we use to control major activities like conducting Process Hazard Analyses (PHAs), management of change, writing operating procedures, training employees, evaluating fitness for duty, conducting incident investigations, etc. If management systems are weak, then layers of protection will fail, and accidents will happen.

III. Human Factor Categories and Typical Impact of Each

To minimize human error, process safety systems should address the **Human Factors Categories** (see various US NRC and US DOE standards from 1980s and 1990s)[1, 2] The following table lists the key human factor categories along with multiplication factors that poor human factors can have on the base human error rates.

TABLE 1: SUMMARY of 10 HUMAN FACTOR CATEGORIES

Based in part on: Gertman, D.; et. al., *The SPAR-H Human Reliability Analysis Method*, NUREG/CR-6883, U.S. Nuclear Regulatory Commission, Office of Nuclear Regulatory Research, Washington, DC, August 2005. PII has modified the list slightly to account for general industry data and terminology and to incorporate PII internal data.

Human Factor Category	Human Factor Issue/Level	Multiplier for Cognitive & Diagnosis Errors
Available Time	Inadequate time	P(failure)=100%
(includes staffing Issues) – ***for responses only***	Barely adequate time (≈2/3 x nominal)	10
	Nominal time (1x what is expected)	1
	Extra time (at least 2x nominal and >20 min)	0.1
	Expansive time (> 4 x nominal and > 20 min)	0.01
Stress/Stressors	Extreme (threat stress)	5
(includes staffing issues)	High (time pressures such as during a maintenance outage; issues at home, etc.)	2
	Nominal	1
Complexity & Task Design	Highly complex	5
	Moderately complex (requires more than one staff)	2
	Nominal	1
	Obvious diagnosis	0.2
Experience/Training	Low	10
	Nominal	1
	High	0.5
Procedures	Not available in the field as a reference, but should be	20
	Incomplete; missing this task or these steps	8
	Available and >90% accurate, but does not follow format rules ***(normal value for process industry)***	3
	Good, 95% accurate, follows >90% of format rules	1
	Diagnostic/symptom oriented	1
Human-Machine Interface (includes tools)	Missing/Misleading (violates populational stereotype; including round valve handle is facing away from worker)	20
	Poor or hard to find the right device; in the head calc	10
	Some unclear labels or displays	2
	Good	1
Fitness for Duty	Unfit (high fatigue level (>80 hrs/wk or >20 hr/day, no day off in 7-day period; or illness, etc.)	20
	Highly degraded fitness (high fatigue such as >15 hr/day, illness, injury, etc.)	10
	Degraded Fitness (>12 hr day and >72 hr/wk)	5
	Slight fatigue (>8 hr per day; ***normal value for process industry***)	2
	Nominal	1
Work Processes & Supervision	Poor	2
	Nominal	1
	Good	0.8
Work Environment	Extreme	5
	Good	1
Communcation	No communication or system interference/damage	10
	No standard for verbal communication rules ***(normal value for process industry)***	3
	Well implemented and practiced standard	1

If all of these human error factors are controlled very well, then the optimized human error rates listed in Section V. of this document are achievable. Alternatively, if one or more of these factors are compromised, the human error rate will increase by the values shown in this table. As an example, an individual whose fitness for duty rating is unfit due to the excessive work hours shown in the table will make errors at a rate 20 times greater than an individual whose fitness for duty is normal.

IV. Incorporating Human Factors into PSM Systems

The basis upon which most companies build their PSM system is their applicable PSM regulation (e.g., the OSHA PSM standard found at 29 CFR 1910.119 in the United States.) Unfortunately, this standard, which was promulgated in 1992 (and has been essentially unchanged since), is devoid of human factor controls with a few exceptions as outlined by Bridges and Tew[3]. The only direct reference to the term "human factors" is mentioned in paragraph (e), Process Hazard Analysis (PHA), which states that the PHA team must consider human factors (presumably in the review of the causes and the quality of the safeguards). The other mention that alludes to human factors, is in Operating Procedures (see paragraph (f)) which states procedures "must be written clearly and understandably." Similar PSM regulations in other countries are likewise silent on human factors.

The next logical place to seek guidance for incorporating human factors into a company's PSM system is industry-specific standards; therefore, some companies may use these to build upon the regulatory framework. In the chemical process industry, such a standard was published in 1985 by the Center for Chemical Process Safety (CCPS), which is a division of the American Institute of Chemical Engineers (AIChE). This standard included a strong emphasis on human factors, since Human Factors was one of its original 12 elements. However, in 2007 the CCPS revised their standard as "Guidelines for Risk-Based Process Safety"[4] (RBPS). This revision abandoned the global element on human factors, and instead spread the direct control of human factors throughout 6 elements. The authors believe that this change not only diminishes the importance of human factors, but also reduces the ability of companies to effectively implement human factors throughout their PSM program.

The CCPS partially addressed this shortcoming by publishing a guideline entitled "Conduct of Operations and Operational Discipline"[5] (COO/OD) in 2011. As outlined in Section VIII of this paper, this newer guideline provides a great deal of practical and useful information for implementing *most* human factors; however, the information contained within it is still incomplete. Therefore, to fully integrate human factors into a PSM program, companies should consult the detailed review of each human factor outlined in Section V of this paper to develop the necessary tools to fully implement human factors into their PSM systems.

V. Detailed Description of the Human Factor Categories

With excellent control of each of the human factors listed in Section III. above, a company can begin to approach the lower limits that have been observed for human error. These lower limits are about:

- **1 mistake in 100 steps for most procedures-based tasks** (such as starting up a process unit), a little less for a routine (daily) task that becomes almost a reflex

- **1 in 10 chance or a little better for diagnosis and response to a critical alarm**

Excellent control requires superior design and implementation of management systems, which is enabled through a thorough understanding of these factors, as outlined below.

A. AVAILABLE TIME and
B. STRESS / STRESSORS

The **Available Time** for the task refers to the time the task is expected to be required for completion. In many cases the time determined for a process is determined using the theoretical times where the conditions are "ideal" and not necessarily realistic. When task and process duration are not realistic people tend to "find the best and easy way" to get it done, ultimately creating latent conditions for error occurrence. **Work Stress** refers to the emotional response that arises when work demands exceed the person's capacity and capability to cope. Some processes have been designed paying little attention to job design, work organization and management systems. Failing to fulfill this requirement eventually results in work stress.

Staffing levels directly impact both factors. Staffing is the process of assessing, maintaining and scheduling personnel resources to accomplish work. An adequately staffed organization ensures that personnel are available with the proper qualifications for both planned and foreseeable unplanned activities. Staffing is a dynamic process in which plant management monitors personnel performance to ensure that overall organizational performance goals are met or exceeded. The result of an effective staffing process is a balance between personnel costs and the achievement of broader organizational goals.

Three key issues must be considered when staffing decisions are made; the first of which is selecting the right staff for a job. Each organization requires the proper amount and type of expertise to safely and competently operate the plant under a variety of conditions. The term "expertise" includes the attributes of talent, effectiveness, knowledge, skills, abilities, and experience necessary to operate and maintain plant systems, structures and components.

The second key issue is avoiding staff overload. Surges in workload, such as during outages, typically require staff augmentation as well as longer work hours for permanent staff. The introduction of contractor personnel or company personnel from other sites may increase the likelihood of errors due to unfamiliarity with the plant, its procedures and hardware, for example. Longer work hours have the potential to increase fatigue, which also contributes to the likelihood of error.

The third key factor is rotating staff every 1 hour or less for tasks that require high vigilance. Humans are inherently unable to remain alert for signals that seldom, if ever, occur. Even a sailor whose life is at stake cannot maintain an effective watch (look-out) for hostile submarines for more than 30 minutes or so. The following figure illustrates the rapid decrease of vigilance with time.

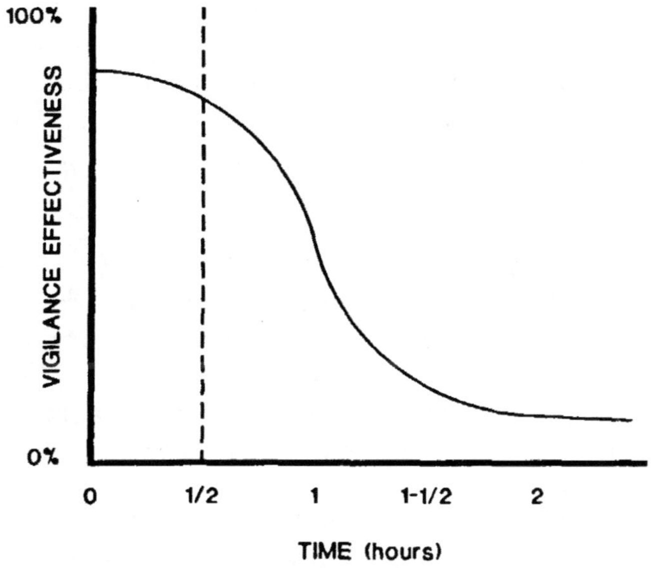

Figure 4: Vigilance as a Function of Time

Placing a worker in situations requiring extended, uneventful vigilance may lead to accidents. Therefore, it is important to design control systems in a manner that requires regular operator interaction so that the operator will remain attentive.

C. TASK DESIGN

A task that is designed with the human limits in mind is much more likely to work effectively than one that assumes humans can and will "always" do what is written. The task must consider that humans think and remember and factor in prior data and prior experiences. Factors to consider in assessing **task design** adequacy (i.e., minimizing the likelihood of human error during its completion) include:

- Complexity of the task
- Probability of repeat errors (coupled error) on redundant aspects of the design
- Error detection and Error Recovery

Complexity of task (procedure-based or a call for action) – If the task is too complex, then humans can forget their place in the task, fail to understand the goal of each step or sub-step, or fail to notice when something isn't going right. Task complexity is a function of:

- number of choices available for making a wrong selection of similar items (such as number of similar switches, number of similar valves, number of similar size and shaped cans)
- number of parallel tasks that may distract the worker from the task at hand (leading to either an initiating event or failure of a protection layer)
- number of staff involved (more staff = more complex)
- number of adjustments necessary to achieve the goal
- amount of mental math required (as a rule, NO math should be required in anyone's head when accomplishing a standardized task)

- how much judgment is required to know when you have accomplished each goal within the task

  For most chemical process environments, the complexity of the task is relatively low (one action per step), but for response actions (in which humans serve as the Independent Protection Layer (IPL)), there are almost always other tasks underway when the out-of-bounds reading or the alarm is activated. Complexity is difficult to predict (since it is not known when a human intervention will be needed), but higher complexity can increase error rates by 2 to 10 times.

<u>Probability of repeat errors</u> (Coupled errors) – For many maintenance tasks, making a repeat error or "common cause error" or "dependent error" can lead to failures of all backup systems. Such errors have led to MANY airplane crashes and major process safety accidents as well.

Coupling represents the probability of repeating an error (or repeating success) on a second identical task, given that an error was made on the first task. The increased probability of failure on subsequent tasks given that an error has already been made is known as dependence. The list below provides some starting point guidance on values to use:

- 1/20 to1/90 – if the same tasks are separated in time and if visual cues are not present to re-enforce the mistake path. *This error rate assumes a baseline error rate of 1/100 with excellent human factors. If the baseline error is higher, then this rate will increase as well.*
- 1/2 – if the same two tasks are performed back-to-back, and if a mistake is made on the first step of two. *This error rate assumes a baseline error of 1/100 with excellent human factors. If there the baseline error is higher, then this rate will increase as well.*
- 8/10 to 10/10 – if the same three tasks are performed back-to-back and strong visual cue is present (if you can clearly see the first devices you worked on), if a mistake is made on the first step of the two or more
- Two or more people become the same as one person (with respect to counting of errors from the group), if people are working together for more than three days; this is due to the trust that can rapidly build.

These factors are based on the relationships provided in NUREG-1278[1] and the related definitions of weak and strong coupling provided in the training course by Swain (1993)[1] on the same topic, as shown here in Table 2. The following relationship is for errors of omission, such as failing to reopen a root valve or failing to return a safety instrumented function (SIF) to operation, after bypassing the SIF. The qualitative values in Table 2 are based jointly on Swain[1] (1993) and Gertman (SPAR-H, 2005 which is NUREG/CR-6883)[2].

One can readily conclude that staggering of maintenance tasks for different channels of the same SIF or for related SIFs will greatly reduce the level of dependent errors. Unfortunately, most sites PII visits do not stagger the inspection, test, or calibration of redundant channels of the same SIF or of similar SIFs; the reason they cite is the cost of staggering the staff. While there is a perceived short-term higher cost, the answer may be different when lifecycle costs are analyzed.

Simple Rule: Staggering of maintenance can prevent significant number of human errors in redundant channels. In fact, the US Federal Aviation Administration (FAA) requires staggering of maintenance for aircraft with multiple engines or multiple control systems (i.e., hydraulics) (*FAA Advisory Circular 120-42A*, as part of Extended Operations (ETOPS) approval).[6]

Table 2: Guideline for Assessing Dependence for a within-SIF Set of Identical Tasks (based partially on SPAR-H, 2005 [1,2], and partially on field observations by PII) Courtesy Process Improvement Institute, Inc., All Rights Reserved

Level of Dependence	Same Person	Actions Close in time	Same Visual Frame of Reference (can see end point of prior task)	Worker Required to Write Something for Each Component
Zero (ZD)	No; the similar tasks are performed by different person/group	Either yes or no	Either yes or no	Either yes or no
Zero (ZD)	Yes	No; separated by several days	Either yes or no	Either yes or no
Low (LD)	Yes	Low; the similar tasks are performed on sequential days	No	Yes
Moderate (MD)	Yes	Moderate; the similar tasks are performed more than 4 hours apart	No	No
High (HD)	Yes	Yes; the similar tasks are performed within 2 hours	No	No
Complete (CD)	Yes	Yes; the similar tasks are performed within 2 hours	Yes	Either yes or no

Once the level of dependence is known, the probability of either repeat success or repeating errors on identical tasks can be estimated. For these probabilities, we use Table 3, which is a re-typing of Table 20-17 from NUREG-1278[1] (and the similar table in SPAR-H [Gertman, 2005][6]).

Table 3. Equations for Conditional Probabilities of Human Success or Failure on Task N, given probability of Success (x) or Failure (X) on Task N-1, for Different Levels of Dependence *Courtesy Process Improvement Institute, Inc., All Rights Reserved*

Level of Dependence	Repeating Success Equations (but shown as error probability)	Repeating Failure Equations
Zero (ZD)	$P_{Success@N} = x$	$P_{Failures@N} = X$
Low (LD)	$P_{Success@N} = (1+19x)/20$	$P_{Failures@N} = (1+19X)/20$
Moderate (MD)	$P_{Success@N} = (1+6x)/7$	$P_{Failures@N} = (1+6X)/7$
High (HD)	$P_{Success@N} = (1+x)/2$	$P_{Failures@N} = (1+X)/2$
Complete (CD)	$P_{Success@N} = 1.0$	$P_{Failures@N} = 1.0$

Error detection and error recovery – Is there enough feedback in the process to allow the worker to realize (in time) that they made a mistake? Have they been trained on how to reason through how to recover from mistakes they or others make? (Sometimes, doing a step too late is far worse than NOT doing the step at all.) Is there enough time available for the type of intervention necessary?

D. EXPERIENCE/TRAINING

Training is necessary for general functioning of management systems and for task-specific skills such as how to start up a compressor, repair a pump, lead a root cause analysis, perform a proper Lock-Out/Tag-Out, etc. Organizations normally do a good job of training, including hands-on training. Training systems can be weak (inadequate) however if the training system does not adequately:

- Address how to troubleshoot the process (i.e., handle process deviations or upsets).
- Provide workers with a mental model(s) of the process so they can perform the assigned tasks correctly, diagnose process upsets properly and quickly, and understand the consequences of their actions.
- Provide workers with enough practice of critical tasks

Ironically, training is many times listed as the "cause" or "root cause" of an accident, when in fact the training is adequate, and some other human factor is the major cause.

E. PROCEDURES

Many procedures do not follow best practices for controlling human error, and so the written procedure actually "contributes" to increased error rates. Further, many organizations are missing guides on how to troubleshoot (what to do when process deviations occur). The best practice rules for writing and validating procedures have been published for many years ***(see Bridges, 1997-2010)***[3, 7, 8]. Below is a checklist based on the current set of best practice rules for developing operating, maintenance, and other work-instructions (procedures):

TABLE 4: PROCEDURE QUALITY CHECKLIST (courtesy PII, 2008)

#	Issue	Response
	Procedure Content Checklist	
1	Is the procedure drafted by a future user of the written procedure? (Engineers should not author procedures to be used by operators or maintenance staff.)	
2	Is the procedure validated by a walk-down in the field by another future user of the procedures?	
3	Is the procedure reviewed and commented on by technical staff (engineers or vendors)?	
4	Is the procedure checked versus the Page and Step format rules below?	
5	Is a hazard review of step-by-step procedures performed to make sure there are sufficient safeguards (IPLs) against the errors that will occur eventually (when a step is skipped or performed wrong)?	
6	Is the content measured using "newly trained operators" to judge the % of steps that are missing, steps that are confusing or wrong, and steps that are out-of-sequence? (A score of 95% accuracy of content is good.)	
	Page Format Checklist	
1	Is the title of the procedure the largest item on the page?	
2	Is the procedure title clear and consistent with other titles, and does it uniquely describes the topic?	
3	Are the document control features the smallest items on the page?	
4	Are temporary procedures clearly identified?	
5	Is white space used effectively? • Is there one blank line between each step? • Does indentation help the user keep their place? • Are the margins large enough to reduce page congestion?	
6	Is type size 12 point font or larger?	
7	Is serif type used (rather than sans-serif)?	
8	Is mixed case used for words of steps, with ALL CAPS used only for special cases (such as IF, THEN, AUTO, and WARNING)?	
9	Is the step number very simple (single level of number used)? Only an integer?	
10	Have sections or information not necessary to performing the steps been moved to the back or to another part of the manual or training guide?	
11	Are section titles bold or larger than the text font? Do sections have clear endings?	
12	Is the decision on electronic presentation versus hard copy correct? Are electronic linkages to procedures clear and accurate and easy to use? If paper is chosen, is the color of the paper appropriate?	
13	Is the overall page format (such as Outline format or T-Bar format) appropriate to the use of the procedure?	
14	Are play script features added for tasks that must be coordinated between two or more users? • Play script is normally used when there are two or more hand-offs of responsibility for steps.	
15	Are rules followed for formatting of Warnings, Cautions, and Notes? (See annotated rules, such as Warnings are for worker safety and Warnings must clearly standout from rest of page.)	
	Step Writing Checklist	
1	Is each step written as a command?	
2	Is the proper level of detail used throughout? This is judged based on: • *Who* will use the procedures • Same level of detail used in similar procedure steps	
3	On average, is there only one implied action per instruction? Best practice is to average 1.2.	
4	Does the procedure indicate when sequence is important? • If sequence matters, each step should be numbered (with an integer or letter) • If sequence does not matter, bullet lists should be used	
5	Are only common words used? Apply "education" level test (5 grade reading level is	

#	Issue	Response
	best)	
6	Do all acronyms, abbreviations, and jargon aid understanding? • Develop a list of such terms for use in procedures *and* communication. • Use terms that users use (within reason)	
7	Is each step specific enough? No room left to guess/interpret: • The meaning of a word or phrase (Check vs. Make sure) • The intent of a step or series of steps • A desired quantity or value • To what equipment the step applies	
8	Is the procedure free of steps that require in-your-head calculations? • Values expressed as ranges rather than targets with error bands • Conversion tables, worksheets, or graphs provided where needed	
9	Are graphics to the user's advantage? • No explanatory paragraphs or lengthy instructions that could be replaced by a picture • No impressive graphics that provide no real advantage	
10	Are references to the user's advantage? • No lengthy explanations or instructions that could be replaced by branching to a reference • No references to a procedure that references still another • No gaps or overlaps between this procedure and a referenced document • If branching, must branch to a procedure, not to a specific step in a procedure	
11	Are rules followed for writing warnings, cautions, and conditional steps? • No more than 2 per page • No actions within a warning or caution (actions must always be numbered steps) • Warnings and Cautions contain descriptions of potential consequences	

For procedures to be effective in ensuring that tasks are performed correctly, they must be used. There are many reasons that workers may not use procedures. ***Deficient Procedures*** are the most prevalent problem in process industries since procedures have not traditionally been developed from the perspective of optimizing human factors; instead, procedures have been traditionally developed to meet a compliance requirement to have written procedures. Examples of procedure deficiencies (inaccuracies) include:

- Incorrect/incomplete/nonexistent. Most procedures we have audited have been only 70-85% accurate; the inaccuracies include missing critical steps, steps as written are not what needs to be done, or the steps are out of sequence
- No/misplaced warnings. For example, a warning should never **contain** the action to take; it should instead **emphasize** the action to take
- Poor format and presentation rules

Other reasons workers may not use procedures include:

- Procedures are out of date
- No procedure has been written for the task
- Users cannot find the procedure they want to use
- Users don't need a procedure because the task is simple
- Users need more information than the procedures contain
- Users see procedures as an affront to their skill
- Procedures are difficult to use in the work environment
- Procedures are difficult to understand

So, in addition to the rules for writing procedures that are shown in the table above, the organization must also address the reasons that cause the worker not to use the written procedure.

F. HUMAN-MACHINE INTERFACE (INCLUDING TOOLS)

The **Human-Machine Interface** (also known as the Human-System Interface, or HSI) is defined as the technology through which personnel interact with plant systems to perform their functions and tasks. The major types of HSIs include alarms, information systems, and control systems. Each type of HSI is made up of hardware and software components that provide information displays, which are the means for user-system interaction, and controls for executing these interactions.

Personnel use of HSIs is influenced directly by (1) the organization of HSIs into workstations (e.g., consoles and panels); (2) the arrangement of workstations and supporting equipment into facilities, such as a main control room, remote shutdown station, local control station, technical support center, and emergency operations facility; and (3) the environmental conditions in which the HSIs are used, including temperature, humidity, ventilation, illumination, and noise.

There are three important goals to be achieved in the design and implementation of the HSI. These are:

- **Design for operability** refers to designing the HSI to be consistent with the abilities and limitations of the personnel who will be operating it. Weaknesses in the design processes can result in an HSI that is not well suited to the tasks that personnel must perform to ensure plant safety, resulting in increased workload, decreased performance by personnel, and an increased likelihood of errors.

  The wide-spread adoption of distributed control systems has provided many advances in process control. However, if implemented improperly, these systems can greatly reduce the operability of the facility. Although they were unable to provide as much information, old analog panels were typically laid out in a manner such that experienced operators were able to monitor the entire process and effectively identify conditions meriting attention. Due to space limitations, and their cost (e.g., hardware and installation), alarms were typically limited to those deemed to be associated with a significant process upset.

  The advent of distributed control systems has eliminated the space limitations associated with alarms, and greatly reduced their costs (in many cases, alarms can be added merely through programming). Without proper alarm configuration structure and discipline, this can lead to "alarm flooding" (an excessive number of simultaneous alarms) during abnormal operating conditions. The human error of failing to respond properly to an alarm increases by a factor of over 1,000 if ten alarms activate simultaneously compared to a single alarm activation.

Therefore, it is critical that control and alarm system designers in facilities with distributed control systems properly prioritize information so that operators can respond appropriately to abnormal conditions. Further, trouble-shooting guides should be readily accessible via the distributed control system to reduce human error in alarm response.

Beyond the topic of alarms, the proper design of the visual displays can also reduce human error. Certain color combinations, such as a light gray background with darker gray equipment and black lines coupled with a flashing red indication for those components in an alarm state effectively draw the operators' attention to the relevant part of the process requiring attention. Further, an effective display should include:

- Safety systems that are bypassed, impaired, or out of service
- Process alarms that are disabled
- Process conditions that are not within established limits
- Backup systems that are unavailable
- Special maintenance or testing activities that are currently in progress

- **Design for maintainability** refers to designing the HSI and associated plant equipment to ensure that personnel are able to perform necessary maintenance activities efficiently. Weaknesses in the design process can result in systems that impose excessive demands on personnel for maintenance and, therefore, are prone to maintenance errors or problems with reliability and availability.

- **Design for flexibility** refers to the way that changes, such as upgrades to the HSI, are planned and put into use. A new HSI component may require the user to perform functions and tasks in new ways. Skills that the user developed for managing workload when using the former design, such as ways for scanning information or executing control actions, may no longer be compatible with the new design. The new HSIs must also be compatible with the remaining HSIs so that operators can use them together with limited possibilities for human error. Also, HSI modifications may not be installed or put into service all at one time, causing the user to adapt to temporary configurations that are different from both the original and final configurations. Weaknesses in HSI implementation can increase operator workload and the likelihood of errors.

Tools are a special category of HSI, which typically refers to hand tools or devices that are generally designed with the user in mind. A big part of human factor consideration is how to make the equipment and process operation mistake-proof (to prevent errors as much as possible). The following are some factors to consider when error proofing in designs:

- **Design for unambiguous assembly.** Design the product or device such that the assembly process is unambiguous (by designing components so that they can only be assembled one way); i.e., design matching parts that are easy to insert and align. For example, use notches, asymmetrical holes, and stops to mistake-proof the assembly process. Products that go together in only one way require less worker training, perform more reliably, and can be repaired more quickly.

- **Consult workers.** Operators, technicians, and maintenance personnel can pinpoint the most troublesome areas.
- **Avoid symmetry.** When a particular orientation is critical to the design, avoid symmetry. For example, use nonsymmetrical hole patterns.
- **Use labels sparingly.** Labels tend to come off equipment too easily and often are wordy.
- **Review the environment.** Environmental problems that encourage mistakes include poor lighting; high/low heat; excess humidity, dust, and noise — anything that distracts workers.

The following are some factors to consider when error proofing in process/operations:

- **Error-proof mechanisms.** Error-proof mechanisms are very powerful in improving system reliability when incorporated into the design. These mechanisms, by design, will not allow a user to perform an illegal operation. For example, if a user enters a value that is outside the accepted range of operation, the control logic will not accept the value.
- **Automatic alerts.** Automatic alerts immediately inform the user of an illegal operation to prompt corrective action. These alerts are particularly useful in critical operations that allow time for corrective action before some adverse consequence occurs. For example, a piping manifold may be designed such that a warning alarm sounds if valves are not opened/closed in the proper, critical sequence, or if a critical valve is left opened/closed.
- **Automatic system shutdown.** Automatic system shutdown should be incorporated into the design when an illegal action is performed during a critical operation and no time is available for corrective action. For example, if an operator uses the wrong sequence in charging a reactor, the reactor will shut down before other materials are charged that may lead to critical temperatures and pressures.

G. FITNESS FOR DUTY

Successful task performance requires that the capabilities that workers bring to the task fall within an expected range. **Fitness for Duty** issues include reduction in an individual's mental or physical capabilities due to substance abuse, fatigue, illness or stress, which increases the likelihood of errors. Types of possible impairments include:

- Physical attributes – strength, reach, eye-sight and color acuity, hearing,
- Mental attributes – drug and alcohol (abuse), mental stress (on and off the job);
- Fatigue – issues from on the job and off the job (especially control of hours per work-day and per work-week)

Safeguards to prevent fitness-for-duty-related errors include company programs for the detection and prevention of potential or actual impairment, as well as the individual responsibility of workers to decline assignments if they are impaired for any reason. The latter safeguard is a weak one, however, because humans are generally over-confident of their capabilities when under the influence of drugs or alcohol, or are stressed, fatigued or ill. Other factors that may discourage self-reporting include the fear of poor performance reports from supervision or not receiving pay for extra overtime.

Company programs that may be implicated in errors caused by personnel impairment include:

- **Fitness-for-Duty Program** – Company fitness-for-duty programs are primarily responsible for detecting and preventing impaired personnel from performing tasks that may affect public health and safety. Medical evaluations of personnel, behavioral observation programs, employee assistance programs and drug and alcohol testing are used to detect impairment. Impaired workers can be prevented from performing tasks by establishing protocols for instances in which a worker is believed to be unfit for duty, training for supervisors on detection of and response to fitness for duty issues, and employment guidelines for personnel who voluntarily enter drug or alcohol treatment programs. Weaknesses in a Fitness for Duty program may allow impaired personnel to have access to vital areas in a plant where they could commit errors. One excellent starting point for a Fitness for Duty sub-element is the guidance provided in ***US NRC's 10 CFR 26 (2005).***
- **Overtime Policies and Practices** – Most companies establish limits for work hours to reduce on-the-job fatigue. It has been shown that 17 hours of work without a break results in the same error rate as being legally drunk. And, at 10 days straight of 12 hours work-days, the error rates for non-routine tasks such as startup of a continuous unit can increase to 1 mistake in 5 to 10 steps (as opposed to the target of 1 mistake in 100 steps). Routine authorization for work hours in excess of those recommended may result in fatigued workers. Further, a practice of excluding training or meetings that occur outside of an individual's normal work schedule from work-hour limitations will also contribute to fatigue. In ***US NRC's 10 CFR 26 (2005)***, the guidance given for control of overtime hours is no more 72 hours of work per 6 day period, no more than 16 hours of work in one day within that period, and a minimum of 24 hours contiguous hours away from work within a 7 day period. The US DOT has even more stringent rules on limiting work hours and establishing required hours for recovery from fatigue.
- **Shift Scheduling** – Shift scheduling may also affect the likelihood that personnel will show performance decrements due to fatigue. A change in the assigned shift or a rotating shift schedule will disrupt circadian rhythms and may increase the likelihood of errors.
- **Safety Culture** – The effectiveness of self-reporting and behavioral observation programs depends greatly upon the safety culture at a site.

H. WORK PROCESSES AND SUPERVISION

Supervision is the process by which work is directed and overseen by first-line management. Successful supervision requires a combination of leadership skills and technical competence. Supervision differs from peer checking or quality control because a supervisor has line management responsibility for the worker(s) as well as responsibility for the work activity.

Supervision is more than the moment-to-moment direction of a work activity. Successful supervision requires the assessment and shaping of worker attitudes and motivation, communication and implementation of management expectations for performing work, the assignment of the best-qualified workers to various tasks, as well as the technical competence to identify incorrect actions and stop improper activities before an error is committed. Effective supervision involves directing the work, overseeing how it is performed, and leadership.

Organizations are typically structured to have sufficient supervision of the job, but many times the delineation between the trainer and the supervisor roles is blurred. Supervision can and normally does play a key role in selecting of the right worker for the job, scheduling of workers to match the required tasks for the day/week, and generally overseeing the task execution to ensure policies and procedures are followed. Supervisors are not always trained on all of their key roles in support of control of human factors, such as detecting issues in workers related to fitness for duty or fatigue.

I. WORK ENVIRONMENT

The **Work Environment** refers to the physical conditions in which work is performed. Environmental conditions that can affect performance include excessive vibration and noise, temperature extremes, and insufficient lighting, as summarized below:

- There are two types of **vibration** that may cause errors. The first is whole-body vibration, in which vibration is transferred to the worker from standing or sitting on a vibrating surface. The second is object vibration, in which a stationary worker interacts with a vibrating object in some fashion. The effects of vibration depend upon its frequency and acceleration. Frequency is the number of oscillations (cycles) that occur in one second. Acceleration is the force, or intensity, of the vibration.
- **Noise** is unwanted sound. Noise can cause errors in several ways. It may disrupt communications, affect the ability to perform tasks and annoy personnel. The effects of noise on communications are complex. Even relatively low levels of noise can mask speech, but only under some circumstances. For example, speakers naturally raise their voices when there is background noise and may be able to overcome some of its effects on communication. Being able to see the speaker's face or using standardized phrases also improves communication in a noisy environment. The type of background noise also affects communication. It is easier to communicate over noise that is steady and uniform than noise that includes sharp tonal peaks, such as background speech.
- **Heat exposure** is a common problem in many areas of a plant, such as the turbine building when the plant is operating. The extent to which workers will be affected by heat depends on many factors. These include physical characteristics, such as age, weight, acclimation to heat, physical fitness and dehydration. Other factors that determine the effects of heat on performance include airflow, humidity, clothing and level of physical activity.
- **Exposure to cold** affects the performance of manual tasks. Decreases in the ability to control hand movements begin at an air temperature of approximately 54° F. The fingers may become numb to pain at this temperature and touch sensitivity is reduced. Performance of gross manual tasks, such as those involving the arms and legs is also degraded at 54° F. The speed at which manual tasks can be performed is affected by the rate of cooling. Slow temperature drops have a greater negative impact on manual dexterity than rapid temperature decreases, during the initial exposure period.

- **Adequate lighting** is required for accurate performance of nearly every task in a unit operation.

The organization must have engineering controls to help control each factor, but sometimes there is no other choice but to rely upon administrative controls.

J. COMMUNICATION (most importantly, verbal communication)

Communication is the exchange of information while preparing for or performing work. Verbal communication occurs face-to-face, by telephone, sound-powered phones or walkie-talkies, as well as over public address systems. Written communication occurs, for example, through policies, standards, work packages, training materials, and e-mail.

Communication involves two sets of behaviors: (1) creating and sending messages and (2) receiving and interpreting them. Communication always involves at least two individuals, the sender and the receiver, and occurs:

- Between individuals
- Within and among work groups
- In meetings
- In pre-job or pre-evolution briefings
- During shift turnover

Successful communication requires several steps. The sender first develops the intention to communicate **either** verbally or in writing. The sender then composes a message that presents the meaning as clearly as possible. The receiver must pay attention to the message and then interpret its meaning. If the communication is successful, the receiver interprets the message consistently with the sender's intended meaning. However, there are many potential errors in both sending and receiving the communication:

- **Sending Errors**
    - *Content wrong*
    - *Content inconsistent*
    - *Content inappropriate for the job*
    - *Content inappropriate for the receiver*
    - *Standard terminology not used*
    - *Familiar terminology not used*
    - *Message production inadequate or interfered with*
    - *Necessary information not sent*
    - *Wrong place or person*
    - *Wrong time*
    - *Sending verification failure*
- **Receiving Errors**

- *Information not sought*
- *Information not found*
- *Information not used*
- *Receiving verification failure*
- *Message misunderstood.*

Communication errors may be reduced in the following ways:

- Structuring messages in a standard format, which alerts the receiver if important information has been skipped
- Providing written instructions or procedures for complicated tasks
- Establishing protocols to repeat key parts of verbal communications
- Preceding special or nonroutine activities with a pre-job briefing
- Using structured protocols, checklists and logs to supplement written instructions

One key opportunity for communication errors is when a new shift takes over responsibilities from the previous shift. An effective tool to reduce these errors is a properly designed shift handover process that includes appropriate checklists and logs of activities, and a structured and consistent method for verbally reviewing this information. Additionally, the logs and discussions need to include:

- Routine operations
- Transition operations (e.g., changes to produce a different product or switch to a different raw material source)
- Nonroutine operations (e.g., maintenance activities or temporary operating conditions)

VI. CONTROLLING HABITS

The direct control of human factors as outlined above is the most effective means to control human error. However, about 15% of human error is due to acquired habits. Some call these Behaviors, but that normally carries negative connotations. Many companies have effective systems for combating bad habits; these systems normally involve peer-to-peer observations and feedback (coaching), and these are many times labeled behavior-based safety management (which is a trademark phrase belonging to BST), or behavior-based reliability, or simply job-observations. A peer-observing-peer system, such as these, can reduce habit-based errors by about 70%.

VII. Applying Human Factors to Risk Reviews and Root Cause Analysis

It is important to note that while not directly controlling human factors, the PHA/HAZOP (hazard identification and risk assessment) and the Incident Investigation management systems help to identify where and how human factor practices are necessary to control certain risks and help to identify weaknesses in human factor practices and implementations.

Risk Reviews (PHA/HAZOP, HIRA, JSA, etc.) – Addressing Human Factors during risk reviews can help identify gaps in all human factor categories since human factor categories are types of safeguards. This is fundamental for predicting where and why humans might make mistakes and for determining (qualitatively, at first) if the protection layers are sufficient if such errors occur and if not, what else is needed. Unfortunately, many organizations do not fully analyze for errors during all modes of operation, and so many (in some cases, most) of the accidents that start with a human error are not predicted, as they should be ***(See Chapter 9.1 of CCPS/AIChE, Guidelines for Hazard Evaluation Procedures, 3rd Edition, 2008[9]; and paper by Bridges, LPS/AIChE, April 2009, Optimizing Hazard Evaluations[10].)*** This practice ultimately strengthens SOPs, human system interface, and most importantly ensures there are sufficient IPLs for scenarios unique to non-routine modes of operation, such as startup, shutdown, and online maintenance.

Incident Investigation/Root Cause Analysis (II/RCA) – II/RCA is necessary to learn from mistakes. Although all companies have an II/RCA system, many companies are lacking in the awareness of where human factors fit into an accident sequence. This leads to the II/RCA missing the human factor weaknesses that led to the human error and so the II/RCA may stop at the cause (and even root cause) being "operator error." Many companies also do not get nearly enough near misses reported (the ratio should be about 20 to 100 near misses reported per actual loss/accident ***(See CCPS/AIChE, Guidelines for Investigating Chemical Process Incidents, 2nd Edition, 2003[12]; and paper by Bridges, 8th conference, ASSE-MEC, 2007, Gains in Getting Near Misses Reported[13].)***

VIII. Closing

An organization implementing PSM (or similar programs such as, SMS or OSHA's 18001 for occupational safety) must develop management systems for optimizing human factors to control human error rates. As noted above, three potential sources for guidance for the chemical process industry are regulatory standards, such as the OSHA PSM regulation, the CCPS RBPS reference, and the CCPS COO/OD guideline. As a matter of closing this analysis, it is helpful to compare how well each of these documents address human factor implementation. The following is a rating scale that the authors have developed to gauge how well human factors are incorporated in these sources.

TABLE 5: DEFINITION OF HUMAN FACTOR RATING SYSTEM

Rating	Description
1	Human Factor is not specifically addressed
2	Human Factor is nominally addressed
3	Human Factor is partially addressed
4	Human Factor is mostly addressed
5	Human Factor is thoroughly addressed

The following table summarizes each document's (management system's) ratings for each human factor:

TABLE 6: HUMAN FACTOR RATINGS FOR PSM REFERENCE DOCUMENTS

Human Factor	OSHA PSM	RBPS Guideline	COO/OD (Component of RBPS)
Available Time	1	1	2
Stress/Stressors	1	1	2
Complexity & Task Design	1	1	2
Experience/Training	2	3	5
Procedures	2	3	2
Human-Machine Interface	1	1	4
Fitness for Duty	1	3	4
Work Processes & Supervision	2	4	4
Work Environment	1	1	1
Communication	1	3	4

As stated above, the OSHA PSM standard does not rate well for incorporation of human factors. It partially addresses them in the area of training, but only nominally at best in the other areas.

The RBPS guideline improves upon the coverage of human factors significantly compared to the US OSHA PSM regulation, particularly in the areas of:

- **Fitness for Duty (FFD)** – specifically by introducing the concept of sleep discipline and overtime limitations, but without specific guidance on limits and without recommended practices for controlling FFD
- **Work Processes & Supervision** – specifically by providing a great deal of emphasis in the areas of process safety culture and workforce involvement
- **Communication** – specifically by providing some (but not all) key communications rules to reduce human error, and by discussing the importance of shift turnover logs, but without enough detail on their content
- **Experience/Training** – specifically by providing a strong structure of the desired training program, albeit at a fairly high level
- **Procedures** – specifically by including most, but not all, of the best practices identified in this paper; the key items missing are the rules to have procedures written by future users of the procedures, and to conduct a hazard analysis of the consequences of human error in following the procedures

The COO/OD guideline further improves upon its parent book, RBPS, in most areas:

- **FFD** – specifically by addressing nearly all the key elements outlined in this paper. One main shortcoming is that it recommends API RP 755 as a guide for Fatigue Management, but this standard is terrible in that it fails to control fatigue that can occur during extended maintenance outages. Further, it does not cover all FFD issues, but rather focuses on fatigue. It would have been better to omit reference to the API RP 755 and instead reference the US NRC standard for FFD, 10 CFR 26. One major advantage of the NRC standard is that it includes the requirements for supervisors to be trained in and search for FFD deficiencies in staff coming onto the shift. A thorough management system for detection and the initial actions and follow-on actions is needed; 10 CFR 26 provides a lot of detail on what is needed.

- **Human-Machine Interface** – specifically by highlighting key DCS design features; however, much of the other factors described in this paper are not discussed.
- **Communication** – specifically by again including key (but not all) communication rules, and by providing specific details about the shift handover log.
- **Experience/Training** – specifically by providing a thorough, and properly detailed, overview of a strong training program.

Taken together, these two CCPS guidelines provide valuable advice on how to incorporate six (6) of the human factors into a PSM program, with notable exceptions in the areas of:

- *Available Time*
- *Stress/Stressors*
- *Task Design and Task Complexity*
- *Work Environment*
- *Procedures - poor/incomplete coverage of best practices*
- *Some gaps within the factors RBPS and COO/OD do cover*

This paper provides sufficient detail for implementing all 10 human factors so that companies following its guidance can be well on their way to building strong human factors elements within their PSM program.

References:

1. Swain, A. D., Guttmann, H. E., *The Human Reliability Handbook - with Emphasis on Nuclear Power Plant Applications*; NUREG/CR-1278, U.S. Nuclear Regulatory Commission, Division of Facility Operations, 1983.

2. Gertman, D.; Blackman, H.; Marble, J.; Byers, J. and Smith, C.; *The SPAR-H Human Reliability Analysis Method,* NUREG/CR-6883, U.S. Nuclear Regulatory Commission, Office of Nuclear Regulatory Research, Washington, DC, August 2005.

3. Bridges, W., and Tew, R., *Human Factors Elements Missing from Process Safety Management (PSM),* 6th GCPS, AIChE, March 2010.

4. *Risk-Based Process Safety*, CCPS/AIChE, John Wiley & Sons, Inc.; Hoboken, NJ, 2007.

5. *Conduct of Operations and Operational Discipline*, CCPS/AIChE, John Wiley & Sons, Inc.; Hoboken, NJ, 2011.

6. *US FAA Advisory Circular* 120-42A (as part of ETOPS approval)

7. Bridges, W. G., Williams, T. R., "Effective Procedure Writing – A Two-Part Series (Part 1: Reducing Risk of Worker Errors; Part 2: Addressing Process Safety Management Requirements for Operating Procedures)," *Chemical Engineering Progress*, 1997.

8. Bridges, W., and Collazo-Ramos, G., *Human Factors and their Optimization,* 8th GCPS, AIChE, April 2012.

9. *Guidelines for Hazard Evaluation Procedures*, 3rd Edition, CCPS/AIChE, John Wiley & Sons, Inc.; Hoboken, NJ, 2008.

10. Bridges, W., *Optimizing Hazard Evaluations,* LPS/AIChE, April 2009.

11. Bridges, W., Lorenzo, D., and Kirkman, J., *Addressing Human Errors During Process Hazard Analyses,* CCPS Conference, AIChE, 1991 (also published in Chemical Engineering Progress May 1994).

12. *Guidelines for Investigating Chemical Process Incidents, 2nd Edition,* CCPS/AICHE, John Wiley & Sons, Inc.; Hoboken, NJ, 2003.

13. Bridges, W., *Gains in Getting Near Misses Reported,* 8th Conference, ASSE-MEC, Bahrain, 2008.

14. Della Rocco, P.S., et al., *The Effects of Napping on Night Shift Performance,* Federal Aviation Administration, U.S. Department of Transportation, 2000.

15. *Fitness for Duty Programs, Preamble, 10 CFR 26, U.S. Nuclear Regulatory Commission, Revised 2005.*

16. US Chemical Safety and Hazard Investigation Board (CSB), *Anatomy of a Disaster: Explosion at BP Texas City Refinery,* published 2006.

17. University of South Australia (UniSA), Adelaide Centre for Sleep Research, *SDA Fact Sheet - AT03*, 2006.

18. Hallbert, B.; Whaley, A.; Boring, R.; MCabe, P.; and Chang, Y.; *Human Event Repository and Analysis (HERA)*, NUREG/CR-6903, U.S. Nuclear Regulatory Commission, Office of Nuclear Regulatory Research, Washington, DC, 2007.

19. Bridges, W., and Clark, T., *LOPA and Human Reliability - Human Errors and Human IPLs,* 7th GCPS, AIChE, March 2011.

20. Bridges, W., and Thomas, H., *Accounting for Human Error Probability in SIL Verification Calculations,* 8th GCPS, AIChE, April 2012.

21. Bridges, W., *The Impact of Human Factors on LOPA,* 9th GCPS, AIChE, April 2013.

22. Bridges, W., and Tew, R., *Best Practices for Writing Operating Procedures and Trouble-shooting Guides,* 13th GCPS, AIChE, March 2017.

23. Bridges, W., *Proven Approaches to Ensuring Operators Can Respond to Critical Process Deviations in Time (Human Response IPL),* 13th GCPS, AIChE, March 2017.

19 AIChE Spring Meeting
15TH GLOBAL CONGRESS ON PROCESS SAFETY
NEW ORLEANS • 2019

Procedure Digitalization
Human Factored Procedure Technology

Presenter/Principal Author
Elliott P. Lander
ATR - Managing Director
6405 Cypresswood Dr #250 | Spring, Texas 77379
Email: eplander@atrco.com

Prepared for Presentation at
American Institute of Chemical Engineers
2019 Spring Meeting and 15th Global Congress on Process Safety
New Orleans, LA
March 31 – April 3, 2019

Procedure Digitalization

Human Factored Procedure Technology

Elliott P. Lander
Managing Director, ATR
eplander@atrco.com

ABSTRACT

The current state of procedure quality and usability tends to be very low across the processing industries. While regulatory compliance has been the primary driver, minimal emphasis on procedure quality, usability and operational discipline results in a poor procedure usage culture. Combined this with the inherent limitations of paper-based procedures and the lack of emphasis on human factors, contribute significantly to increased operational risk.

This paper outlines advantages of Smart Digital Procedures (SDP) and balanced procedure programs to achieve operational excellence and lowered risk driven through mobility, human factors and integrated ecosystems.

INTRODUCTION

Any discussion of procedures must begin with establishing the potential value to the organization. Procedures are a critical "layer of protection" serving to:

- Ensure worker safety
- Ensure consistency of operations
- Minimize risk to people, equipment and environment
- Transfer knowledge from experienced to inexperienced workforce
- Optimize tasks
- Ensure compliance with regulatory standards

Simply having written procedures (with black text on a white background) that are up to date and reasonably accessible are sufficient to be in compliance with Process Safety Management and other regulatory audits, while being insufficient for the intended audience of process operators.

STATE OF THE INDUSTRY

Based on many of the audits conducted by the author, as well as findings from the Texas A&M University Advanced Next Generation Procedure Research Consortium, procedure programs throughout industry share many of these characteristics:

- Mostly done in MS Word by poorly trained or unqualified writers
- Procedure output limited to paper, "one-size-fits-all"
- Inconsistencies between procedures in presentation and structure
- Non-integrated lifecycle components
- No dedicated procedure program managers or executive leadership
- Fair document management controls

- Poor governance documentation
- Poor or non-existent procedure use and adherence policy and usage culture
- Relatively slow change management and continuous improvement
- Little or no human factoring

Procedure administrators generally who perceive themselves as compliant tend to have little interest in migrating to new technologies and procedure writing companies perpetuate "typewriter" technology as tools they are familiar with

PROCEDURE PROGRAM MATURITY FRAMEWORK

Based on years of research, interviews and findings from the Advanced Next Generation Procedure Research Consortium at Texas A&M University, sponsored by ExxonMobil, Chevron, NOVA Chemicals, Shell, NASA and ATR, the authors have developed a Procedure Program Maturity Framework consisting of 4 levels, shown in Figure 1.

The levels progressively measure the potential value and effectiveness of a company's procedure program based on a risk scale.

Figure 1

Procedure Program Maturity Model

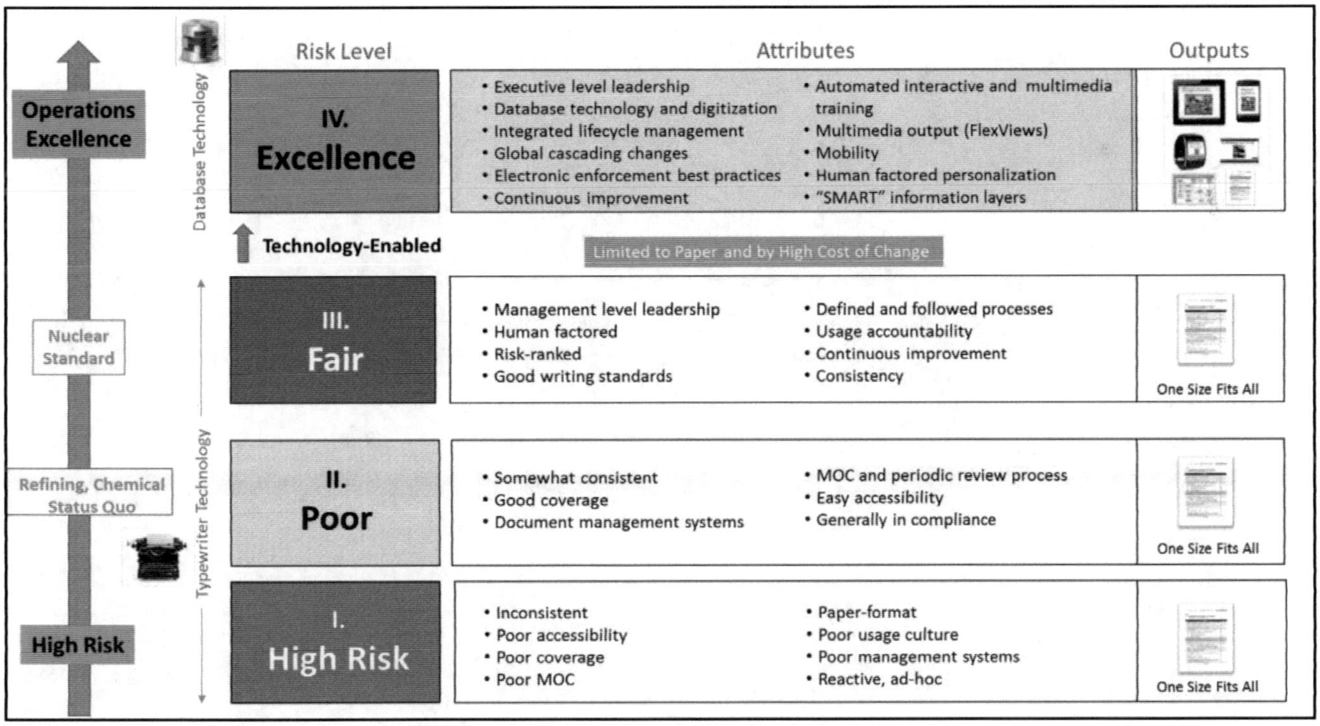

PROCEDURE PROGRAM ASSESSMENTS

Comprehensive procedure program assessments typically focus on 10 critical areas, as shown below. The following table summarizes typical findings for operating facilities, including refineries, chemicals, power, pipeline and exploration and production based on dozens of assessments by the author's company and the Texas A&M Next Generation Procedure Research Project.

Figure 2

General Industry Procedure Program Assessments

Category	Grade	Description
Leadership		Executive leadership not engaged in guiding and enforcing operational discipline, particularly regarding procedure use and adherence. Filters down to lack of supervisory leadership who don't enforce procedure usage. Executive leadership is not generally aware of significant procedure gaps.
Governance		Four pillars of procedure program governance not established: Document Hierarchy, Procedure Management, Procedure Writer's Guide, Procedure Use and Adherence. Lack of governance has significantly limited quality and sustainability.
Technology		Use one-size-fits-all "typewriter technology" to create procedures and link to document repositories. Inefficient processes, lack of enforcement and standardization, and no ability to personalize content or output content "fit for purpose," meaning outputting procedure content to different formats for execution, training and evaluations.
Organization		Documents not properly categorized due to lack of document hierarchy guidelines. Documents not properly prioritized based on risk factors.
Writer Competence		Writers are typically subject matter experts not properly trained or qualified in best practice writing standards leading to inconsistencies and introduction of human error precursors into procedures. Writers often struggle with formatting versus accurate content development.
Content		Best practices in technical writing and human factoring techniques are lacking. Insufficient detail and ambiguous instructions in procedures put workers at risk. Multiple actions in a step are prevalent, challenging usability and human performance.
Format		Multiple formats in use, limiting usability. Front matter not optimized for human performance and important safety information is lost in back of procedure. "One-size-fits-all" not allowing for individual personalization or outputs fit for purpose.
Culture		Operations lacks discipline and does not yet value procedures as a preventative tool for human performance and safety. Technology, governance, and competency infrastructure is missing to allow for a procedure-driven operating culture. Inaccurate content leads to hesitancies to use procedures.
Process		The process is not effectively driving systematic improvement in procedures. Quality controls are not in place to set the standard for reviews. Checklists have been updated without updating associated SOP, resulting in obsolete and conflicting information. Operators and management lack visibility in the process. Operators are not engaging with the process to update and maintain procedures. Procedure reviews and annual recertification are "rubber-stamping" exercises.
Compliance		Despite significant gaps and issues identified above, companies are passing internal and external audits based on relatively superficial and subjective guidelines creating a false impression that "our procedures are good enough."

High risk   Medium risk   Low risk

PROGRAM STATUS

The nuclear power industry, historically considered the gold standard in procedure program excellence, is level III (FAIR). Very good processes, good writing standards and good governance are key elements found throughout the industry. While having good standards, the industry almost exclusively creates procedures in Microsoft® Word and only outputs to a "hard-coded" paper based output.

The "typewriter technology" requires each document to be formatted manually and it is very difficult, expensive and resource intensive to make any substantial changes either in output or to alternative technologies. The reliance on MS Word restricts companies from creating an extensible and scalable human-factored procedure program.

PROCESSING INDUSTRY

While the nuclear industry is clearly at Level III, the processing industries (upstream, downstream, midstream, chemicals and conventional power) are situated at Level II (POOR). The standards, processes and governance are generally nowhere near the nuclear industry, resulting in higher risks, incidents and significant lapses in human reliability.

Unfortunately, while in the POOR classification, these companies may be considered compliant based on the relatively low standards and risk-based audit considerations. From this perspective, *"if it ain't broke, don't fix it"* is the excuse when satisfaction with minimum compliance leads to complacency.

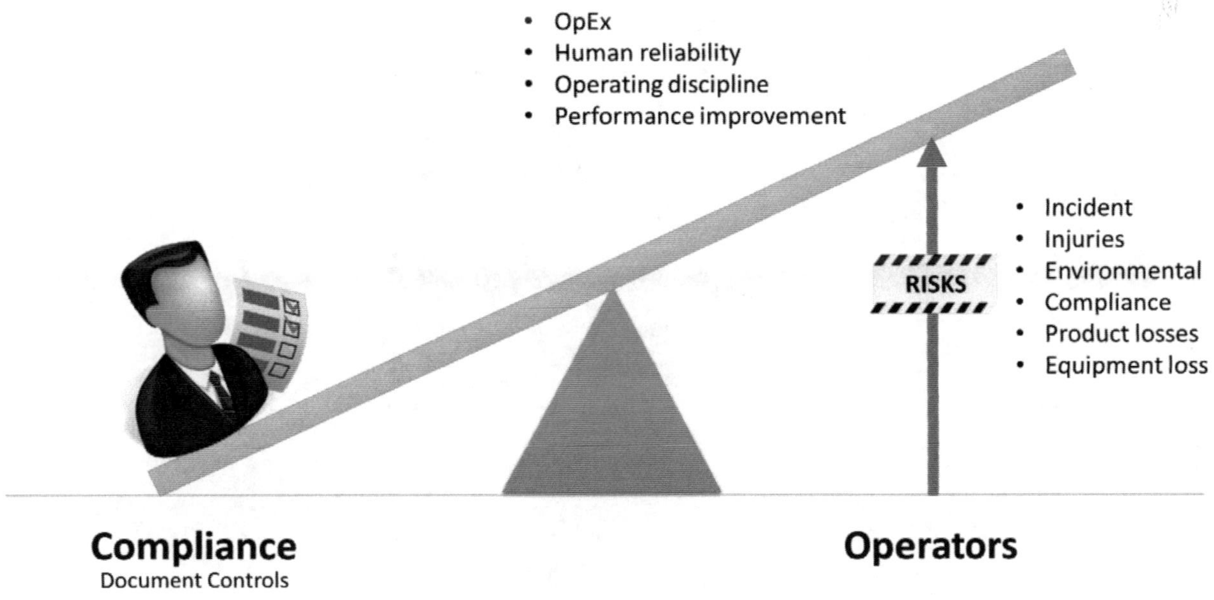

LEVEL IV – OPERATIONAL EXCELLENCE THROUGH DIGITALIZATION

Moving from Level III (FAIR) to IV (EXCELLENCE) requires a Next Generation technology upgrade and requires high standards of governance and accountabilities.

While "typewriter technology" produces an unstructured "blob" of content, Smart Digital Procedure (SDP) technology manages content at the discrete step level with layers of "smart" information, enabling the content to be output to any device in any format.

These SDP procedures have the following characteristics:

- Manage content at discrete step level
- Maintain procedure content in normalized database
- Provide layers of "Smart" information at the step level
- Output procedure content in any format, on any device
- Ability to associate equipment and operating limits to equipment objects
- Personalize the output based on function and user requirements
- Digitize readings and step sign-off for actionable data reporting
- Mobilize the procedures and output to any device
- Enable global changes across entire library of procedures automatically

PROCEDURES AND HUMAN FACTORS

Another important element of Level IV (EXCELLENCE) procedures is the ability to incorporate critical human factors into the procedure presentation. The SDP technology provides the foundation for displaying the content to meet the needs of the user, versus the "one-size-fits-all" limitation of MS Word.

Human factor considerations to improve quality and usability of procedures includes:

- Increase font size to make it easier to read
- Use text-to-speech audio
- Provide options to display or not display graphics based on user experience
- Provide video reinforcement
- Mobilize procedure content
- Create hands-free environment via text-to-speech and speech
- Enable digital sign-off versus pen and ink
- Enable real-time collaborative procedures
- Enable user to change backgrounds in different lighting situations
- Enable user to be remote with immediate access to 100s of procedures via tablet or smartphone
- Different layouts for emergency versus standard operating procedures
- Support for workers with English as a second language

PROCEDURE DIGITALIZATION

Procedure digitalization essentially means that the procedure is broken down into discrete steps, which have unique properties.

The properties define the behavior, and may include such micro-elements as:

- Step type (e.g. action, warning, caution, note, training)
- Step numbered (Y/N)
- Step attachments (e.g. graphics, video, attachments)
- Risk level
- Role(s) assigned to perform step
- Type of sign-off (e.g. role or role + verifier)
- Other types of information
- Equipment and data inserts linked to 3rd party systems

Figure 3 outlines the digitalization process moving from word processor documents to SDP data.

In the digital world, these properties are normalized and enable global changes and guaranteed consistency. No longer are procedure text instructions imprisoned by hard-coded formatting, as digital procedures enable output to any format and any device.

Figure 3

Digital Procedure Transformation Process

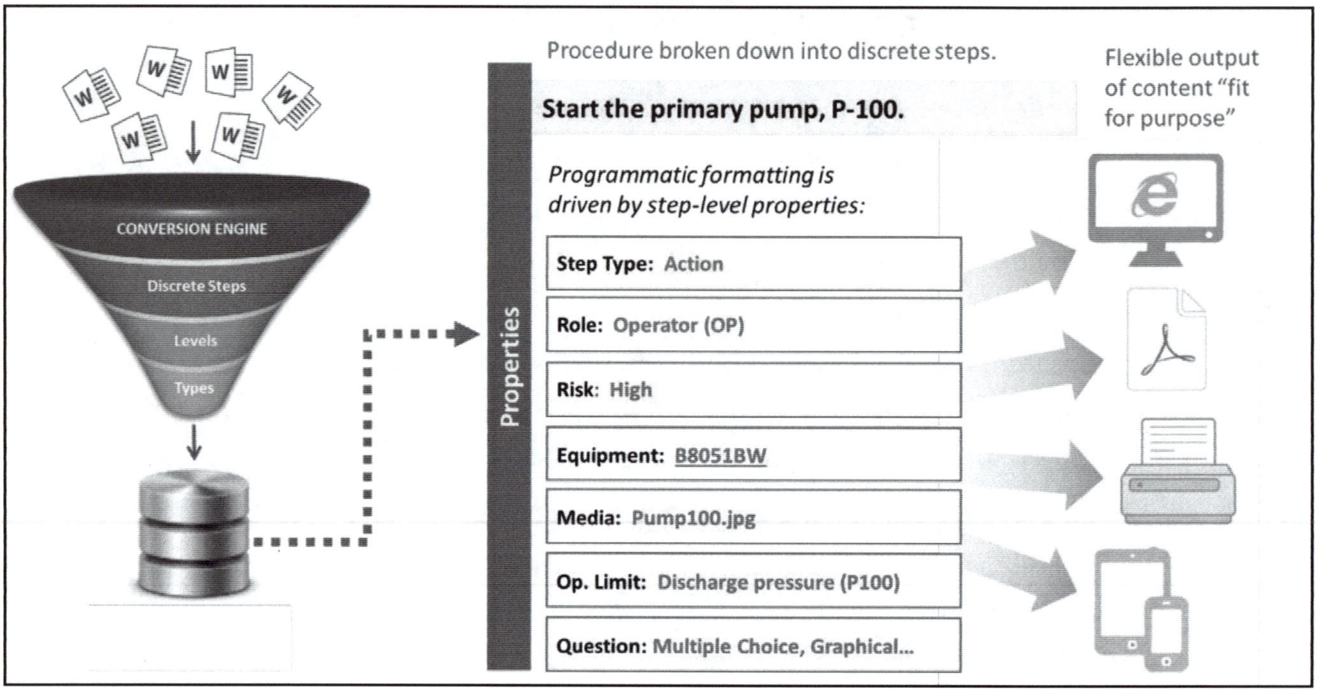

Figure 4 shows multiple automated outputs from a single source SDP database.

Figure 4

Sample SDP Displays

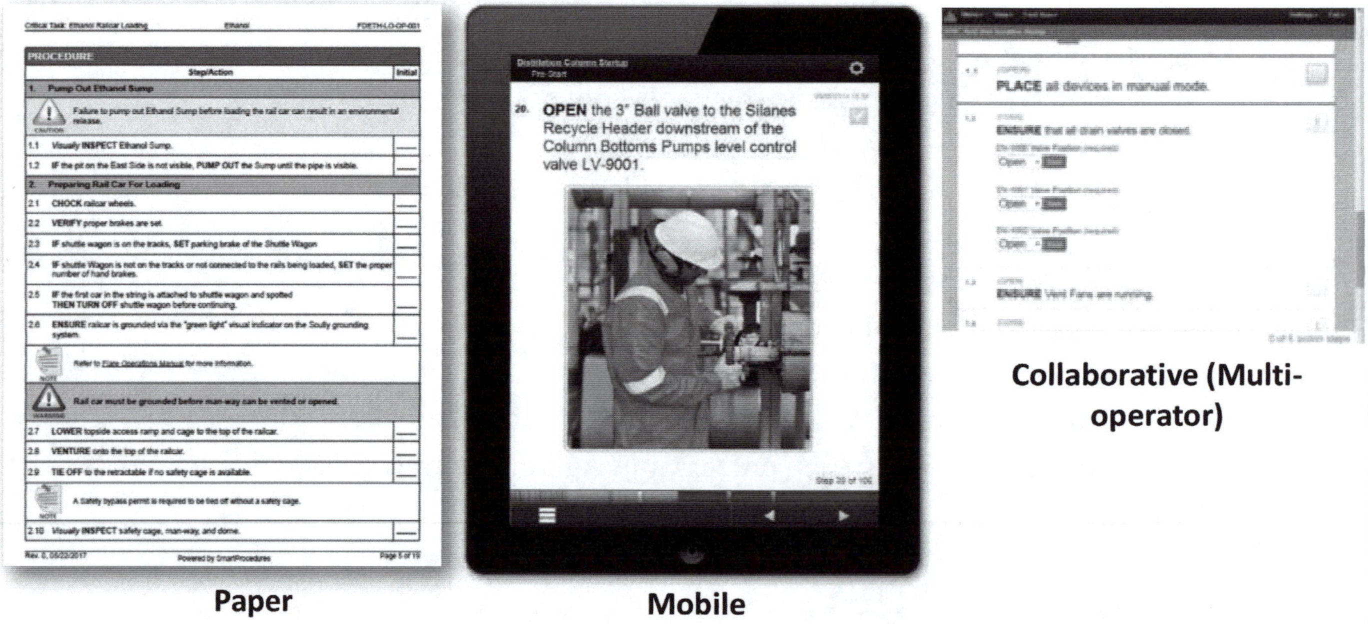

There are 4 key areas where digital procedures enhance operations and worker safety:

- Personalization
- Mobilization
- Fit-for-Purpose
- Integration

Personalization refers to the ability to reformat or restructure the text and associated information for the individual. This could enable individual workers to increase or decrease the font size, show or not show supplementary photos, as well as include or not include additional training or work guidance information.

Mobilization enables workers to display procedure content on mobile devices and interact, sign-off or request additional information. On the devices, workers can enter readings, check-off steps, add notes, photos or videos and automatically track GPS data.

Fit-for-Purpose enables the content to be displayed automatically for different purposes, as shown in Figure 5. This could include normal procedure execution, training, performance demonstration or field validation. This could also mean output to paper as well as interactive mobile devices. In each case the format of the content and potential interaction will change to fit the activity, however the content itself is unchanged.

Figure 5
"Fit for Purpose" Automated Outputs

Digitalization enables content to be output into different formats for different purposes, while maintaining one content repository. The technology also enables extensibility to new technologies and IoT integrations.

Integration is at the heart of digitalization meaning that at the step and procedure level, additional information from other systems, such as distributed control systems, SCADA or equipment databases can be displayed or transferred from one system to another.

This can include dynamic views of equipment or instrument status or transfer of readings to other systems for processing or data storage. Figure 6 shows examples of potential integration points.

Figure 6

Smart Digital Procedure Integration Points

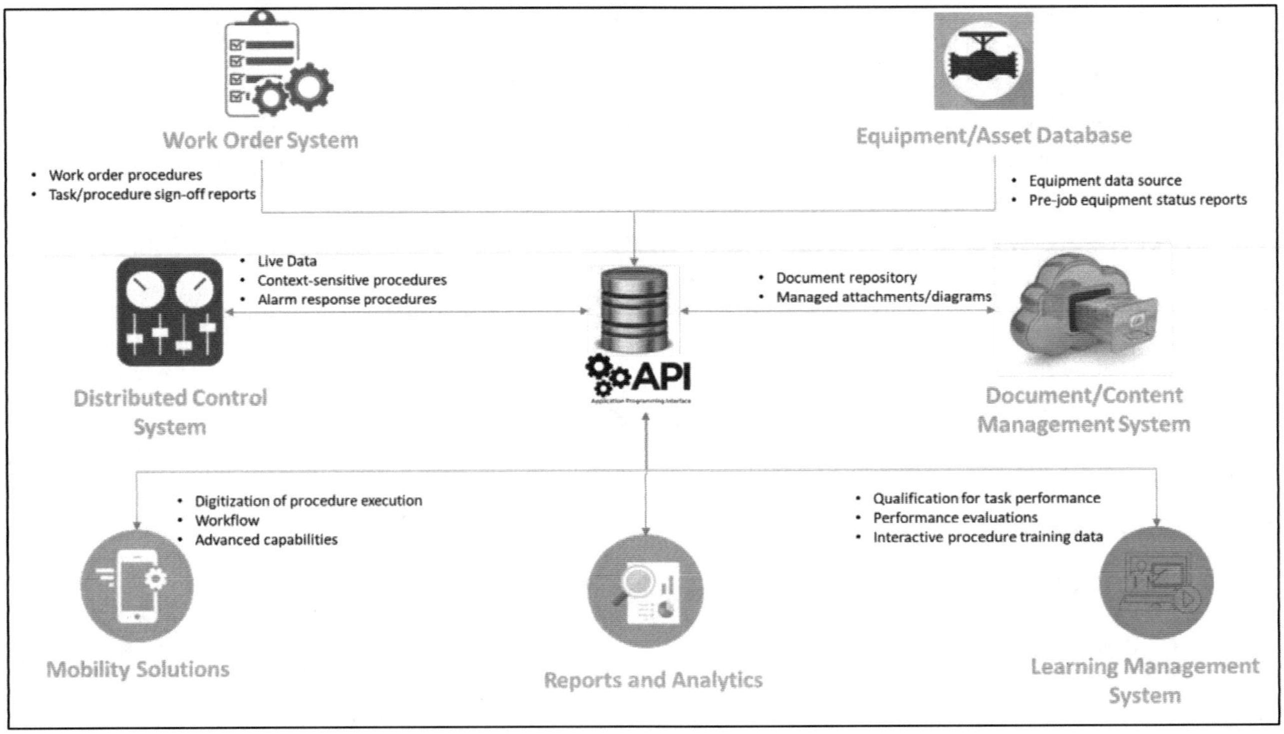

SUMMARY

Compared to conventional paper-based and non-integrated procedure systems, SDP technology proves to have major advantages as shown in Figure7.

To improve bottom line profitability and worker safety, many companies are in serious need to upgrade and modernize their neglected procedure programs. Smart Digital Procedure technology provides an important step to digitize, mobilize and personalize procedure content for greater levels of usability and procedure effectiveness.

Figure 7

Smart Digital Procedure (SDP) Advantages Over Conventional Word Processor-Based Procedures

Current	Future State **Smart Digital**
• Independent files • Manual processes • Use of costly resources	• Relational database • Automated processes • Global changes • Electronic enforcement
• Non-integrated components • Creation, management, workflow and accessibility may be non-integrated • Tasks get dropped, not tracked	• Unified integrated components • Streamline processes • Eliminate redundancies
• Hard-coded paper output • Not extensible to new technologies • Content not mobile-ready	• Any format, any device • Multimedia outputs • Discrete step processing • Extensible
• One-size fits all • Paper document for all procedure tasks • No accounting for human factors, environmental conditions, stress, etc.	• Operational discipline/ reliability • Human factored • Procedure content, "fit-for purpose" • Adaptable to individual needs

CONTACT INFORMATION

Elliott P. Lander
eplander@atrco.com
832-797-6851

Gary Hilberg
ghilberg@atrco.com
281-370-9540

A Practical Approach to Preventing Systematic Error in the Maintenance of Instrumented Safeguards

Dr. Angela Summers, President
SIS-TECH Solutions
12621 Featherwood Dr. Suite 120, Houston, TX 77034
asummers@sis-tech.com

Eloise Roche
SIS-TECH Solutions
12621 Featherwood Dr. Suite 120, Houston, TX 77034
eroche@sis-tech.com

Prepared for Presentation at
American Institute of Chemical Engineers
2019 Spring Meeting and 15th Global Congress on Process Safety
New Orleans, LA
March 31 – April 3, 2019

—

A Practical Approach to Preventing Systematic Error in the Maintenance of Instrumented Safeguards

Dr. Angela Summers, President
SIS-TECH Solutions
12621 Featherwood Dr. Suite 120, Houston, TX 77034
asummers@sis-tech.com

Eloise Roche
SIS-TECH Solutions

Keywords: Instrumentation, safety instrumented systems, electrical systems, maintenance, human factors, human error, systematic

Abstract

Instrumentation and electrical (I&E) maintenance is typically managed using site-wide policies, practices, and procedures. Since I&E equipment is part of the control system and nearly every other layer of protection, the cumulative impact of poor I&E performance can be a significant contributor to major events. Systemic problems in managing I&E equipment reliability lowers process safety performance across a site.

Practical guidance is needed on how to assess the vulnerability of existing sites to instrumented safeguard failure due to maintenance deficits. This paper leverages Reason's organizational accident model as a framework to discuss site-specific factors that impact a site's susceptibility to maintenance error. A table of more than 60 human factors covering I&E maintenance activities was developed and organized by 4 elements of causality: organizational processes, workplace practices, personnel traits, and enabling conditions. The human factors table can be used to rate an industrial site on a negative-to-positive scale, highlighting those areas where systemic changes would likely improve maintenance performance and instrument reliability.

1 Leading or Lagging Indicators

The process industry depends on I&E equipment maintained in a manner that sustains the equipment's ability to act as required, when required, to prevent process safety incidents. While this responsibility is mandated by process safety regulations, it is simply a wise business practice to be proactive in managing instrument reliability. At many facilities, any one of thousands of instruments could cause operational problems. High reliability organizations understand that tackling these challenges head-on yields the best process availability.

Yet, HSE [1] and OSHA [2,3] assessments have found that some companies in the process sector have problems with their maintenance programs, as evidenced by the percentage of maintenance-related findings (Figure 1). These findings echo those published as a series of case studies in Guidelines for Safe Automation of Chemical Processes (Safe Automation) [4].

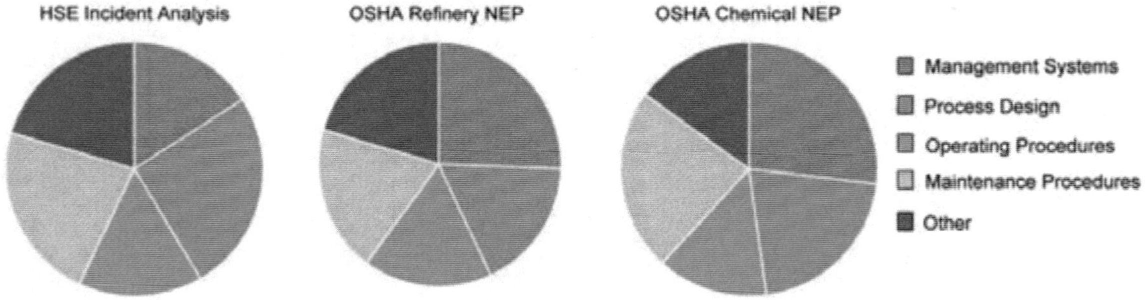

Figure 1. HSE Incident Analysis [1], OSHA Refinery National Emphasis Program (NEP) [2], and OSHA Chemical NEP [3] Findings

Government and industrial organizations have recommended metrics for monitoring the effectiveness of process safety management, including the Health and Safety Executive [5] the Center for Chemical Process Safety [6], and the American Petroleum Institute. API 754 [7] established 4 tiers of indicators (Figure 2). The bottom 2 tiers are leading indicators, because they are measures of the operating discipline, equipment integrity and safety culture. When these tiers are well-managed, it is far less likely that an event will happen. Safe Automation's case studies show that the top 2 tiers are often events where Tier 3 and 4 metrics were not implemented and site practices did not identify and correct systemic problems.

Figure 2 API 754 Metrics

International standards on instrumented safeguards provide more detailed guidance on performance metrics. ISA 61511 [8] requires monitoring the SIS reliability parameters, which are Tier 3 metrics. Quite a few more Tier 3 metrics are recommended by ISA TR84.00.04 in Annex R [9]. Tier 4 indicators, such as using maintenance records as a predictive tool for reliability issues, were introduced into the 2012 edition of ISA TR84.00.03 [10] on asset integrity management of SIS. Recent ISA 84 meetings have included workshops and discussions related to systematic errors during the lifecycle and on reducing systematic errors during maintenance.

2 Incidents are an organization failure

Human factors often play an out-sized role in poor I&E reliability. Safe Automation's case studies demonstrated strong links between the incidents and the failure to manage the on-going reliability of instrumentation and controls [4, 11]. These incidents ultimately were attributed to a series of failures that lined up in a dangerous manner. To prevent incidents, effort is required to sustain the asset integrity of the site's I/E equipment [12]. The errors, violations, and systemic failures in one system, whether control or safety, often repeat in other instrumented systems. If SIS equipment is not maintained, it is highly likely that other I/E reliability deviations are occurring. Systematic issues in maintenance can easily cause a breakdown of multiple IPLs [Independent Protection Layers], even if the IPLs are deemed independent based on an analysis of the equipment and system architecture [13].

The latest industry-published guidance on reducing maintenance errors through diversity is not really helpful. While theoretically attractive, there is no evidence that equipment diversity, staggered testing, or diverse maintenance teams can address the predominant underlying problems that are frequently cited by I&E technicians (Table 1). Rather, the proposed diversity would tend to make many field issues worse by increasing the complexity of the design, installation, testing, and maintenance.

Table 1 Problems commonly cited by I&E Technicians [13]

Unclear roles and responsibilities	Poor procedures
Lack of up-to-date documentation	Lack of warnings or cautions
Poor planning	Poor installation and configuration

3 A Practical Approach to Determining Site Risk for Systematic Error

James Reason, the father of the swiss cheese model, stated "Blaming people for their error is emotionally satisfying but remedially useless [14]." Errors are not inherently bad. There are a multitude of industry stories of errors that birthed significant innovations. Errors can be made by the best people. No one intends to make them. The more competent the person is, the more likely they will commit a very serious error. This is because the most competent people will seek out assignments with the greatest challenges and risk. All employees should be competent enough to do at least the average job, so error management is not equivalent to competency management. Rather, error management is about understanding what promotes errors and changing the situation to discourage them instead. Errors are affected by the mental state of the individual and the situation presented to the individual.

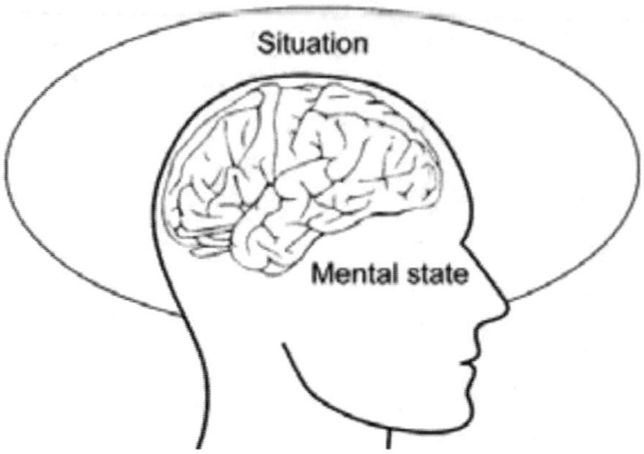

Figure 3 Errors Involve A Mental State and A Situation

Functional safety principles cannot control the individual's mental state, but they can influence the individual's decision-making processes. Everyone weighs the costs versus the benefits of complying to policies, procedures, and practices. The perceived costs might include accidents, injuries, and damages, but these may seem like unlikely future events compared to the benefits of doing something an easier way or saving time today. In most cases of non-compliance, the benefits are easy to see immediately with no obvious negative impact. The cost/benefit balance is rarely shifted by increasing the penalties for non-compliance, because these penalties are already known to be severe in the case of process safety events. Instead, greater influence on decision-making might be achieved by investing more time in acknowledging the important benefits of compliance.

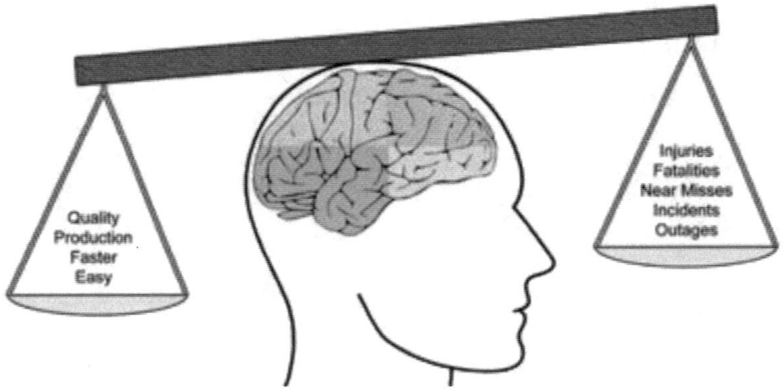

Figure 4 Humans Balance Costs and Benefits When Making Decisions

In contrast to the mental state, the situation can be controlled and managed by functional safety management. The situation can be viewed as what maintenance personnel face when executing the work. The situation involves 4 elements: organizational processes, workplace practices, personnel traits and enabling conditions. These elements are external to personnel, but they influence the decision-making process. The situation can vary significantly in ways that promote or discourage human error. For example, the situation could involve a task that is well-planned with detailed procedures in place to achieve quality work execution. Or the situation could involve troubleshooting unexpected behavior with unfamiliar technology in a poorly lit room with no up-to-date specification.

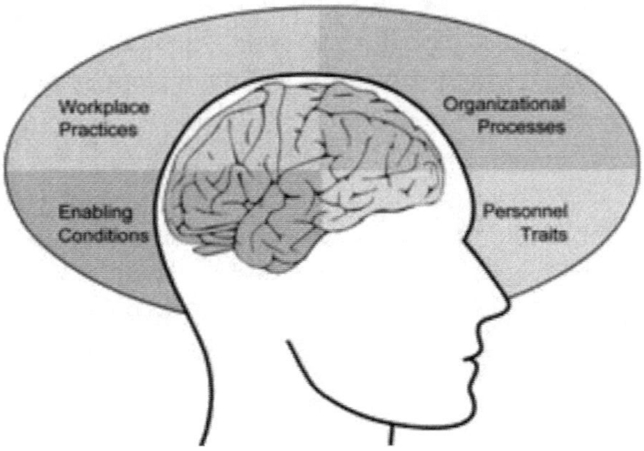

Figure 5 Situations Involve 4 Elements

Reason's organizational accident model (Figure 6) used the 4 elements to illustrate the underlying causality [14] of human errors and incidents. Errors occur when a planned action does not achieve the desired result. For example, the maintenance procedure did not define who takes responsibility for the equipment being returned to service, so this critical task was not done. In contrast, violations occur when deviations from an approved practice are intentionally taken. Violations are rarely malicious acts and are often intended as positive with respect to some aspect of the task. An optimizing violation occurs when someone does something that seems to accomplish the same thing but is easier or faster than the planned way. For example, the deviation meets the deadline or budget, demonstrates a high level of skill, or is simply easier. Routine or optimizing violations can become part of the site maintenance culture when an owner/operator rarely punishes deviations or fails to frequently reward compliance [15].

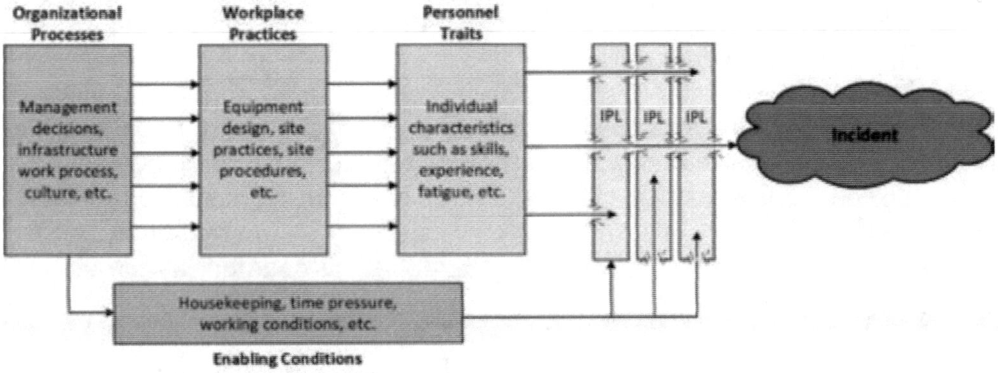

Figure 6 Anatomy of an Organizational Accident (adapted from [14]).

Organizational processes focus worker attention on certain behaviors and metrics, while workplace practices impact the quality and consistency by which maintenance activities are accomplished. Organizational processes determine what is considered important or not. These processes feed into management decisions that directly impact the day-to-day work of planning, forecasting, budgeting, communicating, monitoring, and auditing. Many accidents begin with negative organizational processes that promote poor workplace practices and tolerance of poor instrument reliability. For example, effective communication between maintenance and engineering is critical to ensure that reliability issues are addressed comprehensively. The likelihood of unresolved, long-term reliability problems significantly increases when communication breaks down between engineering, operations, and maintenance.

Workplace practices are the written instructions that govern how maintenance is executed. Poor practices, like inadequate instructions, missing specifications, obsolete procedures, and ineffective interface design, increase the likelihood of errors and violations [16]. For example, a test procedure that does not define the required equipment function becomes a source for maintenance error even though the maintenance person has a high degree of skill with the technology.

Personnel traits are intrinsic to the person executing the task. The most important traits are possessing the skills and experience necessary to complete assigned tasks correctly. Hiring competent personnel is essential, but sustaining competency as I&E technology changes can be challenging. High turnover can make it difficult to ensure that personnel have the in-depth knowledge of the system design needed for troubleshooting problems or evaluating the impact of management of change activities. Competency can also be off-set when fatigue or poor health impacts decision-making. Accidents become even more likely when poor workplace practices are combined with negative personnel traits, such as inexperience, poor skills, or being tired.

Enabling conditions make errors impacting multiple protection layers more likely. These conditions are triggered by organizational decisions that promote errors and violations, such as high work load, time pressure, unreasonable schedules, poor housekeeping, and poor quality tools. Other task-specific conditions, such as ill-fitting personal protective equipment, dim lighting, poor housekeeping and missing labeling, can also exist due to management policies and resource allocations. These enabling conditions potentially increase the likelihood that error will occur regardless of how optimal the organizational processes and workplace practices may be.

The 4 Elements of Causality were used as a framework to develop a list of 61 positive and negative human factors for maintenance activities. The list is based on personal experience and was enhanced by lessons-learned discussions at the Instrument Reliability Network, ISA 84 and ISA 61511 committee meetings. A summary of the systematic error sources, subtopics and human factors is provided in Table 2. The detailed human factors table is provided in Appendix A.

Table 2. Hierarchy of Human Factors in Maintenance

Sources	Subtopics	Specific Human Factors
Organizational Processes	Communications	• Clarity of responsibilities • Engineering and maintenance communications • Operations and maintenance communications • Teamwork and communications • Emergency communications
	Instrument Reliability Program	• Process demand tracking • Maintenance priority • Out of service/bypass management • Repeat failure/bad actor management
Workplace Practices	Maintenance Instructions	• Task complexity • Procedure clarity and detail • Return to service procedures • Change management • Quality control and record keeping
	Maintenance Equipment/Interfaces	• Specification and installation drawing availability • Maintenance feature/facility design
Personnel Traits	General	• Knowledge, skills and experience • Fatigue
	Competency Assessments	• Verification of knowledge and skills
Enabling Conditions	General	• Personal protective equipment • Tools and equipment • Working conditions • Housekeeping • Time pressure

Each human factor listed in Table 2 has multiple prompts in Appendix A. These prompts describe negative and positive human factor attributes. Some negative organizational attributes often cited in incident reports are as follows:

- Instrumented safeguard maintenance is frequently delayed, behind schedule, or not prioritized.
- Frequent bypassing of instrumented safeguards with little oversight, time limits, or risk assessment. Bypasses include operator bypasses, manual operation, changing setpoints, and forces.
- High tolerance for poor process control and upsets leading to frequent demands on the instrumented safeguards.
- High tolerance for poor instrument reliability. Unresolved issues, long-term out-service, and frequent fault conditions accepted.

In contrast, the positive organizational attributes associated with these are:

- Instrumented safeguard maintenance is prioritized, executed as scheduled, and is rarely delayed for operational reasons.
- Instrumented safeguards only bypassed under strict controls, including compensating measures and time limits. Bypasses include operator bypasses, manual operation, changing setpoints, and forces.
- Low tolerance for poor instrument reliability. Proactive attitude to taking action to improve reliability.
- Low tolerance for poor process control; particular focus on reducing frequency of process upsets and process demands on the instrumented safeguards.

The attributes can be rated on any desired scale. This paper takes a binary approach where the site is assessed as displaying either negative or positive human factors. Another approach is to use an analog scale of 1 to 5 with 1 being mostly negative and 5 being the mostly positive. The intent is to provide a means to assess the current status of the strategies, processes and activities used by a site to identify and prevent systematic errors.

4 Applying Human Factors Evaluation to a Case Study

Process safety management provides multiple opportunities to assess the adequacy of I&E equipment, including hazards and risk analysis, risk assessments, maintenance monitoring, management of change, and audits. The use of the positive and negative human factor table in Appendix A can be triggered as a result of findings from these activities. Using the table as part of a corrective process is a good way to get started and to demonstrate immediate benefits. However, it does mean that the site is already experiencing sufficient systemic impact to warrant a deeper dive. This is a classic feedback, or lagging indicator, approach to process safety [16]. Another approach would be to use the table as a self-assessment tool to understand site vulnerability to maintenance error before negative performance data piles up.

To illustrate the methodology, the human factors table was applied to the incident commonly known as "Buncefield." The incident occurred at the Hertfordshire Oil Storage Terminal, which was located in Hemel Hempstead north of London England and was part of a complex of tank terminals known as the Buncefield Depot. The depot had an estimated capacity of 60 million gallons, making it the 5th largest oil depot in the UK [19]. The depot served as a major distribution center for the UK oil pipeline network [19]. It provided fuel to Humberside, Merseyside, as well as to Heathrow and Gatwick airports [17].

An explosion occurred on December 11, 2005, which injured 43 people and devastated the Hertfordshire Oil Storage Terminal, which was jointly owned by Total UK Ltd and Chevron Ltd [17]. Residences and commercial buildings in the area were structurally damaged with some requiring demolition. The economic impact on regional businesses is estimated to be in the range of £130–170 million [17]. Total losses may have been as much as £1 billion [17, 18].

The incident occurred when the Automated Tank Gauging (ATG) system failed, allowing fuel to be fed into a terminal tank for 11 hours [20]. The fuel overflowed through the tank conservation vents for approximately 40 minutes [21] prior to ignition, producing a large vapor cloud estimated to be 8 hectares in size [22]. The vapor cloud ignition resulted in the largest peacetime explosion in European history [19] producing a tremor measuring 2.4 on the Richter scale and blowing out windows five miles away from the site [22].

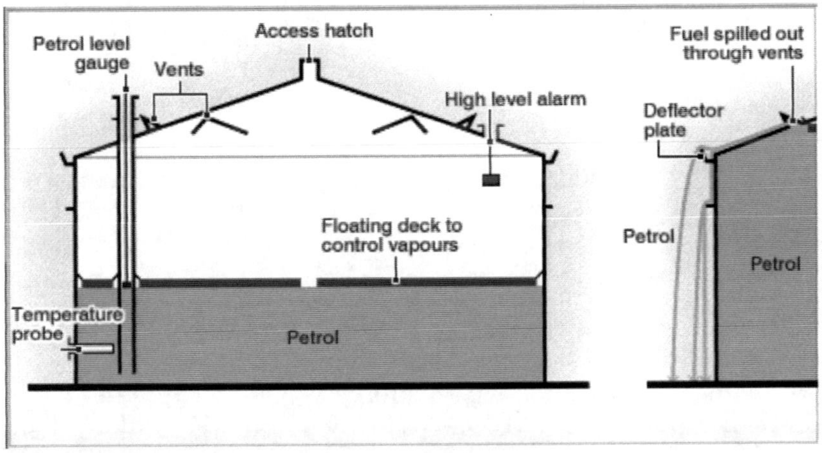

Figure 7. Simplified Graphic of Buncefield Tank

Gasoline was being delivered to the tank on the day before the incident. Early the next morning, the ATG displayed an unchanging level, although the tank continued to fill. By practice, the operator controlled the tank level by terminating transfer upon receipt of the 'user' alarm. However, the 'user', 'high', and 'high high' level alarms used the same transmitter. The failure of the shared transmitter rendered all three alarms inoperative. Since the 'user' alarm never activated, the operator did not take action to terminate transfer.

An independent high-level switch, set above the ATG high-high level, was designed to close inlet valves and activate an audible alarm, but it also failed. The high-level switch had been disabled when the maintenance organization, due to lack of understanding of the relatively new technology and to insufficiently detailed procedures, did not reinstall a lock on the switch test arm. Without the lock, the level switch was not activated when the float was lifted. By late afternoon, the tank overfilled and contents spilled out of tank roof vents. A vapor cloud was formed and noticed by tanker drivers and by people outside the facility. The fire alarm was activated and firewater pumps were started. An explosion occurred a short time later, likely ignited by the startup of the firewater pumps.

Reviews of the readily available literature on the incident identified significant problems with I&E equipment. The analog level involved in the incident had 14 dangerous failures (stuck) in the 3.5 months preceding the incident. It appears that the site had a high tolerance for poor process control and poor instrument reliability. The three "failed" alarm measurements came from the same faulty level device, which is an example of a common cause failure for the intended protection layers. The review also identified quite a few organizational and workplace issues:

- Confusion of responsibilities and expectations
- Poor communication between operations and maintenance
- Lack of consistency on who did what and when
- Lack of timely communication between maintenance technicians and supervision
- No reporting structure for escalation of unresolved problems
- Installation drawings, specifications, and functional requirements were not accessible
- No review of installation and configuration of instrumented safeguards after initial validation

When the entire table in Appendix A was applied to the Buncefield incident, nearly half (28 of the 61) of the negative attributes were identified. It is likely that several other negative human factors were also present on the Buncefield site; however, these potential contributors to systematic error were not specifically addressed in the available literature. The assessment using Appendix A indicates that the Buncefield site had high likelihood of systematic error in I&E maintenance activities. Correcting the problems that led to this incident would have required addressing the site's organizational processes, workplace practices, personnel traits, and enabling conditions. These causal factors for human error made the overfill event much more likely than would have been predicted in a hazards and risk analysis.

5 Summary

Safe Automation's case studies describe incidents where instrumented safeguards should have intervened in the incident propagation but did not. The underlying causes of these failures were often systematic rather than random failures. These underlying causes were potentially impacting the likelihood for incidents across the site, and perhaps the entire organization. A practical first step to correcting systematic error in instrumented safeguard maintenance can be to perform a qualitative evaluation of the existing maintenance human factors. This evaluation can be used to identify the areas in which the organization might be vulnerable to such errors and where there might be more value to focusing additional organizational resources. A table of positive and negative human factors was created to allow assessment of a site's vulnerability to systematic errors during maintenance. As an illustration, the table was used to assess the Buncefield incident. Based on the published investigative reports, nearly half of the 60 negative human factors were present.

6 References

[1] HSE, Health and safety laboratory report, Loss of Containment Incident Analysis, Great Britain (2003).

[2] OSHA, Process Safety Management for Petroleum Refineries - Lessons Learned from the Petroleum Refinery Process Safety Management National Emphasis Program.

[3] OSHA, Process Safety Management for Chemical Plants - Lessons Learned from the Chemical Process Safety Management National Emphasis Program.

[4] CCPS/AIChE, Guidelines for Safe Automation of Chemical Processes, Second Edition, American Institute of Chemical Engineers, NY, (2017).

[5] Health and Safety Executive, "Developing process safety indicators," 1st edition, (2006) ISBN 978 0 7176 6180 0.

[6] CCPS/AIChE, Guidelines for Process Safety Metrics, Wiley-Interscience, New York (2009).

[7] API, Recommended Practice Process Safety Performance Indicators for the Refining and Petrochemical Industries, RP754, API, Washington D.C. (2016).

[8] ISA, ANSI/ISA 61511-1-2018, Functional Safety: Safety Instrumented Systems for the Process Industry Sector - Part 1: Framework, Definitions, System, Hardware and Application Programming Requirements (IEC 61511-1:2016+AMD1:2017 CSV, IDT), Research Triangle Park, NC (2018).

[9] ISA, Guidelines for the Implementation of ANSI/ISA 84.00.01- Part 1, TR84.00.04-2015, Research Triangle Park, ISA, (2015).

[10] ISA, Asset Integrity of Safety Instrumented Systems, TR84.00.03-20012, Research Triangle Park, ISA, (2015).

[11] Summers, Angela, E. Roche, H. Jin, M. Carter, "Incidents That Define Safe Automation," 61st International Instrumentation Symposium, Huntsville Alabama, May 2015.

[12] Summers, Angela, "Risk Assessment Challenges to 20:20 Vision," Process Safety Progress, pages 119-125, June 2015.

[13] Summers, Angela E. and Eloise Roche, "Mobile Interfaces on the Plant Floor," 73rd Instrumentation and Automation Symposium, Texas A&M University, January 2018.

[14] Reason, James and Alan Hobbs, Managing Maintenance Error: A Practical Guide, Ashgate Publishing Company, Hampshire England (2003).

[15] Reason, James, Managing the Risks of Organizational Accidents, Ashgate Publishing Company, Hampshire England (1997).

[16] CCPS/AIChE, Guidelines for Safe and Reliability Instrumented Protective Systems, 1st edition, American Institute of Chemical Engineers, NY, (2014).

[17] Buncefield Major Incident Investigation Board, "The Buncefield Incident 11 December 2005: The final report of the Major Incident Investigation Board," ISBN 978 0 7176 6270 8, (2008).

[18] British Broadcasting Station, "Buncefield blast could cost £1 bn," December 11, 2008, United Kingdom, http://news.bbc.co.uk/2/hi/uk_news/england/7777539.stm.

[19] Wikipedia, "2005 Hertfordshire Oil Storage Terminal Fire," http://en.wikipedia.org/wiki/2005_Hertfordshire_Oil_Storage_Terminal_fire.

[20] British Broadcasting Station, "Buncefield tank was overflowing," May 9, 2006, United Kingdom, http://news.bbc.co.uk/2/hi/uk_news/4752819.stm.

[21] The Daily Mail, "Petrol tank 'overflowing' before Buncefield blast," May 9, 2006, United Kingdom, http://www.dailymail.co.uk/news/article-385607/Petrol-tank-overflowing-Buncefield-blast.html

[22] The Guardian, David Fickling, "Faulty fuel gauge caused Buncefield Explosion," United Kingdom,

http://www.guardian.co.uk/uk/2006/may/09/buncefield.davidfickling

Appendix A - 4 Elements of Causality

Table 1. Organizational Processes

Topic	Positive Human Factors	Negative Human Factors
Communications		
Clarity of responsibilities	Expectations communicated and rules are consistently enforced	Confusion on expectations or inconsistent enforcement
Engineering and maintenance communications	Clear communication between maintenance technicians and engineering to resolve instrumented safeguard issues	Maintenance lacks support from engineering; tolerance for unresolved instrumented safeguard issues
	Timely communication of negative findings by maintenance to I&E engineering/reliability	Poor or lack of communication of negative findings by maintenance to I&E engineering/reliability
Operations and maintenance communications	Clear communication of instrumented safeguard status between operations and maintenance technicians	Poor/unclear communication of instrumented safeguard status between operations and maintenance technicians
	Reliable operator-to-maintenance technician communication equipment (two-way radios, telephone, etc.) with alternative means	Unreliable, no alternative, may not work in an overloaded situation
	Good communication between operations and maintenance technicians	No/not expected communication between operations and maintenance technicians
Teamwork and communications	Formal communication with turnover log when maintenance shift changes occur. Includes communication of active bypasses, overrides, faulted devices, and other issues relevant to safe completion of tasks.	Informal communication when shift changes occur. No defined expectation on what to communicate at shift change.
	Frequent supervisory reviews and quality assurance checks	No/incomplete supervisory reviews or quality assurance checks
	Good communication of new findings related to instrumented safeguard health, such as obsolescence, end-of-life, and any identified installation, commissioning, or functional issues	Lack of timely communication of new findings related to instrumented safeguard health, so that identified issues are not addressed systematically across a site
	Routine reporting of repeat failures (e.g., bad actors) to maintenance supervision	Lack of timely communication of repeat failures to maintenance supervision
Emergency communications	Clear, unambiguous site-wide emergency warning system	No distinction in emergency warnings based on areas or event types. Not audible or reliable in some locations.
Instrument Reliability Program		
Process demand tracking	Low tolerance for poor process control; particular focus on reducing frequency of process upsets and process demands on the instrumented safeguards	High tolerance for poor process control and upsets leading to frequent demands on the instrumented safeguards
Maintenance Priority	Maintenance priority is established based on device criticality and level of functional impairment	Maintenance priority does not consider device criticality; redundant instrumented safeguard equipment essentially treated as "spares;" level of functional impairment not understood
	Spare parts management program considers the time required to acquire specialized instrumented safeguard equipment and the need to remain within the assumed mean time to restoration	Spare parts purchased on failure detection without regard to lead time and planned mean time to restoration
	Instrumented safeguard maintenance is prioritized, executed as scheduled, and is rarely delayed for operational reasons	Instrumented safeguard maintenance is frequently delayed, behind schedule, or not prioritized
	Instrumented safeguards only bypassed under strict controls, including compensating measures	Frequent bypassing of instrumented safeguards with little oversight, time limits,

Topic	Positive Human Factors	Negative Human Factors
Out of service/bypass management	and time limits. Note: Bypasses include operator bypasses, manual operation, changing setpoints, and forces.	or risk assessment. Note: Bypasses include operator bypasses, manual operation, changing setpoints, and forces.
	Low tolerance for poor instrument reliability. Proactive attitude to taking action to improve reliability.	High tolerance for poor instrument reliability. Unresolved issues, long-term out-service, and frequent fault conditions accepted.
Repeat failure/bad actor management	Minimal false or spurious alarms	Many false or spurious alarms or alarms ignored or disabled
	Instrumented safeguards are known to be reliable and effective	Instrumented safeguards are known to be unreliable or ineffective
	Identified failures are investigated and repaired in a timely manner	Failed equipment remains in-service; repeated failures are not investigated
	Failure reporting and escalation notification for unresolved issues is clearly defined	No/poor reporting structure for identified failures; no escalation of unresolved problems

Table 2. Workplace Practices

Topic	Positive Human Factors	Negative Human Factors
	Maintenance Instructions	
Task complexity	Required tasks are well-defined and regularly performed	Infrequently performed or repeated/rapid changes in task expectations
	Manufacturer installation and maintenance manuals are reviewed to ensure that maintenance procedures agree with intended application; manuals are accurately translated, written in native language and written from the perspective of the maintenance technician	Manufacturer installation and maintenance manuals are not reviewed for consistency with application; manuals are not clearly written, in wrong language, or poorly translated
	Units of measure are consistent between provided documents and equipment configuration	Units of measure are not consistent between provided documents and equipment configuration
Specification, installation drawing availability	Procedures include verification of installation and configuration of instrumented safeguards against specification	No review of installation and configuration of instrumented safeguards after initial validation
	Installation drawings, specifications, and functional requirements are accessible when needed	Installation drawings, specifications, and functional requirements are not accessible
Procedure clarity and detail	Procedures are at the right level of detail to ensure consistent execution and record keeping	Too general or too detailed leading to inconsistent maintenance, a tendency to skip steps, or poor maintenance records
	Procedures are written in concise, imperative language	Wordy, inconsistent style
	Procedures include notes, cautions, and warnings where errors could result in impaired equipment	Lack of hazard awareness, unknown impact of error
	Notes, cautions, and warnings set off from procedural steps (e.g., in text boxes placed immediately before applicable steps)	Task criticality not clearly identified
	Procedures/checklists used in the performance of task	Task sequence done by memory
	Procedures contain clear pass/fail criteria	Maintenance determines acceptability based on ad hoc criteria

Topic	Positive Human Factors	Negative Human Factors
Return to service procedures	Maintenance procedures include return to service verification by operations	No operations cross-checking or verification of return to service for instrumented safeguards
Change management	Procedures address change management and version control	Maintenance corrects problems without engineering involvement or change management review
Quality control and record keeping	Procedures include appropriate supervisory checks	No supervisory cross-checking or verification
	Data governance is used to ensure record quality	No data governance
Maintenance Equipment/Interfaces		
Maintenance feature/facility design	Instrumented safeguard equipment is easily accessible for maintenance or accessibility issues are addressed in the maintenance procedures	Instrumented safeguard equipment is not easily accessible; maintenance is known to be delayed by access issues
	Maintenance facilities designed for purpose, arranged in logical order, easy to use, well-labeled, and in-service status is easy to detect	Maintenance facilities are confusing, complicated, unreliable, disablement possible without detection, or in-service status difficult to detect

Table 3. Personnel Traits

Topic	Positive Human Factors	Negative Human Factors
General		
Knowledge, skill and experience	Minimal turnover of maintenance technicians resulting in significant experience with site instrumented safeguards and a high degree of personal knowledge of site systems	High turnover of maintenance technicians resulting in less experience with site instrumented safeguards and less personal knowledge of site systems
	Hiring qualifications are defined and include specific requirements for instrumented safeguards	Hiring qualifications are not defined or do not include specific requirements for instrumented safeguards
	Technicians are well-trained, experienced, and good at troubleshooting the technologies used on site	Technicians are not well-trained, are inexperienced, or lack troubleshooting skills with the technologies used on-site
	Technicians are well-trained on safe work practices, such as lock-out/tag-out, electrical safety, job safety analysis, etc.	No specific/unclear requirements for training on safe work practices, such as lock-out/tag-out, electrical safety, job safety analysis, etc.
	Technicians are well-trained on instrumented safeguard maintenance and required record keeping. Training program includes periodic refresher training.	No specific/unclear training on instrumented safeguard maintenance and record keeping.
Fatigue	Overtime limited by defined policy that ensures reasonable and regular rest breaks	Overtime is extreme and does not ensure sufficient rest
	Permanent shift assignments	Shift rotations
Competency Assessments		
Verification of knowledge and skills	Training verification includes both test and observations	No/inadequate verification of learning

Table 4. Enabling Conditions

Topic	Positive Human Factors	Negative Human Factors
Personal protective equipment	Required PPE does not affect performance of tasks	PPE is heavy, cumbersome, gets in the way of performing tasks.
Tools and equipment	Test equipment is high quality, equipment calibration is verified	Poor quality test equipment; lax tracking of calibration records
Working conditions	Noise level low enough to easily communicate	Hearing protection is required. Noise level hinders ability to hear or use communication equipment.
	Provided with protection from weather; including rain, snow, wind, and sun	Not provided with protection from weather, including rain, snow, wind, and sun
	Task conducted in climate-controlled environment	Task conducted in high temperature and/or humidity extremes
	Clear visibility where task is being executed	Poor visibility where task is being conducted, including fog, smoke, or other sight obscuring element
	Lighting is sufficient to conduct task, including being able to read tags, critical information, procedures, or other documents	Lighting is insufficient to conduct task or to read documents
Housekeeping	Equipment is clearly and uniformly labeled	Equipment is mislabeled or not labeled
	Equipment is installed in the field as would be expected (A to C are upstream to downstream)	Equipment is installed in an unexpected order (C to A are upstream to downstream)
	Equipment criticality is easily distinguished in the documents and in the field	Similar equipment in same area or grouped together without any indication of criticality
	Clearly communicated identifier/location for instrumented safeguard equipment	Ambiguous identifier/location for instrumented safeguard equipment
	Consistent tagging between procedures, P&IDs, and equipment	Inconsistent tagging between installation and documents
	Installation shows discipline toward good labeling, tight wiring connections, and consistent installation practices	Installation shows poor discipline, such as loose wiring connections, lack of consistent labeling, or inconsistent installation practices
Time pressure	Number of tasks well-matched to work force	Required tasks exceed resources
	Pace of tasks is not rushed. Little time pressure on step execution.	Multiple tasks are executed in rapid succession and under time pressure

Qualitative Risk Assessment of a Commercial Refrigeration System

Robert Sladek
Asset Optimization Consultants
16000 Barkers Point Lane, Suite #105
Houston, Texas 77079
rsladek@aoccorp.com

Prepared for Presentation at
American Institute of Chemical Engineers
2019 Spring Meeting and 15th Global Congress on Process Safety
New Orleans, LA
March 31 – April 3, 2019

Qualitative Risk Assessment of a Commercial Refrigeration System

Robert Sladek
Asset Optimization Consultants
16000 Barkers Point Lane, Suite #105
Houston, Texas 77079
rsladek@aoccorp.com

Keywords: risk assessment, risk-based inspection, qualitative, refrigeration, rbi, prioritzation

Abstract

This case study presents the methods that were used to calculate qualitative risk for a critical refrigeration system and the results of the assessment. This case study is intended to demonstrate how a Qualitative Risk Assessment can be a cost-effective and resource-efficient means to prioritize equipment which is often under the radar.

- This assessment was performed to scope a capital, reliability investment by focusing the inspections, improvements and repairs to ensure reliable future operation.
- Probability was assessed using simple hoop stress, historical and expected corrosion rates and also accounted for a unique damage mechanism.
- Consequence was tabulated on Safety, Environmental and Operational factors based on Site SME review and assignment.
- A non-standard risk matrix was developed based on the needs of the site.

1. BACKGROUND

1.1. Commercial Refrigeration System Description

The subject was a 14-year-old commercial refrigeration system which supported four (4) cold rooms in total: two (2) storage rooms; one (1) working room and; one (1) packaging room. Redundancy was required to meet the required temperature tolerance of +/ -5°C. Only one (1) train was required to meet the demands of the system. Mechanical, Electrical and Instrumentation equipment were not part of the scope of this project as these assets were included in the site reliability program. Table 1 lists the equipment which was in the scope of this project.

Equipment Type	Qty.	Function	Skid	Qty. Required
Shell/Tube Exch.	2	Condenser	Evaporator	1
Pressure Vessel	2	Low Temp. Suction Accumulator	Evaporator	1
Pressure Vessel	2	Medium Temp. Suction Accumulator	Evaporator	1
Pressure Vessel	2	Liquid Transfer Vessel	Evaporator	1
Pressure Vessel	2	High Pressure Receiver	Evaporator	1
Pressure Vessel	2	Filter/Dryer	Evaporator	1
Pressure Vessel	2	De-Superheater	Evaporator	1
Shell/Tube Exch.	5	Oil Cooler	Compression	2
Pressure Vessel	5	Oil Separator	Compression	2
Coil Exchanger	4	-35°C Evaporator		2
Coil Exchanger	14	+5°C Evaporator		7
Piping Circuit	7	Process, Insulated C/S		7
Piping Circuit	10	Process, Un-insulated C/S		10
Tubing Circuit	10	Process, Copper Tubing		10
Piping Circuit	4	Cooling Water, Un-insulated C/S		4
Tubing Circuit	2	Cooling water, copper tubing		2

Table 1. Project Scope: Equipment Listing

1.2. Challenges Faced

The value of the product stored in the storage rooms was in excess of $50 million. Any breach of tolerance beyond a defined time period required destruction of the product.

Asset compliance for fixed equipment was required by the state and not Process Safety Management (PSM). As a result, inspection history was only recorded to satisfy the regulatory requirement and was not used to predict asset integrity.

The design data was available from the original project files for the primary assets but not for all assets. Maintenance history was recorded for all assets in the system. Historically, the primary "bad actors" were piping and exchanger components in cooling water service.

1.3. Capital Investment to Improve Near Term Reliability

The client was planning to build a new facility but needed to keep the existing facility in reliable operation until the new facility could be brought on-line. Capital funding was dedicated to improve the near-term reliability of the existing system. Due to the lack of asset integrity history, the client needed a method for identifying which assets required replacement or repair and which assets would sustain the expected run-time without additional maintenance.

2. BENEFITS OF RBI

2.1. Risk Based Inspection (RBI) basics

RBI is defined as "a risk assessment and management process that is focused on loss of containment of pressurized equipment in processing facilities, due to material deterioration. These risks are managed primarily through equipment inspection."1

The probability of failure (POF) and consequence of failure (COF) are calculated individually and the product is the risk. Note that "failure" is defined as loss of containment.

The level of detailed information required to calculate risk can vary depending on the methodology applied. There are three types of methodologies used in industry today: Qualitative, Quantitative and Semi-quantitative. Regardless of the type of methodology chosen, the application of a documented and consistent approach that properly uses relative risk to support the needs will determine the success. The primary output of an RBI assessment is a mitigation plan.

2.2. Qualitative Methodologies

Qualitative methodologies tend to focus on engineering judgement, subject matter expertise and experience for calculating POF and/or COF. The primary benefit of this approach is that they are less costly as they do not require proprietary software or discreet data. The downside is that this approach is less precise than the other methods.

2.3. Quantitative Methodologies

Quantitative methodologies are model-based approaches using discreet data to calculate numerical values for POF, COF and, consequently, risk. This is the most precise methodology and is excellent for the long-term management of assets. This approach typically requires proprietary software program(s) and accurate data from multiple sources (maintenance, inspection, production, engineering and management). Many consider a quantitative methodology as too rigorous for the prioritization of inspection tasks.

2.4. Semi-quantitative

Semi-quantitative models are a blend of the two (2) preceding methodologies. Proprietary software is typically used for this approach. The data required is typically less detailed than that of a quantitative approach.

3. PROJECT CHALLENGES

The initial challenges affecting this project were the limitation of funding and resources, no site endorsed methodology and no software license to use.

- Funding was at a minimum as this project was to precede and support the front-end loading (FEL) of the capital project. The funds allotted for the risk assessment could support approximately 160 hours of consulting time.
- Site resources were available to assist the project.

- The client chose to perform a qualitative RBI assessment to identify which equipment would require funding to achieve the reliability goals of the site.

4. RBI METHODOLOGY - EXCEL

A blended approach was proposed and accepted which used a semi-quantitative methodology for POF and a qualitative methodology for COF. Conservative assumptions were made and combined with field verification where required data was missing.

Risk was determined using a spreadsheet to calculate POF and COF. A non-standard 4x5 risk matrix was used with (4) POF categories (A through D) and five (5) COF categories (1-5). Three (3) color codes were used: Red = High; Yellow = Medium; Green = Low.

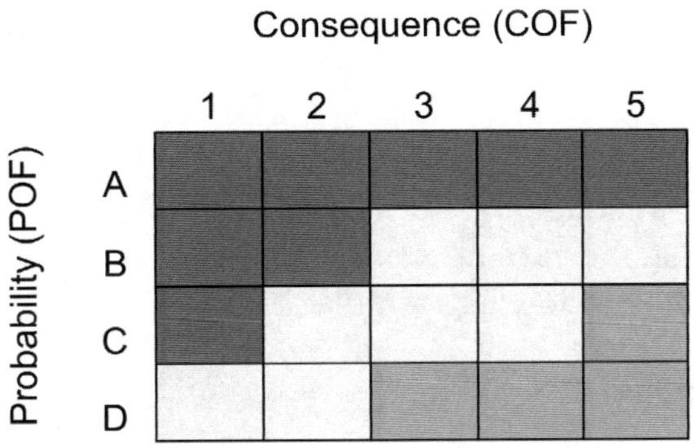

Figure 1. Non-standard Risk Matrix to Meet Site Standard

Damage rates were determined by researching all available records and consulting with site personnel. In the absence of accurate historical data, known values from similar service equipment were applied. A Corrosion Specialist was consulted to confirm/estimate rates based on operating conditions and metallurgy.

A qualitative damage mechanism named "Ice Damage" was assigned to the room evaporators. The evaporators had a history of freezing up due to high humidity year-round. This icing had to be manually removed using picks to clear the coils of ice. The evaporators in the -35°C storage room were assigned a 'B' probability for Ice Damage. The remaining evaporators operate in the +5°C rooms were assigned a 'D' probability for Ice Damage.

For corrosion under insulation (CUI), a corrosion rate of 0.010"/year was used for all insulated, carbon steel components.

4.1. Probability of Failure (POF) semi-Quantitative Methodology

POF was determined by estimating the remaining life (RL) of each component based on the following criteria: A = less than 1 year; B = 1 to less than 5 years; C= 5 to less than 10 years; D = 10 years or longer. For the evaporators, this POF designation was then compared to the Ice Damage mechanism and the more conservative value was used for final POF.

The first step for estimating RL was to determine the minimum thickness required (t_{min}) for each component. The minimum thickness required for pressure (Pt_{min}) was calculated using ASME code formulae for circumferential (hoop) stress. ASME BPVC Section VIII, Division I was used for the fixed equipment while ASME B31.3, Process Piping was used for piping.

Structural minimum thickness (St_{min}) was assigned as follows:

Tubing = 20% of nominal thickness

Fixed equipment and piping ≥ 2" diameter = 0.125"

Piping <2" diameter = 0.100"

De-superheater Plates = 30% of nominal thickness

Pt_{min} was then compared to St_{min} and the greater value was used as t_{min} for further calculation.

The internal (CR_{int}) and external (CR_{ext}) corrosion rates were combined (CR_{total}) and multiplied by the life of the asset to determine the total estimated wall loss. This loss was subtracted from the nominal thickness to provide the current estimated thickness (t_{actual}). Finally, the remaining life estimate was calculated by the following formula: $(t_{actual} - t_{min}) / CR_{total} = RL$

4.2. Consequence of Failure (COF) Qualitative Methodology

COF was determined using site resources to evaluate each component on four (4) independent consequence categories: Safety Impact; Environmental Impact; System Operability Impact; Storage Room Operability Impact. Five (5) severity levels were identified from 1 (High-High) to 5 (Low). Once tabulated, the category with the highest severity (or lowest number) was used to represent the component.

5. PROJECT PLAN AND PROGRESSION

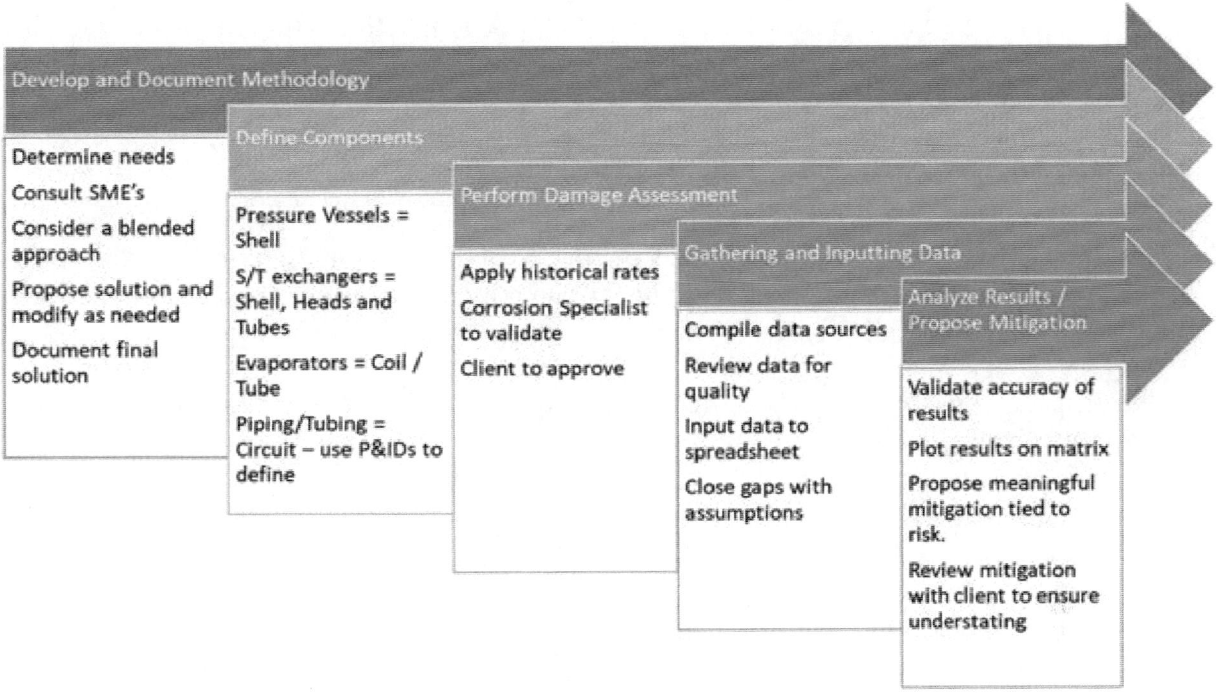

Figure 2. Project Plan and Progression

5.1. Develop and Document the Methodology

The methodology defined above was developed with the client to fit a specific purpose. This included on-site meetings and interviews with site resources. The equipment locations and data were also reviewed to better understand the needs and capabilities. Site subject matter experts (SME) were instrumental in defining the COF severity levels and categorizing damage levels associated with "Ice Damage". This methodology was documented to show transparency of a consistent application and provide for future reference.

5.2. Define components

The fixed equipment was broken into major components, as follows:

- Pressure vessels were assigned one (1) shell component.
- Shell and tube heat exchangers were assigned three (3) components: shell, head and bundle.
- Room evaporators were assigned one (1) tubing component.
- Piping components were defined using the piping and instrumentation drawings (P&IDs). The line numbers were used to identify each component.

5.3. Perform Damage Assessment

All damage values were assigned per the methodology defined above. After the Corrosion Specialist validated and the client approved these values, they were entered into the POF spreadsheet for risk calculation.

5.4. Gathering and inputting data

Site resources were responsible for compiling the data sources needed for the project. Once completed, the project team reviewed the data, then input the required fields into the risk calculation spreadsheet. Where data gaps were discovered, field verification was employed to determine the best conservative assumption.

5.5. Analyze Results and propose mitigation

Once risk was calculated for all components, the methodology was verified for accuracy. The risk results for each component were plotted on the risk matrix (see figures 2-4) and the initial mitigation plans were drafted for client review. The mitigation groupings were designed to fit the client's needs and original purpose of the assessment:

- Inspect high-risk equipment immediately to quantify damage estimates and determine repair/replacement plan.
- Replace equipment based on historical life expectancies.
- Propose damage preventions to extend the life of assets.
- Inspect remaining assets to quantify damage estimates.

The mitigation plans were reviewed with the client detailing the justification of each asset based on the methodology.

6. RESULTS

The results of this risk assessment identified the primary issues affecting the longevity and reliability of this refrigeration system's fixed equipment assets were CUI and cooling water related corrosion. The following matrices in figures 3 and 4 show the risk distribution as a result of this assessment.

ALL EQUIPMENT

PROBABILITY / CONSEQUENCE						
Row Labe	1	2	3	4	5	Grand Total
A		4	3	5		12
B			7			7
C			12	2		14
D		16	33	9	2	60
Grand Total		20	55	16	2	93

Figure 3. Risk Distribution, All Equipment Components

FIXED EQUIPMENT

CONSEQUENCE (columns) / PROBABILITY (rows)

Row Labels	1	2	3	4	5	Grand Total
A		4		5		9
B			4			4
C			7	2		9
D		11	20	5	2	38
Grand Total		15	31	12	2	60

PIPING / TUBING

CONSEQUENCE (columns) / PROBABILITY (rows)

Row Labels	1	2	3	4	5	Grand Total
A			3			3
B			3			3
C			5			5
D		5	13	4		22
Grand Total		5	24	4		33

Figure 4. Risk Distribution, Fixed and Piping/Tubing Components Separated

Based on the results of this assessment, the following mitigation actions were recommended:

1) The following equipment should immediately have the insulation stripped for inspection with expected repairs or replacement due to CUI:
   - The Low and Medium Temperature Suction Accumulators (A-2)
   - The Liquid Transfer Vessels (C-3)
   - Carbon Steel Piping
     - Three (3) circuits (A-3)
     - One (1) circuit (B-3)
     - Three (3) circuits (C-3)
2) All (5) Compressor Oil Coolers (A-4 & C-3) should be replaced in kind immediately due to water side corrosion.
3) The channel heads on the condenser (C-4) should be removed and inspected to quantify water side corrosion rates and repaired and/or coated as required.
   - Sacrificial anode installation should be added to limit corrosion damage of the heads and tubes.
4) The four (4) Evaporators in the -35°C storage room should be replaced due to frequent icing affecting the long-term reliability of this equipment.
5) Due to the high impact related to consequence, all (5) piping circuits (D-2) should be inspected in a high confidence manner to quantify the expected condition.
6) The remaining equipment should be inspected to confirm the findings of this assessment, with priority placed on the higher risk equipment.

7. CONCLUSION

Refrigeration units or "skids" have commonly been managed independently of process equipment integrity programs. Some owner/operators often use special emphasis maintenance teams within the site to manage these assets independently of process related assets and others rely on third party maintenance providers. All too often, refrigeration skids fly under the radar of the site asset integrity program and are not given the same discretion as the process equipment.

This case study provides one example of how facilities can use the fundamentals of RBI to prioritize equipment using minimal effort and resources. The key to success in this approach is to focus the solution based on the specific needs of the facility and resist implementing an over-designed approach in order to minimize the overall effort.

Regulatory compliance is an everchanging target that operators must adjust to. Lessons learned from recent events make their way into our recognized and generally accepted good engineering practices (RAGAGEP). Systems that were once thought of as benign may suddenly become the focus for required asset integrity programs for compliance management. Using a qualitative, focused approach can help streamline the effort in defining which assets to implement into an existing asset integrity program and when.

8. REFERNCES

1. API RP 580 "Risk Based Inspection", 3rd Edition, American Petroleum Institute, February 2016.

19 AIChE Spring Meeting
15TH GLOBAL CONGRESS ON PROCESS SAFETY
NEW ORLEANS • 2019

Managing Reliability of Ethane Crackers using DMR, RBI, and IOWs

Vibha Zaman
Asset Optimization Consultants
16000 Barkers Point Lane, Suite #105
Houston, Texas 77079
vzaman@aoccorp.com

Leomar Penuela
Asset Optimization Consultants
16000 Barkers Point Lane, Suite #105
Houston, Texas 77079
lpenuela@aoccorp.com

Prepared for Presentation at
American Institute of Chemical Engineers
2019 Spring Meeting and 15th Global Congress on Process Safety
New Orleans, LA
March 31 – April 3, 2019

Managing Reliability of Ethane Crackers using DMR, RBI, and IOWs

Vibha Zaman
Asset Optimization Consultants
16000 Barkers Point Lane, Suite #105
Houston, Texas 77079
vzaman@aoccorp.com

Leomar Penuela
Asset Optimization Consultants
16000 Barkers Point Lane, Suite #105
Houston, Texas 77079
lpenuela@aoccorp.com

Keywords: ethane cracker, pyrolysis, carburization, creep, coking, thermal fatigue, reliability integrity operating windows, damage mechanism review, risk based inspection

Abstract

Ethane crackers are subject to severe operating conditions that pose reliability issues and require high maintenance which can have a significant impact on the operational and maintenance cost. Damage mechanisms associated with Ethane crackers and associated piping and fixed equipment include carburization, creep, erosion, thermal fatigue, sigma phase embrittlement, stress relaxation cracking, oxidation, and general corrosion among others. Factors such as high operating temperatures, coke buildup, decoking practices, unexpected shutdowns, and operational breaches can induce or enhance the severity of the damage mechanisms and reduce equipment life. Therefore special attention is needed to avoid unexpected catastrophic failures while optimizing equipment performance.

Risk Based Inspection (RBI) methodology incorporating a damage mechanism review (DMR) can be used to manage the assets in Ethane crackers and reduce the risk of loss of containment by identifying the damage mechanisms that are expected to occur in each fixed equipment and piping component, calculating the risk associated with a potential loss of containment, and clearly defining inspection plans based on the RBI assessment results.

Since day to day changes in process conditions are not fed back into the risk calculation of RBI software, a well-designed and implemented integrity operating windows (IOW) program is required to incorporate these changes by providing real time notification of key process parameters that can accelerate degradation and increase risk. Process data provided by the process engineer is key information that is used when assigning damage mechanisms and developing IOW parameters. This paper discusses the use of the DMR, RBI, and IOW programs to manage reliability in Ethane crackers. It will also touch upon the role that Process Engineers have in the reliability process.

1. MECHANICAL INTEGRITY REQUIREMENTS

OSHA 1910.119 Process Safety Management (PSM) requires facilities to establish and implement written procedures to maintain the ongoing integrity of process equipment and mandates that inspection and testing be performed using procedures that follow recognized and generally accepted good engineering practices. This requires that a mechanical integrity program be in place to assure the continued integrity of process equipment. The frequency of inspection and testing must conform with manufacturer recommendations and good engineering practices, or more frequently if determined to be necessary by prior operating experience. This is important to note for some of the components associated with Ethane crackers where the actual lifespan can be significantly less than design.

Since OSHA PSM is a performance based standard where facilities are free to establish any mechanical integrity program they see fit provided it is based upon recognized and generally accepted good engineering practice, current best practices for mechanical integrity programs include the use of API 580 Risk Based Inspection and API 581 Risk Based Inspection Technology to prioritize and manage the efforts of an inspection program along with API 584 Integrity Operating Windows to provide real time monitoring of process conditions which may increase the risk for equipment damage. The integration of DMR, RBI, and IOWs is illustrated in Figure 1.

Figure 1. Integration of RBI, DMR, and IOWs

2. DAMAGE MECHANISM REVIEW (DMR)

2.1 DMR Overview

The purpose of a DMR is to determine the corrosion rate, expected damage mechanisms, and type of damage (e.g. thinning, pitting, cracking) expected for each asset managed by an RBI program. The results of the DMR are used to obtain the probability of failure which in turn is combined with the consequence of failure to calculate the risk and generate an inspection plan.

The DMR process includes a systematic review of the design data, process information, and materials of construction along with available inspection, monitoring, and sampling data. Typical inputs and outputs of a DMR are shown in Figure 2. Attention to details is critical in achieving a good outcome from the DMR process and it is important not only to understand where the data is coming from, but also do a "sense check" to ensure the information is correct. Errors in the data provided, incorrect interpretation of the data, or incomplete data can result in a failure to identify applicable damage mechanisms and missed opportunities to look for the damage during inspections. Conversely, identification of damage mechanisms that are not realistic can result in increased inspection spending without bringing value or divert efforts from areas where attention is needed.

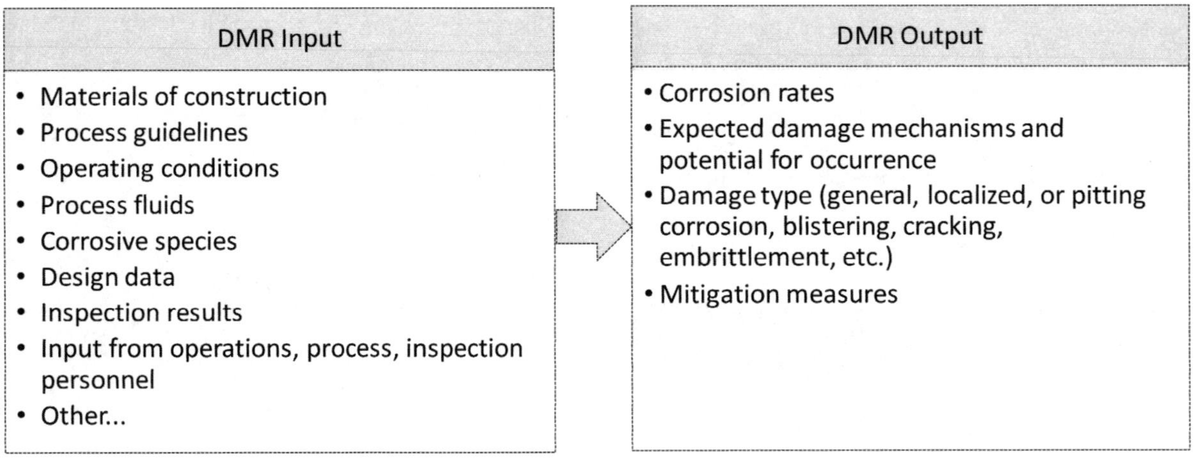

Figure 2. DMR Inputs and Outputs

2.2 Ethane Cracker Process Overview

Ethane pyrolysis furnaces produce ethylene and other byproducts by the thermal cracking of ethane rich hydrocarbons in a mixture of steam and hydrocarbon. There are several pyrolysis furnace designs available in industry depending on the specific feedstock and desired products but most have a configuration similar to that shown in Figure 3. Ethane rich hydrocarbon feed entering the convection section is mixed with dilution steam, injected with dimethyl sulfide (DMS), and preheated prior to entering the radiant tubes where the cracking reactions take place. The cracked gas leaving the radiant tubes is rapidly quenched in the Primary Quench Exchangers (PQE) to stop the reaction in order to obtain the desired products. The heat generated by the pyrolysis furnace is also used to produce high pressure steam in the convection section.

The cracking reactions are endothermic and take place at high operating temperatures (1400-1600°F / 760-870°C) with tube metal temperatures reaching 2100°F / 1150°C due to coke buildup. The dilution steam added to the hydrocarbon feed serves to improve the ethylene selectivity and reduce the coking potential. The DMS is used to inhibit coke formation and control the production of CO by passivating the Radiant coils and prevent rapid coking. Coke does build up over time and reduces heat transfer and increases the pressure drop across the tubes requiring the coils to be regularly decoked. Decoking is started with steam at about 1470°F / 800°C and continued with a steam-air mixture up to about 2010°F / 1100°C.

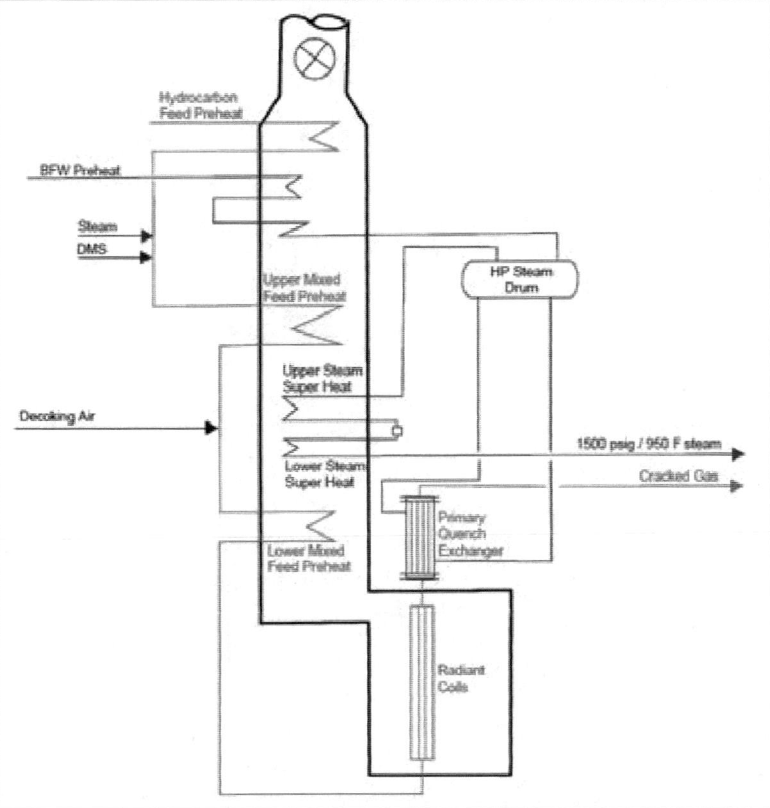

Process	TMT (°F/C)	Metallurgy
Convection Coils		
Hydrocarbon Feed Preheat	555 / 290	Carbon Steel
BFW Preheat	560 / 293	Carbon Steel
Upper Mixed Feed Preheat	900 / 482	1.25Cr-0.5Mo
Upper Steam Superheat	950 / 510	1.25Cr-0.5Mo or 2.25Cr-1Mo
Lower Steam Superheat	1100 / 593	2.25Cr-1Mo, 304H, and 347H
Lower Mixed Feed Preheat	1600 / 870	304H and Alloy 800 HT
Crossovers and Radiant Coils		
Cross Overs	1350 / 732	304H or 347H
Inlet Manifold	1350 / 732	304H or 347H
Cold Pass	1750 / 955	25Cr-35Ni
Hot Pass	2010 / 1100	35Cr-45Ni
Transfer Line and Primary Quench Exchanger (PQE)		
Exit Wye / Transfer line	1650 / 900	Alloy 800 HT or HP Low C (26Cr-35Ni)
PQE Tube Side		Alloy 800 inlet nozzle 1.25Cr-0.5Mo tube
PQE Shell Side		Carbon Steel
PQE Outlet Piping		2.25Cr-0.5Mo

Figure 3. Schematic of a Typical Ethane Pyrolysis Furnace

2.3 Damage Mechanisms in Ethane Crackers

Ethane crackers are susceptible to a number of different damage mechanisms including those that result in the degradation of material properties and those that result in a loss of wall thickness (thinning). A detailed listing of the expected damage mechanisms and damage potentials for each component associated with the Ethane cracker is provided in Table 1. The potential damage mechanisms include the following.

2.3.1 Convection Coils (BFW and Steam Superheat)

The boiler feed water (BFW) Preheat coils produce 640 psig superheated steam and Steam Superheat coils produce 1500 psig superheated steam by recovering waste heat from the flue gases in the convection section. Corrosion rates in the BFW Preheat coils are dependent on the quality of the BFW and are typically low (< 1 mpy) unless there are issues with the BFW treatment. Poor BFW quality can introduce a number of damage concerns including general and pitting type corrosion (due to the presence of acid gases), under deposit corrosion, and fouling.

Corrosion and damage protection in the high pressure Steam Superheat coils is maintained by ensuring the presence of a thin protective magnetite scale (Fe_3O_4). Critical factors that need to be managed include the concentration of dissolved gas (oxygen and/or carbon dioxide), pH, hardness, and the specific BFW treatment system. Poor water chemistry can result in caustic gouging, caustic stress corrosion cracking, scaling, blockage, and steam blanketing which can lead to short term overheat and stress-rupture. In addition, poorly adhering scales and high velocity can result in erosion and erosion corrosion problems.

The upper and lower high pressure Steam Superheat coils have skin temperatures up to 950°F /510°C and 1100°F / 593°C which is above the creep threshold for the typical materials of construction (See Figure 3). Therefore they will accumulate creep over time and are subject to creep-stress rupture as the coils age. Typical coil life is > 20 years. The use of mixed materials (2.25Cr-1Mo ferritic steel and 304H and 347H austenitic stainless steel) can also cause concerns for dissimilar metal weld cracking (DMW) at the high operating temperatures. Additionally, austenitic stainless steel 304H and 347H will develop a sigma phase over time. Sigma is a hard, brittle magnetic phase containing approximately 50% chromium. It results in brittleness and loss of impact strength, which is most apparent at cooler temperatures experienced during shutdowns. The brittleness due to sigma phase formation tends to disappear when the metal is heated above 260°C (500 °F) and reappears on cooling below this temperature. Consequently, the likelihood of an on-stream failure due to embrittlement is low, but failures can occur during cool down or during maintenance activities if high stresses or shock loads are introduced. Presence of a sigma phase also increases the susceptibility to intergranular corrosion. Although the individual potentials for creep, DMW cracking, and sigma phase embrittlement are low, a combination of these damage mechanisms can cause medium or high potential for a loss of mechanical properties over time and lead to cracking damage.

2.3.2 Convection Coils (Hydrocarbon Feed and Mixed Feed Preheat)

The Hydrocarbon Feed Preheat coils operate at low temperatures and contain dry hydrocarbon (typically a mixture of ethane and propane) so there are no internal corrosion or other damage concerns. Since low sulfur fuel is used for the furnace, the potential for flue gas dew point corrosion on the tube OD is also low.

The Upper Mixed Feed Preheat coils are typically constructed of 1.25Cr-0.5Mo material and have skin temperatures up to 900°F / 482°C which is above the creep threshold limit (800°F / 427°C) for the material. Therefore the coils will accumulate creep over time and have a low potential for creep-stress rupture. The coils also have a potential for high temperature sulfidic corrosion due to the presence of sulfur species from the use of DMS. The corrosion potential is low due to the low dosage of DMS (in the range of 100 ppm) that is used.

The Lower Mixed Feed Preheat coils are typically constructed of a mix of 304H, 347H, and Alloy 800HT to provide resistance to high temperature damage mechanisms. Skin temperatures can reach 900°F / 482°C at the coil inlet and 1600°F / 870°C at the outlet. The potential for creep is high, particularly for the coil outlets. In addition, the 304H and 347H materials have a high susceptibility to sigma phase embrittlement. The high skin temperatures also induce coking which promotes carburization.

Carburization results in carbon being absorbed in the steel to form metal carbides. Carburization damage results in a loss of high temperature creep ductility and loss of toughness when the heater is cooled down to ambient conditions. The corrosion resistance is also decreased due to the change in microstructure. Decoking cycles accelerate the carburization process and also introduce concerns with thermal fatigue due to the thermal cycles. Thermal cycles can also be induced from the expansion and contraction experienced from furnace trips and startups/shutdowns. The combination of these damage mechanisms makes the coil butt welds and attachment welds at high risk for cracking. To compound the problem, the change in microstructure results in a loss of weldability. Therefore tube life is typically limited by weld cracking and ability to carry out weld repairs.

Tube replacements driven by the thinning damage mechanisms which include erosion (due to decoking), metal dusting, and external oxidation are less prevalent. Metal dusting is a severe form of carburization that occurs when the carbon diffusion rate is uncontrolled. It results in the formation of a fine powder of iron carbide or pure metal and carbon dust causing rapid metal loss. Once initiated it is difficult to control. Oxidation occurs when oxygen in the combustion air reacts with iron in the steel to form metal oxides. Oxidation corrosion rates depend on the material of construction and TMT and should be less than 5 mpy for the typical combination of materials and TMT.

Other external damage concerns stem from sulfur present in the fuel gas. Although low sulfur fuel is used for the Ethane pyrolysis furnaces, presence of low levels of sulfur species in the fuel can result in the development of iron sulfide scale (over an extended period of time) and lead to concerns for polythionic acid stress corrosion cracking (PTASCC) on the tube OD. PTASCC typically occurs in sensitized austenitic stainless steel materials during downtime when polythionic acids form from the presence of sulfide scale, air, and moisture. Sensitization results in the precipitation of chromium carbides at the grain boundaries and increases the potential for intergranular corrosion. PTASCC has not been a concern for the tube ID's of Ethane crackers. PASCC less likely to occur in the 347H chemically stabilized stainless steel materials since they are less prone to sensitization; the 304H SS and Alloy 800HT coils are the main concern.

2.3.3 Crossovers and Radiant Section Inlet Manifold

Most designs use 304H or 347H SS for the Crossover and Inlet Manifold piping to the Radiant coils. High process temperatures of 1230°F / 666°C are experienced in Crossovers and Inlet Manifold. The corrosion and damage concerns are similar to those experienced in the Lower Mixed Feed Preheat coils and include creep, carburization, sigma phase embrittlement, thermal fatigue, and erosion. Due to the high temperatures experienced and combination of damage mechanisms that result in a loss of mechanical properties, Crossover and Inlet Manifold welds have a high potential for cracking. Poorly supported pipe (e.g. due to improper spring hanger calibration, damaged supports, or poor design) increase the stresses and result in even higher damage potentials. Repair welding becomes increasingly difficult over time due to the change in microstructure. Therefore the useful tube life is often limited by the ability to perform weld repairs.

2.3.4 Radiant Coils

Process temperature is increased in the Radiant coils from about 1230°F / 666°C to 1580°F / 860 °C. Most of the Radiant coils include a cold pass and hot pass connected by welds at the return bends. Current designs use centrifugally cast tubes with 25Cr-35Ni for the cold pass and 35Cr-45Ni for the hot pass. The alloys have an austenitic microstructure, high mechanical strength, and are very resistant to high temperature damage. Several different alloy grades are available. The different alloys employ various additions of carbide forming elements such as niobium, titanium, tungsten and molybdenum to improve the creep strength in the high temperature range. The tungsten, titanium, and niobium alloying elements also improve resistance to carburization.

Skin temperatures are in the order of 1750°F/ 955°C for the cold pass and 2010°F / 1100°C for the hot pass. Due to the very high temperatures experienced and frequency of decoking cycles (about every 20 to 60 days), the radiant tubes can be subject to high potentials for creep, carburization, and thermal fatigue. The welds are particularly sensitive to cracking due to the change in microstructure (which reduces the strength) and higher residual stresses that are present. The precipitation of a sigma phase reduces creep ductility during operation as well as fracture toughness when the tubes are cooled below 500 °F / 260 °C. In addition, solids (spalled coke) present from the frequent decoking cycles result in concern for erosion damage, particularly in areas of turbulence such as elbows and reductions.

The combination of creep and carburization are the most prevalent damage mechanisms that limit Radiant tube life. The use of 25Cr/35Ni and 35Cr/45Ni alloys which readily form chromium oxide scale and DMS injection which promotes the formation of iron sulfide scales, provide carburization resistance. However mechanical damage experienced during the decoking cycles and presence of locally hot areas from the burning of coke deposits cause the protective layers to be disrupted, forming gaps in the scale where the metal base can be exposed to carburization. Cracks formed in the chromium oxide layer as a result of creep or fatigue accelerate the carburization process. Once the carburization process is thus initiated, it is difficult to control and is the primary reason for premature Radiant tube failures.

Industry experience shows the Radiant tubes experience severe levels of carburization and creep damage and generally do not last up to their estimated design life. Therefore, special attention is required to improve their reliability. An IOW program can be used to define the operating limits based on the particular process conditions and materials of construction in order to monitor key process parameters that can influence tube life; a discussion of IOW's is provided later in this report. Additionally, mitigation measures including metallurgy upgrades to provide high temperature damage resistance, use of anti-foulants to resist the formation of coking, use of catalytic or diffusion coatings to reduce carburization, and design changes to reduce tube fouling and improve run lengths have been implemented by many operating units. A discussion of the mitigation measures is outside the scope of this paper.

2.3.5 Radiant Coil Exit Wye / Transfer Line

The pyrolysis furnace effluent leaves the radiant tubes at process temperatures of around 1580°F / 860°C and the Radiant coil Exit Wye / Transfer Line TMT's can reach 1650°F / 900°C at end of run conditions. Materials of construction are typically high chromium nickel alloys (such as Alloy 800 HT or HP Low C) to provide high temperature damage resistance. The main damage concerns are due to creep, carburization, thermal fatigue, sigma phase embrittlement (particularly for HP Low C materials), and erosion due to the high skin temperatures and exposure to the decoking process. As previously discussed, welds are particularly vulnerable to cracking.

2.3.6 Primary Quench Exchangers

The cracked gas is rapidly quenched from 1580°F / 860°C to 830°F / 445°C on the tube side of the PQE's by generating high pressure steam on the shell side. There are several different designs and materials that are currently used for PQE's to provide the rapid quench required while mitigating against the expected damage mechanisms. The designs vary from tube in tube to shell and tube exchangers. The PQE inner tubes and bundles are typically constructed of 1.25Cr-0.5Mo and the shells are usually carbon steel. The material selection for the inlet cone/channel may vary from Alloy 800 HT (tube in tube design) to carbon steel or 1.25Cr-0.5Mo refractory lined (shell and tube design).

- PQE inlet channels/cones constructed of Alloy 800 HT are susceptible to damage mechanisms that are similar to the transfer line (creep, carburization, thermal fatigue, and erosion). Refractory lined channels are susceptible to refractory damage which can result in high skin temperatures inducing creep and high temperature sulfidic corrosion (due to the use of DMS) to the base metal.

- PQE tubesheets (where present depending on the design) are susceptible to thermal fatigue due to the high delta T experienced from the rapid quenching of the cracked gas; seal welds have a high potential for cracking. For tube in tube exchangers, dissimilar metal weld (DMW) cracking at the inlet cone to inner tube joint weld is a concern.

- PQE tubes are subject to fouling and plugging from coke and poly oils as well as erosion corrosion from coke fines, particularly during decoking operations on the process side. Boiler water corrosion is a concern on the shell side. Tube side plugging results in a loss of heat transfer and can lead to short term overheat. The tubes and outlet channels can also experience high temperature sulfidation in the presence of hydrogen (due to the DMS injection). The API 939C Couper Gorman curves are typically used to calculate sulfidic corrosion rates in the presence of hydrogen.

- PQE outlet piping is susceptible to erosion corrosion and very high rates of wall loss (upwards of 40 mpy) have been reported in some cases. High temperature sulfidation in the presence of hydrogen is also a potential damage mechanism but low corrosion rates are expected from sulfidation.

3. RBI AND INSPECTION PLANS

The results of the DMR are used to calculate relative risk and generate inspection plans. Risk calculations can be performed using various methodologies ranging from qualitative to fully quantitative. Due to the availability of, and industry experience with, commercial software that facilitates the RBI implementation process, a semi-quantitative approach using RBI software is often used. This provides a practical approach to developing initial inspection plans based on relative risk as well as facilitating the evergreen process.

A key factor in developing the inspection plan is to ensure that inspection tasks look for damage mechanisms identified by the DMR. Tables 1 and 2 highlight the importance of understanding how the design, process, and operating conditions affect the materials of construction for each component in order to accurately identify the expected damage mechanisms and type of damage expected so that inspections can be carried out at the optimum locations and using the correct inspection technique to look for the expected damage. Incorrect identification of damage mechanisms and their potentials can result in unidentified or increased risk to the unit while overstating the damage potentials can result in increased inspection efforts without reducing risk, or worse – diverting resources from areas requiring attention.

To illustrate the process of identifying inspection methods, we can take the information in Table 2 as an example. Table 2 links the expected damage mechanisms with the type of damage expected (e.g. thinning, pitting, cracking, loss of material properties) and inspection method required to detect damage for the key components in the Ethane crackers. As an example, the Lower Mixed Feed Preheat coil in Table 2 is susceptible to several high temperature damage mechanisms including Creep (High), Carburization (Medium), Sigma Phase (High), and Oxidation (Low). These damage mechanisms can result in a loss of material properties manifesting as bulging, sagging, and cracking. Therefore some of the inspection methods identified to look for this type of damage include Visual inspections of tubes and supports to look for obvious signs of damage, Dimensional inspection (e.g. via strapping, laser profilometry) to identify bulging or changes in dimensions, and PT or UTSW to identify cracking.

The inspection frequency and extent of inspection are derived from the overall risk from the risk calculation and are outside the scope of this paper. The frequency of inspection required to look for damage from each of the different damage mechanisms can vary significantly depending on the damage potential and risk ranking. For example, cracking inspections for the Lower Mixed Feed coils may be required every 2-3 years due to the high potential for cracking but inspection intervals for thinning may be much less frequent if experience shows the potential for thinning (e.g. due to erosion) has not been a concern.

4. INTEGRITY OPERATING WINDOWS

The DMR and RBI efforts provide the following information:

- ✓ Corrosion Rates and type of damage expected
- ✓ Damage Mechanisms and potentials
- ✓ Type of Damage Expected (e.g. thinning, pitting, cracking, etc.)
- ✓ Prioritized risk for each component
- ✓ Inspection Plans – method, frequency, location

This may be sufficient for equipment operating at fairly steady operating conditions, but equipment subject to process fluctuations can benefit from additional real time monitoring via an IOW program to ensure process deviations from set IOW limits are reviewed and taken into consideration prior to the next DMR revalidation effort or turnaround.

A good starting point in determining the process parameters that should be monitored for the IOW program is to review each damage mechanism listed in the "Damage Mechanisms" column in Table 1 and identify the impact process deviations can have on the damage potential. Once the process parameter to monitor has been identified, industry standards, best practices, design values, and corrosion models can be used to identify the values for the IOW limits. When setting the IOW limits, it is important to consider what is an "acceptable" degradation rate for the affected equipment, how quickly personnel are able to respond if the IOW limit is exceeded, and the time required to bring the process parameter within acceptable limits once the response is taken. The actual limits that are set will be based on the risk tolerance of the company combined with the practical operation of the plant. For example, setting a temperature limit for creep that is too low will have operators constantly reacting to alarms and leave little room to operate. Conversely, if the limit is too high the equipment may sustain rapid degradation leading to increased risk and significantly reduced remaining life.

Table 3 illustrates how monitoring of process parameters are linked with potential damage mechanisms that affect the various components associated with the pyrolysis furnace. A discussion of a few of the Ethane Cracker process parameters to monitor and the rationale for monitoring is provided below. The below listing is not intended to be all inclusive, but rather to be used as a starting point for developing IOW process parameters to monitor.

Hydrocarbon Feed Composition — Olefins in the recycle ethane (e.g. ethylene) can increase the coking tendency which can increase the potential for a variety of high temperature damage mechanisms as discussed above. Monitor the olefins in the feed to the pyrolysis furnace and set operational limits.

Hydrocarbon Feed rate — A complete loss of or significant reduction in the hydrocarbon feed can cause short term overheat if the furnace is not promptly shut down since the furnace

	will still be firing while endothermic cracking reactions will no longer be taking place.
Steam Injection rate	Dilution steam reduces the coking tendency. The steam to hydrocarbon ratio is based on process requirements and economics. Monitor the steam injection rate to ensure the correct steam to hydrocarbon ratio required to minimize fouling is achieved. A loss of steam can cause excessive fouling and lead to short term overheat.
DMS Injection Rate	DMS is used to passivate the Radiant coils in order to mitigate against carburization. A low and high limit are both required for the DMS injection rate since too little DMS causes inadequate passivation of the radiant tubes resulting in increased carburization rates and a decrease in tube life. Too high of a DMS injection rate can cause high temperature sulfidic corrosion to carbon steel and low alloy steel heater tubes, heater outlet piping, and quench exchangers. It can also cause an increase in neutralizer demand for the Quench Tower as well as increased demands in the caustic treatment area.
Process Temperature	Process temperature is an important parameter in determining and managing the coking potential. A reduction in the coking rate, and therefore the decoking frequency helps reduce several of the high temperature damage mechanisms including carburization, creep, and thermal fatigue and helps increase the radiant tube life. Monitor the process temperature in several locations in the furnace including radiant coil inlet and outlet.
Tube Metal Temperature	Excessively high skin temperatures can lead to damage due to creep, carburization, and higher rates of coking. Higher coking rates decrease heat transfer, which leads to even higher skin temperatures and shorter runs. Monitor tube metal temperatures online via skin TI's at the Radiant coil inlet (indicates whether incipient hydrocarbon cracking is a concern in the convection section) and at the Radiant coil outlet (indicates the coking severity and whether decoking is required). In addition, periodically use IR Thermography and Pyrometry to monitor for hot spots. Visually look for flame instabilities and impingement once or twice per shift to ensure no hot spots that can cause localized overheating.
Pressure Drop Across Radiant Tubes	Pressure drop across the radiant tubes is an indicator of when decoking is required (outlet to inlet pressure ratio of <0.9 is typically specified to indicate decoking). Excessive pressure drop across the tubes indicates plugging and coking problems which can cause an increase in tube metal temperature (due to endothermic nature of the process) leading to increased creep potential and short term overheat.
PQE Outlet Temperature	PQE tube side outlet temperature is an indicator of whether decoking is required due to fouling and plugging of the PQE tube side.

Pressure Drop Across PQE Tubes	A high pressure drop across the PQE tubes indicates plugging problems from coke fines and/or poly oils. Plugging results in short-term overheat of the tubes and high temperature damage mechanisms in upstream components.
Number of Furnace Trips	Increased furnace trips resulting in increased thermal fatigue potential and decrease the furnace tube life. Record the number of occurrences of furnace trips to determine if the inspection frequency for thermal fatigue needs updating; The information can also be used for fitness for service evaluations.
Spring Hanger Calibration	Improper spring hanger calibration can cause increased stresses on the tubes leading to increased damage from creep and thermal fatigue.
BFW Quality	BFW quality affects multiple aspects of corrosion and damage in BFW and steam service. Corrosion and damage protection is maintained by ensuring the presence of a thin protective magnetite scale (Fe_3O_4). Poor water chemistry can result in caustic gouging, caustic stress corrosion cracking, scaling, blockage, and steam blanketing which can lead to short term overheat and stress-rupture. In addition, poorly adhering scales and high velocity can result in erosion and erosion corrosion problems. Critical factors that need to be monitored in BFW include pH, Conductivity, Silica, Total Iron, Total Hardness, Oxygen, Chloride, Sulfate, Silica, Sodium + Potassium.
H_2S in the Fuel Gas	H_2S in the flue gas can form sulfuric and/or sulfurous acids which can condense on the tube OD (during operation or downtime) and cause corrosion problems in the cooler sections in the convection section. It can also lead to external sulfide scaling over time and introduce concerns for PTASCC of austenitic stainless steel components in the hotter sections of the furnace. The H_2S content in fuel gas should be kept as low as possible.

5. SUMMARY

This paper outlines a methodology for using RBI, DMR, and IOWs to manage the reliability of ethane crackers based on API standards. A well implemented and managed reliability program incorporating these processes can identify potential damage mechanisms, optimize maintenance and inspection, reduce risks and improve the governance in managing Ethane Crackers. The key elements of the program include:

1. An understanding of the damage mechanisms through a DMR to ensure inspections focus on looking for the right type of damage.

2. An RBI program to prioritize risk and ensure components are inspected using the right inspection technique and at the right frequency.

3. An IOW program to provide real time notification of process deviations in order to avoid premature damage that can shorten equipment life.

6. REFERENCES

1. API RP 581, Risk-Based Inspection Methodology, Third Edition, April 2016. API Publishing Services, 1220 L Street, N. W., Washington, D. C. 20005.

2. API RP 580, Risk Based Inspection, Third Edition, February 2016. API Publishing Services, 1220 L Street, N. W., Washington, D. C. 20005.

3. API RP 571, Second Edition, April 2011, "Damage Mechanisms Affecting Fixed Equipment in the Refining Industry", API Publishing Services 1220 L Street. N.W. Washington DC 20005.

4. API RP 584, First Edition, May 2014, "Integrity Operating Windows", API Publishing Services, 1220 L Street, N. W., Washington, D. C. 20005.

5. J. Robinson, 2014, "Avoiding Waterside Corrosion Problems in Ethylene Plant Steam Systems". GE Power & Water Technical Paper GE Water and Process Technologies, Trevose, PA, USA.

6. H.J. Grabke, "Carburization A High Temperature Corrosion Phenomenon", (1998, Materials Technology Institute for the Chemical Process Industries, Inc.). MTI Publication No. 52.

7. C. M. Schilmoller, "HP Modified furnace tubes for steam reformers and steam crackers". Nickel Development Institute. NIDI Technical Series No 10 058.

8. R.C. Yin, A.H. Al-shawaf, B.Y. Al-yami, A.A. Al-Ahmari, "Failure Experiences of Ethylene Cracking Tubes". Corrosion 2009 NACE Paper 09159.

9. Istvan Kucora, Ljiljana Radovanovic, "Pyrolysis Furnace Tube Damaging and Inspection". ACETA Tehnica Corviniensis – Bulletin of Engineering Fascicule 3 [July-September] Tome VII [2014]. ISSN: 2016-3809.

Table 1. Damage Mechanisms in Ethane Crackers

Process	Process Temp (°F/°C)	TMT (°F/°C)	Metallurgy	Damage Mechanisms	Corrosion Rate (mpy)[1]
Hydrocarbon Feed Preheat	In: 140 / 60 Out: 430 / 220	555 / 290	Carbon Steel	• Flue Gas Dew Point Corrosion (Low)	0.5
BFW Preheat	In: 284 / 140 Out: 420 / 215	560 / 293	Carbon Steel	• Boiler Water/Condensate Corrosion (Low)	1
Upper Mixed Feed Preheat	In: 500 / 260 Out: 760 / 405	900 / 482	1.25Cr-0.5Mo	• Sulfidic Corrosion (Low) • Creep (Low)	2
Upper Steam Superheat	In: 620/327 Out: 775/413	950 / 510	1.25Cr-0.5Mo 2.25Cr-1Mo	• Scaling (Low) • Erosion Corrosion (Low) • Creep (Low)	1
Lower Steam Superheat	In: 740 / 395 Out: 950 / 510	1100 / 593	2.25Cr-1Mo 304H 347H	• Scaling (Low) • Erosion Corrosion (Low) • Creep (Low) • Sigma Phase (Low) • DMW (Low) • External PTASCC (Low)	1
Lower Mixed Feed Preheat	In: 630 / 332 Out: 1230 / 665	1600 / 870	304H and Alloy 800 HT	• Creep (High) • Carburization (Medium) • Thermal Fatigue (Medium) • Sigma Phase (304 H: High, Alloy 800 HT: Low) • Erosion (Medium) • Metal Dusting (Low) • Oxidation (Medium) • External PTASCC (Low)	5 (Localized)
Cross Overs	1230 / 665		304H or 347H	• Creep (Medium) • Carburization (Medium) • Thermal Fatigue (High) • Sigma Phase (Medium) • Erosion (Medium) • Metal Dusting (Low)	3 (Localized)

Process	Process Temp (°F/°C)	TMT (°F/°C)	Metallurgy	Damage Mechanisms	Corrosion Rate (mpy)[1]
Inlet Manifold	1230 / 665		304H or 347H	• Creep (Medium) • Carburization (Medium) • Thermal Fatigue (High) • Sigma Phase (Medium) • Erosion (Medium) • Metal Dusting (Low)	3 (Localized)
Cold Pass	In: 1230 / 665	1750 / 955	25Cr-35Ni HP Micro	• Creep (High) • Carburization (High) • Thermal Fatigue (High) • Sigma Phase (Medium) • Erosion (Medium) • Nitriding, external (Low) • Oxidation, external (Low) • Metal Dusting (Low)	3 (Localized)
Hot Pass	Out: 1580 / 860	2010 / 1100	35Cr-45Ni	• Creep (High) • Carburization (High) • Thermal Fatigue (High) • Sigma Phase (Medium) • Erosion (Medium) • Nitriding, external (Low) • Oxidation, external (Low) • Metal Dusting (Low)	3 (Localized)
Exit Wye / Transfer line	1580 / 860	1650 / 900	Alloy 800 HT or HP Low C (26Cr-35Ni)	• Creep (High) • Carburization (High) • Thermal Fatigue (Medium) • Sigma Phase (HP Low C: Medium; Alloy 800 HT: Low) • Erosion (Medium) • Oxidation, external (Low) • Metal Dusting (Low)	3 (Localized)

Process	Process Temp (°F/°C)	TMT (°F/°C)	Metallurgy	Damage Mechanisms	Corrosion Rate (mpy)[1]
PQE Tube Side	In: 1580 / 860		Alloy 800 inlet nozzle 1.25Cr-0.5Mo tube	• Erosion (Medium) • Carburization (High) • Creep (High) • Sigma Phase • (Alloy 800: Low; 1.25Cr-0.5Mo: Not Susceptible) • Thermal Fatigue (High) • Dissimilar Metal Weld Cracking (High) • Metal Dusting (Low) • Stress Relaxation Cracking (Low) • Boiler Water /Condensate Corrosion (Low) • Steam Blanketing (Low)	5
PQE Shell Side	In: 620 / 325		Carbon Steel	• Boiler Water/Condensate Corrosion (Low) • Steam Blanketing (Low)	1
PQE Outlet Piping	830 / 445		2.25Cr-0.5Mo	• Erosion (High)	> 10 (Localized)

Note 1: The corrosion rates in this table are for illustration purposes and are derived from experience with Ethane Crackers in several different units. Actual corrosion rates for a particular unit will be based on the design, materials of construction, process information, and operating conditions, for the particular unit and should be validated using field data.

Table 2. Inspection Methods Based on DMR Results

Process	Damage Mechanisms[1]	Type of Damage Expected	Inspection Method[2]
Hydrocarbon Feed Preheat	• Flue Gas Dew Point Corrosion (Low)	Thinning	Visual (tubes and supports) UT (accessible areas)
BFW Preheat	• Boiler Water/Condensate Corrosion (Low)	Thinning	Visual (tubes and supports) UT (accessible areas)
Upper Mixed Feed Preheat	• Sulfidic Corrosion (Low) • Creep (Low)	Thinning Bulging, sagging, cracking	Visual (tubes and supports) UT (accessible areas)
Upper Steam Superheat	• Scaling (Low) • Erosion Corrosion (Low) • Creep (Low)	Thinning Bulging, sagging, cracking Scaling	Visual (tubes and supports) UT (accessible areas) PT/MPT (outlet header to tube welds) Tube sampling (for scaling)
Lower Steam Superheat	• Scaling (Low) • Erosion Corrosion (Low) • Creep (Low) • Sigma Phase (Low) • DMW (Low) • External PTASCC (Low)	Thinning Bulging, sagging, cracking Scaling	Visual (tubes and supports) UT (accessible areas) PT/MPT (inlet and outlet header to tube welds and dissimilar metal welds) Tube sampling (for scaling)
Lower Mixed Feed Preheat	• Creep (High) • Carburization (Medium) • Thermal Fatigue (Medium) • Sigma Phase (High) • Erosion (Medium) • Metal Dusting (Low) • Oxidation (Medium) • External PTASCC (Low)	Bulging, sagging, cracking Thinning Loss of mechanical properties (strength, toughness)	Visual (tubes and supports) Dimensional inspection of shock tubes (e.g. via strapping, laser profilometry) UT / GUL / RT (thinning and cracking) PT or UTSW (cracking at welds) RT for coking
Cross Overs	• Creep (Medium) • Carburization (Medium) • Thermal Fatigue (High) • Sigma Phase (Medium) • Erosion (Medium) • Metal Dusting (Low)	Bulging, sagging, cracking Thinning Loss of mechanical properties (strength, toughness)	Visual (tubes and supports) PT / UTSW (cracking at welds) UT / RT (thinning)

Process	Damage Mechanisms[1]	Type of Damage Expected	Inspection Method[2]
Inlet Manifold	• Creep (Medium) • Carburization (Medium) • Thermal Fatigue (High) • Sigma Phase (Medium) • Erosion (Medium) • Metal Dusting (Low)	Bulging, sagging, cracking Thinning Loss of mechanical properties (strength, toughness)	Visual (tubes and supports) UTSW (cracking at welds) UT / RT (thinning)
Cold Pass	• Creep (High) • Carburization (High) • Thermal Fatigue (High) • Sigma Phase (Medium) • Erosion (Medium) • Nitriding, external (Low) • Oxidation, external (Low) • Metal Dusting (Low)	Bulging, sagging, cracking Thinning Loss of mechanical properties (strength, toughness)	Visual (tubes and supports) Dimensional inspection of shock tubes (e.g. via strapping, laser profilometry) UTSW (cracking at welds) UT (thinning) Eddy Current (automatized-crawler) Metallurgical analysis and mechanical testing
Hot Pass	• Creep (High) • Carburization (High) • Thermal Fatigue (High) • Sigma Phase (Medium) • Erosion (Medium) • Nitriding, external (Low) • Oxidation, external (Low) • Metal Dusting (Low)	Bulging, sagging, cracking Thinning Loss of mechanical properties (strength, toughness)	Visual (tubes and supports) Dimensional inspection of shock tubes (e.g. via strapping, laser profilometry) UTSW (cracking at welds) UT (thinning) Eddy Current (automatized-crawler) Metallurgical analysis and mechanical testing
Exit Wye / Transfer line	• Creep (High) • Carburization (High) • Thermal Fatigue (Medium) • Sigma Phase (Medium) • Erosion (Medium) • Oxidation, external (Low) • Metal Dusting (Low)	Cracking Bulging Thinning Loss of mechanical properties (strength, toughness)	Visual (for bulging, thinning, refractory damage) PT or UTSW (cracking at welds)

Process	Damage Mechanisms[1]	Type of Damage Expected	Inspection Method[2]
PQE Tube Side	• Erosion (Medium) • Carburization (High) • Creep (High) • Sigma Phase (High) • Thermal Fatigue (High) • Dissimilar Metal Weld Cracking (High) • Metal Dusting (Low) • Stress Relaxation Cracking (Low) • Boiler Water /Condensate Corrosion (Low) • Steam Blanketing (Low)	Thinning Cracking Bulging Loss of mechanical properties (strength, toughness)	Visual (Boroscopy) Bundle (Eddy Current, Remote Field ET, IRIS, and MFL) PT
PQE Shell Side	• Boiler Water/Condensate Corrosion (Low) • Steam Blanketing (Low)	Thinning Pitting	UT Visual (Boroscopy)
PQE Outlet Piping	• Erosion (High)	Thinning	UT

Note 1: The damage potentials in this table are for illustration purposes and are derived from experience with Ethane Crackers in several different units. Actual potentials for a particular unit will be based on the design, materials of construction, process information, and operating conditions, for the particular unit and should be validated using inspection data.

Note 2: These are examples of some of inspections methods used to look for the expected damage. Various other NDT methods are available in industry and this table is not meant to be a comprehensive list of all available inspection techniques.

Table 3. Example of IOW Process Parameters and the Link with Damge Mechanisms

IOW No.	IOW Description	Affected Areas	Process Variable to Monitor	Relevant Damage Concerns	IOW Limits	Monitoring Method	Monitoring Frequency
1.	Hydrocarbon Feed Composition	Ethane Cracker tubes, transfer line, PQE's, and downstream piping.	Olefin content in hydrocarbon feed	Coking in the heater tubes leading to an increase in high temperature damage mechanisms (listed in Table 1) and thermal fatigue.	Consult Process Engineering	Consult Process Engineering	Consult Process Engineering
2.	Hydrocarbon Feed rate	Ethane Cracker tubes, transfer line, PQE's, and downstream piping.	Hydrocarbon flow rate	1. Coking in the heater tubes leading to an increase in high temperature damage mechanisms (listed in Table 1) and thermal fatigue. 2. Short term overheat	Consult Process Engineering	Online flow rate monitoring	Online (Continuous)
3.	Steam Injection rate	Ethane Cracker tubes, transfer line, PQE's, and downstream piping.	Dilution steam flow rate	1. Coking in the heater tubes leading to an increase in high temperature damage mechanisms (listed in Table 1) and thermal fatigue. 2. Short term overheat	Consult Process Engineering	Online flow rate monitoring	Online (Continuous)
4.	DMS Injection Rate	Ethane Cracker tubes, transfer line, PQE's, and downstream piping.	DMS injection rate	Coking in the heater tubes leading to an increase in high temperature damage mechanisms (listed in Table 1) and thermal fatigue.	Consult Process Engineering	Online flow rate monitoring	Online (Continuous)
5.	Process Temperature	Ethane Cracker tubes, transfer line, PQE's, and downstream piping.	Process Temperature	Coking in the heater tubes leading to an increase in high temperature damage mechanisms (listed in Table 1) and thermal fatigue.	Consult Process Engineering	Online temperature monitoring	Online (Continuous)
6.	Tube Metal Temperature	Ethane Cracker tubes, transfer line, PQE's, and downstream piping.	Tube Metal Temperature	Coking in the heater tubes leading to an increase in high temperature damage mechanisms (listed in Table 1) and thermal fatigue.	Based on Process and Reliability Engineering Requirements	Online temperature monitoring	Online (Continuous)

IOW No.	IOW Description	Affected Areas	Process Variable to Monitor	Relevant Damage Concerns	IOW Limits	Monitoring Method	Monitoring Frequency
7.	PQE Outlet Temperature	Radiant coils, transfer line, and PQE's (tube side).	PQE outlet temperature	Coking and plugging of PQE's indicating decoking is required. Failure to decoke increases potential for high temperature damage mechanisms in the piping components upstream of the PQE's.	Consult Process Engineering	Online temperature monitoring	Online (Continuous)
8.	Pressure Drop across the Radiant Tubes	Radiant coils, transfer line, and PQE's (tube side).	Pressure drop across Radiant Tubes	Coking and plugging of the radiant coils indicating decoking is required. Failure to decoke increases potential for high temperature damage mechanisms and short term overheat.	Consult Process Engineering	Online pressure monitoring	Online (Continuous)
9.	Pressure Drop across PQE Tubes	Radiant coils, transfer line, and PQE's (tube side).	Pressure drop across PQE tubes	Coking and plugging of PQE's indicating decoking is required. Failure to decoke increases potential for high temperature damage mechanisms in the piping components upstream of the PQE's.	Consult Process Engineering	Online pressure monitoring	Online (Continuous)
10.	Number of Furnace Trips	Ethane Cracker tubes	Number of Furnace Trips	Thermal Fatigue	Any. Record the number of furnace trips for future engineering evaluations	Record Number of Furnace Trips	Record Number of Furnace Trips
11.	Spring Hanger Calibration	Ethane Cracker tubes	Travel Limits	Creep, Thermal Fatigue	Based on design	Monitor Travel Limits on the Spring Cans	Six monthly
12.	BFW Quality	Steam Coils and PQE Shell side	Multiple (See Comments)	Boiler Water corrosion, fouling, erosion corrosion, fouling, caustic corrosion, caustic SCC, steam blanketing, short term overheat.	Per water treatment vendor recommendations	Per Water Treatment vendor recommendations	Per Water Treatment vendor recommendatio ns

IOW No.	IOW Description	Affected Areas	Process Variable to Monitor	Relevant Damage Concerns	IOW Limits	Monitoring Method	Monitoring Frequency
13.	H_2S in the Fuel Gas	Coils in the convection section	H_2S in the Fuel Gas	1. Flue Gas Corrosion in the Hydrocarbon Feed Preheat and BFW Preheat coils. 2. Polythionic acid SCC on austenitic stainless steel tube OD.	As low as possible	Set a limit for sulfur content in purchased and recycle fuel gas and test periodically	Based on expected variability of sulfur levels in the fuel source.

Relief Valve Optimization Using Adaptive Reliability Based Risk Reduction Targets

Hassan Sadeghi
Lloyd's Register
Address
Shervin.Sadeghi@lr.org

Neil Arthur, Ian Ross
Lloyd's Register
Address (es)
Neil.Arthur@lr.org Ian.Ross@lr.org

[RTAMO is a copy righted methodology]

Prepared for Presentation at
American Institute of Chemical Engineers
2019 Spring Meeting and 15th Global Congress on Process Safety
New Orleans, LA
March 31 – April 3, 2019

Relief Valve Optimization Using Adaptive Reliability Based Risk Reduction Targets

Hassan Sadeghi
Lloyd's Register
1330 Enclave Parkway, Suite 200 Houston, Texas 77077 USA
Shervin.Sadeghi@lr.org

Neil Arthur, Ian Ross
Lloyd's Register
Kingswells Causeway, Prime Four Business Park, Kingswells., Aberdeen, AB15 8PU, United Kingdom
Neil.Arthur@lr.org Ian.Ross@lr.org

Keywords: Maintenance, Optimization, Relief Valves, PRV, PRD, PSV, Reliability

Abstract

Traditional API 581 risk calculation approach can have mixed results for establishing PRVs inspection interval. The problem here is twofold – Probability of Failure (PoF) is low for most cases, therefore Consequence of Failure (CoF) drives risk, leading to impractical inspection frequencies at extreme ends of the spectrum. This coerces the engineering teams to evaluate other more realistic inspection intervals for cost/safety reasons either based on experience or industry data rather than asset information.

An alternative adaptive methodology is discussed in this presentation where PoF and CoF are based on user defined Risk Reduction Targets and Safety Integrity Levels, rather than financial risk basis. This methodology captures how using actual in-service testing results rather than default values in a Weibull distribution can deliver a more representative Probability of Failure on Demand (POFOD) and how it ultimately can be reliably used to deliver an optimum test frequency based on quantitative assessment of failure modes.

1 Introduction

Lloyd's Register (LR) advocates the use of a robust risk based strategy in accordance with API 581 to manage inspection programs in downstream industries. This approach when correctly implemented leads to safe and efficient operation of piping and fixed equipment. In contrast, utilizing the traditional risk based approach for PRVs, instrumentation and rotating equipment can have mixed results leading to impractical inspection frequencies at extreme ends of the spectrum. To understand why this is the case, a discussion of probability and consequence of failure with respect to API 581 [1] is necessary.

This section gives an overview of API 581 with respect to PRVs, followed by an introduction of Real Time Adaptive Maintenance Optimization (RTAMO) philosophy and methodology [2].

1.1 API 581

1.1.1 Overview

API 581 provides direction on the use of risk assessment to determine the maintenance intervals for pressure relief valves, and expresses risk as the product of the probability of a failure event and the financial consequence of a failure event. API 581 introduces the concept of maintaining pressure relief valves at an interval such that a financial risk target is met i.e. a pressure relief device is maintained at an interval such that the residual financial risk exposure is acceptable [1]. No specific guidance is given on acceptable risk targets, and facility owners are directed to consider their own internal risk tolerability (tolerance?) and acceptance criteria when determining an inspection interval. An example is provided showing that the same relief valve can be maintained at maximum intervals of 3 or 5 years if risk targets of $10k and $25k per annum respectively are considered acceptable. Essentially API 581 suggests that pressure relief valves are tested such that a user-defined financial risk reduction is achieved.

1.1.2 Probability of failure

In API 581 the probability of failure can be estimated by considering the probability of a failure event occurring by estimating the likelihood of the pressure relief device being demanded, failing to operate and a subsequent loss of containment being experienced [1]. This is calculated by estimating three independent failure events for each device:

1. Overpressure event.

2. Failure of pressure relief device to operate on demand.

3. Protected equipment loss of containment.

The first step in evaluating the probability of failure is to determine the demand rate placed on the device. This is estimated by referring to guidance where the demand rate is typically once in 10 years to once in 50 years. The next step is to obtain the probability that the pressure relief device will fail to open upon demand in service. API 581 proposes the use of a two parameter Weibull function and modelling to determine this probability of failure to open [1]; however, this may be simplified by considering the exponential distribution (constant rate failure pattern) and API 581 recognizes this in Table 7.4 [1] where it characterizes moderate and severe service relief valves as having predominantly random failure patterns. Similarly, industry standard data sources such as OREDA data sources [3] consider all failure patterns to be random in nature.

API 581 considers the potential loss of containment from the protected equipment item in the case of an initiating event and the pressure relief device functional failure. Provision is made for calculating the probability of the loss of containment event; however, a conservative assumption can be made such that where elevated overpressures occur, and that these overpressures will approach four times the Maximum Allowable Working Pressure (MAWP), generally considered to lead to the rupture case [1]. Given this, it is assumed that the probability of the protected equipment leading to a loss of containment is 1.0 in the event of an initiating event and the relief valve functional failure.

LR has found that using this approach to determine a probability of failure can result in a very low overall probability of failure on demand such that the resulting inspection intervals can be significant.

1.1.3 Consequence of failure

API 581 describes the consequence of failure of pressure relief devices in terms of financial consequences associated with the failure of operation. These financial consequences include, but are not limited to:

1. Equipment repair and replacement.

2. Damage to surrounding equipment in affected areas.

3. Production losses and business interruption as a result of downtime to repair or replace damaged equipment.

4. Personnel injuries associated with a failure.

5. Environmental clean-up.

API 581 provides guidance on the equipment damage and replacement costs, and typical downtime estimates associated with common equipment items through rupture.

1.1.2 Inspection planning

API 581 then proposes that the recommended interval is determined for pressure relief valves by calculation of the risk as a function of time and determining the time at which the risk is equal to the risk target.

1.2 RTAMO Philosophy [2]

The methodology applied by LR RTAMO is similar to API 581 in that a probability and consequence methodology is used to determine an appropriate inspection testing interval based on achieving a user-defined risk reduction. However, LR RTAMO does not use a financial risk basis to determine the testing interval, and instead uses the Risk Reduction Factor (RRF) used in the assessment of Safety Integrity Levels (SIL) for protective systems. LR RTAMO derives the probability of failure from the installed in-service testing results and uses this to calculate an in-service reliability for the pressure relief valves. The testing interval is then determined based on the pressure relief device achieving the required RRF as stipulated in the facility safety studies.

LR's proposed approach is to determine the optimum test frequency for each device based on analyzing plant data rather than using default values in a Weibull distribution. This approach will deliver a more concise representative value for Probability of Failure on Demand (POFOD) and adapts by "learning" about the plant's PRV population as the test history increases over time.

In summary:

- LR reliability based approach assesses PRVs in populations e.g. sites, service categories, types etc.
- Reliability targets based on reliability (risk reduction) required from safety studies e.g. SIL.
- Typical targets 98% for safety critical PRVs; 90% for non-safety critical PRVs
- Individual PRV test results can also be considered in isolation to enable a representative sample and provide a more stringent testing regime for these valves where necessary – 'bad actor' analysis
- Increases population and sample size; reduces sensitivity to individual test results

1.3 Methodology

LR RTAMO methodology consists of three general steps; data preparations, analysis and reporting and finally implementation which are broken down in more details below [2].

1.3.1 Data Gathering

Data gathering and analysis includes the following activities:

- Obtaining data and documentation associated with the PRVs
- Relief valve test results
- Existing benchmarking reports and assessments
- Agreeing the target reliability for the PRVs

1.3.2 Grouping

A key element of the strategy is that PRVs are 'grouped' into collections of PRVs of similar service to obtain a greater population from which to derive reliability and failure data.

1.3.3 Failure Rates

Failure rates are derived for each grouping of relief valve by the review of the historical testing data to determine a dangerous failure rate applicable to each group. This is carried out by assessing the operational life of each grouping of valves and the number of dangerous failures that had occurred in the operational life. This is then used to calculate a relevant failure rate and represents the actual in-service demonstrated reliability of the protective devices.

1.3.4 Reliability Calculations

Using the reliability theory [4] for component found to have a constant failure rate, the in-service reliability of the categories of PRV can be calculated.

It is generally accepted that demand on safety systems and components occurs randomly overtime, therefore it is necessary to evaluate the POFOD function during the fault exposure time T [4]. Assuming the failure can occur with equal probability at any point during the test interval, the probability that should be used in the frequency-weighted POFOD over time T is therefore:

$$\text{PDF}_{\text{Ave}} -= \frac{1}{\mathbf{T}} \int_0^T PDF(t)dt \qquad (1.0)$$

For the constant failure rate this results it:

$$\text{PFD}_{\text{Ave}} = \frac{1}{\mathbf{T}} \int_0^T \left(1 - e^{-\lambda t}\right)dt \qquad (1.01)$$

$$= 1 + \frac{1}{\lambda \mathbf{T}} (e^{-\lambda t} - 1)$$

The simplification of the reliability function above results in the well know reliability formula:

$$\text{PFD}_{\text{Ave}} = \frac{\lambda \text{T}}{2!} - \frac{(\lambda T)^2}{(3\,!)} + \frac{(\lambda T)^3}{(4\,!)} - \ldots.$$

$$\approx \frac{\lambda \text{T}}{2} \qquad (1.02)$$

Where λ represents unrevealed/dangerous failures.

1.3.5 Test Interval

The testing interval for each grouping of PRV will then be determined by considering the reliability demonstrated, how this compares with the required reliability of the protective device and how this in-service reliability will be affected by modulating the testing interval. The existing testing interval must be considered in determining the new testing interval, as it may be imprudent to extend the intervals to the maximum limit. Similarly, in addition to the application of a testing interval per fluid service, each individual PRV should be assessed in turn to ensure that specific 'bad actors' are dealt with appropriately.

Using the above and given the reliability target for a given PRV, we can calculate the required test interval based on the failure history:

$$Reliability = 1 - \text{PFD}_{\text{avg}} \qquad (1.03)$$

$$= 1 - \frac{\lambda \text{T}}{2}$$

2 Results

2.1 Upstream Oil and Gas Installations

This methodology has been proven in the complex upstream Oil & Gas assets as well as single process train FPSOs over the past 3 years and the results have been positive:

- Success in identifying opportunities to extend testing intervals
- Quick response times in analyzing data
- Analysis tools have led to speedy identification of optimized maintenance intervals
- These tools provide detailed analysis reports that can be used to support Management of Change (MoC) for maintenance management systems and maintenance strategies

Results for offshore and onshore installation are captured in the subsequent sections.

2.1.1 Fixed Offshore Installation

Pre & post optimization testing distributions for all PRVs:

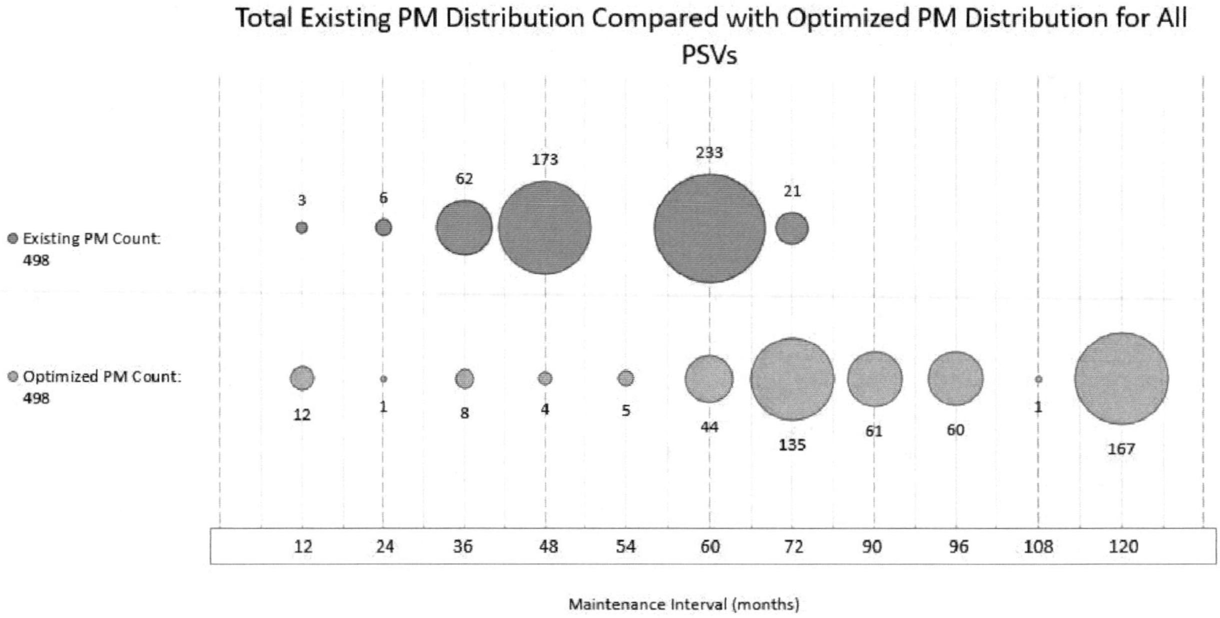

Figure 1. Existing PM and Optimized PMs for all PRVs

Pre & post optimization testing distributions for Air Service PRVs:

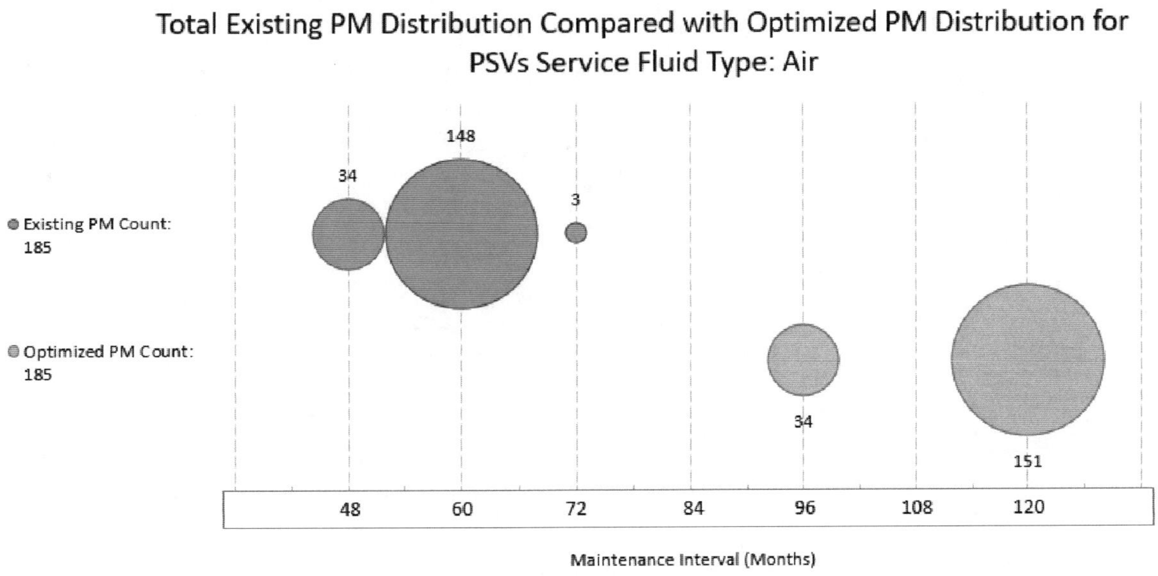

Figure 2. Existing PM and Optimized PMs for Air PRVs

Pre & post optimization testing distributions for Hydrocarbon Gas Service PRVs:

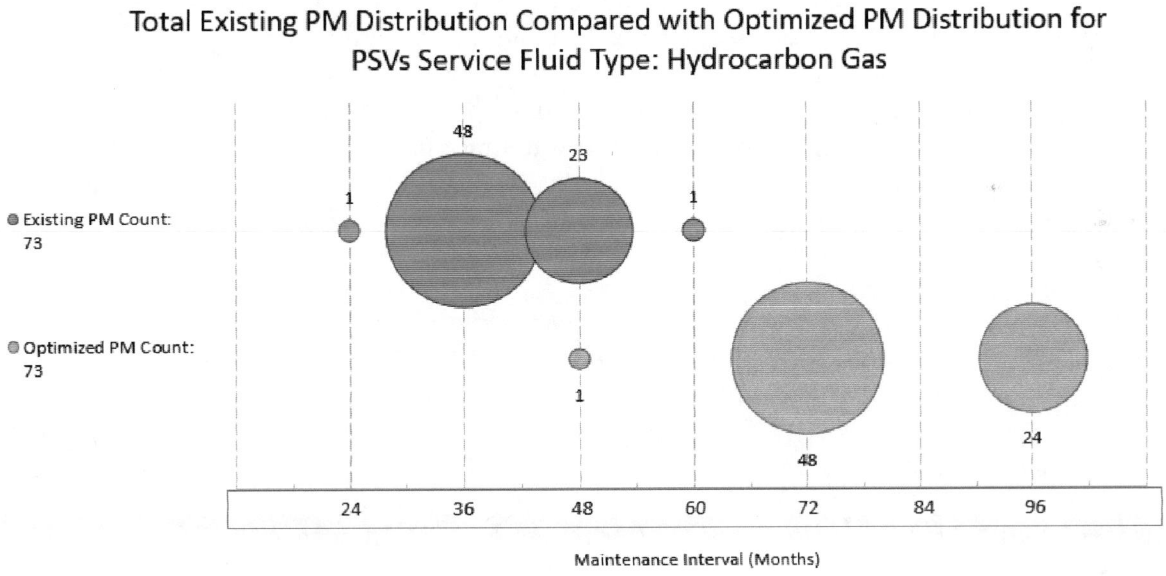

Figure 3. Existing PM and Optimized PMs for HC Gas PRVs

2.1.2 <u>FPSO</u>

Pre & post optimization testing distributions for Hydrocarbon Gas Service PRVs:

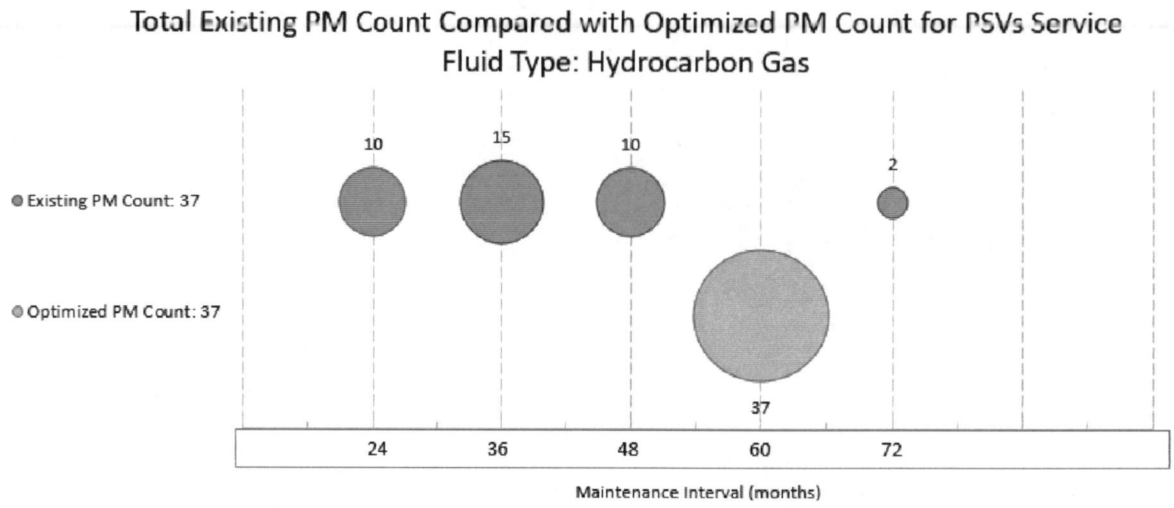

Figure 4. FPSO Existing PM and Optimized PMs for HC Gas PRVs

2.2 Downstream Oil and Gas

Significant reductions in maintenance activity and materials costs were observed by applying the methodology in comparison to the existing planned maintenance. Majority of intervals were extended and some maintenance test intervals were reduced based on actual operational performance. Whilst maintenance interval extensions tended to be more limited for safety critical service valves, the reduction in maintenance activity and materials costs are nonetheless significant. Furthermore, greater asset size and population of PRVs in comparison to upstream applications creates a powerful potential for optimization as seen in the results captured below.

Pilot studies were initiated for:

- Major onshore gas processing plant with approx. 1,000 PRVs
- Refinery with a total of 1,000+ PRVs

2.2.1 <u>Gas Processing Plant</u>

Previous results with Air PRVs allowed a 40-50% reduction in maintenance

Oil & Gas Major gas plant with 961 PRVs, 391 valves are air PRVs. This suggested high potential for maintenance optimization based on previous results seen in upstream applications.

Table 1. Gas Processing Plant Analysis Results

PRV Service Fluid	PRV Count	PRV Count (with Test Cert)	Test Certs Analyzed	Dangerous Failures	PM Annualized Hours (Before)	PM Annualized Hours (After)	% Change
AIR	391	73	74	2	665	448	-33%
OIL	95	78	80	1	474	262	-45%
TOTAL	486	151	154	3	1139	710	-38%

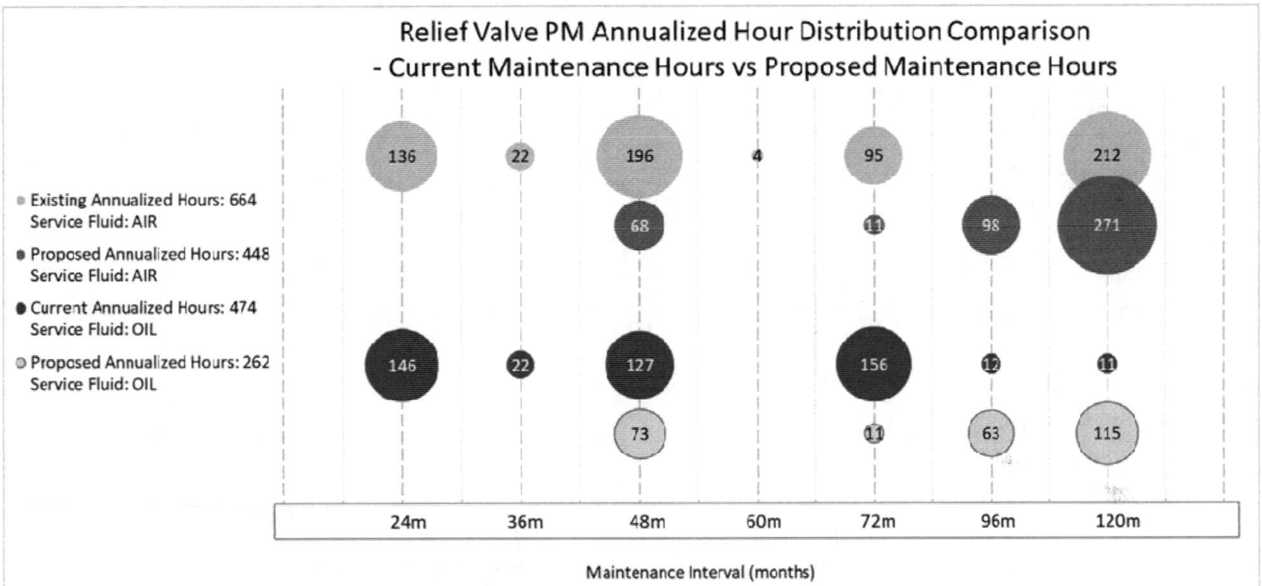

Figure 5. Downstream application Existing PM and Optimized PMs for AI and Oil PRVs

2.2.2 <u>Refinery – Placeholder Analysis ongoing…</u>

- Indicative Failure rate derived for 2 safety critical PRV types
  - Steam 90% Reliability Target

    Significant optimization potential from upstream results
  - HC Gas 95 - 98% Reliability Target

    6 different categories of HC gas to be applied to refinery

    Previously treated as a single service category in upstream

    Sample pool used to generate indicative failure rate

This was tested against a 98% reliability target

Table 2. Downstream refining project PRV results

	PSV	Test Certs	Failure/ month	Current Interval	RTAMO Optimized Interval (months)	Recommended Interval	Annualized PM Change
Steam	221	625	0.0012	36.00	179	96	-62.50%
Gas	186	275	0.0020	36.00	52	48	-25.00%
Water	51	64	0.0022	36.00	98	72	-50.00%
Heavy HC	51	31	0.0025	36.00	41	36	0.00%
Inerts	47	71	0.0012	36.00	179	96	-62.50%
Caustic	12	22	0.0102	36.00	10	12	+200.00%
Amine	10	4	0.0046	36.00	23	24	+50.00%
BFW	4	2	0.0078	36.00	28	24	+50.00%
Totals	582						-35.82%

These failure rates compare favourably with industry comparators e.g. SINTEF failure rate of 0.0044 failures/month [5], providing confidence in the interval extensions.

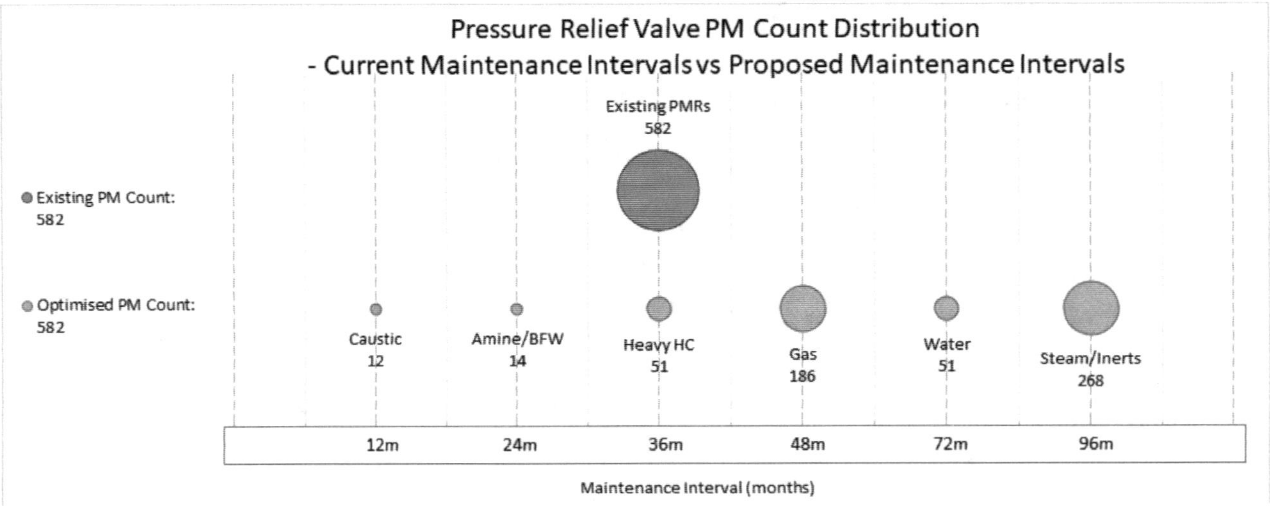

3 Conclusions

- Agile approach that offers a sound alternative to the traditional RBI approach
- Reliability centred analysis of PRV testing intervals can be conducted quickly and effectively based on actual asset performance data
- Reliability targets can be set appropriately for safety critical and non-safety critical services
- This Reliability centred approach uses grouping of valves to obtain a greater sample size when determining reliability data to support review of the testing regime
- This approach allows a deterministic link between relief valve testing and risk
- Software is available which supports this analysis
- Proven approach with strong track record in offshore oil & gas platform and processing facilities, very applicable to downstream as shown via results
- This methodology supports timely and effective reduction in OPEX, through reduced maintenance and materials costs, delivering maintenance capacity back to operators
- Similar methodology can be applied to optimize PM of Rotating equipment and instrumentations such as pumps and fire and gas detection systems

4 References (examples given to show format).

[1] American Petroleum Institute. "Risk-based Inspection Methodology, 3rd ed," API RECOMMENDED PRACTICE 581, pp 34 - 83, April 2016

[2] Lloyd's Register, Real Time Adaptive Maintenance Optimization "RTAMO". Available at https://www.lr.org/en/rtamo/ Accessed on Feb 08, 2019.

[3] OREDA. "Offshore and Onshore Reliability Data, 6th ed. 'Volume 1 – Topside Equipment'", pp 240 – 260, 2015

[4] J.C.H. Schüller, J.L. Brinkman, P.J. Van Gestel and R.W. van Otterloo. "Methods for determining and processing probabilities 'Red Book'" The State Secretary of Housing Spatial Planning and the Environment (VROM), pp 5.10-5.28, 1997

[5] SINTEFF Technology and Society. "Reliability Data for Safety Instrumented Systems, PDS Data Handbook", SINTEF Research, pp 30-40, 2013

548430 Forecasting of Major Incidents Based on Big Data of Near Misses and Minor Incidents: Heinrich's Vs. Deep Learning

Forecasting of coming major incidents, though challenging, would be useful in preventing accidents in advance, allocating resources, and proactive management of process safety. This research compares approaches of advanced analytics such as deep learning as well as applying the traditional Heinrich's law, in forecasting major incidents in the areas of national fire occurrences and operation of LNG gasification and distribution plants. Heinrich's law shows a ratio distribution of near misses, minor incidents, and major accidents, and it also proposes a periodicity of all those incidents. Using Heinrich's law, the occurrence of a coming major incident may be predicted in advance, if a number of near misses and minor incidents before the next major incident are accumulated a specific number of times and monitored carefully. In conventional research, experts artificially classified the 3-level category of incidents, but natural classification using a clustering method of machine learning performs better, without biases, along with the era of big data. The proposed method is applied to 1) the past 6-year investigation data (about 40,000 occurrences/year) of National Fire Data System (NFDS) of Korea and 2) collected operation data of LNG gasification and distribution plants. 3-level or 4-level distribution of past incidents depending on the damage and loss shows a similar tendency as Heinrich's law. Forecasting models predicting occurrences of major incidents in 1) national fire occurrences and 2) LNG gasification plant operations are developed using deep neural networks (feedforward as well as recurrent), and their applications in real-time risk management are discussed.

Opportunities to enhance barrier management through incident analysis

Presenter/Principal Author: Fabian Zitman
Affiliation: CGE Risk Management Solutions
Address: Vlietweg 17v, Leidschendam
Email address f.zitman@cgerisk.com

Other Author(s): Emma Verschoor
Affiliation(s): CGE Risk Management Solutions
Address (es): Vlietweg 17v, Leidschendam
Email address (es) e.verschoor@cgerisk.com

Prepared for Presentation at
American Institute of Chemical Engineers
2019 Spring Meeting and 15th Global Congress on Process Safety
New Orleans, LA
March 31 – April 3, 2019

Opportunities to enhance barrier management through incident analysis

Presenter/Principal Author: Fabian Zitman
Affiliation: CGE Risk Management Solutions
Address: Vlietweg 17v, Leidschendam
Email address f.zitman@cgerisk.com

Other Author(s): Emma Verschoor
Affiliation(s): CGE Risk Management Solutions

Keywords: barrier management, near miss reporting, incident analysis

Abstract

Many high hazard industries use bowties to analyze and assess risks. In the bowtie risk analysis model, barriers are identified which aim to prevent, control or mitigate unwanted scenarios.

However, despite best effort risks assessments, incidents and near misses may still occur. When they do, organizations typically conduct investigations to come up with recommendations to avoid similar situations in the future. Unfortunately, in many cases these investigation reports end up as a collection of largely unstructured documents. They are often treated as single events, making it difficult to clearly identify trends over multiple events.

Scenario-based Incident Registration (SIR) aims to provide a more structured process. A user selects a specific scenario from a set of bowties and assesses the effectiveness of the barriers identified on the original bowtie. Specific questions are asked per barrier about the reasons of the failure. These findings are stored on a barrier level. If this process is repeated, the bowtie will accumulate barrier failure data of multiple incidents in a single diagram. This information helps the organization to identify the strength of their barriers, which in turn facilitates risk-based decision-making.

Some incidents require in-depth investigations and a registration only is not enough. However, this registration can be used as a starting point for the incident analysis. Doing barrier-based incident analysis allows you to learn from barrier failures and improve their performances in order to prevent similar incidents from happening. This incident data together with the SIR data can be linked back to the bowtie diagram in order to spot the strengths and weaknesses of your visual risk assessment.

SIR is a continuous process which is more than a reporting system. The reporter does not require knowledge about the bowtie methodology, while incident managers receive valuable insight to their pro-active risk assessment.

1 Introduction

Through the years, as organizations have focused more on safety, accidents have become less frequent. Overall, this is a good thing. However, for safety managers, this makes things more complicated due to the way safety is currently measured: the absence of incidents and accidents. Paradoxically, fewer incidents also means less data to measure whether the organization is safe or not. With the increasing complexity which hides the underlying factors in organizations, merely the absence of accidents no longer guarantees that we are safe (Hollnagel, 2014). Many examples of organizations with incidents after a long absence of incidents exist: the 1986 Space Shuttle Challenger explosion, the 1997 Hindustan refinery explosion in India, 1999 Paddington train crash, the 1998 Morton explosion and the Three Mile Island nuclear plant accident. Therefore, rather than relying on the number of accidents as an indicator of safety, we need a more proactive metric.

A proposed solution to this paradoxical problem of too few incidents is a larger focus on lower severity incidents as these generally happen more frequently, e.g. near misses. Near miss reporting can quickly build up a large database of would-have-been accidents.

In this paper, a near miss reporting system will be presented. The effectiveness of near miss reporting in general will be examined, before exploring how the bowtie method can be used to enhance the effectiveness of near miss reporting.

The bowtie methodology is a risk analysis tool and if used well, can help to answer at least the first two questions from the three questions raised by Gordon MacDonald of the UK Health & Safety Executive from the aftermath of the Buncefield incident in 2005 (ODN, 2010).

"Lessons must be learned from this incident. From the boardroom down, companies must ask themselves these questions:

1. *Do we understand what can go wrong?*
2. *Do we know how what our systems are to prevent this from happening?*
3. *Do we have information to assure us they are working effectively?"*

"*Many organizations are able to answer these questions to some extent*" (Smit, 2017). Various risk management tools and methodologies can be used to answer these questions, where the bowtie methodology is one of them. However, to be able to answer the third question, organizations should use more than only the bowtie methodology. Audit methodologies or incident analysis techniques are often used to verify whether the barriers are working effectively. SIR is one methodology that can help to answer MacDonald's third question.

1.1 Evidence in favor of near miss reporting

Critiques of incident reporting have questioned whether organizations actually become safer after implementation of an incident reporting system. As such, it may still seem unclear whether near miss reporting should be pursued. These critics have suggested that incident reporting can be more successful if there is an emphasis on "analysis of incidents and organizational learning" but can fail if there is too big of an emphasis on just the counting of incidents (Vincent et al., 2008; Clarke, 2008; Davies, 2000; Braithwaite et al., 2011). In order to properly refute these critiques, the causal link between near misses and accidents must be demonstrated through the use of the iceberg model.

The basic premise that more information regarding (minor) errors can hold predictive value for larger accidents rests on the assumption that near misses and accidents have the same set of root causes. This is also known as the 'iceberg model' (after Hyden, 1987). The iceberg model shows a "continuum of events ranging from normal behavior via conflicts and deviations to actual accidents" (Van der Schaaf, 1992, p21.). Van der Schaaf (1992) has simplified this model into the one seen in Figure 1.

The events at the bottom of the triangle occur first, i.e. the earlier you detect the behavioral acts and near misses, the higher the chance you can prevent an accident from occurring. Another assumption here is that the behavioral acts and near misses have the same root causes as the accidents. Therefore, if the detection of a certain event happens later 'up' the triangle, the analysis of that event should lead to largely the same root causes as if the event was detected earlier 'down' the triangle. Likewise, the recommendations should be of a similar nature.

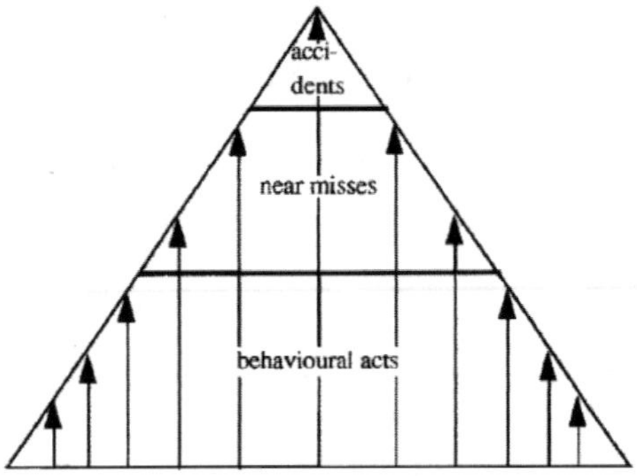

Figure 1 - Simplified qualitative version of the iceberg model by Schaaf

Supportive evidence for the qualitative assumptions of the model comes from numerous studies across multiple industries (construction, aviation, railway and oil and gas among many others) that indicate that an increase in near miss reporting decreases not only the occurrence of minor injuries, but also that of larger so-called OSHA (Occupational Safety & Health Administration) injuries (Mckay, (2013). Thus, the pursuit of near miss reporting can make an organization safer as the information resulting from near misses holds predictive value for preventing accidents.

In conclusion, this paper proposes that near miss reporting can help make organizations safer. This paper explores how a near miss reporting system can be used to support the process of learning from near misses. The foundation of the design is the bowtie method and must be looked at in more detail.

2 Method

CGE Risk Management Solutions has developed a tool, Scenario-based Incident Registration (SIR), to improve the quality of incident reporting and allow trend analysis for improved risk-based decision-making. SIR uses the bowtie methodology and bridges the gap between bowtie risk analysis and incident analysis.

2.1 The bowtie method

Using SIR in the organization requires bowtie knowledge and expertise from the middle to low management of an organization, usually the safety and risk department. Operational personnel, who eventually report the incidents, do not necessarily require this knowledge and expertise.

The bowtie methodology, presented in Figure 2, is a proactive risk assessment tool to analyze potential incident scenarios and assess risks. "Once constructed, the bowtie purpose is best used to support risk management and risk communication" (the American Institute of Chemical Engineers, 2018). The bowtie also fits as a framework to structure incident data if the incident is analyzed using a barrier-based incident analysis technique, such as Barrier Failure Analysis, Barrier-based Systematic Cause Analysis Technique (BSCAT) (DNV GL, 2015) or Tripod beta (Stichting Tripod Foundation, 2014).

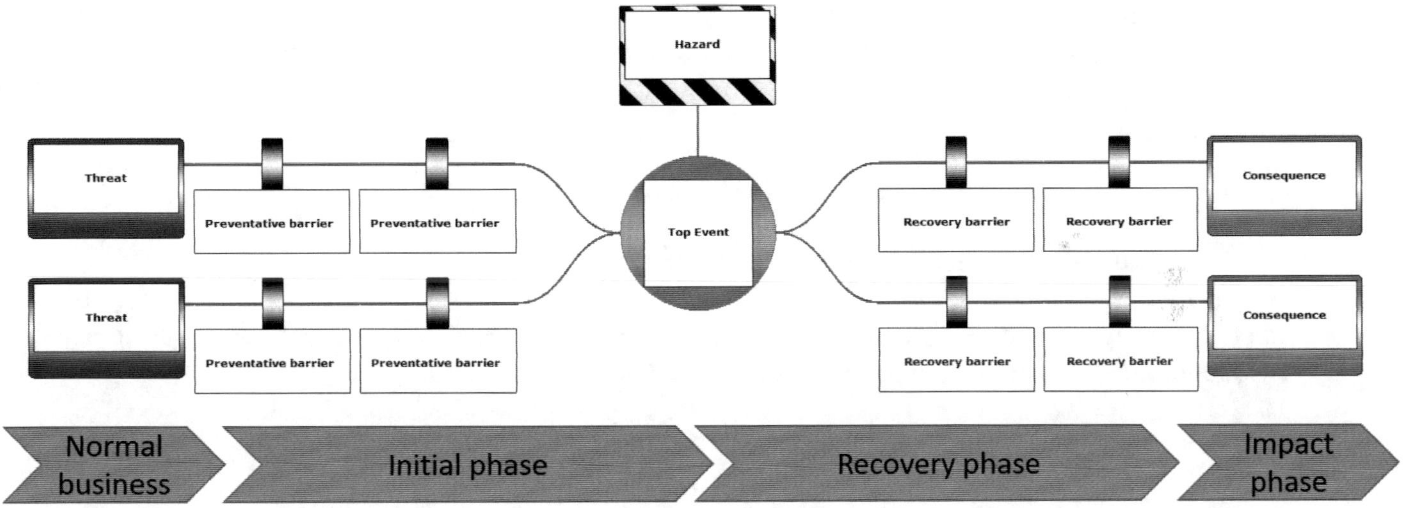

Figure 2 - The bowtie model

The bowtie methodology is a combination of two risk analysis tools (FTA, preventative and ETA, reactive) and James Reason's Swiss Cheese theory.

The Swiss Cheese metaphor is developed by James Reason in 1990. The model represents layers of defenses to stop an incident. However Swiss Cheese has holes in it and so does each layer of defense or barrier. The holes are created by active failures and latent failures (Reason, 1997). SIR helps to identify active failures during an (near-) incident scenario with the help of the bowtie.

The bowtie model is represented in Figure 2. The hazard defines the operation of where the risks are analyzed. Examples are 'Working at height' or 'Storage of hydrocarbons'. The hazard should be described in a controlled state, as it is part of an organization's normal business. The reason of referring to normal business is to be able to link risks back to process optimization.

The top event related to the hazard is the moment of loss of control. Examples are 'Person falls from scaffold' or 'Loss of containment of hydrocarbons'. It is an unwanted event but not the actual incident or accident. When the top event is understood completely, it can help an organization to recover faster from a loss of control moment to prevent the actual incident or accident.

On the left side of the top event, the threats, are the causes of the loss of control (top event). Examples are 'slippery surface' causing a person to fall or 'corrosion' causing a loss of containment. Threats may lead directly to the top event.

On the right side of the top event, the consequences, are the actual incidents like a 'Fire', 'Hydrocarbon spill resulting in environmental damage', or 'Person falls on hard surface resulting in fatality'.

The next most important step is to define the barriers. "*Safety barriers are physical and/or nonphysical means planned to prevent, control, or mitigate undesired events or accidents*" (Sklet, 2006 2.2). Barriers consist of multiple activities that maintain the barrier function and effectiveness, such as training or maintenance. Barriers on the left side, preventative barriers, may stop the threat from occurring or prevent the threat leading to the top event. Examples of preventative barriers are 'anti-slip surface' or 'anti-corrosion painting'. Barriers on the right side, recovery barriers, may stop these consequences or reduce the impact on these consequences. Examples of barriers are 'fall harnesses' or 'fire extinguisher'.

SIR helps to understand what went wrong in the barrier supporting activities.

2.2 Scenario-based Incident Registration (SIR)

SIR is a methodology which collects meaningful data about barrier failures. This data can be collected through incident investigations. However not every incident can be completely investigated, due to time, resources and expertise constraints.

Figure 3 is an illustration of the usual follow up process of an incident. Organizations normally try to learn from accidents by time-consuming investigations. Incidents with a lesser severity, e.g. near misses, are stored individually in a database, with limited to no learning.

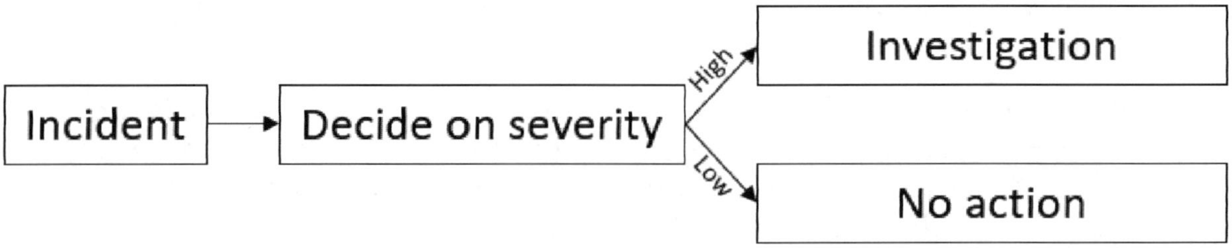

Figure 3 - Incident follow up process

In Figure 4 the SIR-analysis compliments the incident decision process by storing low severity incidents on a bowtie. Additionally, SIR helps to decide whether an investigation is needed during the follow up process.

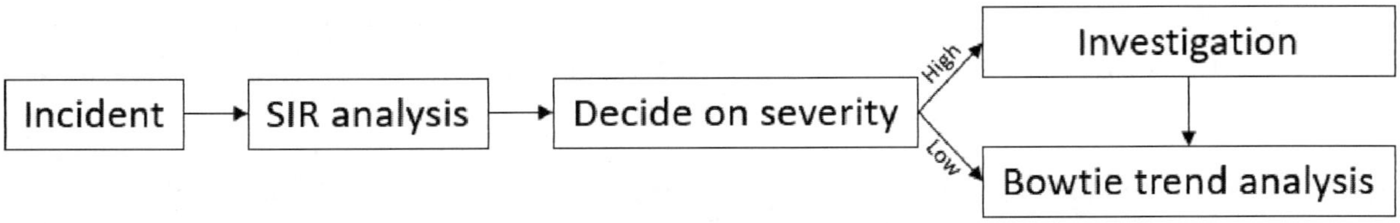

Figure 4 - Incident follow up process including SIR

In the initial form, the reporter enters the details about the location, date, time, and a short description of what happened. SIR starts after the initial incident form is completed. SIR is conducted in three steps.

1. Reporter selects the operations at the time of the incident
2. Reporter selects the scenario of what happened
3. Reporter assesses each barrier in that scenario

Each step is based on existing bowties and links data back to these bowties.

1) Reporter selects the operations at the time of the incident

In this step, the reporter chooses the operation he or she was doing at the time of the incident from a set of pre-built bowties. Behind the scenes, this will select a hazard and top event combination. The reporter does not necessarily need to know this, as knowledge of the bowtie methodology or how the bowtie is constructed is optional. Examples of hazards are working at height or lifting objects with a crane.

2) Reporter selects the scenario of what happened

From the selected bowtie in the previous step, the reporter chooses the relevant scenario. The relevant scenario starts on the left side, from a threat, all the way to the right to a consequence. It is optional to choose multiple threats or multiple (potential) consequences, as there might be multiple causes or multiple (potential) outcomes.

3) Reporter assesses each barrier in that scenario

The reporter is then taken through all the barriers that are in the selected bowtie scenario. Each barrier is assessed, and the primary cause is identified.

The reporter first decides on the barrier state and chooses one of the following options per barrier:

- This safety measure functioned correctly
- This safety measure failed
- This safety measure was not applicable in this scenario.

When the reporter indicates a barrier as 'failed', a question appears about why the barrier failed, i.e. the primary cause of the barrier failure. These questions can be customized and will differ per barrier.

As an example, the reporter selects the barrier: 'Restricted access to lifting area'. The primary cause can be all types of human error, i.e. slips, lapses and mistakes (Reason, 1990). Optional answers could be that the restricted area was not set up or the restricted area was not adhered to.

The barrier specific questions indicate the primary cause of the failure of the barrier. According to the Tripod Beta incident analysis methodology (Stichting Tripod Foundation, 2014), the primary cause is "An action or omission of a person, or group of people, that causes a barrier to fail."

Thus, by performing SIR analysis for every low severity incident, time can be saved for the follow up investigator and if not further investigated this data is now permanently captured onto the bowtie.

3 Discussion

In this section, using three purposes defined by Reason (1997), academical critiques of near miss reporting are examined and how they relate to the presented SIR design. Limitations of the design are mentioned.

3.1 Can SIR be an effective near miss management system?

Reason (1997) identifies three purposes of collecting and analyzing near miss data:

1. to gain qualitative insight into how (minor) errors can result in near misses or even major accidents

2. to gain enough data for statistically correct trend analyses to find out the combinations of factors leading to accidents

3. to remind organizations of the hazards challenging their systems and so "slow down the process of forgetting to be afraid" (Reason, 1997, p119), especially very safe organizations which have low occurrences of incidents

3.1.1. Qualitative insight – Does near miss reporting prevent larger accidents?

As we saw, using the iceberg model, studies in different industries (McKay, (2013)) found evidence in favor that near misses can hold predictive value for larger accidents. However, Reason (1991) critiques that information regarding near misses and accidents, only relates to the last steps of a long chain of causal events. The information is therefore not useful enough as it is both too noisy—it is difficult if not impossible to find the real root cause of the incident—and too late, as it encourages reactive rather than proactive safety management. Instead, Reason (as read in Van der Schaaf, 1992) proposes to use auditing with a focus only at the earliest events; management decisions. In response, Van der Schaaf (1992) agrees that any near miss reporting system by itself, would always be lacking considering the complexity and difficulty of incident analysis. This is where the bowtie methodology lends itself extremely well, as the barrier construct allows for both incident and audit data to be mapped onto itself. Thus, the limitations outlined by Reason above can be overcome through combining near miss results with audit results.

3.2.1 Trend analyses – How can trend analyses be ensured?

The presented SIR design tried to tackle two challenges in getting enough statistical data for effective trend analyses. The first challenge is getting enough data. By facilitating near miss reporting, SIR can quickly contain enough data, as: "Near misses are usually estimated to occur one or two orders of magnitude more frequently than actual accidents" (Van der Schaaf, 1992, p25).

The second challenge is standardizing the data, as incidents are often investigated in isolation. In many cases investigation reports of near misses or accidents end up as a collection of largely unstructured documents. They are often treated as single events, making it difficult to clearly identify trends over multiple events. As we saw, using the bowtie method, standardization becomes much easier through the barrier constructs. The barriers serve as a common denominator for more effective trend analyses. A shortcoming, though, is that bowties are usually of a much more generic nature, whilst incidents are much more specific scenarios. As such it is difficult to capture all possible events and some information might be lost in the process of switching from specific scenarios to generic ones.

3.3.1 Awareness - Does more information through near misses lead to a safer organization?

A challenge that organizations face when implementing or improving an incident reporting system, such as near miss reporting, is that there can be both 1) a lack of feedback on how the reported information is used and 2) the perceived difficulty of collecting enough data (Van Der Schaaf and Kanse, 2004). A lack of feedback in how the reports are handled can demotivate reporters to report again in the future, which can also result in more difficulty in gathering data. Therefore, SIR immediately feeds the reported information back onto the bowtie, for all to see. Moreover, according to Phimister (2003), a well-functioning system should also give feedback of the results and improvements of the near miss investigation to the reporter. In addition, these results should be communicated to as many other employees as appropriate in order to raise awareness levels. Thus, SIR can be set up to allow the reporter and related users to be given updates of the recommendations. Furthermore, to address the perceived difficulty of data collection, the reporting forms should be easy and convenient and not require lengthy analysis for every near miss reported. Again, using the bowtie methodology, SIR can save a lot of time and avoid lengthy analyses, which should stimulate employees to report more. A limitation here, however, is that SIR might be confusing at first for employees.

In line with the importance of feedback, Swain (as read in Van der Schaaf, 1992) suggests that just the act of observing and reporting a near miss can be considered employee engagement, which should result in higher levels of awareness. Raised levels of awareness can in turn lead to a safer organization. Indeed, Cooper (2000); Glendon and Stanton (2000) and Jones, Kirchsteiger and Bjerke, (1999), have found that near misses can boost safety culture. Additionally, near miss reporting can help show that employees are exhibiting learning, as it demonstrates how employees have recovered and prevented errors from developing into incidents (Van der Schaaf, 1992). Therefore, organizations can become safer not only by analysis of the near misses through SIR but also by the increased feedback that SIR provides, stimulating further reporting of near misses.

4 References (in order of appearance)

[1] Hollnagel, E. (2014). Safety-I and Safety-II: the past and future of safety management. CRC Press.

[2] ODN. (2010). Reaction to Buncefield explosion decision. Retrieved from https://www.youtube.com/watch?v=E_daM9tM4bs

[3] Smit, J. (2017). Advanced barrier management: Linking Bowtie Analysis, Barrier-Based Incident Analysis and Risk Based. Leidschendam, The Netherlands.

[4] Vincent C, Aylin P, Franklin BD, Holmes, A., Iskander, S., Jacklin, A. and Moorthy, K. Is health care getting safer?, BMJ , 2008, vol. 337 pg. a2426 https://doi.org/10.1136/bmj.a2426

[5] Clarke l. Learning from critical incidents, Adv Psychiatr Treat, 2008, vol. 14(pg. 460-8)

[6] Davies, HTO, NutleySM.Developing learning organisations in the new NHS, BMJ, 2000, vol. 320(pg. 998-1001)

[7] Braithwaite J, Westbrook MT, Robinson M, Michael, S., Pirone, C. and Robinson, P. Improving patient safety: the comparative views of patient-safety specialists, workforce staff and managers , BMJ Qual Saf , 2011 , vol. 20 (pg. 424 -31)

[8] Hyden C.(1987) The development of a method for traffic safety evaluation. The Swedish traffic conflicts technique. Bulletin 70, University of Lund.

[9] Schaaf, van der, T.W. (1992). Near miss reporting in the chemical process industry Eindhoven: Technische Universiteit Eindhoven DOI: 10.6100/IR384344

[10] Mckay, B. (2013). Measures of effect: Near miss reporting on construction site injuries.

[11] Pitblado, R., Potts, T., Fisher, M., & Greenfield, S. (2015). A method for barrier-based incident investigation. Process Safety Progress, 34(4), 328-334.

[12] Stichting Tripod Foundation. (2014). Guidance on using Tripod Beta in the investigation and analysis of incidents, accidents and business lesses. London: Energy Institute.

[13] The American Institute of Chemical Engineers; the Energy Institute. (2018). Bow ties in risk management. Hoboken: the American Institute of Chemical Engineers.

[14] Reason, J. (1990). *Human error*. Cambridge university press.

[15] Reason, J. (1997). Managing the risks of organizational accidents. Routledge.

[16] Sklet, S. (2006). Safety barriers: Definition, classification, and performance. *Journal of loss prevention in the process industries*, *19*(5), 494-506.

[17] Reason, J. (1990). *Human error*. Cambridge university press.

[18] Stichting Tripod Foundation. (2014). Guidance on using Tripod Beta in the investigation and analysis of incidents, accidents and business lesses. London: Energy Institute.

[19] Reason, J.T. (1991). Too little and too late: a commentary on accident andincident reporting systems. In: T.W. van der Schaaf, D.A. Lucas and A.R.Hale (eds). Near Miss Reporting as a Safety Tooi. Butterworth-Heinemann. Oxford.

[20] Van Der Schaaf, T., & Kanse, L. (2004). Biases in incident reporting databases: an empirical study in the chemical process industry. Safety Science, 42(1), 57-67.

[21] Cambraia, F. B., Saurin, T. A., & Formoso, C. T. (2010). Identification, analysis and dissemination of information on near misses: A case study in the construction industry. Safety Science, 48(1), 91-99.

[22] Phimister, J. R., Oktem, U., Kleindorfer, P. R., & Kunreuther, H. (2003). Near-miss incident management in the chemical process industry. Risk Analysis: An International Journal, 23(3), 445-459.

[23] Cooper, M.D., 2000. Towards a model of safety culture. Safety Science 36 (2), 111–136.

[24] Glendon, A.I., Stanton, N.A., 2000. Perspectives on safety culture. Safety Science 34(1–3), 193–214.

[25] Jones, S., Kirchsteiger, C., Bjerke, W., 1999. The importance of near miss reporting to further improve safety performance. Journal of Loss Prevention in the Process Industries 12 (1), 59–67.

[26] Swain, A.D. (1974). The Human Element in Systems Safety: a guide for modern management. InComtec Ltd., Camberley.

Development, Implementation and Use of a Process Safety Incident Identification and Capture System within a Utility Context

Brendan Bleasdale
Engineering Authority
Contact Energy
Wellington
New Zealand
brendan.bleasdale@contactenergy.co.nz

Prepared for Presentation at
American Institute of Chemical Engineers
2019 Spring Meeting and 15th Global Congress on Process Safety
New Orleans, LA
March 31 – April 3, 2019

Development, Implementation and Use of a Process Safety Incident Identification and Capture System within a Utility Context

Brendan Bleasdale
Engineering Authority
Contact Energy
Wellington
New Zealand
brendan.bleasdale@contactenergy.co.nz

Keywords: Incident reporting, incident categorization, KPIs, leading and lagging indicators

Abstract

This Paper sets out the background to Contact Energy's journey to enhancing process safety awareness and performance, looking at key milestones and success factors providing the context within which we adapted the approach set out in API754 to identify and classify process safety incidents for use within a utility context. The importance of good engagement and communication, including how Contact raised staff awareness of what constitutes a process safety incident, and how we engaged them to identify and report incidents are described. The Paper demonstrates that while being able to identify and report incidents is important, it's only part of what is required and that real improvement comes from collecting and using the right data to drive actionable insight, using technology to make process safety visible, and integrating process safety across the business. We outline the use of 'lagging' incident data, (including minor incident data) to provide insights and learnings on the strength of our process safety barriers, how we integrate insights into our Process Safety Management Framework, how we utilise a near-real-time dashboard to display 'leading' indicators of barrier strength, and how we report on process safety, including incident data, at various levels within the Company.

1 Background and Introduction

Contact Energy is a New Zealand based vertically integrated utility company, operating a number of physical assets which require effective process safety management. The physical assets cover a wide range of technologies and also vary in age from plant commissioned in the 1950s to recently commissioned power plants. The electricity generation portfolio includes:

- Geothermal plant, drawing on sub-surface geothermal steamfields accessed via 300+ in-service wells for steam extraction, fluid re-injection and monitoring. These wells vary in depth down to approximately 3500m and are connected to the generating units by approximately 115 km of above surface pipelines. Four of the generating units use pentane as the motive fluid in their Rankine Cycle turbines and are classed as 'Upper-Tier Major Hazard Facilities' subject to regulatory Safety Case acceptance

- Hydro plant with three dams classed as having 'High' Potential Impact
- Fossil fuel plant, including plant with high pressure gas compressors, hydrogen cooled generators, and high energy steam pipework systems. To support these plants the company, until late 2018, owned and maintained an underground gas storage facility (with a storage capacity of up to 18 PetaJoules of natural gas) classified as an 'Upper-Tier Petroleum Production Facility' and subject to regulatory Safety Case acceptance.

Until December 2018 Contact also owned and operated an LPG storage and distribution (bottled and reticulated) business with one facility designated an Upper-Tier Major Hazard Facility and five locations designated as Lower-Tier MHF facilities.

With such a wide range of process hazards and potential major accident scenarios, back in 2013 Contact Energy became concerned about the number of process safety events happening globally. While our industry had an excellent record in occupational safety, the increase in significant asset failures worldwide was a leading indicator for deteriorating process safety and pointed to the need for enhanced and systematic understanding within the Company of our own performance in this area. There were other drivers for process safety, including evolving international awareness of, and guidance around process safety, legislative changes and stronger business imperatives.

The process safety improvement journey progressed through a number of stages:

1. Establishing a system for identifying and recording process safety incidents
2. Creating a common understanding of what was needed to ensure the integrity of our operations and to allow us to systematically answer three key questions.
    - Do we understand what can go wrong in our facilities?
    - Do we know what our systems are for preventing this?
    - Do we have information to assure our systems are working effectively?
3. To provide a structure for assessing and monitoring the performance of our risk control areas and processes we produced a Process Safety Management Framework (adapted from a framework initially developed by Scottish Power and based on multiple inputs, including the Energy Institute framework[3]).
4. Visualizing our Major Accident Hazards using bow-tie methodology
5. Establishing an Operational Safety Improvement Project to close gaps identified in our process safety management culture, leadership and processes. This Project focused on our people and on their role in understanding and enhancing process safety.
6. Provision of high quality, easily accessible information and insights on our performance through a near-real-time dashboard which brought together lagging data (from process safety incidents) and leading indicators (such as the effectiveness of our Management of Change process) and displayed these against the key risk control areas within our process safety management framework.
7. Embedding the output of the earlier stages into a culture of continuous improvement and process safety awareness.

2 Identification, Classification and Recording of Process Safety Incidents

Identification, classification and recording of Process Safety incidents was the first step on the journey and was intended to provide systematic information on our performance in this area. Incident identification was also fundamentally intended as a mechanism to raise awareness of the process safety controls that we had in place (many of which at that time were not recognized by staff as process safety related) and the need to identify and report where these controls had failed or been weakened.

The system to be used had to be simple to understand and apply, relevant to our business context by covering our wide range of potential accident scenarios and controls, allow easy analysis of the types of incident that were being captured, and have a basis in and consistency with international practice.

API Recommended Practice 754 "Process Safety Performance Indicators for the Refining and Petrochemical Industries" [1] was identified as a sound foundation to a tiered approach to incident identification and classification. Other guidance, such as that issued by the UK Energy Institute [2] and the UK HSE [3] were also taken into consideration.

2.1 Adaptation of API754 for use within Contact Energy

In its introduction, API754 states "*This RP was developed for the refining and petrochemical industries but may also be applicable to other industries with operating systems and processes where loss of containment has the potential to cause harm*" and as such is strongly focused on loss of containment of flammable and toxic material. API 754 utilizes a 4-tier approach to incident identification with:

- Tiers 1 & 2 definitions being linked to loss of primary containment of material leading to one or more of the specified consequences for that Tier, or loss of greater than threshold values quantities of hazardous materials. Loss of containment of non-toxic or non-flammable materials (such as steam, hot water, compressed air…) are only included in the definition where they result in one or more of the consequences specified for that Tier.
- Tier 3 incidents are challenges to Safety Systems and,
- Tier 4 incidents relating to "Operating system and management system performance".

Strict application of the API definitions would, within the Contact Energy context, have excluded some scenarios related to process safety, in particular those not related to loss of toxic or flammable material.

2.1.1 Adaptation of Tier 1 and Tier 2 definitions

Use of standard API754 definitions for Tiers 1 and 2 within Contact would have excluded scenarios related to loss of containment of:

- mechanical energy (for example catastrophic failure of a turbine due to overspeed),
- electrical energy (for example significant arc-flash events) or,
- ambient temperature water (for example major failure of hydro dam containment)

A decision was therefore made that we would include loss of containment of mechanical, electrical and bulk water energy sources as Tier 1 or Tier 2 incidents if they resulted in the specified consequences for that Tier.

2.1.2 Adaptation of Tier 3 and Tier 4 definitions

- Tier 3 indicators are intended for internal company use, with API754 providing guidance on example indicators. This means there is a lot of flexibility in how Tier 3 indicators can be used, but direct application of the guidance on the examples relating to demands on Safety Systems means only demands on Safety Systems with related loss of primary containment protection are counted. Contact Energy therefore decided to classify activation of any process safety related protection as a Tier 3 Incident, irrespective of the process safety hazard they are providing protection against.

- Tier 4 indicators under API754 are very much leading indicators and within Contact are covered in some detail within our Process Safety Dashboard KPIs. Use of Tier 4 Process Safety Incident indicators was therefore discontinued a short time after the introduction of our categorization process in July 2012. Tier 3 Process Safety Incidents are, however, separated into sub-categories to aid in trending and identification of the nature of the challenges to process safety as shown in Tables 1 and 2 below.

Table 1. Sub-categorization of Tier 3 Process Safety Incidents

Tier	Descriptor
3.1	**Plant operating outside its safe design operating limits.**
3.2	**Safety Critical equipment failing to operate on demand or found compromised* during inspection or test** (*compromised means that the inspection or test indicates that if called upon the equipment would not operate with the required level of reliability).
3.3	**Activation of a process safety protective device.** Activation is only counted as a process safety incident if the equipment is responding to a genuine exceedance of the activation parameter. Spurious trips caused for example by a faulty sensor sending an incorrect trip signal are therefore excluded. Note: this criterion is the one that generally requires the greatest guidance on interpretation.
3.4	**Loss of primary containment** where the loss of material is less than the threshold for a Tier 2 incident.
3.5	**Error or failure in applying a process safety management system.** See sub-categorization in Table 2.

Table 2. Sub-categorization of Tier 3.5 Process Safety Incidents

Tier	Descriptor
3.5	**Error or failure in applying a process safety management system.**
3.5.1	Management of Change Process error or failure.
3.5.2	Incorrect application of the Permit (tag out/ lock out) process.
3.5.3	Delayed/ missed maintenance or test of safety critical equipment without a formal risk assessment in place.
3.5.4	Override or bypass of safety critical equipment without a documented risk assessment in place.
3.5.5	Prohibited item taken into hazard area.
3.5.6	Inadequate, or failure to follow, a safety critical procedure.
3.5.7	Failure to meet competency requirements for work on safety critical equipment.
3.5.8	Cyber threat to a safety critical control system.
3.5.9	Other process safety management system requirements not met.

– Contact has implemented a process for escalating some Tier 3 incidents to Tier 2 even in the absence of the specified Tier 2 consequences. This escalation occurs if the 'near-miss is deemed to be very significant and a Major Accident has only been avoided by chance, rather than engineered protection. There are two clear scenarios where this approach would be applied.

The first relates to safety critical equipment being found compromised on inspection or test where no other lines of protection are in place and a credible initiating event would have resulted in a major accident. For example, on one plant, routine testing identified a common mode failure affecting actuation of the steam admission valves on a steam turbine. This fault meant that neither the control nor emergency stop valves would have fail-safe closed on a trip signal. A loss of grid-connection fault would have meant steam continuing to be supplied to a turbine with no connected load, with resulting overspeed and catastrophic failure potential.

The second scenario is similar but relates to plant operating outside its safe design parameters where escalation to Tier 2 is applied if there were no barriers in place to prevent a credible plant disturbance leading to a major accident. For example, a major steam pipeline had a build-up of a significant quantity of condensate water within the pipeline, outwith the operating parameters for the system. Whilst that water was present, any change in system flows or pressure on either side of the condensate would have moved the water into the steam flow and caused rapid condensation induced water hammer (CIWH) pressure transients with a very high risk of pipeline rupture. CIWH risks are described in many publications including US DOE report 'Condensation Induced water hammer safety' [4]

2.2 *Recording of Process Safety Incidents*

Contact Energy's existing 'Risk Manager' incident recording software was modified to include selection of 'Process Safety Incident Yes/No' and 'Tier' as an additional characteristic available to be attached to an incident, with those entering incidents into the system encouraged to consider any potential process safety implications. To assist front line staff in this, a simple flow chart was produced which guided the user through the decision making process. A copy of the current flow-chart is attached in Appendix A. Because of the prevalence of LPG, natural gas, and pentane use within the Company, these are specifically captured in the flow chart. If other materials are involved, the user is pointed to the API754 document for threshold quantities.

One of the key considerations when developing the flowchart and associated guidance was to keep things as simple as possible and recognize that in some situations this could result in lack of consistency in identification or categorization. This wasn't seen as an issue as the benefits of engaging staff through a simple-to-use process was seen to outweigh any disbenefits over consistency.

There were, and still are, two main areas where guidance on interpretation is most often sought. The first is when more than one category could be applied to an incident. For example, inadequate application of our Permit (lock-out, tag out) system (Tier 3.5.2) may result in loss of primary containment of pentane from a system opened up for maintenance (Tier 3.4 if loss is below the Tier 2 threshold). Two options are then available. Firstly, an incident could be assigned multiple categories and, in the example above, it would be both 3.4 and 3.5.2. This maximizes our understanding of the incident but would skew our view of the total number of process safety incidents occurring so currently only one category is assigned to an incident with the 'most significant' being selected. A judgement then needs to be made on the relative severity of each of the challenges. In the example above, the system failing in our permit process would dominate unless there were specific aspects of the nature of the release which made that more significant.

The other area often requiring interpretation is 'Activation of a process safety protective device'. For example, on a power plant a steam turbine trip may be initiated by a number of parameters, some of which will be safety related (such as overspeed protection) and others are to prevent commercial damage to the machine (for example water carry-over protection). So a steam turbine trip may or may not be a process safety incident depending on the purpose of the protection which operates. Discussion with site staff on this issue is always helpful as it raises the awareness of process safety and the systems and controls which keep the plant safe. Activation of a process safety protective device is not captured if the activation is 'spurious' and directly activated, for example, by a faulty speed sensor. Less obvious however is where a faulty sensor results in a plant transient which causes, say, a safety relief valve to lift. In this case whilst the initiator was a faulty instrument the protective device was responding to an actual plant parameter exceeding the point at which the protection operates and is a genuine process safety incident.

All incidents within the Company are reviewed by corporate staff on a weekly basis and this review was extended to include process safety expertise review to look at the accuracy of capture of process safety incidents. In the first months after introduction of process safety incidents, about 20% were being identified by front-line staff when entering the incident with the remainder being picked up as part of the weekly review. In those circumstances there would be a coaching conversation with the appropriate business unit to take them through the process safety aspects of the incident and the basis of its categorization. Updates to the system would then be made by the initiator and not by the reviewer. Within about 12 months the ratios had switched with approximately 80% of incidents being identified and classified by the initiator and 20% following review. At our current state of maturity, almost all incidents are correctly categorized at the time of input into the system.

3 Raising staff awareness and engagement

Implementation of process safety incident reporting was only one part of an overall program to raise awareness of process safety and the controls and barriers that need to be in place. We engaged our people on the importance of always having a heightened sense of awareness of safety and of communicating effectively, particularly when involved in hazardous operations. We emphasized the important role they play in developing our process safety culture by ensuring they had the competencies needed to do the job, with a focus on skills which are safety critical. We encouraged our people to understand the processes, procedures and instructions that apply to the activities they carry out and to question anything that didn't seem right or was hard to follow. Establishment of a Process Safety Management Framework and using front-line staff to identify process safety barriers and controls and capture these within bow-tie diagrams were key activities which helped the understanding of when there were challenges to barriers which needed to be captured as process safety incidents.

Table 3. Contact Energy Process Safety Management Framework

4 Integration of Process Safety Leading and Lagging Indicators

The process safety incident identification process provides a series of lagging indicators but, to get the full picture on process safety performance, it is necessary to integrate leading and lagging indicators in a systematic way which gives visibility of where:

- leading indicators should have provided advance warning of failures captured as lagging indicators, or
- leading indicators are positive but associated lagging indicators are negative (suggesting inadequate or non-effective leading indicators).

To achieve this, lead KPIs were developed to show the strength of the controls within each of the Risk Control areas (the 7 vertical columns) within the process safety management framework. A process safety dashboard was then developed to display these in near-real-time with 'donut' graphics being used to show lead and lag indicators together. To achieve this, further characteristics needed to be added to each process safety incident to link the incident to the appropriate box on the process safety management framework. Activation of a fire suppression system for example would be allocated to the 'Fire Systems' risk control system within the Safety Critical Elements column on the matrix and would result in the centre of that site's donut graphic being shaded light grey. Hovering over this area would then allow the details of the incident to be displayed.

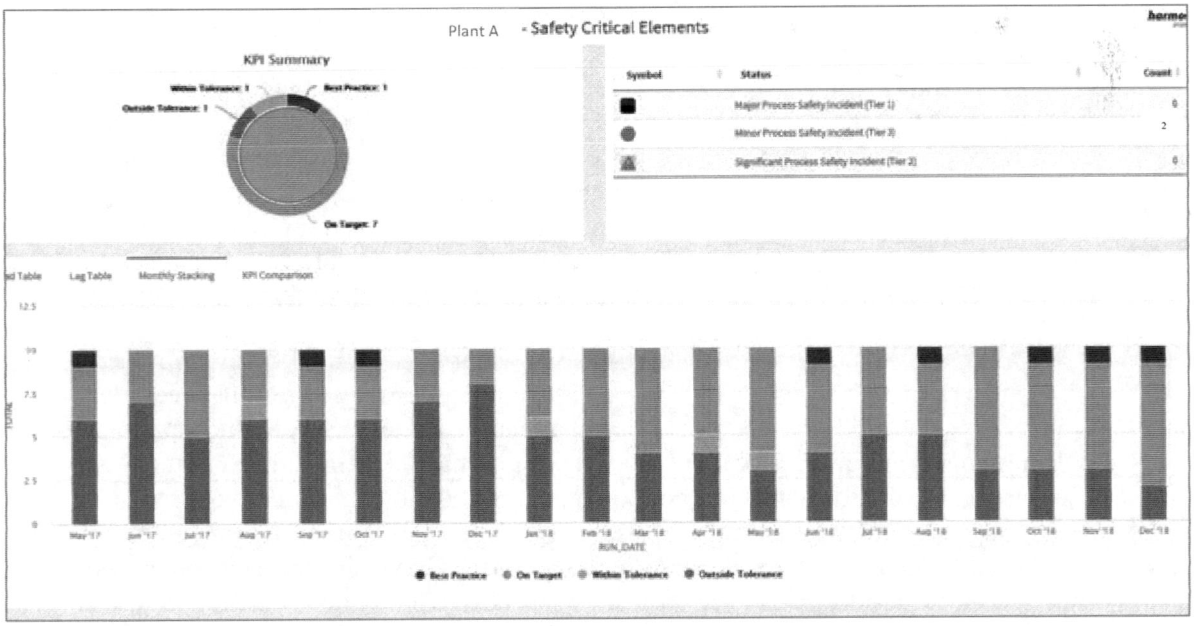

Figure 1. Example Process Safety Dashboard screenshot

5 Trending, follow-up and Governance of Process Safety Incident data

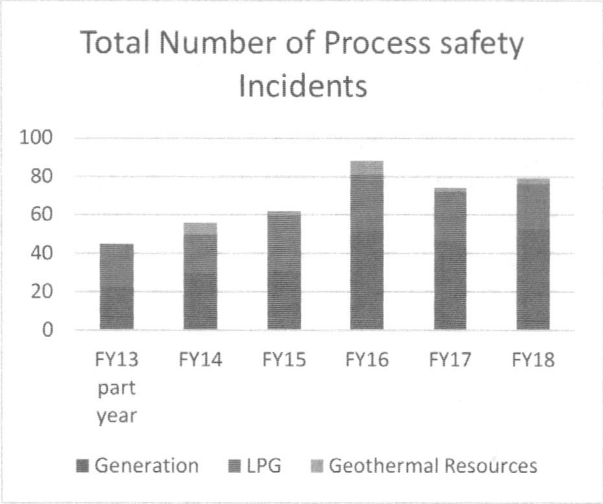

Figure 2. Trend in number of incidents reported

Having established a systematic process for identifying and recording process safety incidents it was important that maximum use was made of the insights this information can provide. The data can be cut and trended in many different ways such as by number of incidents over time, by technology (e.g. LPG), by location/plant, or by type of incident (e.g. Permit lock out/tag out). Some examples of these are shown below. Reporting commenced in July 2012 with the first years showing an increasing trend in number of incidents which was attributed to the ongoing improvements in process safety awareness as a result of our operational safety improvement program. There is some evidence of a slight downward trend post July 2017 which could be the result of improved strength of our barriers, or could be a fall-off in reporting. The latter is thought to be unlikely given the strength of process safety awareness and culture being seen in the business.

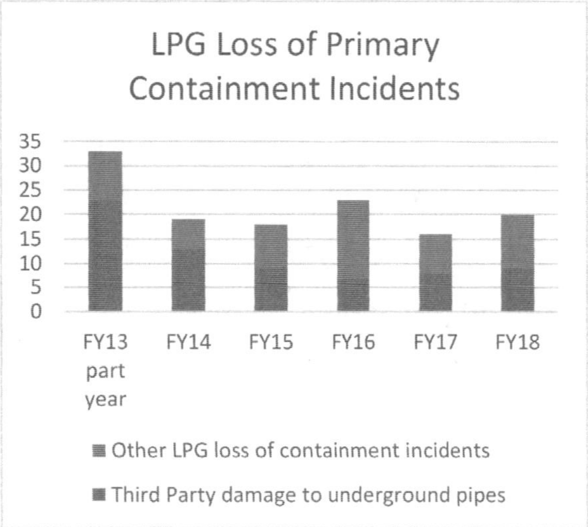

Figure 3. LPG Loss of primary containment incidents

Figure 3 shows the large number of loss of primary containment incidents in the 2012/ 2013 financial year caused by third parties damaging buried LPG domestic distribution pipework. Many of these were in Christchurch and were related to the very high volume of post-earthquake reconstruction work. Contact's only two Tier 1 incidents, and the majority of the Company Tier 2 incidents, were as a result of third-party damage. A concerted campaign, including billboard and radio advertising, on the need to 'check before you dig' resulted in a significant decrease in this type of event.

The pie-charts in Figure 4 below provided insights into areas requiring enhanced focus and attention. As an example, the data related to 'Safety Rules non-compliance' (lock-out tag-out) is automatically transferred to the Safety Rules Assurance team who then analyze the root causes of the incidents and adjust the process or training as required. Where there has been activation of a process safety device this can be viewed as the plant simply operating as designed. However, these events provide an opportunity, not just to record that the protection has been called upon, but also to follow up on the message of 'leaving nothing to chance' and to follow up on the maintenance records on protective devices (to confirm that all routine maintenance, inspection and tests are up to date) and that it was systematic maintenance rather than 'chance' which underpinned its successful operation.

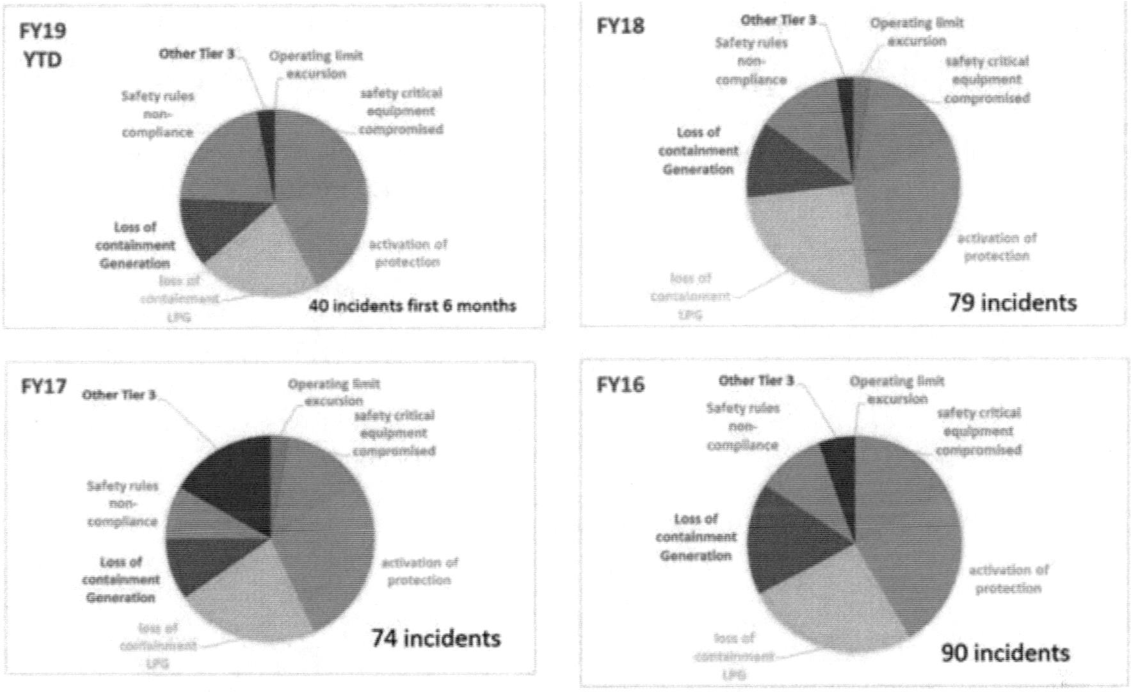

Figure 4. Comparison of relative contributions to the total number of Tier 3 Incidents

5.1 Investigation of Process Safety Incidents

Investigation of Process Safety Incidents follows the same 'Learning Team' approach used for all incidents in the Company. Whilst not eroding people being accountable for their actions, the Learning Team approach explores, with all those involved, the factors resulting in the adverse event. This approach has been in place since 2015 and is very successful in moving from a potential adversarial and blame seeking approach, to a facilitated approach which calls on the experiences of all involved to really draw out the learnings about where our defenses are weak and how can we fail safely. Worksafe, has published the Contact approach in a good practice case study on their website [5] together with Contact's guidelines for running successful learning teams.

5.2 Oversight and Governance

Oversight and governance take place at all levels within the organization. The process is overseen by the Company Engineering Authority supported by site Technical Authorities who provide close-in guidance and coaching on the processes. Formal review of the incident data together with the associated leading indicators is carried out as follows:

Table 4. Process Safety Leading and Lagging Indicator Governance Reviews

FREQUENCY	OVERSIGHT BODY	ATTENDEES
Weekly	Site Management Meeting	Site Head of Generation, Departmental Heads, HSE Adviser
Monthly	Site Management Meeting	Site Head of Generation, Departmental Heads, HSE Adviser
Monthly	Operations Leadership Team	General Manager Operations, Site Heads of Generation, Engineering Authority, Heads of Technology, People, Finance
Monthly	Generation & Development Leadership Team	Chief Generation & Development Officer, Business Unit General Managers, Engineering Authority, Head of People
Every 4 months	Executive HSE Meeting	Chief Executive, Company Chief Officers, Engineering Authority, HSE Adviser, Head of Business assurance & Audit, General Managers Operations and Geothermal Resources
Every 4 months	Tri-annual Risk Review for each location	GM Operations, Site Head of Generation and site Departmental Managers, Engineering Authority, Risk Manager
Every 4 months	Board Health & Safety Committee	Company Chairman, Board members, Chief Executive, Chief Officers, General Manager Operations, Engineering Authority

6 Conclusions

Simple changes to interpretation of API751 have allowed Contact Energy to deploy this approach to identification and capture of Process Safety Incidents within a Utility context. Deployment of the incident reporting process has been a key part of the overall drive to raise awareness of process safety, the barriers that keep us safe, and the importance of recognizing any challenges, failures or weaknesses in the barriers. Integration of the incident 'lag' indicators with our online dashboard of lead indicators, within a process safety management framework, has allowed visibility of the overall picture with respect to process safety and provided a simple process for identifying trends with respect to barriers or groups of barriers. Staff engagement and buy-in has been an essential element in making the process work and has relied on making the identification, reporting and trending activities as simple as possible.

7 References

[1] Process Safety Performance Indicators for the Refining and Petrochemical Industries. ANSI/API Recommended Practice 754, April 2016

[2] High level framework for process safety management.1st Edition December 2010. Energy Institute, London. ISBN 978 0 85293 584 2

[3] Developing process safety indicators. A step-by-step guide for chemical and major hazard industries. HSG254. UK Health and Safety Executive. ISBN 978 0 7176 6180 0

[4] Gintner, M.A. USDOE Office of Environmental Restoration and Waste Management, Washington, DC (US) report HNF-SA-3179-FP March 1997. Link via https://www.osti.gov/biblio/16909/ accessed January 2019

[5] "Worker engagement and participation case studies – Involving everyone in learning reaps benefits (Worksafe New Zealand)" Sept 2017, https://worksafe.govt.nz/the-toolshed/case-studies/wepr-case-studies/involving-everyone-in-learning-reaps-benefits/ accessed January 2019

Process Safety Incident Categorisation

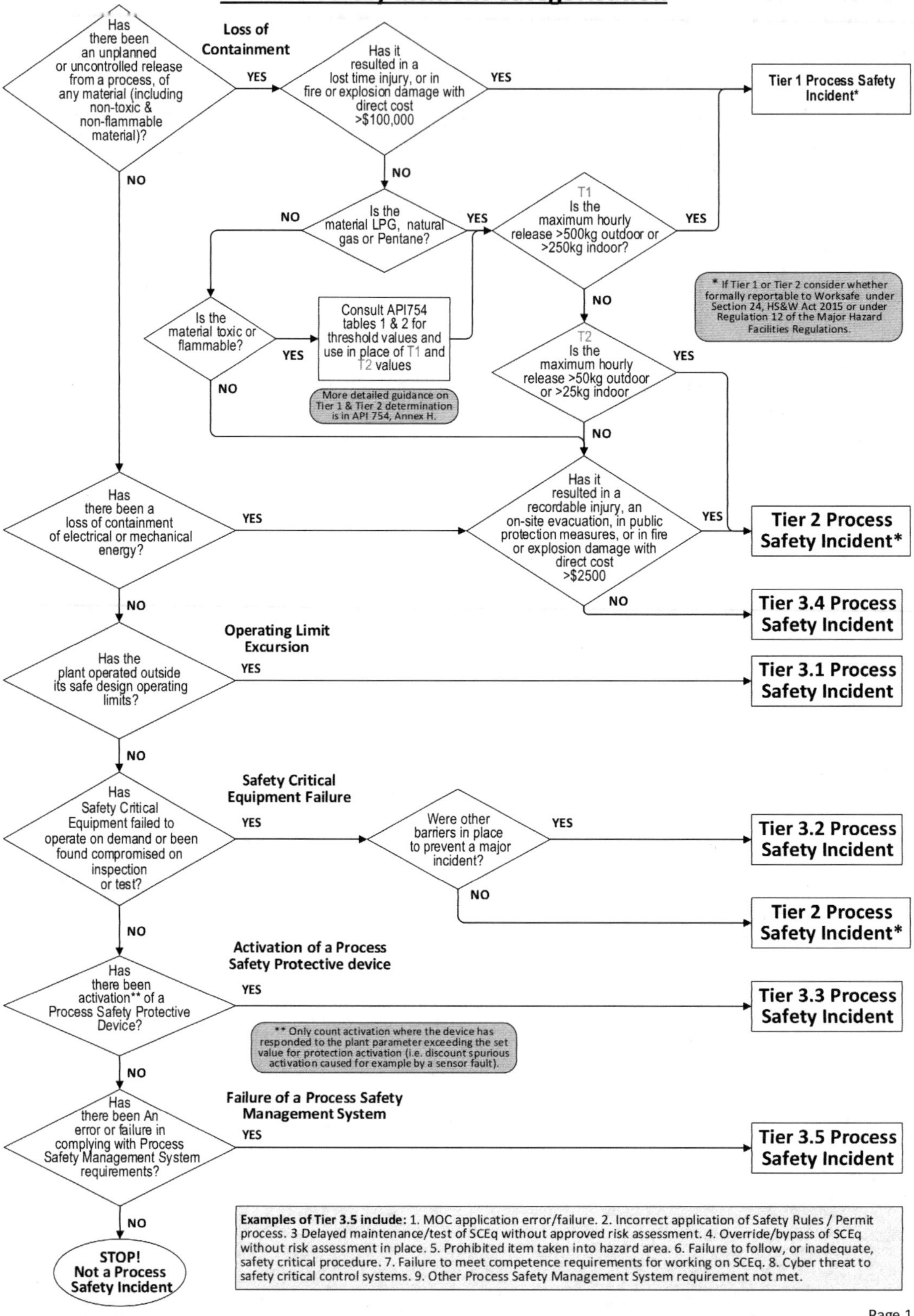

When Emergency Response Goes Fatally Wrong: Lessons Learned from Investigating Two Major Incidents

Scott Davis
Gexcon US
4833 Rugby Avenue Suite 100
Bethesda, MD
sgdavis@gexcon.com

Prepared for Presentation at
American Institute of Chemical Engineers
2019 Spring Meeting and 15th Global Congress on Process Safety
New Orleans, LA
March 31 – April 3, 2019

When Emergency Response Goes Fatally Wrong: Lessons Learned from Investigating Two Major Incidents

Scott Davis
Gexcon US
4833 Rugby Avenue Suite 100
Bethesda, MD
sgdavis@gexcon.com

Keywords: emergency response planning, lessons learned, case studies, major explosion incidents

Abstract

Emergency response plans are an essential, yet oftentimes overlooked, layer of protection for facilities where all other safeguards failed to prevent an incident. While catastrophic accidents such as large releases of chemicals, fires or explosions are devastating for the process industry, experience investigating numerous incidents has shown that a lack of an effective emergency response plan can lead to an unnecessary and tragic escalation of the initial incident. More specifically, investigation of two recent incidents: (1) 2013 West Texas Explosion and (2) 2015 explosion on the FPSO Cidade São Mateus; will demonstrate how fatal consequences resulted from the lack of emergency planning in the first case study and from deficiencies in the emergency response plan or implementation of the plan in the second case study. More specifically, the FPSO explosion resulted in 9 fatalities, eight of which were responding to a leak of a light-end hydrocarbon, and the West Texas explosion that resulted 15 fatalities, most of whom were responding to the fire that preceding the devastating explosion. Lessons will show that the fatal consequences could have been avoided with the proper emergency planning. In addition, while emergency response plans consider "maximum credible" scenarios, past events have shown that low probability high consequences should also be at least considered.

1 Introduction

Emergency response plans are an essential, yet oftentimes overlooked, layer of protection for facilities where all other layers of protection failed to prevent an incident. An emergency plan specifies procedures for handling sudden or unexpected situations, where the ultimate objectives are to be prepared to: (1) prevent fatalities and injuries; (2) reduce damage to buildings, stock, and equipment; (3) protect the environment and the community; and (4) accelerate the resumption of normal operations [1]. Developing the emergency response plan consists of five major steps (see Figure 1). The first step is to assess the vulnerability of the site and determine what hazards pose a threat to the installation. Such an exercise will help determine not only what "credible" events can occur at a given facility, but also what catastrophic or major events can occur. It should be noted that major emergencies are generally rare events.

Next, it is necessary to develop your emergency response plan and determine what means are currently available to execute the plan and what is still needed. At the planning stage, it is important that several groups be asked to participate, such as: subject matter experts and process engineers; the health and safety group; and even appropriate municipal officials should also be consulted since control may be exercised by the local government in major emergencies and additional resources may be required. Once the emergency response plan is developed, the necessary facilities and equipment need to not only be provided but tested and maintained. Communication, training and periodic drills will ensure adequate performance if the plan must be carried out. Lastly, there should be a periodic review of the emergency response plan.

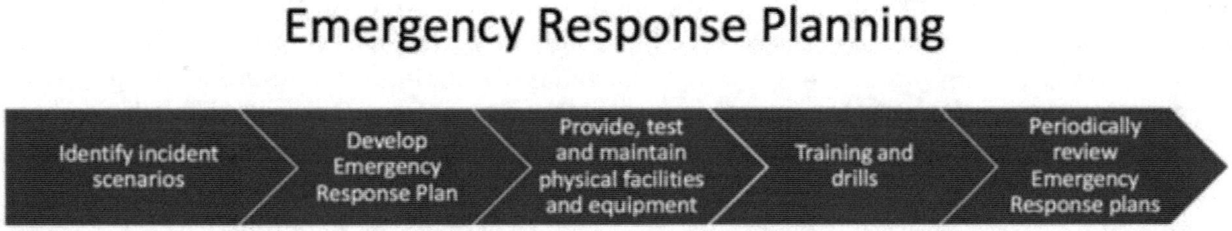

Figure 1: Steps of Emergency Response Planning

While the major benefit is to provide guidance during an emergency, developing the plan has other advantages. For example, previously unrecognized hazardous conditions, which would aggravate an emergency situation, could be discovered during the emergency planning and means put in place to mitigate or eliminate these hazards. The planning process may also identify deficiencies, such as the lack of resources (equipment, trained personnel, supplies), which can be easily corrected before an emergency occurs.

While catastrophic accidents such as large releases of chemicals, fires or explosions are devastating for the process industry, experience investigating numerous incidents has shown that a lack of an effective emergency response plan can lead to an unnecessary and tragic escalation of the incident. Chaos can quickly occur during an emergency if there are lack of resources and trained personnel coupled with an urgent need for rapid decisions and a very short period of time. The stress of the situation can lead to poor judgment resulting in severe losses, and normal channels of authority and communication cannot be relied upon to function routinely. Therefore, it is imperative to eliminate the "it can't happen here" mentality, as emergencies can occur and preplanning is necessary.

The importance of emergency planning will be discussed within the context of two recent incidents: (1) 2013 West Texas Explosion and (2) 2015 explosion on the FPSO Cidade São Mateus. These two case studies will demonstrate how the lack of emergency planning resulted in devastating consequences in the first case study and from deficiencies in the emergency response plan or implementation of the plan in the second case study.

2 Background of Incidents

2013 West Texas Explosion

On April 17th, 2013, an explosion occurred at the West Fertilizer Company storage and distribution facility in West, Texas. The explosion at West Fertilizer resulted from an intense fire in the seed and fertilizer area of the facility that led to the detonation of approximately 30 tons of ammonium nitrate stored inside a wooden receiving bin. The explosion occurred while emergency services personnel were responding to a fire at the facility. Fifteen people were killed, more than 250 were injured, and numerous buildings were damaged or destroyed.

The seed and fertilizer building (Figure 2) was built in 1958 and stored fertilizer and seed, which were distributed to farmers. Figure 3 shows the layout of the seed and fertilizer building, which included an ammonium nitrate (AN) storage bin neighboring the seed room, as well as an empty bin used for ammonium nitrate storage, and bins containing ammonium nitrate, K-Mag (fertilizer mix of potassium, magnesium and sulfur), Potash, ammonium sulfate, DAP (diammonium phosphate), and a bin for potassium chloride along the west wall. This building was capable of receiving rail shipments directly from the rail car by means of a rail spur junction and an underground conveyor belt and bucket elevator; the bucket elevator dumped directly into the northern AN storage bin, or onto a conveyor belt feeding the bins along the west wall.

Figure 2: West Fertilizer Facility – viewed from the west (upper) and east (lower)

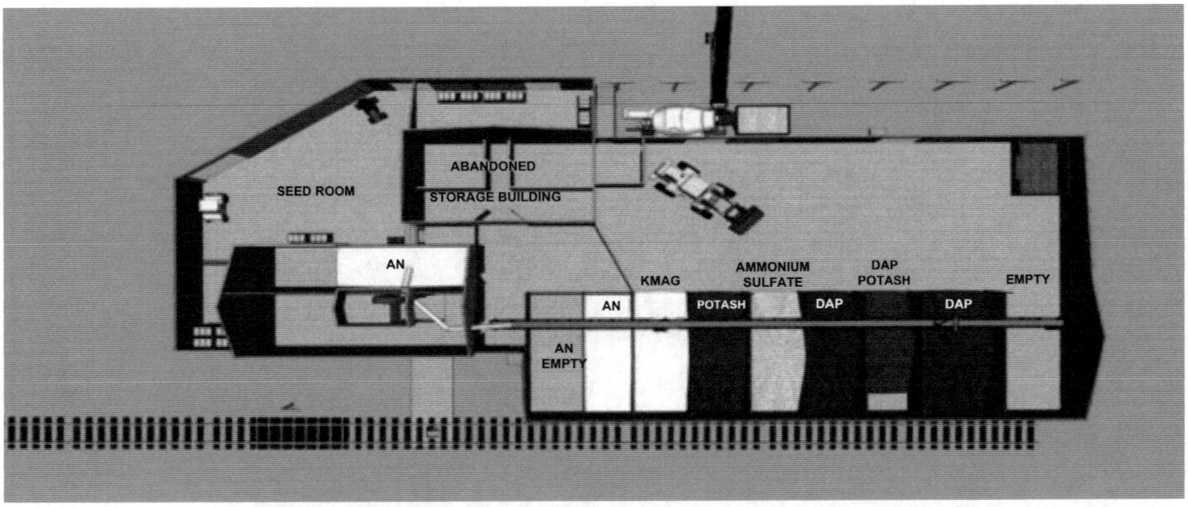

Figure 3: Layout of the Seed and Fertilizer Building.

On April 17th, 2013, at approximately 7:29 pm, a fire was reported, and fire fighters responded to the fertilizer plant [1]. All operations were already stopped for the day and employees had left the plant at about 5:30-6:00 pm. GexCon obtained several videos that recorded the progression of the initial fire [3]. Figure 4 shows the initial fire as viewed from the east and west; note that the highest point on the building represents the location of the bucket elevator by the northern AN storage bin. The image shows that the fire is isolated to the northern section of the building, which includes the seed room and portions of the northern AN storage bin containing about 30 tons of ammonium nitrate.

At approximately 7:50 pm, a massive explosion occurred at the fertilizer plant. Figure 5 shows images from a video taken from the high school parking lot to the southeast and Figure 6 the ensuing blast damage and crater from the detonation. Based on the videos from the day of the incident, there were actually two explosions: one smaller blast followed by a larger blast roughly 300 msec later.

Figure 4: Close up of the Seed and Fertilizer Building during fire as viewed from east (left image) and viewed from the west (right image)

Figure 5 – View of the explosion (viewed from southeast)

Fifteen people were killed as a result of the explosion, which included voluntary firefighters, and over 250 people were injured [3]. Damage was extensive throughout the neighboring community, which included homes, schools, an apartment building and a nursing home.

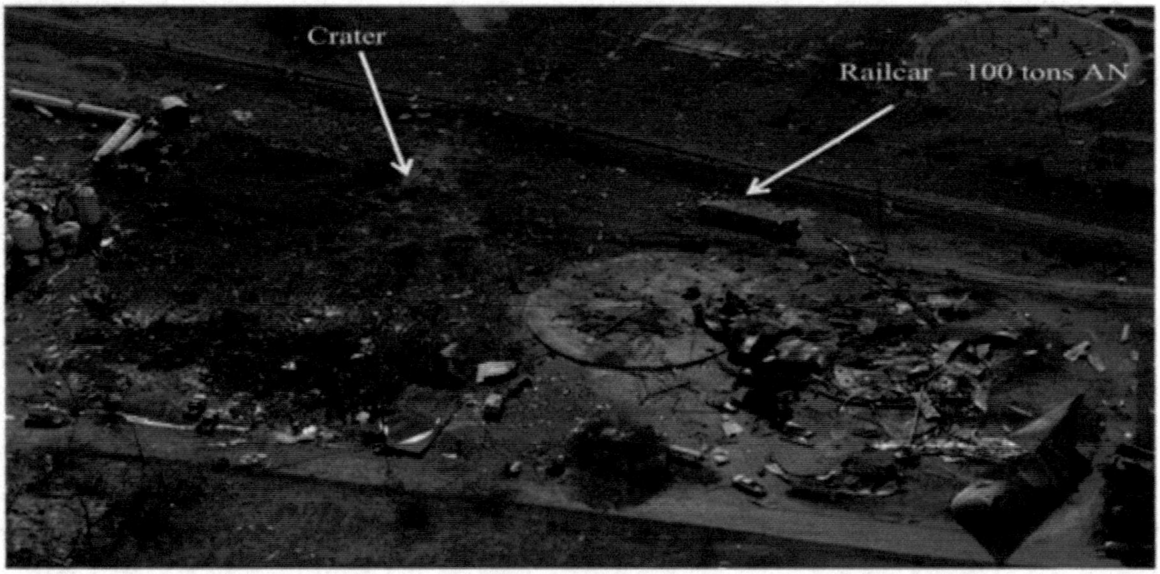

Figure 6: Blast damage and resulting crater

2015 explosion on the FPSO Cidade São Mateus

Due to the confidentiality of matters in this case, all references are made to two publicly available reports regarding the incident [5], [6]. The Cidade São Mateus was a Floating, Production, Storage and Offloading of Oil and Gas Export (FPSO), with a capacity to produce 4,000m³/day of oil and 10,000,000m³/day of gas as shown in Figure 7.

Figure 7: 2015 FPSO Cidade São Mateus.

On February 11, 2015 there was a leak of condensate (light-end hydrocarbon liquid) discovered in the pump room at a spade blind flange connection. Figure 8 shows that the pump room is situated just forward of the engine room and aft of the storage tanks. The layout and levels of the pump room is given in the left image of Figure 9, while the leaking spade blind flange is in the right image.

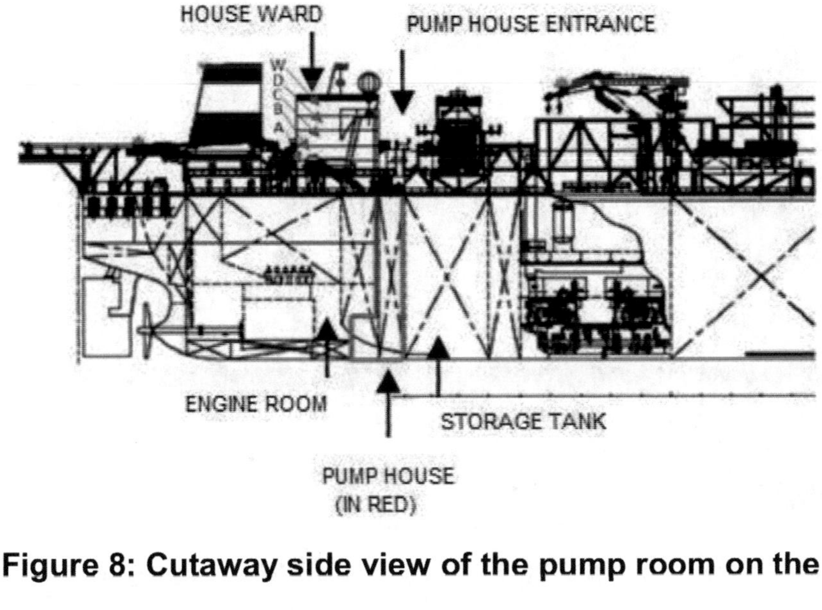

Figure 8: Cutaway side view of the pump room on the FPSO Cidade São Mateus.

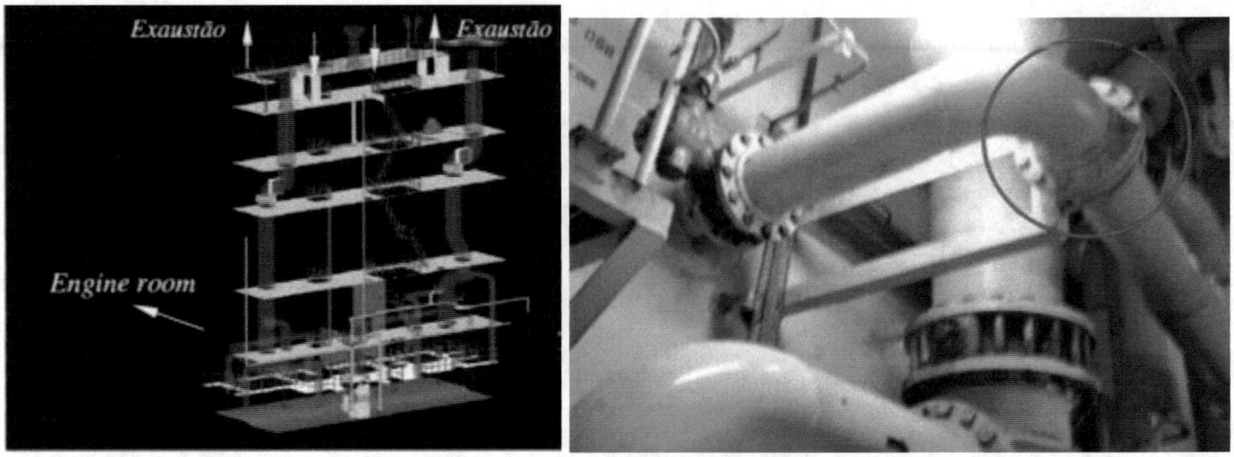

Figure 9: Model of the pump room on the FPSO Cidade São Mateus (left) and spade blind flange (right).

Just prior to the leak, the operations crew was attempting to drain condensate center cargo tank 6 (6C) into an alternate tank and they used an alternate pump alignment (stripping pump). However, the valve on discharge side of stripping pump was closed and they were pumping against a spade blind that was inadequate for such pressures, which ultimately failed the blind and seals. The gas detectors in the pump room began to alarm, and ultimately all three were alarming.

The ventilation was interlocked to shutdown per "confirmed gas" in the pump room. Emergency response teams entered the pump room to attempt clean up and repair the flange. There were three separate entries, and on during the third entry the explosion occurred. Figure 10 through Figure 12 show the level of blast damage to the pump room. Nine people were killed, of which eight were first responders.

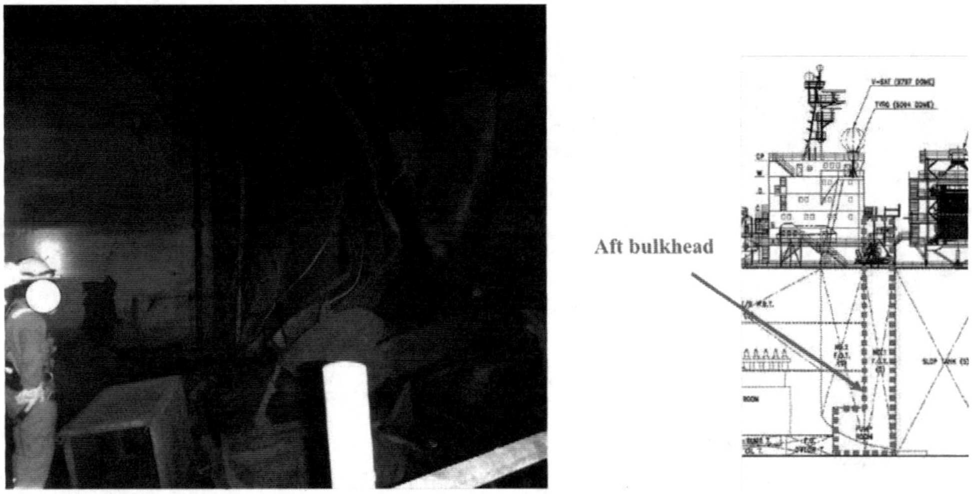

Figure 10: Damage to level above the pumps – complete bulkhead failure.

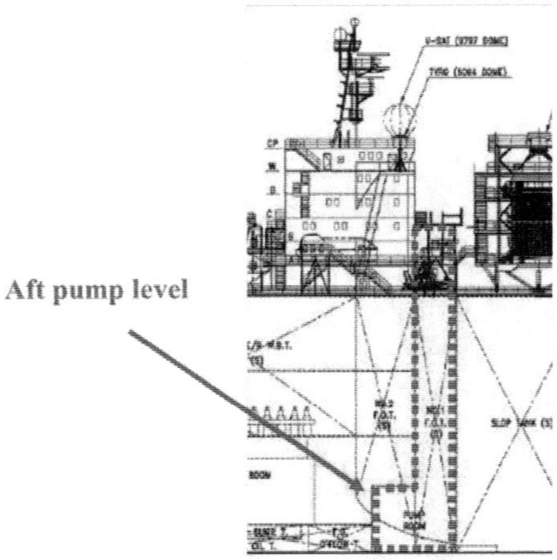

Figure 11: Damage to the engine room.

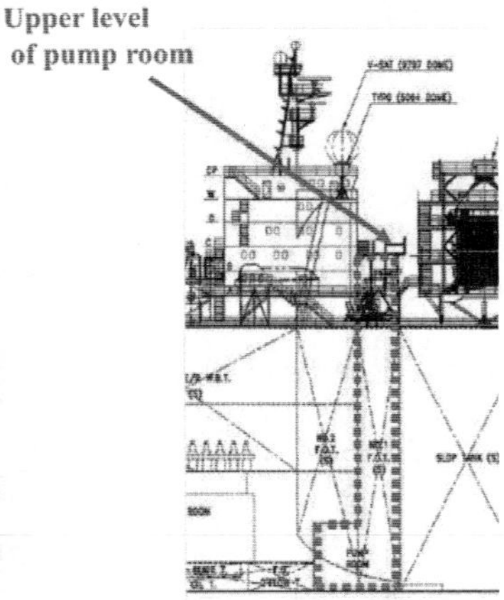

Figure 12: Damage to upper level of pump room.

3 Emergency Response

Response during the 2013 West Texas Explosion

As the fire occurred 20 minutes prior to the explosion, fire personnel were deployed to scene. Of the 15 people killed, 12 were first responders (8 volunteer firefighters and emergency services), and the other individuals were good Samaritans who supported the emergency response at the fertilizer plant. One of the volunteer fire fighters, who died in the blast, was employee of West Fertilizer. In fact, he was the plant foreman and knew what was stored in the facility. The emergency response is outlined in Figure 13.

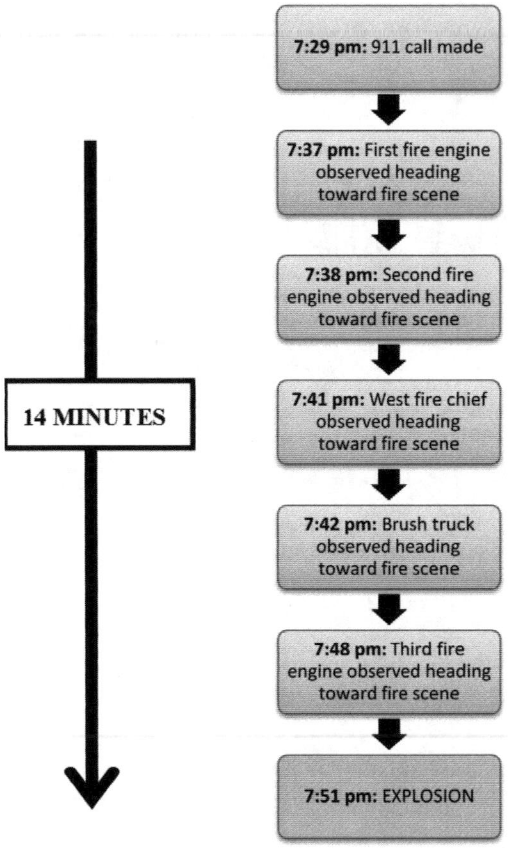

Figure 13: Emergency Response Plan at West Texas

Plant personnel and management did not understand the unique explosive nature of ammonium nitrate. More specifically, while the conditions necessary for detonation are not precisely known, the data suggest that detonations are promoted by severe shock or by heating with confinement, such as the massive fire at the plant. In fact, pools of molten ammonium nitrate may form in the event of fire, and if the molten mass becomes confined (e.g., in the elevator pit) it could explode. These hazards were not known nor understood by plant personnel and emergency responders. Therefore, responders were unaware of the potential of ammonium nitrate to detonate, had no knowledge of how long the fire had been burning, no robust incident pre-planning process in place, and no previous ammonium nitrate-related fire emergency training or drills. Hence, the strategy and tactics utilized by the West Volunteer Fire Department were not appropriate for the rapidly developing situation and they were exposed to unnecessary and extreme risk that proved fatal.

Response during 2015 explosion on the FPSO Cidade São Mateus

The first detector alarmed at 11:31, and by 11:38 all three detectors reached the over limit levels. The 1st team entered the pump room between 11:49-11:54 to assess the gas alarms, and they reported a leak at the flange and a 2-meter diameter pool under the flange. A 2nd team entered the pump room around 12:07 to assess the damage and whether repairs could be made. Despite the confirmed gas on all three detectors, the 2nd team still entered, and a personal gas detector alarmed at 100% LEL. The 2nd team exited around 12:15 and the ship was partially demobilized around 12:20 for lunch. The 3rd team enters around 12:27 to clean and repair, with ladders, wrenches, fire hose and absorbent blankets. The explosion occurred around 12:38 and killed four members in the pump room and four members at the top of the pump room on watch. Due to the failure of the wall into the engine room, the utilities operator was fatally injured in the engine control room.

There was a detailed Emergency Response Plan on the FPSO, along with the muster points for each person involved in the emergency response as shown Figure 14.

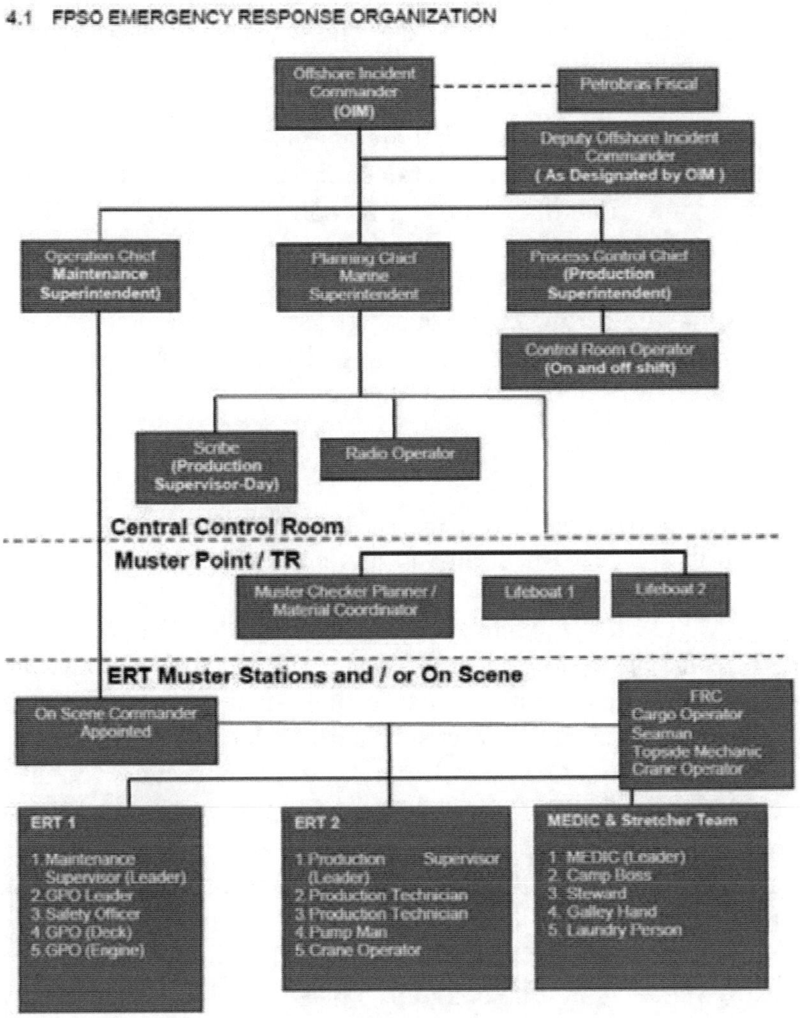

Figure 14: Organization Chart and muster points of the emergency response team

The Offshore Incident Commander is the Offshore Installation Manager (OIM) and would have the roles of implementing a command structure that would satisfy the organizational needs to protect life, environment and the installation, as well as deploying quick and efficiency actions to control incidents, as well as managing response actions to ensure that the escalating potential, risks of injury to personnel and damages to the installation are minimized. In the event of a leak in potentially explosive atmospheres, the emergency response must consider the possibility of the event escalating to a fire or explosion scenario. Thus, systems and procedures should be provided for the crew to control the situation and mitigate the risks of escalation from explosive atmospheres.

Investigation showed that the vessel was not designed for condensate, but for crude and gas. Yet despite the fact that there was confirmed gas detection within the pump room and on personal gas detectors, the crews entered three times. On the third entry, the explosion occurred. It was also reported that an explosive atmosphere in the pump or engine room was not included as an accident scenario in the emergency response plan, and there were no clear instructions of this particular scenario. Despite the knowledge that there was confirmed gas detection on the fixed sensors installed at the bottom of the pump room and also on personal detectors, Emergency Response Teams were sent into the pump room to evaluate the scenario in an explosive atmosphere.

The actions of Command essentially played down the risk of the condensate leak and allowed personnel to enter the pump room on multiple occasions to clean the spill without first performing a risk analysis, and without additional monitoring the room's atmospheric conditions. This noncompliance with the applicable emergency procedure was not only observed during entry operations, but also through the partial or total demobilization of the people irrespective of their muster points with an ongoing confirmed gas detection in the pump room. This demobilization resulted in two members being injured as they were found in a damaged elevator shaft after the blast.

4 Analysis

While deficiencies in the emergency response plans resulted in unnecessary and tragic consequences, the reasons each were lacking are very different and distinct.

For the West Texas Explosion, the Texas State Fire Marshal incident investigation [2] concluded that neither West Fertilizer nor West Fire Department had pre-fire and pre-incident plan at the facility. -Therefore, the recommendations provided were to ensure that local fire departments implement an Emergency Response Plan, with a particular focus on high-hazard occupancies. It was identified that fire departments visit and pre-incident plan responses to commercial structures in their jurisdiction. Further, they concluded that it is imperative that high-risk occupancies containing hazardous materials are identified and appropriate emergency response strategies and tactics are developed through pre-plan documents and SOP directives. Despite the extremely low likelihood of ammonia nitrate detonations, the recommendations state that high-hazard and high-risk occupancies containing heavy fire loads and hazardous materials should be identified by Fire Departments and all members made aware of their potential dangers. This will better prepare them for not only credible hazards, but also those high consequence hazards which may be found in the event of a fire and also provide them with insight as to when it is prudent to attack a fire or evacuate the area. More specific recommendations regarding the emergency response planning included:

- Develop a detailed Emergency Response Plan and Evacuation Plan for the area within 1 mile of existing ammonia nitrate buildings
- Develop standard operating procedures/guidelines to train and educate members on firefighting and emergency scene strategies and tactics for high-hazard occupancies
- Develop pre-incident plans by visiting the commercial and high-risk properties within the response jurisdiction.
- Utilize risk management systems on all emergency responses to determine when to employ alternative tactics.

- Incorporate incident management systems in every emergency incident, training exercise and basic operational process of the fire department to provide appropriate command and control for conducting operations.
- Implement a strategic incident action plan (IAP) for all emergency operations through a coordinated application of appropriate emergency scene tactics for the presented dangers
- Ensure that an incident safety officer (ISO), a person who is independent from the incident commander, is appointed and effectively utilized at every incident that would require an ISO.
- Establish a hazardous materials program and actively train members on the tactics to use when an incident involving hazardous materials occurs

In contrast, for the 2015 FPSO event, there was a detailed Emergency Response Plan for the vessel. Despite this plan, investigations concluded that there were significant deficiencies in the plan that needed to be remedied, and largely included reference to personnel exposure, failure to identify the accident scenarios for the emergency plan, training and mobilization. The execution of the plan failed to properly identify the potential high consequence and catastrophic result of the actions of the emergency responders. More specifically:

- Exposure to personnel
    - In the event of a leak in potentially explosive atmospheres indoors, the plan must consider the possibility of event escalation to a fire and/or explosion. Thus, systems and procedures should be provided once there is the formation of an explosive atmosphere. To ensure the success of mitigation actions, emergency procedures and resources for emergency response should be pre-planned, drawing on the use of all available resources, including the time of scenario assessment.
    - To determine that all functions involved in emergency response, including land-based support, have periodic theoretical training in emergency response procedures.
    - Set in the emergency response plan the available material and human resources to the response of each accidental
- Failure to identify accident scenarios or not included from the risk scenarios
    - Establish criteria and procedures for the migration of risk scenarios identified in the risk analysis and safety studies for emergency response plans. In addition to the selected scenarios, the procedures should contain scenarios that represent challenges to response actions, such as fatalities, and multiple injuries, among others.
    - Establish periodically simulated emergency covering all identified accident scenarios, including scenarios with multiple injuries and fatalities. Such simulations should include the activation of external resources to inform platform that response structure is effectively tested and that corrective measures arising from simulated ensure continuous improvement of the response actions of all entities involved.
- Demobilization of meeting points
    - Periodically coordinate all staff regarding the hazards involved in the assumptions made in the response procedure to the emergency unit.

  - Prohibit the release of people to normal activities before the full control of the emergency.
- Ignition of the explosive environment
  - As the ignition was introduced by the action of the emergency responders within explosive atmospheres, review the procedures related to emergency response and safe work practices considering the possible ignition sources identified in existing rules

5 Conclusion

The lessons from two specific incidents show that critical deficiencies in the emergency response plan or implementation of the plan can result in fatal consequences that could have been avoided. The 2013 West Texas explosion shows how important it is to not only recognize high risk events, but also have the appropriate plans in place to avoid unnecessary escalation of an already tragic incident, where 12 of the 15 people fatally injured were first responders. By preparing first responders for not only credible hazards, but also those high consequence hazards, can provide them with crucial insight as to when it is prudent to attack a fire or evacuate the area. In addition, not properly following or playing down the risk of a high consequence event was observed in the 2015 FPSO event, where eight of the nine crew members killed by the blast were first responders. The failure to recognize the potentially high consequence and catastrophic event, led to an emergency response that unnecessarily exposed responders and ultimately led to an exacerbated event. This study confirms that emergency response plans are an essential layer of protection for facilities where all other safeguards failed to prevent an incident, which could have prevented unnecessary and tragic escalation of the incident.

6 References

[1] Preparedness for Emergency Response - FEMA Training - FEMA.gov

[2] Texas State Fire Marshal, "Firefighter Fatality Investigation FFF FY 13-06," West, TX, 2013

[3] Davis, SG, DeBold, TF, & Marsegan, C (2017). Investigation Findings and Lessons Learned in the West Fertilizer Explosion. Journal of Fire Science. Volume 35, Issue 5, September 5, 2017: pp. 379-395.

[4] "A Public Health Report on Injuries Related to the West (Texas) Fertilizer Plant Explosion," June 24, 2014.

[5] Brazilian Navy Directorate of Ports and Coasts, Maritime Accident Safety Investigation Report, FPSO Cidade São Mateus - Explosion followed by flooding and casualties, February 11th, 2015.

[6] Investigation report of the explosion incident occurred on 11/02/2015 in the FPSO Cidade São Mateus, Brazilian National Agency of Petroleum, Natural Gas and Biofuels, Superintendence of Operational Safety and the Environment, August 2015.

19 AIChE Spring Meeting
15TH GLOBAL CONGRESS ON PROCESS SAFETY
NEW ORLEANS • 2019

Major Explosion at a Chemical Company in China: the Causes and Lessons Learned

Wei Wang
China University of Petroleum (East China)
No. 66, West Changjiang Road, Huangdao District, Qingdao, China
15621097585@163.com

Yi Liu
China University of Petroleum (East China)
No. 66, West Changjiang Road, Huangdao District, Qingdao, China
liuyi@upc.edu.cn

Dongfeng Zhao
China University of Petroleum (East China)
No. 66, West Changjiang Road, Huangdao District, Qingdao, China
zhaodf@vip.sina.com

Yanchun Wang
China University of Petroleum (East China)
No. 66, West Changjiang Road, Huangdao District, Qingdao, China
Wychun25@163.com

Jian Chen
China University of Petroleum (East China)
No. 66, West Changjiang Road, Huangdao District, Qingdao, China
836771528@qq.com

Hanhai Dong
China University of Petroleum (East China)
No. 66, West Changjiang Road, Huangdao District, Qingdao, China
donghanhaiupc@126.com

Prepared for Presentation at
American Institute of Chemical Engineers
2019 Spring Meeting and 15th Global Congress on Process Safety
New Orleans, LA
March 31 – April 3, 2019

Major Explosion at a Chemical Company in China: the Causes and Lessons Learned

Wei Wang
China University of Petroleum (East China)
No. 66, West Changjiang Road, Huangdao District, Qingdao, China
15621097585@163.com

Yi Liu
China University of Petroleum (East China)

Dongfeng Zhao
China University of Petroleum (East China)

Jian Chen
China University of Petroleum (East China)

Yanchun Wang
China University of Petroleum (East China)

Hanhai Dong
China University of Petroleum (East China)

Keywords: Fishbone Diagram; Analytic Hierarchy Process; Process Safety Management; Incident Investigation

Abstract

On July 12, 2018, a devastating explosion incident occurred in Yibin City Hengda Technology Co., Ltd. in Sichuan Province, resulting in 19 deaths and 12 injuries, and impacting an area of about 1,800 square meters. This was one of the most disastrous explosions that occurred in a chemical plant in China in the past five years. The incident reflects the general deficiencies of process safety management in small and medium-scale companies. Therefore, it is significant to analyze the causes of the incident and put forward countermeasures to avoid such kind of incidents.

According to the characteristics of complexity, diversity, and uncertainty about the influencing factors of the explosion, this paper presented a systematic analysis from qualitative and quantitative perspectives. The fishbone diagram model was used to identify influence factors from six aspects of people, design, management, environment, safety prevention, and materials. Furthermore, an assessment index including 26 factors was determined. The analytic hierarchy process method was used to analyze the importance degree of each factor quantitatively, and the key factors leading to the incident were obtained. The results showed that the sum of weights of three main aspects, including people, design, and safety prevention is nearly80%, so they were the dominant causes leading to this incident. Lack of chemical industry experience, non-standard process design, and incomplete automatic control system were the focus of prevention among 26 factors.

Process safety management is the application of management systems to the identification, understanding, and control of process hazards to prevent incidents from occurring. This analysis considered process hazard analysis, pre-startup safety review, operation procedures, training, and emergency plan and response, which were the main elements in process safety management framework. Finally, we proposed safety precautions for the identification, storage, production, and use of hazardous chemicals, as well as recommendations for process equipment management.

1 Introduction to the event

At 18:42:33 on July 12, 2018, an explosion occurred in Yibin Hengda Technology Co., Ltd. in the industrial park of Jiang'an County, Sichuan Province, causing 19 fatalities and injuring 12 people. The explosion caused the second and third workshops to catch fire, and the combustion area is $1800m^2$ [1]. The first workshop, warehouse, drying room, analysis room, office building, and other buildings had been damaged to some extent. The fire department dispatched 42 fire engines and more than 200 firefights. It took more than two hours to extinguish the fire on the spot as shown in Figure 1. It was estimated that the total energy of the explosion in this incident was equivalent to 235 kg of TNT. The direct economic loss was 41.42 million yuan.

At the time of the incident, the company was producing imazamox and triazole. Triazole is a pesticide, and imazamox is an herbicide. The process of the incident was the dehydration of butyramide in the process of synthesis of imazamox. At about 11:30 on July 12, the company collected 2 tons of chemicals labeled as raw materials (actually the explosive chemical sodium chlorate, not the raw material required for the production of butyramide). At around 14:00, the deputy director of the company's second workshop received raw materials from the warehouse. At around 15:30, the workers in the second workshop raised the raw materials for production through the lift to the third floor of the second workshop. At around 16:00, the 2R301 kettle for dehydration of butyramide was completed, and the feed was in an empty state. Before 17:20, the 2R301 kettle was completed. At about 6 pm, the dehydration completed. At 18:42:33, it was due to the on-site handover time. A chemical explosion occurred in the 2R301 kettle on the third floor of the second workshop. The direct cause of the incident was that the operator replaced the unlabeled sodium chlorate with butyramide and put it into the 2R301 kettle for dehydration, which caused the explosion and fire. Sodium chlorate is a kind of explosive chemical, which is easy to form an explosive mixture when mixed with butyramide. The explosion led to the disintegration of 2R301 kettle. The high-temperature toluene vapor rushed out rapidly formed an explosive mixture with the external air and produced a secondary explosion, which also caused the explosion and combustion of the materials such as sodium chlorate, toluene, and methanol stored at the workshop site.

Figure 1. Extinguishing at the scene.

According to the investigation, in this incident, the second workshop was the core area of the explosion, and the overall structure was severely damaged as shown in Figure 2. The platform on the third floor collapsed. The reactor with a height of nearly two meters and a diameter of about one meter fell from the upper level.

Figure 2. Severely damage to the second workshop.

2 Incident evaluation index system based on the fishbone diagram

Fishbone Diagram is also named as cause and branches figure analysis. Kaoru Ishikawa, a Japanese management master, invented it for finding out the root cause of incident [2]. It indicates the relationship between a problem and its underlying causes. It is a qualitative analysis tool to help identify incident causes [3]. With step by step in-depth study to analyze its influencing factors,

Fishbone Diagrams can systematize complicated incident causes. When starting the specific report, we first analyze the factors which affect the safety most and then followed by departure from the overall reason to find the basic reason and to detect and determine the main reason [4,5,6].

After the incident, the government set up a team of experts to identify risk factors from six aspects: human, design, management, environment, safety prevention, and material and establish an incident evaluation index system with 26 elements. The fishbone diagram that induced the explosion is shown in Figure 3.

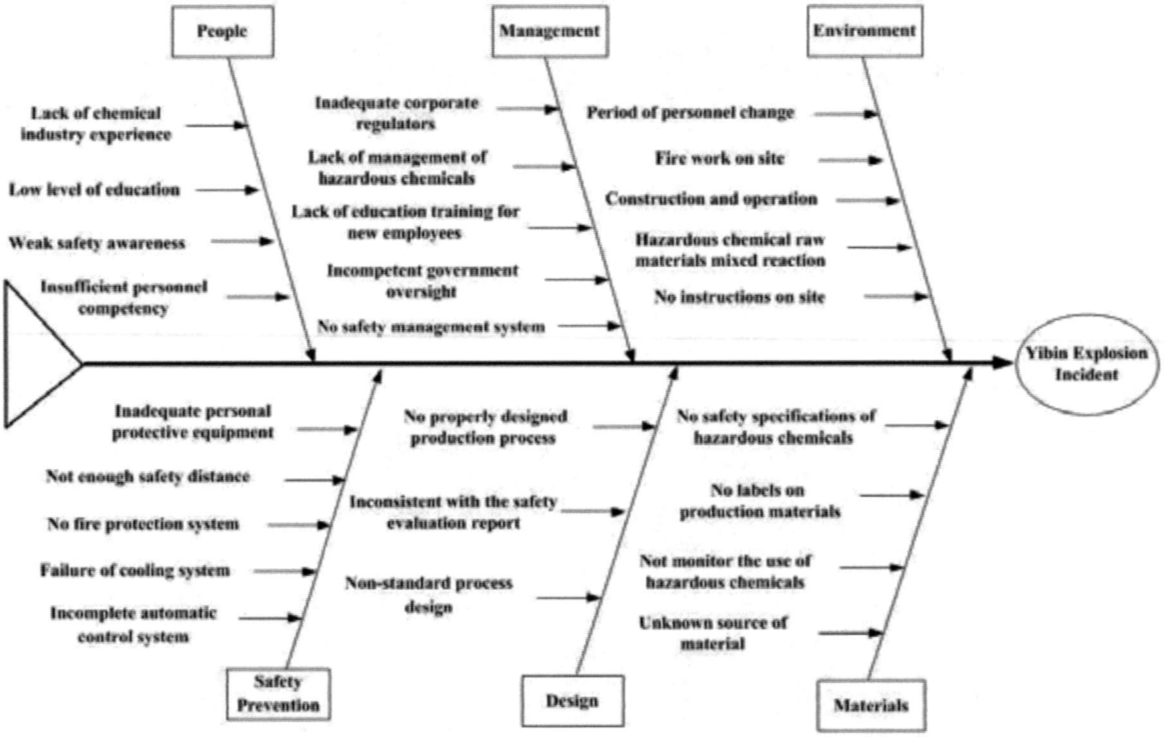

Figure 3. Fishbone diagram of Yibin explosion incident [7].

2.1 People

Neither the safety managers nor the operators demonstrated proficiency in the requirements of process safety competency. There were two main aspects below:

- Corporate management - The general manager of the enterprise had responsibility for the mechanical processing and production contract. However, he did not have appropriate experience in the chemical industry. The technical director did not have experience with management of process safety and production systems. The corporate legal person responsible for finance and logistics also had no experience in chemical engineering.

- Operators - Among the 19 people who died in the incident, 16 were operators, and most of these operators were local farmers. They lacked the basic knowledge of chemical process safety production, and they did not understand the process safety risks in the production process.

2.2 Design

The second workshop where the incident occurred was not constructed and installed strictly by the design drawings, and it was originally planned to have nine sets of 5000L tanks and one set of 3000L tanks and produce 5-nitroisophthalic acid. As built, the workshop had three sets of 5000L tanks, seven sets of 3000L tanks, and eight sets of 2000L tanks were set up. The product was also different from the original, but imazamox and triazole. A similar situation existed in the third workshop. The production technology of imazamox and triazole used by the company had no formal technical source, and the enterprise also did not entrust professional institutions for process calculation and construction drawing design. The technical director of the enterprise only planned the required production equipment and operation process according to his experience, but he knew nothing about the process safety risks of related technologies. It was during the production of imazamox that the distillation kettle used for the dehydration of butyramide by toluene first exploded. According to the *Notice on Further Strengthening the Safety Design Management of Hazardous Chemicals Construction Project*, Hengda company needed to start construction in April 2018, but the company had already started construction in July 2017. Also, the fire separation design of each building did not meet the requirements of the *Petrochemical Enterprise Design Fire Code (GB50160).*

2.3 Management

The lack of process safety culture is the main cause of management failure. Safety culture is developed through the interactions of factors such as safety policies and safety goals, the discipline of the organization, the behavior and attitude of the company staff, and the competence of workers. Improper operating discipline and procedures can lead to a negative process safety culture and an increased likelihood of process safety incidents. The enterprise had not formulated the post safety responsibility system, safety management rules, and regulations. Furthermore, the enterprise manager and the person in charge of each department only had the oral authorization, and there was no formal appointment document to clarify the post responsibility. The safety management organization was not complete. The company's safety and environmental protection department had only one safety officer who did not have safety management capabilities. There were also no full-time or part-time safety officers in each workshop. There was a lack of management of hazardous chemicals in the process safety management. The technology provider concealed some chemical name marks for confidentiality, and there was no safety specification. Required knowledge and competence connected with safety for workers were the pivotal content of safety training. New employees did not have security education directly work or internship. The government had not strictly regulated the safety departments of local key hazardous chemicals companies.

2.4 *Environment*

The factory built the workshop while producing, and used the polypropylene pipe which was easy to produce static electricity to transport flammable and explosive liquid materials. The building materials, hardware, and chemicals in the warehouse were mixed and piled with hidden dangers. The incident occurred during shift time, resulting in a large number of casualties. In the on-site operation workshop, there was neither operation instructions nor warning signs for hazardous chemicals. At the time of the incident, the second workshop produced two products at the same time. This practice of producing multiple produced at the same time in the same workshop has excellent security risks.

2.5 *Safety prevention*

Compared with the 20 elements of CCPS, China drew on the process safety management experience of the Occupational Safety and Health Administration (OSHA) and further optimized the formation of 12 elements for the production of Chinese chemical companies. After that, a series of process safety technologies were developed and promoted for the key elements of risk management and equipment integrity. In 2013, the SAWS issued *Guidance for enhancing chemical process safety management.* The Ministry of Emergency Management, which was integrated by the State Administration of Work Safety(SAWS) in 2018, attaches great importance to the supervision of hazardous chemicals, hazardous chemical processes, and major hazardous sources, referred to as "Two Priorities and One Major." If an enterprise violates The relevant provisions, it shall be punished according to the upper limit of economic punishment according to law.

In the production process of imazamox and triazole, the nitrification, oxidation, diazotization, hydrazine hydrate, hydrogen peroxide, methanol, and other dangerous chemicals and processes involved were all "Two Priorities and One Major" projects, which needed strict supervision. However, the company's automatic control system, flammable and toxic gas alarm system and fire water system and other safety facilities were not installed. Once there was a danger, the enterprise itself could not deal with an emergency. The lack of an automated control system led to more than ten operators performing manual operations around the reactor, which was an important cause of major casualties.

2.6 *Materials*

According to the investigation, the safety evaluation report made by the enterprise before the incident did not fully consider the process safety information, such as the lack of description of dangerous chemicals and the risk identification of the process, so the enterprise had great potential risks. Workers didn't know the source of raw materials, and many chemical production raw materials were not marked. Workers knew nothing about the risks of these dangerous chemicals, and company officials failed to communicate information about the hazards of the process to employees. The workers compiled codes to replace the names of products and raw materials, and these numbers were not abbreviated as specifications. The raw materials used by Hengda Company, such as hydrazine hydrate, nitric acid, toluene, hydrochloric acid were dangerous chemicals or precursor chemicals, but the company still purchased them illegally without relevant licenses. In the production process, regulators did not supervise the management of hazardous chemicals.

3 Determination of the weights of the incident incentive index based on the analytic hierarchy process

The Analytical Hierarchy Process (AHP) is a decision-making method proposed and used by American operations researcher T.L.Saaty in the mid-1970s [8]. AHP is a process that breaks down a problem in a hierarchical matrix with increasing level of detail to determine the most significant causes of an incident. According to the obtained fishbone diagram of the Yibin explosion incident, an incremental layer structure for analytic hierarchy analysis was constructed, as shown in Fig. 4, where the A, B, and C layers illustrated the primary effect and hierarchy of causes.

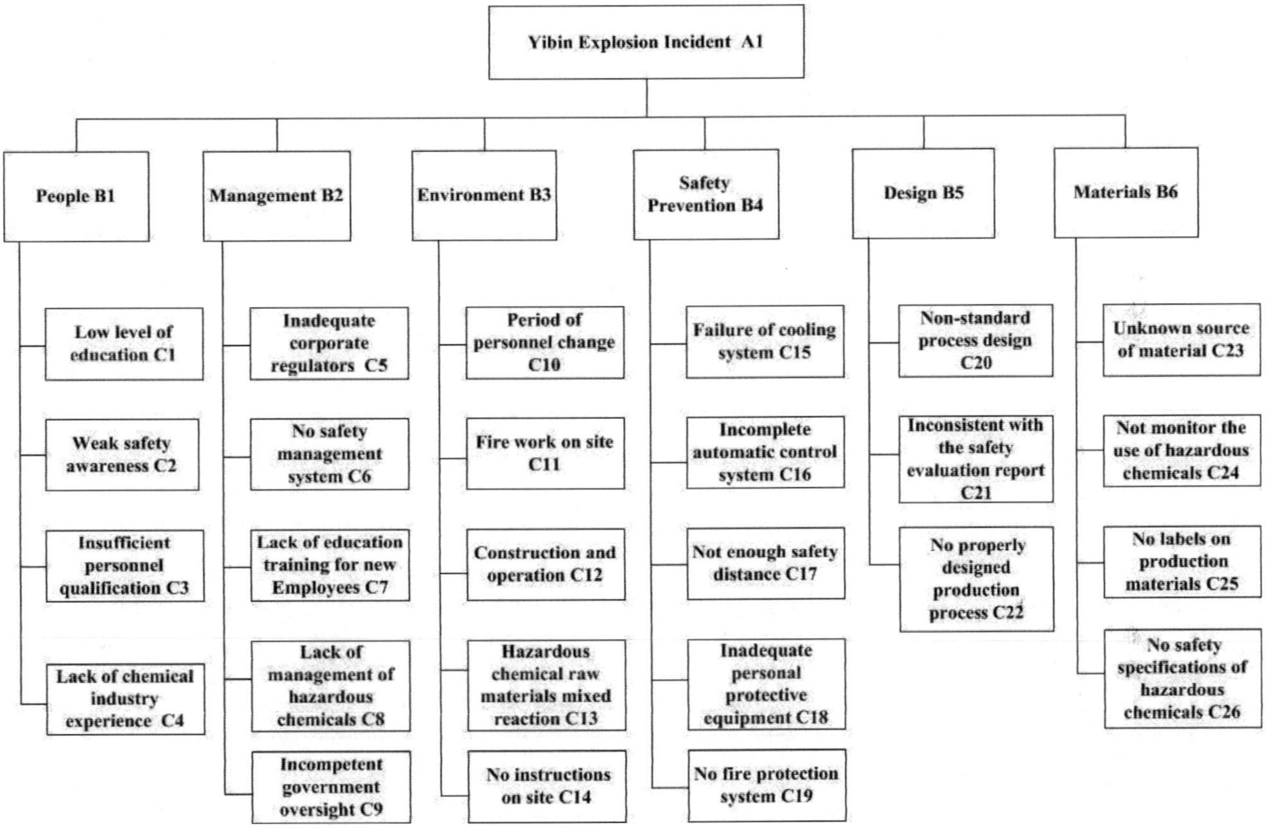

Figure 4. Hierarchical structures of safety evaluation system for Yibin explosion incident.

3.1 The single-layer index weights of the influencing factors

According to the actual data collected and expert opinions, we used 1-9 scale to construct the judgment matrix (A) of the first-level evaluation index, as shown in table 1, and obtained the influence of various factors in layer B on layer A. λ_{max} is the maximum eigenvalue, CI is the consistency index, RI is the average random consistency index, and CR is the random consistency ratio; When $CR<0.1$, it passed the consistency test [9]. According to table 1, the weights of design is the largest, followed by people and safety prevention, with the sum of the weights of the three as high as 0.8. It can be seen that the leading cause of this incident is due to the low level of process design, operators' insufficient qualification of chemical safety production and non compliant protection measures. Therefore, enterprises should pay attention to intrinsic safety design, improve the safety awareness of staff by popularizing safety knowledge and operational guidelines, and finally achieve the goal of safe production.

Table 1. Judgment matrix of primary evaluation indicators and weights results

A	B1	B2	B3	B4	B5	B6	Weights	Consistency	Ranking
B1	1	2	3	1/2	1/3	7	0.1645	λ_{max}=6.4698 CI=0.094 RI=1.26 CR=0.0746<0.10 Satisfy consistency	3
B2	1/2	1	2	1/3	1/4	6	0.1060		4
B3	1/3	1/2	1	1/4	1/5	5	0.0700		5
B4	2	3	4	1	1/2	8	0.2545		2
B5	3	4	5	2	1	9	0.3810		1
B6	1/7	1/6	1/5	1/8	1/9	1	0.0239		6

3.2 The comprehensive weights of the influencing factors

We built each judgment matrix of layer C to layer B and calculated the single layer weights of layer C, and finally, we obtained the comprehensive weights of incidents caused by bottom influencing factors. We considered the product of these two layers as the comprehensive weights of the sub-factor. For example, the comprehensive weights of C1 is equal to the single-layer weights of C1 to B1 multiplied by the single-layer weights of B1 to A. The comprehensive weights result and ranking of each sub-factor were shown in table 2.

Table 2. Single and synthetic weights of C layer indexes

		Single weights	Consistency	Synthetic	Ranking
B1	C1	0.0843	λ_{max}=4.1375 CI=0.0458 RI=0.9 CR=0.0515<0.1 Satisfy consistency	0.0139	16
	C2	0.2120		0.0349	8
	C3	0.1544		0.0254	12
	C4	0.5492		0.0904	3
B2	C5	0.1556	λ_{max}=5.2550 CI=0.0638 RI=1.12 CR=0.0569<0.1	0.0165	15
	C6	0.0566		0.0060	21

	C7	0.4735	Satisfy consistency	0.0502	5
	C8	0.2578		0.0273	11
	C9	0.0566		0.0060	22
B3	C10	0.6253	λmax=5.3598 CI=0.08995 RI=1.12 CR=0.0803<0.10 Satisfy consistency	0.0438	7
	C11	0.0901		0.0063	19
	C12	0.0435		0.0030	23
	C13	0.1976		0.0138	17
	C14	0.0435		0.0030	24
B4	C15	0.0781	λ_{max}=5.1717 CI=0.042925 RI=1.12 CR=0.0383<0.1 Satisfy consistency	0.0199	13
	C16	0.5895		0.1501	2
	C17	0.1271		0.0324	9
	C18	0.0781		0.0199	14
	C19	0.1271		0.0324	10
B5	C20	0.6833	λ_{max}=3.0246 CI=0.0123 RI=0.58 CR=0.0236<0.1 Satisfy consistency	0.2604	1
	C21	0.1168		0.0445	6
	C22	0.1998		0.0761	4
B6	C23	0.1242	λ_{max}=4.1891 CI=0.0630 RI=0.9 CR=0.0708<0.1 Satisfy consistency	0.0030	25
	C24	0.2626		0.0063	20
	C25	0.5465		0.0131	18
	C26	0.0667		0.0016	26

As can be seen from table 2, the full weights of C16 (Incomplete automatic control system) and C20 (Non-standard process design) was greater than 0.1. The full weights of C4 (Lack of chemical industry experience) and C22 (No properly designed production process) were very close to 0.1. The combined weights of these four factors reached 0.577, which was more than half of the total. Therefore, these four influencing factors were the primary causes of the explosion incident in Yibin, Sichuan, as strict control and prevention were given priority.

Among them, both C20 and C22 belong to the design aspect, and enterprises should entrust qualified formal companies to design the process and production plan. In the method of safety design, the operation and design should be analyzed in real time to check the status of dangerous materials, effectively reduce the operating temperature and pressure, and control the influence of hazards factors in the process on chemical equipment. The equipment used in production shall meet the strength and anti-corrosion performance required by the process, and shall be designed and manufactured by qualified factories.C4 is a human problem. Therefore, it is necessary to conduct chemical safety education and operation training for operators before entering the plant. Many of Hengda's workers were the relatives and friends of management personnel. They were generally poorly educated and skilled in operation, and had not received any training in chemical safety. The workers did not go through systematic learning, but the workshop director directly taught them operating skills. C16 is a problem of safety prevention. The automatic control system can not only improve the operation efficiency of production equipment but more importantly, it can protect the safety of chemical personnel and reduce the loss caused by incidents. Enterprises should follow the safety production regulations, install automatic control systems, toxic and harmful gas alarm systems, and fire protection systems to effectively prevent incidents.

The weights of C7 (Lack of education training for new employees), C10 (Period of personnel change) and C21 (Inconsistent with the safety evaluation report) were all around 0.05. The weights of C2 (Weak safety awareness), C19 (No fire protection system) and C17 (Not enough safety distance) also exceeded 0.03. These five factors were important causes of incidents and should be effectively controlled.

C7 is a new employee training management problem. Enterprises should pay attention to the educational training of employees before entering the plant and implement the three-level safety training education of the plant, workshop, and team. C10 involves change management. At the time of the incident, it was during the shift time of employees, the change of personnel and the change of materials and equipment, which had great potential risks. Enterprises should establish a change management system, conduct comprehensive training for the staff who play a vital role in the change management process and clarify the position and responsibility of each person in the change system.

4 Lessons learned from the causes of the incident

The Yibin explosion was a pharmaceutical chemical incident. Through the investigation and analysis of historical disasters in the chemical industry in China, it was found that the proportion of pharmaceutical chemical incidents in the chemical industry was huge [10]. It was characterized by a random change of chemical process, small scale, low level of process control, involving dangerous materials and complex chemical reaction. Especially in small and medium-sized chemical enterprises, the lack of process safety management is an essential cause of such incidents. From the perspective of process safety management, this paper summarized and put forward suggestions for preventing such incidents.

- Process Hazard Analysis - In the design stage, the risk identification and assessment tool system should be applied to identify the risks of materials in the process of technology. In the review stage, HAZOP, SIL are used to review the process, layout and material flow. Companies should use scientific models to calculate the explosion-proof grade and safety distance of buildings. By adding the explosion-proof wall and the explosion-proof film, the building close to the installation area can reach the required safety factor. After the device is put into operation, it needs to be compared with the hazard analysis in the design stage, and the hazard that is inconsistent with the design should be analyzed again. Hengda's actual operating equipment, raw materials, and products were inconsistent with the original plan, and the company had not discovered and controlled the real hazards [11].

- Pre-startup Safety Review - The company shall organize a safety inspection team, whose task is to confirm whether the improvement measures and safety guarantee measures in the process hazard analysis report have been implemented as required; Whether adequate training of personnel in operation, maintenance, and management of process equipment has been completed; Whether necessary safety information of process equipment is completed and updated; Whether necessary equipment testing and checking have been completed and confirmed; Whether the process equipment is constructed and installed according to the design requirements [12].

- Operation Procedures - The incident occurred during the change management phase, and there was no review or other risk assessment during the shift. Staff should review the operating procedures as often as possible to ensure that the operating procedures reflect current operating conditions, including changes in chemicals, process equipment, and facilities. The enterprise shall confirm the adaptability and effectiveness of the operating procedures and ensure that the operators understand and master the operating systems [13].

- Training - A complete process safety management is inseparable from the participation of operators, and the human factor is the most unstable factor in process safety management. Therefore, Enterprises should establish strict production operation disciplines and require operators to master process safety information, equipment integrity management points, and emergency response policies and finally improve workers' ability to control equipment safety operations [14].

- Emergency Plan and Response - Enterprises need to establish a set of overall emergency plans, which usually stipulate in written form how the factory should deal with abnormal or urgent situations. Also, enterprises should establish emergency response teams, traditionally composed of enterprise personnel, but also include external staff; The responsibilities of each team member should be clear, ensuring that members have no questions about accountability and delegation. In case of an emergency, the person in charge may determine the safe area according to the emergency response manual and direct the person to evacuate to a safe place [15].

5 Conclusion

The circumstances of the incident described in this article are typical in China, and the incident reflects the common safety problems in small and medium-sized chemical enterprises. Managers should understand these lessons we learned from this incident. By using fish-bone diagram method, the causes of the explosion incident were obtained through qualitative analysis, including six aspects, such as people, design, management, environment, safety prevention and materials, and 26 factors influencing the incident were further analyzed. Then the analytic hierarchy process was used for quantitative analysis, and the results showed that lack of chemical industry experience, non-standard process design, and incomplete automatic control system were the main factors affecting the incident, so the enterprise should strictly control and prevent these factors. Finally, we summarized the causes of such incidents and put forward the measures from five aspects of process safety management, including process hazard analysis, pre-production safety inspection, operation procedures, training, and emergency plan and response plan. An incident brings not an only economic loss but also bad social influence to the factory. It is our hope that this paper will help prevent future accidents.

6 References

[1] Sichuan Provincial Emergency Management Office, Yibin Hengda Technology Co., LTD. "7.12" Major Explosion Fire Accident Investigation Report, Available at http://www.chinasafety.gov.cn/gk/jsts/201902/t20190214_224397.shtml, Accessed on February 14, 2019.

[2] Tongyuan Luo, Chao Wu and Lixiang Duan, Fishbone diagram and risk matrix analysis method and its application in safety assessment of natural gas spherical tank, Ind. Eng. J. 25(02), 2007, 138-141.

[3] Xing, W., QC Team Based Materials, China Society Press, Beijing, China, 2004.

[4] T Deng, X., Chen, Q. and Jiang, D., Application of fishbone chart analysis method in prevention for steam turbine overspeed incident, Guangdong Electr. Power, 20(02), 2013, 73-77.

[5] Changlong Zhu, Jiayong Zhu, Li Wang and M. Sam Mannan. Lessons learned from analyzing a VCE incident at a chemical plant, Journal of Loss Prevention in the Process Industries,50,2017, 397-402.

[6] Dennis P. Nolan, Chapter 5: Specialized Reviews—CHAZOP, EHAZOP, Bow-Tie Analysis, Layers of Protection Analysis, Safety Integrity Level, FishboneDiagram, and Cyber Security Vulnerability Analysis, Safety and Security Review for the Process Industries (Fourth Edition), Gulf Publishing Company,2015,Pages 17-27.

[7] Xinhua News Agency, Emergency management department: 7·12 incident enterprises have no law and discipline, and government supervision is lax, Available at http://www.gov.cn/xinwen/2018-07/17/content_5307196.htm?_zbs_baidu_bk, Accessed on July 17, 2018.

[8] Seong KonLee, GentoMogi and Jong WookKim. Decision support for prioritizing energy technologies against high oil prices: A fuzzy analytic hierarchy process approach, Journal of Loss Prevention in the Process Industries,22(6), 2009, 915-920.

[9] Saaty T.L., A scaling method for priorities in hierarchical structures, Journal of Mathematical Psychology, 15 (3),1977, 234-281.

[10] Heda Zhang and Xiaoping Zheng, Characteristics of hazardous chemical incidents in China: A statistical investigation, Journal of Loss Prevention in the Process Industries, 25(4), 2012, 686-693.

[11] Hao WU, Igor PEÑARRUBIA, Lin CUI and Jinsong ZHAO, Process safety management considerations for biofuel production, Higher Education Press, 4(3), 2017, 357-367.

[12] Yuan Gao, Yunxiao Fan, Jing Wang, Xi Li and Jingjing Pei, The mediating role of safety management practices in process safety culture in the Chinese oil industry, Journal of Loss Prevention in the Process Industries, 57, 2019, 223-230.

[13] Paul R.Amyotte and Cathleen S.Lupien, Chapter Three - Elements of Process Safety Management, Methods in Chemical Process Safety, 1, 2017,87-148.

[14] T. Kontogiannis, M.C. Leva and N. Balfe, Total Safety Management: Principles, processes and methods, Safety Science, 100(B), 2017, 128-142.

[15] Hao WU, Igor PEÑARRUBIA, Lin CUI and Jinsong ZHAO, Process safety management considerations for biofuel production, Front. Eng. Manag, 4(3), 2017, 357- 367.

Taming the Wild River Rapids – How Process Safety Can Apply Outside the Process Industries

Trish Kerin
IChemE Safety Centre
Level 7, 455 Bourke St, Melbourne, Victoria, Australia
tkerin@icheme.org

Prepared for Presentation at
American Institute of Chemical Engineers
2019 Spring Meeting and 15th Global Congress on Process Safety
New Orleans, LA
March 31 – April 3, 2019

—

Taming the Wild River Rapids – How Process Safety Can Apply Outside the Process Industries

Trish Kerin
IChemE Safety Centre
Level 7, 455 Bourke St, Melbourne, Victoria, Australia
tkerin@icheme.org

Keywords: lessons learned, process safety, design, maintenance

Abstract

On the 25th October 2016 the Thunder River Rapids ride at Dreamworld on Australia's Gold Coast suffered a pump failure, resulting in the tragic deaths of four riders. This incident was investigated by the police and later a coronial inquest to determine how a tragedy like this could occur. As a process safety professional, it was interesting to follow the coronial inquest findings and then think about how process safety management may have helped to prevent this incident. Assuming that a process safety event can be defined as a high consequence or potential consequence low likelihood loss of control event [1], this incident had all the hallmarks. Therefore, it is possible to apply the IChemE Safety Centre (ISC) framework [2] for process safety to identify where the system failed and understand how it could be averted in future. This paper will explore various elements of the incident and how process safety management techniques may have helped, including that the ride had been modified to be different from all other similar rides, that the staff did not know how to stop the ride in an emergency and that all means to stop the ride required manual detection and intervention, with no alarms or shut down systems installed.

1 The Incident

The Thunder River Rapids ride at Dreamworld in Queensland, Australia, was a water ride, simulating white water rafting. Up to six riders sit in a circular raft, which flows along a turbulent river, including travelling up a conveyor to ride down a waterfall. The water flow is driven by pumps and made turbulent by the installation of underwater obstacles. It was in full operation on 25th October 2016, when the pump managing the water flow tripped out. This resulted in the water level on the ride dropping. This drop in water level caused a raft to become stuck on the rails at the conveyor leading to the disembarkation platform. This raft remained stationary for some 57 seconds while the ride continued to run before it was impacted by the next raft in the sequence. This following raft contained 6 passengers, four adults and two children. When this raft struck the stationary one, it flipped over, killing the four adult passengers. The two children managed to escape the raft [3]. This incident is

—

currently under a Coronial Inquest. As such this paper is written from several media outlets that have been reporting the proceedings and based on the evidence reported, not the findings of the Coroner. The Coronial Report is due to be published in mid 2019.

2 History of the Ride

The Thunder River Rapids ride was completed in 1986 and was likely an in-house design and construction [4]. Sometime after construction, the ride was modified to remove some of the timber slats that form the base of the conveyor [4]. Two in every three slats were removed, though is unclear why this was done. This did create significant gaps in the base of the conveyor belt. The ride was also refurbished some months prior to the incident. [5].

Over the history of the ride there had been a number of incidents that had occurred with the rafts and the pumps. This in itself is not unusual, in that pumps can trip out from time to time, however there had been an incident in 2001 where the pump had tripped, and a raft became stuck [6]. This was then struck by a second raft, which flipped over. In this incident there were no passengers in the raft, and therefore no injuries. The investigation found that the ride operator had failed to adequately monitor the ride and he was dismissed. Details of this incident were not shared broadly and the operators and maintainers of the ride in 2017 had no knowledge of it. At the time an email was written by a park engineer discussing the incident and stated "I shudder when I think if there had been guests on the ride" [3]. This demonstrates that someone at some time recognized the significance of this incident, but it was then lost in history.

3 Technical and Organizational Causes of the Incident

There is one key immediate cause of this incident, however there are many different and wide-ranging causal factors. The immediate cause has been determined to be the tripping of the water pump, which in turn allowed the water level to drop, causing the first raft to become stuck. This provided the obstacle that the second raft, with six passengers, struck, causing it to then flip over. But let us examine the broader aspects of this incident, from the pump reliability to the safety systems and organizational capability. The ISC framework [2] provides a useful template to examine the causes. This framework in represented by Figure 1. Fundamentally leadership underpins all aspects of process safety. There are six functional elements that must be robust in an organization to manage major hazard safety. These are; knowledge & competence; engineering & design; systems & procedures; assurance; human factors and culture. In this incident there were clear failures in each of these elements.

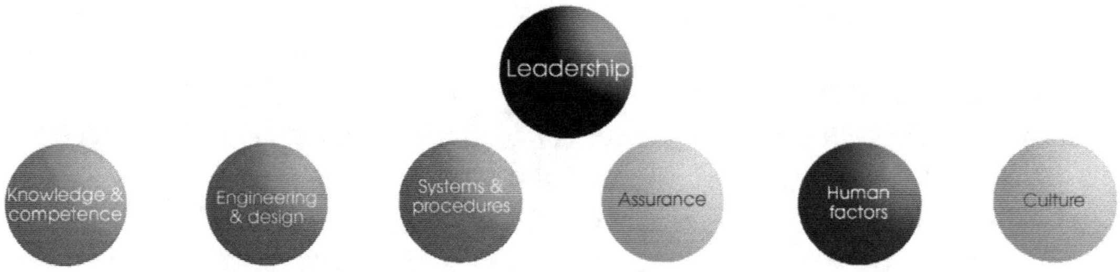

Figure 1. ISC Framework for Process Safety

3.1 Knowledge & Competence

The training of the ride operators took only 90 minutes and was unstructured with no evidence of competence consistently recorded [7]. There was a procedure manual however it was thought to be confusing by the ride operators [8]. The operator running the ride on 25th October 2016 had been trained for 90 minutes that morning, as it was her first day running the Thunder River Rapids. The junior operator stated that the trainer "showed me the front page of the manual and just flipped through it. I had to go through it myself. She said just sign off on it when you are happy." [9] The trainer stated that she sometimes had to "invent" training techniques because of the complexity of the control panel [8]. There was no clear guidance on how and when to use the emergency stop for the ride. The new operator testified that she was told "don't worry about it, you don't need to use it" with regard to the emergency stop button [10]. This is disputed by the person who delivered the training. The new operator believed it was the role of the senior operator to shut the ride down. It should be noted that the emergency stop would shut down the moving equipment with in two seconds, where the normal shutdown took around eight seconds. The ride operator that day stated that she "would have waited for the senior operator to tell her to hit the e stop" [10]. The senior operator that day was "surprised" to learn there was an emergency stop that would have shut the ride down within two seconds [7].

3.2 Engineering & Design

Intamin Amusement Rides designed and built the first river rapids rides in the world, and following this tragedy released a statement stating that the ride was "neither built by Intamin nor supplied by Intamin". The ride at Dreamworld was a custom build, and this was said to be common for rides built in the 1980's in the Gold Coast[4].

3.2.1 The Timber Slats

—

A few years after the ride was opened, a modification was made to the timber slats in the conveyor system [4]. The slats appear to serve at least two safety purposes. The first was to provide a solid platform for a passenger if they fell from the raft, essentially isolating them for the moving machinery and providing a platform to stand on [11]. The second was so that the rafts did not get caught in gaps and become stuck or indeed flip over. The Thunder River Rapids ride had two of every three slats removed. This provided large gaps between the remaining slats. There does not appear to be a theme park anywhere else that has made this change. There also appears to be no evidence as to why the change was made, nor was there a considered Management of Change (MoC) review undertaken. It is not clear if the organization had a structured and systematic MoC process. The safety regulator did not approve the modification and believed it added in additional hazards [4, 12]. In this incident, the raft became lodged in the gap, and as the conveyor continued to move forward, the raft was driven up vertically. When the passengers then fell from the raft, there was no solid base and they landed in the mechanism of the conveyor. There had also been conveyor framework modifications made to the ride in early 2016 in an attempt to stop rafts flipping over [13]. It is unclear what this modification was or indeed why it was made.

Figure 2. Aerial view of Thunder River Rapids showing the large gaps in the timber slat conveyor after the incident [10].

Figure 3. Aerial view of Infinity Falls, Orlando, showing close spacing of the boards on the conveyor [14].

3.2.2 The Pump

The water pumps had failed three times in the days leading up to the incident, and then twice on 25th October 2016, before the final failure that lead to the tragedy [15]. The pump's purpose was to keep the water flowing through the ride to keep the rafts moving. If the pumps stopped the water not only stopped flowing, it also drained back, and the water level dropped such that rafts could become stuck on the underwater obstacles that produced the rapids [16]. The reason for the pumps tripping out was not determined, the protocol seemed to be to just reset them to get the ride operating again [3]. There also is some confusion as to whether an alarm was sounded when the pumps tripped out, with an old procedure having the comment about an alarm crossed out [17]. At the time of the incident

—

the pumps were approximately 10 years old, and the manufacturer stated that they had a life expectancy of about a decade. There did not appear to be any plans to replace the pumps, however there was a maintenance request into the pump supplier in the week prior, but this visit did not eventuate [18].

3.2.3 Low Water

Low water in the ride was a critical hazard, as the rafts could become stuck or indeed flip. There had been previous incidents where low water level hazard led to rafts flipping, notably in 2001. There was no low water level switch that would shut down the ride if the water level became too low [3]. The only indication the operators had was to notice the water stain level on the wall of the ride canal to determine if the water was too low [19]. When the water level dropped on 25th October, one empty raft became stuck on the rails near the base of the conveyor. As the second raft containing four adults and two children engaged with the conveyor it was then driven into the stuck raft. A safety review conducted after the event stated that the primary cause was a lack of a safety rated water level detection system. It was believed that such a device could have been fitted for as little as AUD 3,000. The ride had been refurbished some seven months prior to the incident, but a device was not fitted. The nearby log ride at the park had such a safety device fitted [18].

3.2.4 Emergency Stop

Around two years prior to the incident an external safety report was written suggesting the emergency stop system should be simplified, however this was rejected by the company as the emergency system in place was seen to be an "acceptable risk". The stopping procedure for the ride contained several individual steps. There was a single button which would stop all components of the ride, however this was located at the unloading area and not the control panel where the operators worked. [20]

There was a great amount of confusion regarding the emergency stop button and what it actually did. The senior operator stated that they were "surprised" to learn there was an emergency stop that would stop all components and the junior operator on her first day was unsure what the button did [7]. She stated she was told to not worry about the button because she did not need to use it and testified, she "would have waited for my senior operation to tell me to press it". She also did not know the button would stop the conveyor from moving. [10] The maintenance workers who performed the daily checks on the ride also stated they did not know what the emergency stop button did as they never tested it [21].

3.2.5 Maintenance

Seven months prior to the incident a directive to cut all repairs and maintenance spending was ordered by the owners. At the time of the incident the facility was approximately AUD125,000 over maintenance budget for the year to date. Capital expenditure could continue to be spent during this time [16]. One maintainer stated that he performed comprehensive daily, monthly and annual checks on the ride, which include consumable

—

replacements such as nuts and bolts [22]. When the regulator inspected the ride after the incident, he noted it was in "generally poor condition" with excessive corrosion and rust on the electrical cables [23].

The maintenance planner noted that if spending could be deferred it would be. He also noted that the ride was shut annually for review, but on the nine occasions he witnessed a shutdown, he never saw an engineer inspect the ride nor were any risk assessment carried out [24].

3.2.6 Seatbelts

The Thunder River Rapids rafts had seat belts fitted for the safety of the riders. It should be noted that in May 2017, some seven months after the Dreamworld incident, that an 11 year old girl died on a similar ride, Splash Canyon, at Drayton Manor in the United Kingdom. The Splash Canyon ride rafts had no seatbelts installed at all, and as the child stood to swap seats with a fellow rider she fell overboard and struck her head. She died days later in hospital. This was not the first time a rider had fallen from the Splash Canyon ride [25]. This incident highlights the importance of proper restraints on these types of rides.

The seatbelts provided on the Thunder River Rapids rafts consisted of Velcro straps. It was vital that these were engaged correctly to provide the necessary restraint. Unfortunately, the Velcro was worn, and this made correct engagement more difficult. It is thought that at least one of the four riders may have survived the incident if the rafts had a better quality seatbelt, such as a mechanical buckle type that are on aircraft. This is because as the raft was shaken by the mechanism the belt failed to hold and the rider dropped to the conveyor system below [23, 26].

3.3 Systems & Procedures

Dreamworld has a series of operating procedures for the Thunder River Rapids ride however they were thought to be confusing [8]. During the investigation into the 2001 incident where empty rafts flipped, it was determined that the operator failed to follow the startup and emergency procedures. The investigation also reviewed the operating procedures and determined that the chance of a similar incident occurring with passengers on the rafts "is nil" [6]. A senior engineer believed that the cause of the 2016 incident was the operator not following procedures and that if procedures were followed the "system is a safe system" [27, 28].

There was also a lack of adequate emergency response procedures. Operators had discussed wanting better procedures and undertaking emergency drills, however, even two years after the incident, these have not been conducted at all [29].

3.4 Assurance

—

Over the history of the ride there had been a number of third-party reviews, inspections or audits carried out. Three years prior the incident a safety audit was conducted, and it recommended replacing the complicated emergency shut down process with a single emergency stop button, but this recommendation was never actioned [30].

There was also a legal requirement for the rides to have a "certificate to operate" and the Thunder River Rapids ride certificate was due to expire in January 2016. The company sought at extension as they were unable to find a "competent person" as defined under the legislation to inspect the ride [16]. Eight day before the incident the ride was inspected by a third-party engineer. He declared the ride as "structurally sound and properly functioning" even though he did not complete the full safety audit and merely did a visual inspection. He also did not inspect the log book as was required by law in a push to have the ride certified before it expired, though he admitted that he could have refused to certify the ride because of a "lack of documentation" [31, 32].

3.5 Human Factors

The control panel for the ride was confusing [3] and the operators needed to perform 36 different tasks in a one minute period, consistently [10]. This made for a system that encouraged human error, as it was not possible for the operators to be effective under these conditions. In addition to this, the engineering team believed the operators needed to simply follow procedures for the ride to be safe, not taking into account how difficult the procedures were or the actual workload. A senior engineer stated "In my view if procedures are followed the system is a safe system, you may as well not open the park. Procedures are there to be followed, if we don't follow procedures in everyday life, things happen." This blind focus on procedures does not take into account how the human interfaces with the machine and how the system advantages or disadvantages human performance.

3.6 Culture

A maintenance worker stated that he believed that the "safety culture at Dreamworld is very sound" [22] however this is in contrast to other workers who testified that familiarity had bred contempt with respect to the Thunder River Rapids ride [16], where new attractions were given more focus that existing ones [17]. Junior operators believed there was a culture of fear and had engaged with their union over their working conditions and the demands put on them regarding lack of rest breaks [33].

The culture also appeared to not learn from incidents, with the 2001 incident of flipping rafts said to be a one off and having "nil" chance of occurring with riders on board [6]. even though the consequence appeared to be understood at the time with an engineer writing in an email "I shudder when I think if there had been guests on the ride." [3].

The park relied on its apparent good safety history rather than challenging how and when an incident could occur [11]. They did not have an effective incident tracking system to

—

help with the learning process. They introduced a database but there was "general reluctance to use what was a fairly complex database" by the operators [34].

Safety did not appear to be a value in the company culture. The former safety manager stated that "everyone thought it was someone else's responsibility". He believed that deficiencies in the corporate structure led to a lot of problems. In 2018 his corporate safety role was made redundant.

Around five months prior to the incident, a memo was issued to staff to 'remain vigilant when monitoring conveyor movements" for items that might interfere with the rafts. This suggests there was some understanding of hazards and consequences, but it does to appear to have driven any improvement activities or modifications to prevent an incident [24].

4 Ramifications

The full ramifications of this incident are not yet clear. The Queensland government has initiated some legislative improvements to manage theme parks, but there is likely to be more recommendations coming from the Coronial Inquest when it releases its findings. Aside from the tragic loss of four lives, from the Dreamworld perspective, their guest numbers have remained low since the incident occurred, and it is struggling to attract patrons. This has led to their parent company suffering financial losses for the past two years and 50% decreases in the share price over the two year period [35].

The incident also provided an interesting lesson in crisis management. The company was scheduled to hold their annual general meeting in the days following the incident. The meeting went ahead and in it the CEO was awarded a performance bonus of $840,000 including both shares and a cash component despite the four deaths days before [36]. From a public relations perspective this was not a successful event. The CEO also stated they had been in contact with the victim's families, but this was publicly disputed by the families [37]. This further damaged the company reputation at a difficult time.

This incident will not only have ramifications for the families of the deceased, but for the survivors and the staff who responded due to the nature of injuries sustained by the victims. Four former staff members are suing Dreamworld for psychological damage [38].

These ramifications show how a process safety type incident can have significant impact on a business and its ability to continue operations in a positive way.

5 Could the Principles of Process Safety Have Helped?

As had been shown, there were clear and systematic failures across the whole organization that led to this tragedy. Starting with knowledge and competence, which relates to

—

organizational capability, had the operators been trained better in the safe operation of the ride as well as in emergency response, their performance that day may have been improved. From an engineering and design perspective, the design and installation of the ride was inadequate, it underwent undocumented changes which introduced hazards, and it was not adequately maintained. The systems and procedures for the operation and maintenance the ride was confusing and did not appear to contain necessary information. The assurance processes conducted did not seem to be thorough, rather they focused on keeping the ride operating and recommendations were not actioned. The ride was not designed with human factors in mind, from the high workload to the confusing operating panel. The organization culture did not seem to place a value on safety, it only focused on cost reduction. It also did not appear to be a learning culture, where lessons from the past are understood and embedded. Lastly at the time of crisis, the overall leadership displayed by the parent company was publicly found to be lacking.

There are transferrable learnings for the process industries from this incident, with some clear questions to ask. How are you managing your competency program to ensure you have the right capability? Are your designs adequate, maintained and any changes managed? Are your procedures thorough, identifying hazards, consequences and controls and easy to comprehend? How are you monitoring the operation of your facility? Does the design and operation of your facility enhance human performance or make it more difficult? Does your culture support safe operations?

6 Conclusion

A focus on the elements of process safety can help identify what is important in a business. In this instance there had been recommendations for a low-cost instrumented safety system to be installed that would have prevented the incident from occurring. The importance of this was not understood by the organization, and the opportunity to upgrade the ride during its recent refurbishment was ignored. It is clear that the hazard and potential consequences and controls were not understood. As is often the case, when the hazard is not understood and therefore not managed, people suffer tragic consequences.

7 References

[1] Kerin, T. *Managing Process Safety. In the Core Body of Knowledge for Generalist OHS Professionals*. Safety Institute of Australia, Tullamarine, VIC. 2017

[2] ISC (IChemE Safety Centre). Safety Centre Framework, Available at https://www.icheme.org/knowledge/safety-centre/framework/, Accessed on January 2, 2019

—

[3] ABC News, Available at https://www.abc.net.au/news/2018-06-18/dreamworld-fatal-accident-ride-had-broken-down-inquest-told/9871898, Accessed on January 2, 2019

[4] Australian Financial Review, Available at https://www.afr.com/news/dreamworlds-rapids-ride-different-from-others-around-the-world-20161101-gsfd07, Accessed on January 2, 2019

[5] 9 News, Video 11, Available at https://www.9news.com.au/2018/12/06/03/33/dreamworld-inquest-enters-final-days, Accessed on January 2, 2019

[6] news.com.au, Available at https://www.news.com.au/national/queensland/courts-law/coroner-james-mcdougall-requests-proof-all-dreamworld-rides-were-certified/news-story/e77275b97b0bbee9c1c541bda6df0035, Accessed on January 2, 2019

[7] ABC News, Available at https://www.abc.net.au/news/2018-06-19/dreamworld-operator-on-first-day-when-four-people-died-on-ride/9885214, Accessed on January 2, 2019

[8] ABC News, Available at https://www.abc.net.au/news/2018-06-27/dreamworld-trainer-promoted-since-four-killed-on-ride/9914188, Accessed on January 2, 2019

[9] 9 News, Video 5, Available at https://www.9news.com.au/2018/12/06/03/33/dreamworld-inquest-enters-final-days, Accessed on January 2, 2019

[10] ABC News, Available at https://www.abc.net.au/news/2018-06-20/dreamworld-operator-impossible-load-responsibilities-manage-ride/9889776, Accessed on January 2, 2019

[11] 9 News, Video 1, Available at https://www.9news.com.au/2018/12/06/03/33/dreamworld-inquest-enters-final-days, Accessed on January 2, 2019

[12] 9News, Available at https://www.9news.com.au/2018/11/20/03/32/dreamworld-ride-passed-wphs-safety-audit, Accessed on January 2, 2019

[13] ABC News, Available at https://www.abc.net.au/news/2018-10-08/dreamworld-inquest-hears-total-failure-to-ensure-safety/10350034, Accessed on January 2, 2019

[14] Intamin Worldwide, Available at https://www.intaminworldwide.com/infinity-falls-orlando/ , Accessed on December 12, 2018

—

[15] 9 News, Video 8, Available at https://www.9news.com.au/2018/12/06/03/33/dreamworld-inquest-enters-final-days, Accessed on January 2, 2019

[16] ABC News, Available at https://www.abc.net.au/news/2018-06-25/dreamworld-inquest-told-of-ride-maintenance-cutbacks/9906214, Accessed on January 2, 2019

[17] Australian Financial Review, Available at https://www.afr.com/real-estate/commercial/hotels-and-leisure/dreamworld-prioritised-new-attractions-over-maintenance-of-old-rides-inquest-20180626-h11urb, Accessed on January 2, 2019

[18] 9 News, Video 10, Available at https://www.9news.com.au/2018/12/06/03/33/dreamworld-inquest-enters-final-days, Accessed on January 2, 2019

[19] ABC News, Available at https://www.abc.net.au/news/2018-06-21/dreamworld-ride-operator-highly-distressed-give-evidence-inquest/9893308, Accessed on January 2, 2019

[20] news.com.au, Available at https://www.news.com.au/national/queensland/courts-law/dreamworld-engineer-tells-families-i-apologise-for-not-being-to-make-eye-contact-with-you/news-story/40545dc747c95eb96b0293c888d77d96, Accessed on January 2, 2019

[21] Sunshine Coast Daily, Available at https://www.sunshinecoastdaily.com.au/news/dreamworld-thunder-river-rapids-button-wasnt-regul/3543997/, Accessed on January 2, 2019

[22] ABC News, Available at https://www.abc.net.au/news/2018-06-26/dreamworld-ride-operator-says-conveyor-was-never-stopped/9910192, Accessed on January 2, 2019

[23] news.com.au, Available at https://www.news.com.au/national/queensland/courts-law/safety-report-shows-simple-change-on-dreamworld-ride-could-have-saved-lives/news-story/9c1fa2c195efd12f86a3e8ad516ea1b8, Accessed on January 2, 2019

[24] 9 News, Video 20, Available at https://www.9news.com.au/2018/12/06/03/33/dreamworld-inquest-enters-final-days, Accessed on January 2, 2019

[25] news.com.au, Available at https://www.news.com.au/travel/travel-updates/incidents/girl-11-dies-after-falling-into-the-water-on-splash-canyon-ride-during-school-trip-to-uk-theme-park/news-story/f83330b0ce12ea11df2d089ece676519, Accessed on January 2, 2019

—

[26] 9News, Available at https://www.9news.com.au/2018/12/06/03/33/dreamworld-inquest-enters-final-days, Accessed on January 2, 2019

[27] 9News, Available at https://www.9news.com.au/2018/10/16/14/43/dreamworld-inquest-thunder-river-rapids-ride-engineers-give-evidence, Accessed on January 2, 2019

[28] 9News, Available at https://www.9news.com.au/2018/10/17/10/49/dreamworld-inquest-engineer-to-appear-again, Accessed on January 2, 2019

[29] Australian Financial Review, Available at https://www.afr.com/news/dreamworld-ride-trainer-promoted-after-tragedy-staff-wanted-emergency-training-20180628-h11yke, Accessed on January 2, 2019

[30] 9 News, Video 12, Available at https://www.9news.com.au/2018/12/06/03/33/dreamworld-inquest-enters-final-days, Accessed on January 2, 2019

[31] ABC news, Available at https://www.abc.net.au/news/2018-10-09/dreamworld-ride-declared-safe-days-before-tragedy/10352164, Accessed on January 2, 2019

[32] ABC News, Available at https://www.abc.net.au/news/2018-10-17/dreamworld-supervisor-sympathy-has-families-in-tears-inquest/10387186, Accessed on January 2, 2019

[33] 9 News, Video 9, Available at https://www.9news.com.au/2018/12/06/03/33/dreamworld-inquest-enters-final-days, Accessed on January 2, 2019

[34] 9 News, Video 16, Available at https://www.9news.com.au/2018/12/06/03/33/dreamworld-inquest-enters-final-days, Accessed on January 2, 2019

[35] Australian Financial Review, Available at https://www.afr.com/research-tools/AAD/share-prices?days=1095, Accessed on November 26, 2018

[36] The Guardian, Available at https://www.theguardian.com/australia-news/2016/oct/27/deborah-thomas-set-to-receive-840000-in-bonuses-despite-dreamworld-deaths, Accessed on December 19, 2018

[37] ABC News, Available at https://www.abc.net.au/news/2016-10-27/dreamworlds-shareholders-ask-parent-company-to-support-victims/7970104, Accessed on January 2, 2019

—

[38] news.com.au, Available at https://www.news.com.au/lifestyle/real-life/news-life/dreamworld-first-responders-to-sue-theme-park-for-psychological-damage/news-story/e8fd7399eec9d165c657dbdb3ac35968, Accessed on January 2, 2019

19 AIChE Spring Meeting
15TH GLOBAL CONGRESS ON PROCESS SAFETY
NEW ORLEANS • 2019

Fatal Toxic Chemical Release at DuPont

Tamara Qureshi
U.S. Chemical Safety and Hazard Investigation Board
1750 Pennsylvania Ave, NW Suite 910 Washington, DC 20006
tamara.qureshi@csb.gov

Dan Tillema, Steve Cutchen, Lauren Grim, and Chris Lyon
U.S. Chemical Safety and Hazard Investigation Board
1750 Pennsylvania Ave, NW Suite 910 Washington, DC 20006

Prepared for Presentation at
American Institute of Chemical Engineers
2019 Spring Meeting and 15th Global Congress on Process Safety
New Orleans, LA
March 31 – April 3, 2019

Fatal Toxic Chemical Release at DuPont

Tamara Qureshi
U.S. Chemical Safety and Hazard Investigation Board
1750 Pennsylvania Ave, NW Suite 910 Washington, DC 20006
tamara.qureshi@csb.gov

Dan Tillema, Steve Cutchen, Lauren Grim, and Chris Lyon
U.S. Chemical Safety and Hazard Investigation Board

Keywords: DuPont, toxic release, gas detection, alarm

Abstract

On November 15, 2014, four workers were killed after approximately 24,000 pounds of highly toxic methyl mercaptan released from an insecticide production unit (Lannate® Unit) at the E.I. du Pont de Nemours chemical manufacturing facility in La Porte, Texas. This paper will discuss the events leading to the methyl mercaptan release, the deficiencies in DuPont La Porte's toxic gas detection system, and offer lessons on how to prevent some of the deficiencies in the toxic gas detection system from occurring. An effective toxic gas detection system can protect workers by alerting them of a potentially dangerous situation and preventing them from unknowingly entering a potential toxic chemical release.

1 Introduction

On November 15, 2014, approximately 24,000 pounds of highly toxic methyl mercaptan released from an insecticide production[1] unit (Lannate® Unit) at the E.I. du Pont de Nemours chemical manufacturing facility in La Porte, Texas. Four workers at the DuPont La Porte facility entered the Lannate® Unit's manufacturing building (manufacturing building) and the highly toxic methyl mercaptan release. Three of these workers were responding to a coworker's distress call from inside the manufacturing building. The fourth worker entered after all the workers inside the manufacturing building stopped communicating with workers inside the control room. The Lannate® Unit's methyl mercaptan toxic gas detection system did not effectively warn workers of the large chemical release occurring within the manufacturing building. Additionally, this toxic gas detection system did not provide emergency responders with sufficient information to effectively evaluate the potential harm to the public.

[1] The Insecticide Business Unit (IBU) was divided into two primary insecticide product lines: Lannate®, which is DuPont's trade name for methomyl; and Agricultural Products Intermediates (API), which produced insecticides and nematicides known as Vydate®, the trade name for oxamyl [3, p. 11].

The DuPont La Porte methyl mercaptan toxic gas detection system had several deficiencies. First, the methyl mercaptan detector system did not meet its design intent. The alarm set point for methyl mercaptan concentration and the method of conveying alarms did not forewarn workers of the immediately dangerous to life and health (IDLH) conditions they would encounter by entering the manufacturing building. The methyl mercaptan gas detectors only alarmed on the Lannate® control room operator's computer screen. There was no audible or visible alarm provided in the field for outside personnel. Second, the placement of the methyl mercaptan toxic gas detectors did not fully portray the extent of the November 15, 2014 methyl mercaptan release. Third, the DuPont La Porte site normalized the hazards associated with methyl mercaptan. Finally, the alarms did not prompt protective action or investigations into the sources of alarms. During the hours prior to the November 15, 2014 incident, multiple highly toxic chemical gas detectors alarmed (sounded). However, the DuPont emergency response team was not notified nor was the area cleared of personnel. In addition, methyl mercaptan releases on November 13 and 14, 2014, were identified by methyl mercaptan detectors, but were never reported as releases nor investigated as serious process safety incidents. Personnel believed they knew the source of the releases and normalized their responses to these alarms. Consequently, there was a delayed awareness of the release by operations, and workers who entered the manufacturing building after the release began were not alerted of nor prepared for the lethal concentrations of methyl mercaptan being released, resulting in the death of two of these workers.

The DuPont La Porte incident is the CSB's third investigation of a fatal DuPont incident in five years.[2] On September 30, 2015, the CSB issued a presentation, safety video, and an Interim Recommendations report [1],[3] which detailed needed safety improvements at the facility and issued recommendations to the DuPont La Porte facility to be implemented prior to resuming operations of the Lannate® Unit.[4] Due to the amount of methyl mercaptan stored at the DuPont La Porte site, the Lannate® process was covered by Occupational Safety Health Administration's (OSHA) Process Safety Management (PSM) program and the Environmental Protection Agency's (EPA) Risk Management Plan (RMP) Rule [2].

An effective toxic gas detection system can make the difference between personnel being prepared for a potentially dangerous situation or entering a potential toxic chemical release. To be effective, a toxic gas detection system needs to consider the chemical's hazards, the gas detector's

[2] The CSB also investigated an incident at a DuPont plant in Belle, West Virginia (three incidents that occurred in a 33-hour period over January 22-23, 2010, resulting in one fatality), and another in Buffalo, New York (a single incident that occurred on November 9, 2010, resulting in one fatality and one serious injury) [16], [17].

[3] In the Interim Recommendations, the CSB identified significant process safety management deficiencies at the DuPont La Porte facility, including delayed maintenance of safety-critical equipment, lack of written procedures, poor hazard analysis practices, lack of implementation of important inherently safer design concepts, and a lack of hazard recognition.

[4] In response to the CSB findings and recommendations, DuPont committed to making changes at the DuPont La Porte facility. DuPont La Porte, however, was not able to address all of these findings and recommendations, because in the spring of 2016 DuPont announced that it would not reopen the Lannate® Unit. The company had determined that "significant changes in market conditions during the period of the shutdown [would] persist over the long term [14]" and the cost required for the restart "[was] not a long-term viable and cost-efficient option for the DuPont Crop Protection business [15]." In 2017, DuPont dismantled and removed the buildings and equipment from the DuPont La Porte facility that were associated with its insecticide and herbicide business units, which included the Lannate® Unit.

capabilities, the location of the gas detectors, and the alarms associated with the gas detectors. This paper will discuss some of the events leading to the methyl mercaptan release, the deficiencies in DuPont La Porte's toxic gas detection system and offer lessons on how to prevent some of the deficiencies in the toxic gas detection system from occurring.

2 Events Leading to the Incident

On Monday, November 10, 2014, an inadvertent chemical dilution caused operating difficulties that forced a shutdown of DuPont La Porte's Lannate® Unit [3, pp. 20-21]. On Wednesday, November 12, 2014, operators tried to restart the Lannate® process. During the shutdown, the methyl mercaptan piping to the reaction section of the process plugged due to the formation of a clathrate hydrate (hydrate),[5] halting restart. At that time, operators did not know the source of the plugging and began troubleshooting the process to try to clear the piping, unaware that the solid hydrate had formed.

Two days later, on Friday, November 14, 2014, the troubleshooting efforts to clear the plugging were still ongoing. That morning, a technical team composed of engineers and other employees experienced in the Lannate® process, met with operations personnel to discuss troubleshooting options to clear the plugging. This meeting identified the likely scenario that water had entered the methyl mercaptan system and a solid hydrate formed as a result. Based on their understanding of the hydrate,[6] the technical team asked operators to put hot water on the outside of the methyl mercaptan piping, under the insulation, to warm the piping and its contents, to break-up (dissociate[7]) the plugging. The team also realized that when heated, methyl mercaptan would expand, requiring a safe place to vent to avoid over-pressuring the piping. To address this concern, operations personnel opened valves between the methyl mercaptan piping and a waste gas vent header, not recognizing that this alignment created a pathway for liquid methyl mercaptan to eventually release from the piping inside the manufacturing building when drain valves were opened. In addition, when operations personnel opened other valves while troubleshooting, such as drain valves, methyl mercaptan released to the atmosphere causing a strong methyl mercaptan odor[8] that could be smelled by site personnel.

[5] At low temperatures (≤ 52 °F) water and methyl mercaptan form a solid, ice-like material called methyl mercaptan clathrate hydrate [10]. Temperatures in the Houston area for the 24 hours preceding the incident averaged approximately 40 °F and had been consistently below 55 °F since Tuesday, November 11, 2014. Although the potential for methyl mercaptan hydrate formation had been identified in DuPont's methyl mercaptan technical standard and in a process hazard analysis years earlier, DuPont never implemented safeguards, such as heat tracing, or developed a procedure to dissociate the hydrate safely [3, p. 22].

[6] DuPont's methyl mercaptan technical standard identifies the potential to form a hydrate at low temperatures. The technical standard states, "[methyl mercaptan] will form a hydrate with water, which is a solid below 40 deg F [degrees Fahrenheit] per information provided by a [methyl mercaptan] supplier."

[7] Methyl mercaptan clathrate hydrates dissociate back to the liquid phases of methyl mercaptan and water at approximately 52 °F [10].

[8] Methyl mercaptan has a low odor threshold. It gives off a smell like rotten cabbage. It can be smelled at only two parts per billion (ppb) [9, p. 1]. OSHA's permissible exposure limit (PEL), however, is 10 parts per million (ppm), which is 5,000 times the minimum concentration it can be smelled. The low odor threshold of mercaptans makes them useful as fuel gas odorants [9, p. 2].

During this troubleshooting process, methyl mercaptan released both outside and inside the manufacturing building, triggering 32 methyl mercaptan gas alarms on the control panel throughout the 17 hours preceding the incident. While methyl mercaptan is a toxic and flammable chemical, and at high enough concentrations lethal, operations personnel did not respond in accordance with the nature of the emergency circumstances, because they associated the alarms with ongoing troubleshooting efforts. They did not perceive the methyl mercaptan alarms as signifying a serious hazard because they normalized[9] both the methyl mercaptan odor within the Lannate® Unit as well as the detector alarms associated with them.

Early in the morning on Saturday, November 15, 2014, the hot water from the hoses warmed the hydrate plugging, causing it to dissociate, clearing the plugging. Due to the valve alignment, liquid methyl mercaptan flowed into the waste gas vent header, located inside the manufacturing building [3, pp. 25-26]. This vent header, however, was not intended for liquid methyl mercaptan.

At 2:51 am alarms began to sound on the control system indicating high pressure in equipment located inside the manufacturing building. Operations personnel did not realize that liquid methyl mercaptan was causing the high pressure. Instead, they attributed the high pressure to a common, long-standing problem with process condensate (that DuPont La Porte personnel believed to be mostly water) accumulating in the vent header piping [3, p. 28]. The Lannate® control room operator (Board Operator) contacted the night shift supervisor (Shift Supervisor) and another operator (Operator 1) separately, asking them to help troubleshoot the high-pressure situation. The Shift Supervisor and Operator 1 each went separately to the manufacturing building, likely with the intent to drain liquid from the vent header piping manually.[10] Sometime between 3:01 am and 3:13 am, two sets of drain valves on the vent header piping, located on the third floor of the manufacturing building were opened (Figure 1). But instead of being comprised mostly of water as workers believed, liquid methyl mercaptan that had filled the piping escaped out of the valves, vaporized, and killed the Shift Supervisor.

Between 3:24 am and 3:26 am, three methyl mercaptan detectors inside the manufacturing building sensed at least 25 parts per million (ppm) methyl mercaptan[11] and triggered alarms at the control panel. The Board Operator—still focused on reducing the high pressure—did not realize a major chemical release inside the manufacturing building was occurring.

---

[9] Normalization of deviance is a phenomenon in which individuals or work teams gradually accept a lower standard of performance until the lower standard becomes the norm. It is typically the result of conditions slowly changing and eroding over time [11, p. 4].

[10] In 2011, DuPont installed a NOx (nitrogen oxides) reducing scrubbed (NRS) incinerator system. The waste gas vent header piping to the NRS Incinerator was designed without sufficient consideration of liquid accumulation. It contained low points where liquid could accumulate, and there was no engineered equipment provided, such as a knock out drum, to remove liquid safely from the waste gas vent header piping system. The liquid accumulation in the piping would create pressure in process equipment. Pressure buildup within the equipment was resolved by manually draining liquid from the waste gas vent header. See DuPont La Porte, Texas Chemical Facility Toxic Chemical Release: Interim Recommendations [3, p. 18].

[11] The methyl mercaptan detectors alarm at 25 ppm–well above the National Institute of Occupational Safety and Health (NIOSH) recommended exposure limit (REL) of 0.5 ppm and the OSHA enforceable (PEL) ceiling limit of 10 ppm. OSHA recommends that employers consider using the alternative occupational exposure limits because the agency believes that exposures above some of these alternative occupational exposure limits may be hazardous to workers, even when the exposure levels are in compliance with the relevant PELs [5]. The Cal-OSHA permissible exposure limit (PEL) is an 8-hour time-weighted average (TWA) of 0.5 ppm and NIOSH recommended exposure limit (REL) is a 15- minute ceiling limit of 0.5 ppm [7], [6].

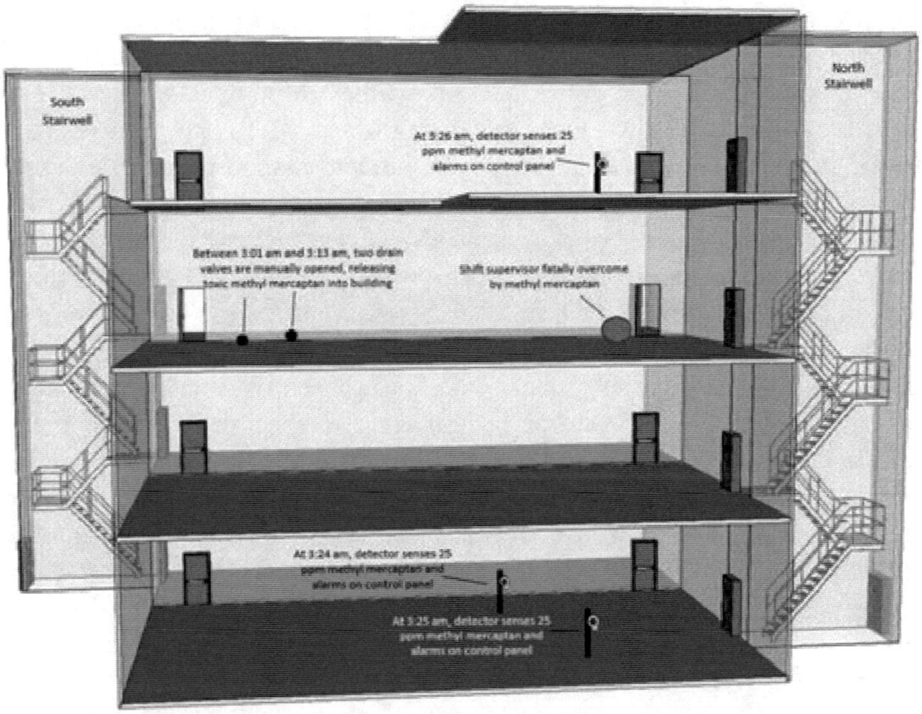

Figure 1. Depiction Showing Location Where Drain Valves Were Opened. These drain valves released toxic methyl mercaptan into the manufacturing building. Methyl mercaptan detectors on the first and fourth floors detected high concentrations of methyl mercaptan shortly after the release began. This depiction is a simplified graphic of the manufacturing building and does not show the location of Operator 1. Source: CSB

At about 3:30 am, Operator 1 made an urgent call for help over the radio. Various personnel interpreted her communication in different ways, generally as either "We need help!" or "I need help on the fourth floor!" The Board Operator tried to get more information through radio communication, but neither the Shift Supervisor nor Operator 1 ever responded [3, p. 29].
Two operators (Operator 2 and Operator 3) who were in the control room and heard the distress call ran to the manufacturing building to help. Another operator (Operator 4) saw Operator 2 and Operator 3 running to the manufacturing building, and he followed them. None of these operators knew there was a major release of toxic methyl mercaptan inside of the manufacturing building, and therefore they did not wear any respiratory protection when they ran into the manufacturing building. The manufacturing building lacked automatic visual or audible alarms to alert field workers or prevent them from entering a potentially toxic atmosphere.

Operator 2, Operator 3, and Operator 4 each entered the south stairway and took different routes inside the manufacturing building to try to find Operator 1. Operator 2 went to the third floor, where he was fatally overcome by methyl mercaptan [3, p. 29]. Operator 4 went to the second floor. He walked about ten feet and hit what he later described as a "wall" of methyl mercaptan, but he managed to retreat to the stairwell [3, p. 29]. Operator 3 went to the fourth floor but did not find anyone. He then announced on the Lannate® Unit's public-address system that he did not see anyone on the fourth floor. The Board Operator responded that the Shift Supervisor and Operator 1 may be on the third floor. Operator 3 then began to feel light headed. He made his way to the stairwell and lost consciousness while descending from the fourth floor [3, p. 29].

When Board Operator could not reach people inside the manufacturing building, an operator in the control room, Operator 6[12] (the brother of Operator 2), then grabbed three 5-minute escape respirators. Other operators in the control room warned him not to enter the manufacturing building because they did not know what was going on or where the non-responsive operators were. Operator 6, however, took the escape respirators and rushed to the manufacturing building [3, p. 29].

At about 3:40 am, Operator 6 encountered Operator 4 in the south stairway of the manufacturing building and put an escape respirator on him. The breathing air helped Operator 4 recover, and he exited the manufacturing building safely. A worker—likely Operator 6—then manually activated the manufacturing building fume release alarm, an alarm intended to alert area workers of a toxic chemical release in the building. Operator 6 then went to the third floor of the manufacturing building. At some point inside the manufacturing building after helping Operator 4, Operator 6 retrieved a self-contained breathing apparatus (SCBA). He tried to rescue his brother (Operator 2) by putting one of the escape respirators on his face. Operator 6 was then fatally overcome by methyl mercaptan before he could connect his own respirator mask to his SCBA tank.

3 Lessons from the DuPont La Porte Incident

3.1 Ensure Toxic Gas Detector Systems Operate as Intended

DuPont La Porte's methyl mercaptan technical standard stated, "[a]utomatic or continuous specific chemical detection systems *must* be installed where a [methyl mercaptan] release could have any off-site impact [...] [Emphasis in the original]." DuPont La Porte's methyl mercaptan toxic gas detection system consists of methyl mercaptan detectors, alarms, distributed control system (DCS), control panel, and a control room operator. The manufacturing building is equipped with gas detectors for methyl mercaptan in various locations. The third floor, where the release occurred, however, did not have any methyl mercaptan detectors. The detectors inside the manufacturing building were also set to alarm on the control panel when they detected a 25 ppm concentration of methyl mercaptan and were intended to protect workers from exposure to methyl mercaptan.[13]

DuPont used hydrogen sulfide (H_2S) detectors to identify methyl mercaptan. DuPont records indicate that the manufacturer has stated the detector should be capable of detecting methyl mercaptan within 1 ppm.[14] The alarms activated when the detectors sensed the hydrogen sulfide equivalent to 25 ppm methyl mercaptan.[15] The implementation of the methyl mercaptan toxic gas

[12] Operator 6 was a board operator for another process in the Lannate® Unit.

[13] The 25 ppm alarm set point is 50 times the NIOSH's REL of 0.5 ppm and more than double OSHA's ceiling limit of 10 ppm [7]. Therefore, alarming at this detection level could not protect workers from exposure to methyl mercaptan.

[14] DuPont La Porte's detectors for methyl mercaptan are calibrated for 0-50 ppm using a hydrogen sulfide (H_2S) calibration gas.

[15] The methyl mercaptan detectors were evaluated post-incident. Although the detectors inside the building have the ability to span 0-50 ppm, these detectors were set to span 0-25 ppm. Post-incident, the span was increased to 50 ppm. This configuration of the alarms limited the amount of information communicated to DuPont La Porte personnel.

detection system inside the manufacturing building, however, did not meet the design intent, federal safety regulations, or DuPont corporate standards.

The methyl mercaptan detectors outside and inside the manufacturing building were installed for two different purposes—the outside detectors were geared towards an offsite release and the inside detectors were meant to protect workers from a hazardous environment. DuPont La Porte documentation stated that the basis for the alarms outside the manufacturing building was to provide an early warning mechanism for a potential offsite methyl mercaptan release. The alarm setpoint was based on the ERPG (Emergency Response Planning Guidelines)-2[16] concentration[17] for methyl mercaptan which is 25 ppm. DuPont La Porte installed the methyl mercaptan detectors inside the manufacturing building after the methyl mercaptan detectors outside the building. Unlike the outside detectors, there was no documentation for the basis for setting the alarm setpoint to 25 ppm for the methyl mercaptan detectors inside the manufacturing building nor was there evidence that these setpoints were set at 25 ppm to prevent methyl mercaptan alarms from becoming a nuisance. It appears that the setpoints were copied from the outside methyl mercaptan detectors even though the inside detectors served a different purpose—worker protection. The methyl mercaptan detectors alarmed at 25 ppm—well above the NIOSH recommended exposure limit (REL) of 0.5 ppm and the OSHA enforceable PEL ceiling limit of 10 ppm.

Depending on the purpose of a toxic gas detection system alarm, there are different sets of values for facilities to consider. For instance, ERPG values are designed for the protection of the public during an emergency and not intended for the everyday routine protection of workers [4].[18] For worker protection, OSHA recommends that employers consider using the lower alternative occupational exposure limits because the agency believes that exposures above some of these alternative occupational exposure limits may be hazardous to workers, even when the exposure levels are in compliance with the relevant PELs [5]. For example, the Cal-OSHA permissible exposure limit (PEL) is an 8-hour time-weighted average (TWA) of 0.5 ppm and NIOSH's REL is a 15- minute ceiling limit of 0.5 ppm [6], [7].

Additionally, information from the methyl mercaptan detectors inside the manufacturing building only displayed on the DCS monitor and was not recorded by the site's DCS and data historian [3, p. 37]. CSB investigators, however, were able to obtain information from alarm logs about the detectors.

DuPont La Porte's management of change documentation for installation of methyl mercaptan detectors inside the manufacturing building stated the condition prompting the change was the

---

[16] ERPG-2 is the maximum airborne concentration below which nearly all individuals could be exposed for up to 1 hour without experiencing or developing irreversible or other serious health effects or symptoms which could impair an individual's ability to take protective action. At concentrations above ERPG-2, some members of the community may experience significant health effects which could impair them from taking protective action. Emergency responders often alert the community to take protective action in areas susceptible to the ERPG-2 concentration and above [19].

[17] The ERPG values for methyl mercaptan in parts per million (ppm) are: ERPG-1, 0.005; ERPG-2, 25; ERPG-3, 100 [18, p. 6].

[18] Emergency Response Planning Guidelines (ERPGs) are chemical concentration levels that are used during emergency response and when preparing for emergencies. They are designed to anticipate public health effects not occupational health effects from exposure to airborne chemicals [4].

possibility that personnel could enter a potentially hazardous location without knowing methyl mercaptan had been released:

> *Present Conditions and Description of Proposed Change:*
>
> *Personnel can enter a potentially hazardous location without knowing MeSH [methyl mercaptan] has been released to the environment.*

The methyl mercaptan gas detectors, however, only alarmed on the Lannate® control room operator's computer screen.

DuPont La Porte site did not have audible or visual alarms on or in the manufacturing building to alert field personnel of a methyl mercaptan release automatically. DuPont relied on verbal communication of alarms that automatically displayed on a continuously manned control board. Without being made aware of the major chemical release inside the building, Operator 2, Operator 3, and Operator 4 entered without respiratory protection. As a result, Operator 2 was killed from methyl mercaptan exposure, and Operator 3 and Operator 4 were exposed to dangerous concentrations of methyl mercaptan.

Facilities should make strategic decisions when choosing a toxic gas detector to ensure that a toxic gas detection system operates as intended. Considerations should include regulatory requirements, NIOSH recommendations, chemical interferences, and utility (worker protection, release modeling, etc.). The methyl mercaptan detectors at the DuPont La Porte facility had a span of 0-50 ppm. This range may be appropriate for early detection of a toxic leak when coupled with alarms and clear instructions to evacuate the premises at lower concentrations. If the goal is to understand other aspects of a release, like potential lethal concentration (150 ppm is methyl mercaptan's IDLH concentration) or modeling an offsite release, then a wider concentration range should be considered.

Additionally, facilities should also consider the use of automatic alarms designed to alert personnel of hazardous conditions—such as a release of a toxic chemical or explosive atmosphere—which can relay critical safety information immediately to personnel, without potential delays inherent in a system that solely relies on a control room operator to verbally communicate safety alarms to operators in the field.

3.2 Ensure Toxic Gas Detector Placement is Strategic

Although the DuPont La Porte site had 13[19] fixed methyl mercaptan detectors inside and outside the manufacturing building, there were no fixed methyl mercaptan detectors located on the third floor inside the manufacturing building where a majority of the November 15, 2014 methyl mercaptan release occurred. Additionally, no workers were provided with either portable or personal methyl mercaptan monitors. Part of the Lannate® process occurred inside the closed manufacturing building, a four-story building containing a large amount of piping and process equipment, dividing walls, multiple stairwells, and an elevator. The Lannate® Unit had methyl

[19] The site had a total of 15 methyl mercaptan detectors—there were two additional methyl mercaptan detectors installed in another process building adjacent to the manufacturing building.

mercaptan process piping both inside and outside the manufacturing building. A facility siting study and a process hazards analysis had identified the necessity of having methyl mercaptan detectors in the Lannate® manufacturing building. These studies identified the need for protecting workers from toxicity and explosivity hazards related to potential methyl mercaptan releases.

DuPont La Porte installed 10 methyl mercaptan detectors near process piping outside the manufacturing building. The design basis for the outside methyl mercaptan detectors was to provide an early warning of a methyl mercaptan release. DuPont La Porte placed these detectors around its methyl mercaptan storage tank and the area where methyl mercaptan could be unloaded. There were, however, no fixed methyl mercaptan detectors along the fence line that could relay whether there was an offsite release.

DuPont La Porte installed three methyl mercaptan detectors inside the manufacturing building. The stated design purpose for the methyl mercaptan detectors inside the building was worker protection. The inside detectors were located within the building on the first and fourth floors. These detector locations were determined based on potential release scenarios from a 2001 process hazard analysis (PHA).

The methyl mercaptan release occurred on the third floor where there were no methyl mercaptan detectors. Although the 2001 PHA did not identify methyl mercaptan detector locations on the third floor, DuPont La Porte routed the waste gas vent header piping along the third floor and this piping was directly connected to the top of the methyl mercaptan storage tank. During abnormal or upset conditions, the waste gas vent header could have been a potential source for a methyl mercaptan release. For example, ineffective hazard management while troubleshooting the plugged methyl mercaptan feed piping during the November 15th incident allowed liquid methyl mercaptan to flow into the waste gas vent header piping towards the NRS incinerator—a location where it was never intended to go.

Additionally, the routine draining of liquid from the waste gas vent header should have triggered a management of change review and considered whether additional safeguards like methyl mercaptan detectors were necessary. As discussed earlier, DuPont La Porte never fully resolved liquid accumulation in the waste gas vent header piping to the NRS incinerator with an engineered solution. Instead, operators were directed to drain the system by area management through written instructions in a daily instruction log book to operations staff [8, p. 18].[20] This routine draining presented potential dangers because DuPont La Porte personnel never sampled or analyzed the liquid workers routinely drained from this system [8, p. 18]. Daily instructions before the incident requested the operators to sample the uncharacterized liquid. DuPont La Porte management wanted samples of the liquid for a subsequent laboratory analysis to identify its chemical composition. Workers reported that the liquid had a strong, noxious odor of sulfur compounds [8, p. 18]. Even though DuPont La Porte was aware that the liquid likely contained more than just water and was unaware of the actual chemical composition [8, p. 18], instructions to workers did

---

[20] Beginning January 14, 2014, and continuing until the incident, the daily instructions directed operators to drain the methyl mercaptan waste gas vent header's "low point" on the 3rd floor once per shift. After the installation of a flexible hose from the waste gas vent header drain valve to the floor, the instructions changed to include directing the liquid through the hose under a running safety shower until the liquid stops flowing from the hose [8, p. 18].

not specify that additional precautions—like wearing additional PPE—were needed when draining the liquid from the waste gas vent header to the floor inside the manufacturing building.

Consequently, DuPont La Porte workers dealt with the common problem of liquid accumulation in the waste gas vent header on a routine basis by draining the liquid (line breaking) without engineered solutions or ensuring the use of safety procedures or personal protective equipment. On November 15, 2014, when the liquid drain valves were opened this time, flammable and highly toxic methyl mercaptan flowed onto the floor and filled the manufacturing building with toxic vapor and there no methyl mercaptan detectors on the third floor to alert personnel of the hazardous conditions that existed there.

As with toxic detector selection, the placement of toxic gas detectors must be strategic in order to capture the potential areas where a release could occur and the extent of the release. The methyl mercaptan detector placement both externally and internally at the DuPont La Porte facility did not operate as intended in part because the detectors did not fully cover the areas where the majority of the methyl mercaptan was released. For example, there was no real time data for emergency responders to use to evaluate the potential for offsite concentrations during the methyl mercaptan release. Although emergency responders conducted offsite monitoring hours after the initial release, the information obtained from air monitoring could not give an accurate portrayal of the methyl mercaptan release. Furthermore, the methyl mercaptan monitors inside the building did not provide accurate information about the release because they were not located near the open drain valves.

Facilities can adopt a variety of methods to select the location of toxic gas monitors with the understanding that they should be prepared for scenarios where the monitoring location may not be effective. One technique that a facility could use is to develop a series of release scenarios and based on those scenarios place monitors in the path of those releases (including by the potential sources of a leak). This technique, however, may not cover all potential release areas. Additionally, facilities should evaluate additional locations for toxic gas detectors if there is a change in process that creates new potential release scenarios. Another technique, if focusing on worker protection, is to give workers personal monitors when they enter toxic gas hazardous areas, especially if these areas are not normally occupied or congested with equipment. At DuPont La Porte, commercially available personal hydrogen sulfide monitors could have been issued to workers. Additionally, if using fixed toxic gas monitors for personnel protection, they should be installed near the breathing zone.

3.3 Ensure Personnel Understand Toxic Gas Hazards

DuPont La Porte personnel did not fully appreciate the effects of methyl mercaptan's physical characteristics. Although methyl mercaptan[21] is a highly toxic, flammable chemical, its low odor

[21] Personnel at DuPont La Porte would refer to methyl mercaptan as "MeSH." MeSH is shorthand based upon methyl mercaptan's chemical formula: CH_3SH. "Me" represents the methyl group (CH_3) and "SH" represents the mercaptan group (SH)—and when combined, these initialisms forms MeSH.

threshold[22] and ability to cause olfactory fatigue in personnel can generate a false sense of security when working with the chemical.

Although DuPont's methyl mercaptan technical standard states, "[a]ll DuPont sites *will* protect the site and neighboring businesses and communities from the escape of a highly toxic material [,]" methyl mercaptan detectors provided essentially no value to DuPont personnel [Emphasis in the original]. DuPont operations staff became accustomed to the odor of methyl mercaptan being present in their work environment. The low odor threshold of methyl mercaptan resulted in a false perception of safety. Consequently, the workers did not associate detecting the odor of methyl mercaptan as representing an immediate hazard.

> *"And I smell MeSH [methyl mercaptan] all the time ... but it never ... I smelled MeSH but I never felt I couldn't breathe. Like whether it's, there's -- whether it's a small leak or anything, it just never -- I didn't think MeSH was that dangerous. I thought it, you know, but it is."*

As seen by this comment from a DuPont La Porte worker, smelling methyl mercaptan at low concentrations without having any observable, adverse effects can potentially normalize the danger of a highly toxic chemical.

Furthermore, extended exposure to methyl mercaptan could potentially eliminate the perception of a methyl mercaptan release because olfactory fatigue will prevent workers from being able to smell the chemical. The odor threshold for methyl mercaptan is approximately two parts per billion (2 ppb). However, olfactory fatigue is known to occur with methyl mercaptan, resulting in the exposed person to no longer being able to detect the chemical by smell. As a result, using odor as an indicator of a release may not provide adequate warning of hazardous concentrations [9].

In the events leading to the major methyl mercaptan release, there were many smaller releases, filling the area with methyl mercaptan odor. In part because personnel believed they understood the source of releases and normalized the risk associated with these smaller releases, personnel did not apply additional safeguards like clearing the area or wearing the required personal protective equipment required by the site when dealing with a methyl mercaptan release.

Facilities should ensure that employees understand the hazards of chemicals that they are working with and may be potentially exposed to if there is a release. In cases where there are multiple hazardous chemicals, such as methyl isocyanate[23] (MIC) and methyl mercaptan (as was the case at DuPont La Porte), it is important that companies effectively train workers on the hazards of each chemical at their facility. Furthermore, companies should train workers that handle multiple highly toxic chemicals, such as methyl mercaptan and MIC, to treat these chemicals with equal

---

[22] Odor threshold, in general, is the lowest concentration of a gas or a material's vapor that can be detected by odor. Odor threshold values are not fixed physiological parameters or physical constants but are statistical points representing the best estimate value from a group of individual responses [12, p. 2].

[23] Methyl isocyanate (MIC) is a raw material used in the production of insecticides at La Porte including the Lannate® process. In December 3, 1984, a methyl isocyanate release at the Union Carbide insecticide plant in Bhopal, India resulted in an estimated 3,800 people that died within days, and tens of thousands that were injured. Eventually, the release killed tens of thousands of people. See CSB Safety Message on 30th anniversary of fatal chemical release that killed thousands in Bhopal, India [13].

importance. This training is important for workers to help ensure they have a working knowledge of those chemical hazards.

3.4 Ensure Response to a Toxic Gas Detector Alarm Is Appropriate

The toxic gas detection system at DuPont La Porte had a history of false readings. Because of these false readings, personnel supplanted their ability to observe and detect a leak over the toxic gas detection system. Personnel adapted their responses to a potential toxic gas release to include determining whether the toxic gas detectors were functioning properly by using their perception of smell before implementing the site's leak response procedures.

For example, the Lannate® Unit used chlorine in addition to methyl mercaptan. Workers could observe from a distance a chlorine leak by the green color of the gas or the immediate adverse effects (i.e., difficulty breathing) of the gas. Workers, similarly, would also use their sense of smell to detect methyl mercaptan. Methyl mercaptan gas, however, is colorless and the odor is not a reliable indicator of a release (for the reasons noted above). In both situations where workers used their physical observations to determine where there was a leak, workers were at risk for being overcome by a toxic chemical release because they were not using additional safeguards like personal protective equipment (PPE).

DuPont's highly toxic material standards require that toxic gas detection systems be treated the same as safety interlocks. DuPont defines a safety interlock as a system or function that detects an out-of-limits (abnormal) condition or improper sequence and brings it to a safe condition. In this case, an alarm should trigger a predetermined response from operators to bring the system to a safe condition. In practice, the methyl mercaptan detectors were not treated as a safety interlock.

Multiple DuPont employees informed the CSB that odor was the primary method used to locate a methyl mercaptan leak. DuPont personnel routinely responded to methyl mercaptan detector alarms without any additional respiratory protection. When they responded, if a significant leak was identified, these personnel consistently stated they would leave to get a SCBA and then return.

On the night of the incident, multiple methyl mercaptan detector alarms activated, but because the workers were aware that methyl mercaptan was being released from their activities associated with clearing the hydrate, these alarms were normalized and not treated as a safety interlock. The alarms were not treated as a near miss and there was no actual investigation into the source or how to prevent these releases. No additional personal protective equipment, such as air-purifying respirators or SCBAs was used. In addition, DuPont did not order a unit evacuation and did not notify the emergency response team until hours later, after Operator 1 made the distress call.

When a toxic detection system has a history of false readings it creates an opportunity for workers to normalize toxic gas detector readings and their response, making the toxic gas detection system ineffective. Facilities should take steps to address the potential for false readings and reduce opportunities for workers to normalize false readings and their alarms. One method is to treat toxic gas detector alarms as a near miss and investigate them even if there are no injuries or the readings are determined to be false. This method should trigger an effort to understand alarm management and safety management system issues and address them. Additionally, facilities should ensure that workers are using leak response procedures to prevent personnel from entering into potential hazardous situations without proper safeguards, like PPE.

4 Conclusions

The Lannate® Unit's toxic gas detection system did not effectively warn workers of the November 15, 2014 methyl mercaptan release occurring inside the manufacturing building. The manufacturing building, including the building's overall construction, detectors, and alarms, was not designed such that personnel could be readily notified of a major chemical release in the manufacturing building. Additionally, the methyl mercaptan toxic gas system did not provide real-time, accurate information to warn of potential offsite consequences.

There are several lessons that facilities can learn from the incident to help improve their toxic gas detection systems. Gas detectors must operate as intended to prevent workers from normalizing the hazard the detectors are protecting workers from. For toxic gas detection system to be effective, the detector selection and placement have to be strategic to identify and quantify releases. Finally, personnel have to understand and appreciate the hazards associated with the toxic gases they work with to help ensure they respond appropriately to potential toxic gas releases.

5 References

[1] U.S. Chemical Safety and Hazard Investigation Board (USCSB), "DuPont La Porte Facility Toxic Chemical Release," [Online]. Available: https://www.csb.gov/dupont-la-porte-facility-toxic-chemical-release-/. [Accessed 1 March 2019].

[2] U.S. Environmental Protection Agency (USEPA), "List of Regulated Substances under the Risk Management Plan (RMP) Program," [Online]. Available: https://www.epa.gov/rmp/list-regulated-substances-under-risk-management-plan-rmp-program. [Accessed 2017 August 28].

[3] U.S. Chemical Safety and Hazard Investigation Board (USCSB), "DuPont La Porte, Texas Chemical Facility Toxic Chemical Release: Interim Recommendations," 30 September 2015. [Online]. Available: https://www.csb.gov/assets/1/20/dupont_la_porte_interim_recommendations_2015-09-30_final1.pdf?15526. [Accessed 28 February 2019].

[4] National Oceanic and Atmospheric Administration (NOAA) Office of Response and Restoration, "Emergency Response Planning Guidelines (ERPGs)," [Online]. Available: https://response.restoration.noaa.gov/oil-and-chemical-spills/chemical-spills/resources/emergency-response-planning-guidelines-erpgs.html. [Accessed 10 March 2019].

[5] United States Department of Labor (USDOL) Occupational Safety and Health Administration (OSHA), "Permissible Exposure Limits - Annotated Tables," [Online]. Available: https://www.osha.gov/dsg/annotated-pels/. [Accessed 28 February 2019].

[6] United States Department of Labor (USDOL) Occupational Safety and Health Administration (OSHA), "Annotated Table Z-1," [Online]. Available: https://www.osha.gov/dsg/annotated-pels/tablez-1.html. [Accessed 28 February 2019].

[7] Centers for Diseases Control and Prevention National Institute of Occupational Safety and Health (NIOSH), "NIOSH Pocket Guide to Chemical Hazards - Methyl Mercaptan," 11 April 2016. [Online]. Available: https://www.cdc.gov/niosh/npg/npgd0425.html. [Accessed 28 February 2019].

[8] U.S. Chemical Safety and Hazard Investigation Board (USCSB), "DuPont La Porte, Texas Chemical Facility Toxic Chemical Release: Interim Recommendations," 30 September 2015. [Online]. Available: https://www.csb.gov/assets/1/20/dupont_la_porte_interim_recommendations_2015-09-30_final1.pdf?15526.

[9] Agency for Toxic Substances and Disease Registry (ATSDR), "Methyl Mercaptan," [Online]. Available: https://www.atsdr.cdc.gov/MHMI/mmg139.pdf. [Accessed 28 February 2019].

[10] A. H. Mohammadi and D. Richon, "Equilibrium Data of Sulfur Dioxide and Methyl Mercaptan Clathrate Hydrates," *Journal of Chemical & Engineering Data,* vol. 56, pp. 1666-1668, 2011.

[11] Center for Chemical Process Safety (CCPS), Recognizing Catastrophic Incident Warning Signs in the Process Industries, Hoboken: Wiley, 2012.

[12] A. H. Lehocky, S. S. Murnane and D. Patrick, Odor Thresholds for Chemicals with Established Occupational Health Standards, Falls Church: AIHA, 2013.

[13] USCSB, "On 30th Anniversary of Fatal Chemical Release that Killed Thousands in Bhopal, India, CSB Safety Message Warns it Could Happen Again," [Online]. Available: http://www.csb.gov/on-30th-anniversary-of-fatal-chemical-release-that-killed-thousands-in-bhopal-india-csb-safety-message-warns-it-could-happen-again-/?pg=4.

[14] K. Baker, "DuPont Closing Texas Plant Where 4 Died," The News Journal, 1 April 2016. [Online]. Available: https://www.usatoday.com/story/money/2016/04/01/dupont-closing-texas-plant-where-4-died/82512270/. [Accessed 29 June 2017].

[15] DuPont, "Form 8-K," 31 March 2016. [Online]. Available: https://www.sec.gov/Archives/edgar/data/30554/000003055416000115/dd33116form8-k.htm.

[16] U.S. Chemical Safety and Hazard Investigation Board (USCSB), "E.I. DuPont de Nemours & Co., Inc. Belle, West Virgina Investigative Report," September 2011. [Online]. Available: https://www.csb.gov/assets/1/20/csb%20final%20report.pdf?13966. [Accessed 1 March 2019].

[17] U.S. Chemical Safety and Hazard Investigation Board (USCSB), "E. I. DuPont De Nemours Co. Fatal Hotwork Explosion," [Online]. Available: https://www.csb.gov/e-i-dupont-de-nemours-co-fatal-hotwork-explosion/. [Accessed 1 March 2019].

[18] American Industrial Hygiene Association (AIHA), "Current ERPG Values," 2014. [Online]. Available: https://www.aiha.org/get-involved/AIHAGuidelineFoundation/EmergencyResponsePlanningGuidelines/Documents/2014%20ERPG%20Values.pdf.

[19] F. Cavender, S. Phillips and M. Holland, "Development of Emergency Response Planning Guidelines (ERPGs)," *Journal of Medical Toxicology,* vol. 4, no. 2, pp. 127-131, 2008.

553063 Hurricane Harvey 20 Months Later - Insights on the Arkema Crosby Flooding

On August 24, 2017, Hurricane Harvey, a Category 4 hurricane, made landfall in southeast Texas producing an unprecedented amount of rainfall. Hurricane Harvey killed 68 people and flooded over 300,000 structures, forcing roughly 40,000 people to evacuate their homes. Hurricane Harvey required a large emergency response effort supported by local, state, and Federal officials.

At Arkema's Crosby site, Harvey delivered an unprecedented amount of rainfall. The National Oceanic and Atmospheric Administration estimates show that the rainfall during Harvey had less than a one-in-one-thousand-year likelihood of occurring. Harris County has estimated that at Crosby, Harvey was a 5,000 year to 20,000 year event. The plant's hurricane preparedness procedure was followed and resulted in a safe shutdown and hurricane preparedness of the manufacturing units. However, finished goods inventory remaining in the warehouses was overwhelmed by high water and overcame multiple layers of protection used to safely store organic peroxide finished goods.

This presentation will explain the sites preparedness ahead of time, the events leading up to the flooding, the incident itself, and the emergency response.

The presentation will focus on the impacts, the main lessons learned and the key recommendations from the CSB report.

AUTHOR INDEX

AUTHOR INDEX